Practical Engineering Perspectives
The Environment:
Air, Water, and Soil

Gail F. Nalven, Editor

American Institute of Chemical Engineers
345 East 47th Street • New York, New York 10017-2395

Copyright © 1997
Practical Engineering Perspectives The Environment: Air, Water, and Soil
ISBN: 0-8169-0713-7
American Institute of Chemical Engineers
345 East 47th Street
New York, New York 10017-2395

Library of Congress Catalog Card Number: 96-41927

AIChE shall not be responsible for statements or opinions advanced in its papers or printed in its publications.

cover design: Joseph Roseti

PREFACE

This compendium of recent articles is aimed at chemical engineers and other technical professionals in the chemical process industries (CPI) who are involved with environmental aspects of chemical processes and plants. With the recent and growing emphasis on environmental issues, this is indeed a very large group, including individuals in virtually all chemical engineering functions—research and development, design, construction, operations, management, consulting, academia, and so on.

The environmental field is extremely broad, and the subjects of interest to chemical engineers are many and diverse. This volume takes an in-depth look at the specifics for the various environmental media—air, water and wastewater, and soil. (A companion volume covers the key topics of pollution prevention and waste minimization, and environmental management, which apply to any CPI process.)

The Air Pollution section, with 24 articles, is by far the largest, reflecting the tremendous amount of activity in this field in recent years as a result of the Clean Air Act Amendments of 1990. Developing an air emissions inventory—the first step in understanding and controlling a plant's air emissions—is the subject of the first article. The next seven articles discuss options for controlling volatile organic compound (VOC) and nitrogen oxide (NOx) emissions. Additional articles cover emissions testing and monitoring, odor control, fugitive emissions, and several technologies for air pollutant control, such as carbon adsorption and biofiltration.

The Water and Wastewater Treatment section contains articles on: developing treatment strategies; zero discharge and water reuse; using biomonitoring data to reduce effluent toxicity; stormwater management; and specific treatment technologies, such as steam stripping, carbon adsorption, Fenton's chemistry, and liquid/liquid extraction.

The third section, Site Remediation, begins with a discussion of information resources on innovative cleanup technologies available from the U.S. Environmental Protection Agency. The next article provides an overview of the various technologies available for remediating hydrocarbon contamination. The remaining articles look at specific technologies in more detail, including bioremediation, "pump-and-treat" systems, soil vapor extraction, and soil washing.

Articles have been selected based on their usefulness to practitioners, and their emphasis on practical and generic guidance. The roster of authors includes many of the world's foremost authorities in the field.

All the articles have been published by the American Institute of Chemical Engineers over the past five years. Three articles originally appeared in *Environmental Progress;* all others were originally published in *Chemical Engineering Progress,* the broad-based monthly magazine of AIChE.

Cynthia Fabian Mascone
Technical Editor
Chemical Engineering Progress
August 1996

ACKNOWLEDGMENTS

Creating a new publishing program is a daunting task that can only be made possible with the help of a group of talented publishing professionals who share their expertise.

Thanks go to Mark D. Rosenzweig, Cynthia Fabian Mascone, and Rich Greene for their guidance in developing the text content, and to Stephen R. Smith for his vision in creating this program. Special appreciation goes to Chung S. Lam whose assistance in handling the "details" is really what made this series possible!

Gail F. Nalven
June 1996

CONTENTS

Water and Wastewater Treatment

Site Remediation

CONTRIBUTORS

William V. Adams, Vice President, Durametallic Corp., Kalamazoo, MI

Todd A. Baesen, Professional Engineer, James M. Montgomery Consulting Engineers, Inc., Walnut Creek, CA

Joseph N. Bays, Advanced Chemical Engineer, Eastman Chemical Co., Kingsport, TN

Dannelle H. Belhateche, Project Engineer, Environeering, Inc., Houston, TX

Richard J. Bigda, P.E., President, Technotreat Corp., and Richard J. Bigda and Associates, Tulsa, OK

Robert Block, Geotechnical Engineer, Remediation Technologies, Inc., Concord, MA

Jerold W. Boddy, P.E., Associate and Coordinator, Air Quality Services, Malcolm Pirnie, Inc., Pittsford, NY

Hinrich Bohn, Ph.D., Professor, Department of Soil and Water Science, University of Arizona, Tucson, AZ

José L. Bravo, P.E., Vice President, Jaeger Products, Inc., Houston, TX

Ronald Brestel

Leigh Ann Carroll, Air-Pollution Control Engineer, Burns & McDonnell Engineering Co., Kansas City, MO

Nick Chadha, PE, Senior Technical Associate, IT Corp., Knoxville, TN

Soung M. Cho, Ph.D., Director of Engineering, Energy Applications Division, Foster Wheeler Energy Corp., Clinton, NJ

Richard S. Colyer, Environmental Engineer, U.S. Environmental Protection Agency, Research Triangle Park, NC

T. Ted Cromwell, Director, Air and Water Programs, Regulatory Affairs Division, Chemical Manufacturers Association, Washington, DC

Stanleigh R. Cross, Design Coordinator, Phoenix Engineering, Inc., Houston, TX

Roger W. Cusack

Karen S. Eble, Project Engineer, Betz Industrial, Betz Laboratories, Inc., Trevose PA

Jennifer E. Feathers, Project Engineer, Betz Industrial, Betz Laboratories, Inc., Trevose PA

Robert W. Fitzgerald, Sales Manager, 3M Corp., Research Triangle Park, NC

Ashutosh Garg, P.E., Manager, Thermal Engineering, Kinetics Technology International Corp., Houston, TX

Deepak Garg, ENSR Consulting and Engineering

Paula S. Gess, P.E., Associate, James M. Montgomery Consulting Engineers, Inc., Metarie, LA

William K. Glynn, P.E., Senior Chemical Engineer, Camp Dresser & McKee, Inc, Cambridge, MA

Mike E. Goldblatt, P.E., Project Engineer, Betz Industrial, Betz Laboratories, Inc., Trevose, PA

Dirk Gombert, Westinghouse Idaho Nuclear Company, Inc., Idaho Falls, ID

Kimberly A. Groff, Ph.D., Director, Environmental Technology, List Inc., Acton, MA

Robert M. Howe, AWARE Engineering Inc., Houston, TX

Marc Karell, P.E., Senior Project Engineer, Malcolm Pirnie, White Plains, NY

Michael C. Kavanaugh, James M. Montgomery, Consulting Engineers, Inc., Walnut Creek, CA

J. David Keating, Jr., AWARE Engineering Inc., Houston, TX

Gerald L. Kirkpatrick, Partner, Environmental Resources Management, Inc., ERM Group, Exton, PA

Gary N. Lawrence, P.E., Vice President, Phoenix Engineering, Inc., Houston, TX

Ramon Li, P.E., Associate and Manager, Air Quality Engineering, Malcolm Prine, Inc., White Plains, NY

Sydney Lipton, Consultant, Port Townsend, WA

Gilbert M. Long, Certified Hazardous Materials Manager and General Manager, ENSR Consulting Engineering, Somerset, NJ

Alison M. Martin, Ph.D., P.E., Supervising Engineer, James M. Montgomery Consulting Engineers, Inc., Metairie, LA

Robert G. McInnes, Principal Air Quality Engineer, ENSR Consulting and Engineering, Acton, MA

Lisa A. McLaughlin, President, Waste Min Inc, Groton, MA

Hugh S. McLaughlin, Ph.D., P.E., Vice President, Waste Min Inc., Groton, MA

Jan Meyer, Ph.D., Environmental Engineer, U.S. Environmental Protection Agency, Research Triangle Park, NC

Michael E. Miller, Ph.D., Environmental Chemist, Camp Dresser & McKee, Inc., Cambridge, MA

Edward C. Moretti, Engineer, Baker Environmental, Inc., Coraopolis, PA

Nikhiles Mukhopadhya, Manager, Process and Environmental Services, Baker Environmental, Inc., Coraopolis, PA

Sharon L. Nolen, Ph.D., Group Leader, Eastman Chemical Co., Kingsport, TN

David C. Noonan, P.E., Associate, Camp Dresser & McKee, Inc., Cambridge, MA

Edgar J. Oubre, AWARE Engineering Inc., Houston, TX

R. Benson Pair, Jr., Chief Environmental Technology Engineer, The M.W. Kellogg Co., Houston, TX

Charles S. Parmele, Engineering Manager, IT Corp., Knoxville, TN

Daniel M. Powell, Ph.D., P.E., Director, Center for Environmental Engineering and Science, New Jersey Institute of Technology, Newark, NJ

Jean A. Rogers, James M. Montgomery, Consulting Engineers, Inc., Mannheim, Germany

Robert M. Rosain, Senior Project Manager, CH2M Hill, Bellville, Washington

Edward N. Ruddy, P.E., Project Manager, Industrial Air-Pollution Control, Burns & McDonnell Engineering Co., Kansas City, MO

M. John Ruhl, P.E., Manager, Solvent Recovery Division, Dedert Corp., Olympia Falls, IL

Joanne R. Schaich, Senior Environmental Affairs Representative, Eastman Chemical Co., Kingsport, TN

Jeffrey H. Siegell, Ph.D., Senior Staff Engineer, Exxon Research and Engineering Co., Florham Park, NJ

Mark H. Stenzel, Manager, Equipment Product Management, Calgon Carbon Corp., Pittsburgh, PA

Hans Stroo, Ph.D., Manager, Biotreatability Lab, RETEC, Seattle, WA

Geoffrey H. Swett, Senior Program Manager, RETEC, Tucson, AZ

Stacey E. Swisher, Mechanical Engineer, Eastman Chemical Co., Kinsport, TN

Dante J. Tedaldi, Bechtel Environmental, Inc., San Francisco, CA

Karen Walker, Marketing Planning Specialist, Horiba Instruments, Inc., Irvine, CA

Charles Wells, Ph.D., Vice President, Rotoflex, Inc., Pleasanton, CA

Stephen C. Wood, Southern Regional Manager, Energy Technology Consultant, Woodward-Clyde, Inc., Houston, TX

Helen Yoest, Vice President, Sales and Marketing, Entropy, Inc., Research Triangle Park, NC

Elaine G. Zoeller, Ph.D., Environmental Representative, Eastman Chemical Co., Kinsport, TN

The Editor

Gail F. Nalven, Acquisitions Consultant, American Institute of Chemical Engineers, and Technical Editor, New York, NY

INTRODUCTION

Industrial activity inevitably has an impact on the environment. However, we have come a long way since the Industrial Revolution. Environmental standards have risen and continue to rise. In part, this can be attributed to improvements in technology, which have enabled us to be more efficient in the use of raw materials and fuels. Improved efficiency means less waste, which in turn means a reduction in environmental releases. This also can have a direct, and very positive effect on profitability.

However, changing public perceptions have also contributed to the drive for higher standards. Global issues such as acid rain, deforestation, depletion of the ozone layer, and the growing numbers of extinct and endangered spices, have a very high profile internationally. At the local level, communities are less and less tolerant of possible adverse effects on health, safety, and quality of life due to discharge from nearby factories. This is true not only in North America and Europe, but increasingly in the developing countries, too. Disasters such as Chernobyl and Bhopal have further fueled public concern about the effects of industry on the environment.

Industry is responding—partly out of a desire to be a good neighbor and enhance good will, and partly because of growing regulatory pressure. It is no longer socially, politically, or legally acceptable to be seen as a polluter, and the high visibility of environmental issues has made this a matter of great importance to managers at every level in industry.

Against this background, the need to design, build, and operate greener, cleaner, and more efficient plants has challenged process engineers in the past decade, and it seems set to remain that way for the foreseeable future.

Environmental management is a complex subject. It has an important technological dimension. Contamination of land and water is the legacy of excessive emissions in the past, and remediation is now receiving a great deal of attention. A large number of methods are in use or under development, including both physico-chemical and biological processes.

However, our goal for the future must clearly be to prevent unacceptable levels of contamination. New equipment and processes continue to be developed and improved to minimize emissions from factories. These include, for example, flue gas desulfurization technologies and selective catalytic reduction to remove SO_X and NO_X, respectively, from stack gases; and improved treatment processes to reduce organic loadings in aqueous effluents. However, these technologies focus primarily on end-of-the-pipe removal or destruction of contaminants. Increasingly, the emphasis is shifting away from this type of pollution control to more fundamental structural changes in industrial processes. These aim to eliminate emissions by either not creating pollutants in the first place, or by recovering and reusing materials that would otherwise be discharged. This approach is known by many different names, including pollution prevention, waste minimization, and conservation-based environmental compliance.

Progress in materials and equipment design is an important factor in this change of emphasis—for example, improved membrane separation technologies enable the separation and recycle of some organic materials that have hithero been discharged in aqueous effluents. Another important part of the response to this challenge has been the development of new systematic process design methods that embody waste minimization concepts. Advances in process integration techniques have been particularly important in this regard.

However, technology is only one dimension of the field. Continuous improvement in environmental performance requires the right focus at all levels within industrial organizations, and this change of emphasis provides the second dimension. This calls for changes in attitudes, and in the way we work. It demands awareness of the environmental impacts of the things we do. Many companies have implemented programs to involve their workforce in this continuous improvement process, and a number of notable successes have been reported.

The third dimension of environmental management is the expanding body of environmental regulations. Some of these appear contradictory, and it has been argued that in many cases regulations stand in the way of genuine environmental improvements. The complexity of this field has created a need for specialists in environmental law, and necessitates close interaction between technical and legal people as environmental management activities are pursued.

Environmental management is a multifacited, global problem. It is dynamic, responding to rapid changes in technology, law and public perception.

Alan P. Rossiter
Linnhoff March, Inc.
August 1996

The Environment:
Air, Water, and Soil

Develop a Plantwide Air Emissions Inventory

Compiling an emissions inventory is the first step in reducing or controlling emissions. It can also help in achieving waste minimization, process safety, and quality goals.

**Ramon Li,
Marc Karell, and
Jerold W. Boddy,**

Malcolm Pirnie, Inc.

Increasingly stringent and technical air emission regulations are being issued under the Clean Air Act Amendments of 1990 (CAAA) at the federal, state, and local levels. To respond to these regulations in a responsible and cost-effective manner, companies will need to develop a complete engineering approach, including compiling a reliable emissions inventory, investigating plant practices, and implementing optimal strategies to minimize air emissions.

Many CAAA regulations contain emission thresholds allowing smaller facilities to avoid potentially onerous and expensive requirements. Many regulations are and will be written with precise emission reduction goals that must be achieved. For these reasons, the effort necessary to develop a reliable emissions inventory will pay off in either demonstrating that certain regulations are not applicable to a facility or by defining more clearly how reduction goals should be achieved.

Before an evaluation of which regulations are applicable to a plant and what their implications are, an Emissions Characterization and Assessment Program (ECAP) should be undertaken. Such a program emphasizes the use of a multidisciplinary task force to set emission reduction goals, review data, and plot strategies to attain the goals. The intent of the program is to integrate process and environmental engineering experience in order to institute operational changes to minimize discharges. While offering long-term benefits, ECAP is a dynamic practice that requires periodic review.

ECAP encourages a process-oriented assessment of pollutant emitting steps in order to calculate emissions and determine which approach can most effectively reduce emissions. Accessible process-related information, such as standard operating procedures (SOPs) and process flow diagrams (PFDs), are used with the proper methodology to estimate emissions. Use of computer software to estimate emissions and to manage the database is encouraged.

This approach can have ramifications beyond estimating emissions and reducing them. This can be the first step toward assembling and managing multimedia environmental information. As a result, other goals, such as waste minimization, process safety management, ISO 9000, and materials recovery, may also be realized.

This article outlines the steps involved in compiling a reliable emissions inventory. It reviews the methods available and the data needed for determining emissions rates, and discusses how to collect and manage the data. It also offers some insights on what to do with the emissions inventory.

Getting started

An ECAP program introduces a stepwise organized plan to inventory and reduce air emissions. It is generally recognized as an effective approach to demonstrate and maintain compliance and maximize benefits.

Given the far-reaching goals of determining emission rates of every process step, a phased project approach is preferred. The development phase of an ECAP Plan should involve the following components:

1. Form a task force. The first step is the organization of a task force. Ideally, its members should represent different divisions within the industrial facility, including engineering, environmental, manufacturing, legal, and administration. Including members from outside the company, such as an environmental consultant, may give the task force an additional independent, experienced voice.

2. Understand existing facility conditions. The task force should begin by discussing the plant's products, equipment, and chemicals used as they relate to emissions and discharges. Air emission regulations that may be applicable should be considered and evaluated to determine regulatory impacts.

3. Set goals. Overall project goals could include decreasing air emissions, reducing the risks of accidental discharges, optimizing process operations, and improving production. In order to provide focus, the task force should also propose preliminary emission reduction goals. Meeting regulatory requirements should be the top priority. However, additional reductions should be encouraged if economically acceptable. Any corporate environmental standards that exist should be met, as well.

4. Identify the emission sources to be investigated. Before beginning the effort, it is important to define what will be investigated. For a large chemical process plant, it may be better to begin performing the inventory on a few targeted significant sources or processes rather than get overwhelmed with data for every point in the plant. Depending upon the specific industrial source category, the task force may also need to study fugitive emissions.

Methods to determine emission rates

The ECAP team must decide what methods it will use to estimate emission rates from the selected sources. There are several acceptable methods

for computing emission rates. The team must weigh the cost expended vs. the resulting accuracy in deciding which methods to use for specific points.

Stack testing/monitoring. The most accurate means of measuring emissions is stack testing or monitoring, where an actual sample of exhaust is taken at the stack and pumped to the appropriate analyzer for chemical analysis. Real-time measurement of total volatile organic compounds (VOCs), NO_x, SO_2, CO, and other pollutants is possible. EPA Method TO-14 *(1)* can provide ppb concentrations of a wide variety of specific VOCs in a sample of gas.

Emissions can be estimated through stack testing, emission factors, material balances, or engineering equations.

Stack testing results should not necessarily be taken as absolutely accurate. As with any field work, errors can occur that affect results. However, stack testing does provide information under actual operating conditions. While useful, stack testing can be expensive, given the cost of the sampling equipment, analyzers, and labor. Therefore, it is generally used sparingly on selected high-priority emission points or to verify other calculations.

Emission factors. While it may be expensive to stack test many emission points, there is a growing body of knowledge of emissions from various industrial processes. The U.S. Environmental Protection Agency (EPA) has been compiling records of stack tests and provides emission factors for specific operations, on a pound of pollutant per operation unit basis, in its AP-42 reference manual *(2)*.

An example of the usefulness of emission factors is stationary combustion equipment. AP-42 provides emission factors for pollutants such as hy-

drocarbons, CO, NO_x, SO_2, and particulate matter on a fuel-usage basis for different types of boilers combusting different types of fuels. Given the lack of emissions data for most boilers (particularly older units), emission factors can be useful in estimating emissions of different pollutants.

Because emission factors generally represent compilations of many tests on different types of equipment, a particular emission factor may not accurately represent the actual unit emissions. The rule of thumb is that compilation emission factors tend to be worst-case (conservative) because they include emissions from older, less-efficient units in the average. Many combustion equipment vendors are now performing their own in-shop testing on specific models and can provide their customers with more accurate emission factors (generally, in lb/million Btu) for use in determining regulatory impacts. Engineers should use caution in determining when and how to use emission factors.

Material balance. Another technique to estimate emissions is the use of a material balance. The quantity of a compound exiting a piece of equipment can be estimated by adding the amount entering the equipment plus the amount formed from chemical reactions minus the amount accumulated within it or converted to another compound. If the compound is not recovered as product, the quantity leaving may be further refined to exit pathways, such as water, solid, and air. If a certain compound undergoes no reaction, is not formed in any additional quantities, and does not accumulate in the equipment, then the compound is either in the process' wastewater or solid waste, or is discharged to the air. The quantity of that compound emitted in the air can be estimated by subtracting the estimated amount that is lost in both the wastewater and solid waste from the total quantity used in the process.

Material balance can best be utilized in estimating emissions from operations such as painting or plating. Commercially available coating material contains a certain percentage of a pollutant, such as organic solvent, which can be assumed to be airborne once applied to a surface. Any addition of a diluent, such as methyl ethyl ketone (MEK), should be noted and computed into the organic total. Emissions on an hourly or annual basis may be estimated based upon either the appropriate total quantity of coating used, or the total number of objects coated if a usage factor is known. In some cases, coated samples must be cured. Therefore, a certain percentage of the emissions would occur in the paint booth, while the remainder occurs in the curing oven. The ECAP team will need to estimate the appropriate emissions split.

While relatively inexpensive to perform, material balance can produce erroneous information. For instance, even a small error in measuring a concentration in another medium, such as wastewater, may have a large impact given the large volume of wastewater produced over a period of time. Air emissions estimated by material balance should be used with care and verified by other means.

Engineering equations. Another relatively inexpensive way to estimate emissions is to use engineering equations that can compute the volatility of compounds under varying conditions and from a variety of process steps. Equations can be found in such references as *(3–6)*. While the emission rates determined by this manner are defensible, they represent theoretical treatment of actual conditions.

Overall, this approach represents, for most facilities, the best balance between cost and accuracy, particularly for large numbers of emission points. The remainder of this article focuses on how to produce an air emissions inventory using engineering equations

with particular emphasis on chemical batch processing.

Information needed

It is easiest to perform an emissions inventory on a process-by-process basis, because equipment is generally located near each other, process steps are linked together, and annual emissions of many steps are related to the number of times a process is run. To begin an emissions inventory, it is crucial for the engineers gathering the data to familiarize themselves with the process, equipment, and steps. They should understand the process steps, the chemicals used, and the intermediates, byproducts, and end products formed.

One of the most important tasks in a data gathering project is to properly define the plant's processes into workable components. The multistep chemical process must be broken down into definable air emission operations. While seemingly daunting, this is generally not as time-consuming as would be expected, because this information may have been prepared for other purposes.

Probably the best source for such information is the plant's SOPs, which contain step-by-step instructions to operators on how to use the equipment. During the review of this document, the team can select which steps may be producing air emissions. While most steps are process-related (for example, "turn on steam jacket"), many can result in the volatilization and emission of compounds, such as charging chemicals to a reactor, raising the temperature, and pulling a vacuum. These air emitting steps should be numbered and identified.

A critical task in the data gathering project is breaking down the plant's processes into workable components.

After chemical process steps are identified from the SOPs and other sources, they should be analyzed to gain a better understanding of the nature of emissions. The best approach is to study the state of the volatile compounds within the vessels during the process steps, and follow them out to the stack. This should provide the most accurate description of emissions. Some common unit operation steps and the fate of volatile compounds from chemical batch processes are described below.

• *Fill or charging* — When a substance is added to a vessel, such as a reactor or a tank, an equal volume of vapor is displaced and considered emitted from the vessel. If the substance added is a volatile compound or the vessel already contains a volatile substance, then the displaced vapor will contain some quantity of that compound. In most cases, the volatiles will be saturated based upon the temperature of the system at the partial pressures in the displaced vapor. Note that charging of solids can also result in emissions of volatiles already present, with emission rates dependent upon the amount of solids charged and their charging rates.

• *Tank breathing* — In addition to losses during tank filling, storage tanks containing volatile compounds lose small amounts of their contents throughout the year because of tank "breathing." This phenomenon, caused by normal diurnal temperature variations affecting the tank vapor space expansion and contraction, usually accounts for much lower emissions than filling. Emissions from tank breathing have been characterized and can be calculated using published equations, such as in *(2)*.

• *Gas sweep* — Equipment is often swept with an inert gas, such as air or nitrogen, before reuse in a new process operation, to remove residual chemicals. This action sweeps the residual chemicals out of the vessel and, potentially, into the atmosphere.

- *Evacuation* — A vessel may need to be placed under vacuum during a process to remove its vapor contents. The contents of the vessel are emitted during this step.
- *Heating* — When a reactor is heated, thermal expansion of the liquid phase causes a volume of the vapor phase to be displaced. The displaced vapor often contains relatively high concentrations of volatile compounds because of the high temperatures. Therefore, heating could contribute significantly to total emissions.
- *Vacuum operations* — Sometimes a piece of equipment is kept under vacuum. The degree of volatilization of a compound is affected by its vapor pressure relative to the reduced system pressure. Examples are vacuum distillation and vacuum drying. Vacuum drying is unique because its emissions profile is not constant — most of the volatile compounds are emitted during the early part of a run, and emission rates generally decay at roughly exponential rates.
- *Gas evolution* — During a reaction, new compounds will form. These compounds may be volatile and, therefore, have an equilibrium ratio in the liquid and gaseous phases. The formation of new compounds in itself could cause increased pressure and, as a result, release head space vapor containing volatile components. The rate of formation or evolution of the compounds and their volatilities determine their emission rates. Because these gases compete with others already in the head space, the relative volatility and emission rate of other compounds would be affected, as well.

A critical requirement in order to develop emission rates is the volatility of compounds in question. A chemical's volatility can be best determined from vapor pressure data found in engineering handbooks or by the use of Antoine's Equation:

$$\log p = A - B/(C + T) \qquad (1)$$

where p = vapor pressure, atm; T = boiling point, K; and $A, B,$ and C are

Table 1. Information needed for various process operations.

Operation Type	Parameter Needed
Charging	Liquid Pumping Rate (gpm)
Evacuation	Time to Evacuate (min) Free Space of Vessel (gal)
Gas Sweep	Rate of Gas Sweep (cfm)
Heating	Time to Heat (min) Free Space of Vessel (gal) Beginning and Final Temperatures (°C)
Gas Evolution	Rate of Gas Evolution (lb/h)
Vacuum Operation	Air Leak Rate (lb/h) Absolute Pressure (mm Hg) Free Space of Vessel (gal)
Vacuum Drying	Batch Weight (lb) Percent Wet Material (%)

empirical dimensional constants. Many chemical engineering references [such as *(4)*] provide values for the constants A, B, and C used to determine the vapor pressure at a wide variety of temperatures for many commonly used compounds.

Equations are available from a variety of sources, including standard chemical engineering references and EPA publications, to determine the behavior of chemicals during different chemical process steps. One of the best sources for chemical batch processes is an EPA publication entitled "Control of Volatile Organic Compound Emissions from Batch Processes" *(5)*.

Data gathering

The task force must now go out to the plant and the records room to gather the information described above for the air emitting steps. It is recommended that the ECAP team take a tiered approach, first identifying the processes being investigated. This should be followed by identifying the equipment and emission points involved in the process, including any air-pollution control equipment. Then for each piece of equipment or emission point, the process steps that could potentially result in air emissions should be identified

and numbered (discussed below). Finally, the potential pollutants (volatile compounds) of each process step should be recorded. In total, this will provide the facility with a complete listing of emission points, process steps, and chemicals potentially emitted.

A good source that can help identify process steps that may result in air emissions is the plant's SOPs for the process. Most SOPs provide numbered, specific procedures that operators must perform for the process to work, and can be used to identify air emitting steps in the process and what type of step it is.

After these qualitative items are catalogued, the team must obtain specific information concerning the nature of each process step to determine its emissions. Besides the SOP, useful information may be obtained from batch records, plant flow diagrams, material safety data sheets (MSDSs), and material balances.

Another very good source of information is interviews with the operators. The workers in the plant who operate the processes on a daily basis often know best about the equipment. They are often eager to share their knowledge and feel a part of the team. It is not unusual that such interviews result

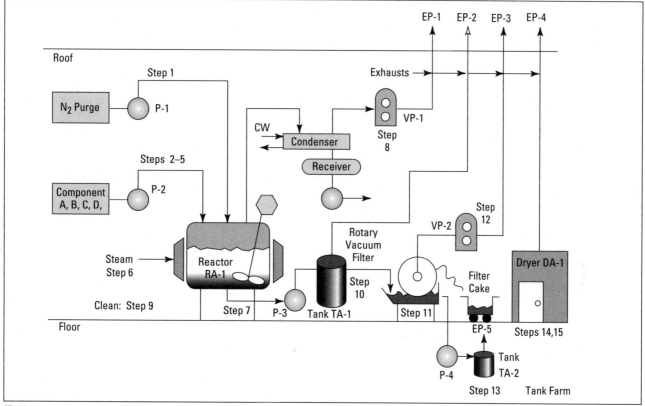

■ *Figure 1. Process flow diagram for hypothetical "pirniedine" process.*

in discrepancies being found and corrected between operator information and SOPs.

In all cases, there is common information needed to estimate emissions. This includes the potential pollutants present in the system, their quantities, the operating temperature, and the types of air-pollution control equipment and their effects on pollutant removal. The specific information needed for each type of air-emitting process step is described in Table 1.

Compiling the data

One problem that the team must overcome during data gathering is the handling of large amounts of information concerning pollutants and process parameters for many process steps and pieces of equipment. Therefore, it is best to be organized before data gathering begins.

Process steps should be numbered for easy tracking. Data sheets should be prepared so that the team can gather the specific data needed for a given process step. Notebooks should be prepared to provide handy access to this information. It is important to include on data sheets any assumptions used to estimate emissions, as they may be important in years to come when engineers review the data.

After emission rates are estimated, the ECAP team must keep track of the information and report it in an easy-to-read, accessible format. The plant may choose to minimize paperwork by keeping all data in a computer database or may choose to print out information in accessible books.

It is crucial that while compiling information, notes be included describing the origin of process data used so that there is no confusion when referred to in the future. Examples of clarifying notes are "heating duration from 6/16/90 version of SOP, pg. 4," "pressure recorded from pump dial by L.A. 5/20/94," "assumed pure exponential decay in absence of data," or

"leak rate provided by John Doe of XYZ Co., conversation of 4/13/92." Because of the large quantity of critical data that must be handled, recording such references and notes — even in handwritten form — is critical.

Computer software can be used to create and store a database of environmental information. Software exists that store relevant process and environmental data for an emission point, and print the information on appropriate permit application forms.

Computer programs can save the user time in estimating emission rates. They contain many of the engineering equations, emission factors, and chemical information discussed above. The user can input process data for storage and computation of emission rates.

[See *CEP's Software Directory* for information about commercially available software packages. — *Editor*]

After gathering the data and calculating emission rates, the ECAP team should review the data from a quality point of view. Process data should be reviewed for their accuracy and reliability; approaches, calculation methods, and the calculated emission rates should be examined to determine if they "make sense." As discussed earlier, engineering calculations, while based upon known theoretical concepts, may not apply to specific process situations, and alternative methods need to be developed. Therefore, qualified team members should review the data.

An example of a potential misapplication of engineering equations is in vacuum operations. As a vacuum pump reduces the pressure in a vessel, the sum of the pure components' vapor pressures times their liquid mole fractions (partial pressures) at a given temperature are also reduced. Many equations, using the basis of ideality, can result in the sum of the partial pressures exceeding total system pressure. However, in reality, this cannot happen. Under vacuum conditions, actual partial vapor pressures are reduced along with the component's activity coefficient. Those most familiar with the

chemistry of the compounds should study this matter in more detail to estimate what the reduced activity coefficients or vapor pressures may be.

Therefore, it may be useful for the plant to conduct additional studies, where feasible, to clarify input data and substantiate emission rates. For instance, the plant may wish to conduct selected stack testing or ambient monitoring studies to determine the concentration of VOCs in an exhaust stack or unit, particularly if calculations show that it has a high emission rate. Another example is a simple test to determine the decay factor of a dryer — if it does not interfere with production, runs can be interrupted periodically (perhaps hourly during the key beginning of a run) to measure the weight loss of selected trays with time, and the weight loss can be graphed and its characteristics determined.

Example

As an illustration of how to estimate emissions, consider the simplified process to make the mythical compound

"pirniedine" shown in Figure 1. It consists of four major pieces of process equipment, exhausting through four emission points. Various batch process steps take place in each vessel, which is typical of activities at a chemical plant involved in the formation and purification of a compound.

To determine emissions from this process, we first review pirniedine's SOPs, and identify the following 15 process steps capable of emitting air pollutants:

• Reactor RA-1: a gas sweep to clear the vessel space (Step 1), four successive fills of solvent and reactants (Steps 2–5), heating (Step 6), the reaction (which is accompanied by gas evolution) (Step 7), a vacuum distillation to purify the product (Step 8), and a fill with a cleansing solvent solution after the product has been removed (Step 9).

• Tank TA-1: fill by wet product preceding filtration (Step 10).

• Filter FA-1: fill by wet product (Step 11), and filtration/evacuation (Step 12).

• Tank TA-2: fill by filtrate (Step 13).

Table 2. Sample data input sheet for emissions inventory.

Process Step:	
Process:	Exit Temperature:
Step Number:	Step Type:
Fill Rate:	gal/min
Total Volume Transferred:	gal
Exit Temperature:	°C (Control Option 1)
Condenser Selected For Multiple Control Strategy:	(Control Option 2, 3, or 4)
Selected Control Option:	(1-7)

Pollutant	Quantity Present in Equipment, lb

Comments: _____

• Dryer DA-1: evacuation of dryer (Step 14), and vacuum drying of wet product to final dry state (Step 15).

We then review the SOPs and batch records and interview operators to determine the precise operating parameters, such as charging rate for fills, free space for the vacuum distillation, and so on. In order to more efficiently handle data, process data are stored on data sheets, such as the one for the fill step presented in Table 2.

Proposed process data can be analyzed using various computer programs to estimate emission rates. An excerpt of a typical output printout is shown in Table 3. Emission rates for each step are presented on hourly and per batch bases, and are compared for different control equipment scenarios.

A benefit of using an emissions estimating program is that all the process and emissions data can be reorganized and presented in various types of management reports, which can be used to help make key decisions concerning compliance, pollution prevention, and waste minimization.

Literature Cited

1. **U.S. Environmental Protection Agency,** "Compendium of Methods for the Determination of Toxic Organic Compounds in Ambient Air," U.S. EPA, Research Triangle Park, NC, Document No. EPA 600/4-89/018 (June 1988).

2. **U.S. Environmental Protection Agency,** "Compilation of Air Pollutant Emission Factors," U.S. EPA, Research Triangle Park, NC, Document No. AP-42, Supplement D (Sept. 1991).

3. **Perry, R. H., and D. Green, eds.,** "Perry's Chemical Engineers' Handbook," 6th ed., Section 13 — Distillation, McGraw-Hill, Inc., New York (1984).

4. **Dean, J. A., ed.,** "Lange's Handbook of Chemistry," 13th ed., McGraw-Hill, Inc., New York, pp. 10-26–10-53 (1985).

5. **U.S. Environmental Protection Agency,** "Control of Volatile Organic Compound Emissions from Batch Processes" U.S. EPA, Washington, DC, Document No. EPA 450/2-78-029 (1994).

6. **Buonicore, A. J., and W. T. Davis, ed.,** "Air Pollution Engineering Manual," Van Nostrand Reinhold, Inc., New York (1992).

PROJECT: Analysis of VOC Emissions
PROCESS: Pirniedine
BATCH SHEET/REF.: January, 1995

EQUIPMENT: RA-1-Reactor   **EQUIPMENT SIZE:** 1500

Step Number and Type	Solvent	Temp., °C	V.P., mm Hg	Emissions lb/h	Cycle, h	Total lb/run	(1) UNCTRL	(2) COND 20.0C
1, GS	Hexane	27.0	165.0	183.07	1.17	213.58	183.07*	125.44*
2, F	Hexane	27.0	165.0	3.81	0.58	2.22	2.22	1.67
3, F	Hexane	27.0	165.0	2.10	0.30	0.63	0.63	0.48
4, F	Hexane	27.0	165.0	1.79	0.50	0.90	0.90	0.67
5, F	Ethanol	27.0	67.0	1.39	0.11	0.15	0.15	0.11
5, F	Hexane	27.0	165.0	5.01	0.11	0.56	0.56	0.42
							0.71	0.53
6, H	Ethanol	35.0	103.9	0.89	1.00	0.89	0.89	0.33
6, H	Hexane	35.0	229.7	2.95	1.00	2.95	2.95	1.37
							3.83	1.71
7, GE	Eth Acet	35.0	150.9	21.68	1.00	21.68	21.68*	9.14
7, GE	Ethanol	35.0	103.9	2.13	1.00	2.13	2.13	0.79
7, GE	Hexane	35.0	229.7	45.32	1.00	45.32	45.32*	20.64*
							69.13*	30.57*

(*This chart continues on next page.*)

What to do with an emissions inventory?

What do you do with an emissions inventory after it is complete? The first use of an inventory is as a tool for a regulatory compliance evaluation. With more accurate information concerning what and how much is being emitted into the air, the ECAP team can identify which current or upcoming regulations affect them and the most cost-effective measures to comply. Thus, the emissions inventory should address both current and anticipated future conditions.

In fact, examining future operations in the plant is a critical part of evaluating regulatory compliance, and any anticipated changes in the plant should be accounted for in the evaluation. For example, if a piece of process equipment is currently used in the manufacturing of two different products, future demand may dictate the increased production of one at the expense of the other. If such a change in production patterns is anticipated, this should be noted in the inventory, because it may result in greater (or at least different) emissions, which

LOCATION: Malcolm Pirnie, Inc.
INT.: Step #1
CYCLE: 35.0 h PAGE 1

⎯ LB IN AN HOUR ⎯							
(3) COND 10.0C	(4) COND −10.0C	EFF	SCRUBR	CONDS COLMN	(6) SCRUBR & COND	CONTRL COLMN	Emissions lb/run
71.42*	21.47*	0.00	183.07*	2	125.44*	1	213.58*
1.08	0.40	0.00	2.22	2	1.67	1	2.22
0.31	0.11	0.00	0.63	2	0.48	1	0.63
0.44	0.16	0.00	0.90	2	0.67	1	0.90
0.06	0.02	0.90	0.02	2	0.01	1	0.15
0.27	0.10	0.00	0.56	2	0.42	1	0.56
0.33	0.12		0.58		0.43		0.71
0.17	0.04	0.90	0.09	2	0.03	6	0.03
0.82	0.27	0.00	2.95	2	1.37	6	1.37
0.99	0.31		3.04		1.40		1.40
4.85	1.26	0.90	2.17	2	0.91	6	0.91
0.38	0.08	0.90	0.21	2	0.08	6	0.08
11.61*	3.45	0.00	45.32*	2	20.64*	6	20.64*
16.85*	4.79		47.70*		21.63*		21.63*

could influence the regulatory evaluation. Once future operating scenarios are identified, regulatory compliance strategies can be finalized and implemented and modifications of permits completed.

The ECAP team can also use the emissions inventory as a start toward an emissions reduction program, as a vehicle for prioritizing emission points for reductions. For instance, the team may wish to select those emission points that have high emission rates, are relatively easy to control, emit toxic compounds, or are regulatory concerns.

The team can then assess different options to reduce emissions from each point. Strategies include recycling, process modifications, solvent substitutions, and installation of air-pollution-control equipment. For example, operation of equipment such as condensers and carbon adsorption units can pay for itself in terms of solvent that may be recovered and reused by the plant.

After the team has selected the strategies, it should follow up on their implementation and revise the emissions inventory. To be most useful, the inventory should remain a "living document" that is automatically updated whenever any process or environmental changes are implemented in the plant.

In conjunction with available database management software programs, process and emissions information gathered and produced by the ECAP team can be used to determine compliance with other regulations without having to search for the data again. The process-related information developed in the emissions inventory can also be used to help prepare process safety management plans. A database management software system makes the transfer and handling of data more convenient. CEP

R. LI, P.E., is an associate and manager of air quality engineering at Malcolm Pirnie, Inc., White Plains, NY (914/641-2648; Fax: 914/ 641-2455). He has 16 years of experience in air-quality permitting, emissions inventories, air-pollution control, and monitoring. He received a BS in chemical engineering from Columbia Univ. and an MBA from Pace Univ. He is a licensed professional engineer in New York and New Jersey, is a member of AIChE, and has published many articles on air-pollution control.

M. KARELL, P.E., is a senior project engineer at Malcolm Pirnie, White Plains (914/641-2653; Fax: 914/641-2455). He has 10 years of experience in air-quality permitting, emissions inventories, air-pollution control, and monitoring for a variety of chemical process industries, and has worked in industry, consulting, and the government. He has a BS in biochemistry from New York Univ., an MS in biochemistry from the Univ. of Wisconsin, and an MS in chemical engineering from Columbia Univ. He is a licensed professional engineer in New York, is a member of AIChE, and has published many articles on industrial air-pollution issues.

J. W. BODDY, P.E., is an associate and coordinator of air quality services for the Upstate New York offices of Malcolm Pirnie, Inc., Pittsford, NY (716/248-5161; Fax: 716/248-2332). He has over 21 years of experience working with industry with a broad background in process systems design. Air project experience includes audits, emissions inventories, permitting, and emissions control evaluations. He has a BS in thermal and environmental engineering from Southern Illinois Univ., and is a member of AIChE. He has recently published papers on "best management practices" for air emissions reductions.

VOC Control: Current Practices and Future Trends

Users and suppliers of abatement technology project where VOC control is headed over the next several years.

Edward C. Moretti and Nikhiles Mukhopadhyay,
Baker Environmental, Inc.

One of the most formidable challenges posed by the Clean Air Act Amendments of 1990 (CAAA) is the search for efficient and economical control strategies for volatile organic compounds (VOCs). VOCs are precursors to ground-level ozone, a major component in the formation of smog. Under the CAAA, thousands of currently unregulated sources will be required to reduce or eliminate VOC emissions. In addition, sources that are currently regulated may seek to evaluate alternative VOC control strategies to meet stricter regulatory requirements such as the maximum achievable control technology (MACT) requirements in Title III of the CAAA.

Because of the increasing attention being given to VOC control, the American Institute of Chemical Engineers' (AIChE) Center for Waste Reduction Technologies (CWRT) initiated a study of VOC control technologies and regulatory initiatives. A key objective of the project was to identify and describe existing VOC control technologies and air regulations, as well as emerging technologies and forthcoming regulations. That work is the basis for this article. (Two other objectives of the study were to present economic analyses for commonly used VOC control technologies and to provide guidance on selecting abatement technologies, but those aspects of the project are not covered here.) Further details of the project can be found in *(1)*. [For information on obtaining the report, circle No. 121 on the Reader Inquiry card in this issue.]

The following technologies were identified as existing or emerging "end-of-pipe" VOC abatement controls:
- thermal oxidizers (thermal incinerators);
- catalytic oxidizers (catalytic incinerators);
- flares;
- condensers;
- adsorbers;
- absorbers;
- boilers/process heaters;
- biofilters;
- membrane separators;
- ultraviolet oxidizers;
- corona destruction reactors; and
- plasma technology devices.

Table 1 summarizes most of these technologies and their applicability ranges. Corona destruction and plasma technology are not included because of their relative early development status, but they are described briefly in the next section.

Current practices

Because of their broad applicability to a wide variety of VOC emission streams, thermal and catalytic oxidizers, flares, condensers, and adsorbers are the most popular VOC controls in use today. Oxidizers and flares are destruction devices, where VOCs are combusted, reduced, or otherwise destroyed without being recovered. Adsorbers and condensers, on the other hand, are recovery devices, where VOCs are recovered and often reused. The selection of a destruction vs. a recovery device is usually based on the economic benefit of recovering the VOCs in the waste gas stream.

Absorbers and boilers/process heaters are used less often for VOC control. In general, absorbers are less effective than other abatement technologies and are more commonly used to recover inorganic compounds. Boilers and process heaters are not used solely for VOC abatement; however, existing boilers or process heaters may be used to destroy VOCs as long as the safety and reliability of the primary process is not adversely affected.

Three of the technologies identified by survey participants — biofiltration, membrane separation, and ultraviolet (UV) oxidation — have only recently become commercially available for VOC control. Biofiltration involves the use of soil or compost beds containing microorganisms to convert VOCs into carbon dioxide, water, and mineral salts. [For further discussion of biofiltration, see (2,3).] Membrane separation refers to the use of a semi-permeable polymeric membrane to separate VOCs from a waste gas stream (4). UV oxidation uses oxygen-based oxidants, such as ozone, peroxide, and OH^- and O^- radicals, to convert VOCs into carbon dioxide and water in the presence of UV light (5).

Two of the technologies identified —

The survey

Some of the information necessary to identify current practices and possible trends was obtained via surveys of: users and suppliers of VOC abatement technologies, for detailed technical information and unique applications of current and emerging technologies; university-based research centers, which may be at the forefront of VOC technology development; and CWRT members and other cooperating organizations, for information about the impacts of forthcoming VOC regulations on technology development and corporate environmental policymaking.

In all, 294 surveys were conducted, either by telephone or by mail. Eighty-six responses were received for an overall response rate of 29%. The table below profiles the respondents. The chemical process industries (CPI) were represented by 22 responses from operating companies and 7 from consulting firms that serve the CPI.

Group	Number of Responses
Equipment Suppliers	26
Operating Companies	21
University-based Research Centers	15
AIChE/CWRT Members	17
Cooperating Organizations	7
Total	**86**

Table 1. Summary of VOC control technologies.

Technology	Emission Source	VOC Category	Emission Stream Flow Rate, scfm	VOC Concentration, ppmv
Thermal Oxidation	PV, ST, TO, WW	AHC, HHC, A, K	<20,000 without with heat recovery ≥20,000 with heat recovery	20–1,000 without heat recovery ≥1,000 with heat recovery
Catalytic Oxidation	PV, ST, TO, WW	AHC, HHC*, A, K	Unlimited	50–10,000
Flaring	F, PV, ST, TO, WW	AHC, A, K		
Condensation	PV, ST, TO	AHC, HHC*, A, K	≤2,000	5,000–12,000
Adsorption	PV, ST	AHC, HHC, A	≥300	20–20,000
Absorption	PV, ST, TO	A, K	≥1,000	1,000–20,000
Boilers and Process Heaters	PV	AHC, A, K		
Biofiltration	PV, WW	AHC, HHC*, A, K	Unlimited	500–2,000
Membrane Separation	PV, TO	AHC, HHC, A, K		0–1,000
Ultraviolet Oxidation	PV	AHC, HHC, A, K		

Legend:
Emission Source: F = Fugitives; PV = Process Vents; ST = Storage Tanks; TO = Transfer Operations; WW = Wastewater Operations
VOC Category: AHC = Aliphatic and Aromatic Hydrocarbons; HHC = Halogenated Hydrocarbons (* = Limited Applicability); A = Alcohols, Glycols, Ethers, Epoxides, and Phenols; K = Ketones and Aldehydes

corona destruction and plasma technology — are not yet commercially available. In corona destruction, energetic electrons are generated in a high-intensity reactor, where they collide with VOCs to produce nonreactive compounds such as carbon dioxide and water (6). The exact mechanism by which the excited electrons react with VOCs is not well known at this time. Early results indicate that corona destruction appears to be effective for lean VOC streams. It does not appear to produce intermediate hazardous compounds that would require disposal as hazardous waste. It operates at ambient temperature, does not require auxiliary fuel, and can treat halogenated and nonhalogenated compounds.

Plasma technology is not yet well known as a VOC abatement technology, although research and development are proceeding in the United States and Europe. [More information on plasma technology can be found in (7,8).]

The evaluation and selection of an appropriate VOC abatement technology depends upon factors such as the environmental, economic, and energy impacts of installing, operating, and maintaining the equipment. While the priority of these selection criteria may vary among companies or even individual process units within the same company or plant, the fundamental decision-making philosophy remains the same. The article following this one, by Ruddy and Carroll, covers the selection procedure in detail and provides additional information for evaluating the technologies.

Impacts of regulations

According to the survey participants, the stringency of VOC regulations that will emerge from the CAAA will be driven by the real and publicly perceived health effects of VOCs and ozone. The nonattainment provisions (Title I) and the hazardous

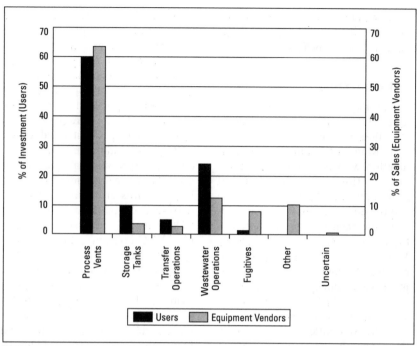

■ *Figure 1. More than half of the capital expenditures are expected to be for the control of process vents.*

air pollutant provisions (Title III) of the CAAA are expected to have the greatest impacts on regulated sources.

Most survey respondents feel that regulatory compliance is frequently the driving force behind VOC abatement technology developments or modifications. Many of the design or operating improvements that have been made to VOC control equipment to optimize performance were made in response to stringent state and federal restrictions on allowable emissions. As a result of these technological improvements, some VOC abatement equipment can achieve control efficiencies exceeding 99%.

However, there is an additional cost associated with achieving incrementally higher efficiencies, and these economic impacts are often the primary concern of regulated companies. The costs associated with regulatory compliance not only include the capital and annualized costs of abatement strategies but also the costs for required actions such as record keeping and reporting, emissions monitoring, air permitting, and

voluntary actions such as routine inspections or periodic training programs to ensure compliance.

Most participants feel that VOC regulations will become increasingly more stringent in the future. They believe that the long-term effect of the U.S. Environmental Protection Agency's (EPA's) regulatory agenda will prompt most operating companies to evaluate either volume reduction or pollution prevention alternatives to VOC abatement.

Reducing the volume of VOC-laden exhaust streams (for example, by concentrating the VOCs in an adsorber prior to catalytic or thermal oxidation) may lower capital costs by reducing the size of control equipment. Annualized operating costs would also be expected to decrease due to lower fuel and other operating expenses.

Preventing pollution can be even more cost-effective, since cost savings can be realized from the elimination of expensive pollution control equipment. Additionally, a company's public image may be enhanced by its decision to implement a pollution prevention strategy. [For further information on minimizing VOC emissions through process changes, see (9).]

Future trends

The incentive to reduce or eliminate VOC emissions has evolved from three driving forces:

1. Many VOC emitters, in conjunction with their associated trade organizations (*e.g.,* the Chemical Manufacturers Association), have responded to an imperative to preserve the environment and minimize the adverse effects of atmospheric VOC releases. This proactive commitment to environmental management has resulted in significant and permanent reductions in VOC emissions over the past decade.

2. Many emitters are faced with the economic imperative to recover and reuse VOC-containing solvents, especially if the VOCs are expensive commodities.

3. Many emitters are required to comply with government regulations in order to avoid costly penalties.

These three forces will continue to drive the efforts of regulated sources toward pollution prevention and VOC emissions control.

In the CWRT project, the future trends in VOC control were characterized based on the predictions by users and suppliers of VOC control equipment. Users were asked about their approximate capital budgets for VOC control, and equipment suppliers were asked about projected U.S. sales, both over the next five years. Expenditures are characterized by five criteria: type of emission source, VOC category, exhaust stream flow rate, VOC concentration, and VOC abatement technology.

Type of emission source. Process vents will, by far, require the largest share — almost two-thirds — of the projected investment in VOC abatement technologies within the next five years, as illustrated in Figure 1. The main reason for this may be that many forthcoming VOC regulations will focus entirely or partly on process vents, including the Hazardous Organic National Emission Standard for Hazardous Air Pollutants (NESHAP) [or HON, for Hazardous Organic NESHAP] and the petroleum refining NESHAP.

Users and suppliers both project that VOC controls for wastewater operations will account for the second-largest portion of the total reported investment. VOC emissions from wastewater operations are expected to receive increased focus from regulatory agencies, especially after promulgation of the HON and the petroleum refining NESHAP, both of which will include specific control requirements for wastewater operations.

Relatively small amounts of investment will be directed toward controlling VOC emissions from storage tanks, transfer operations, and fugitive sources (mainly equipment leaks). While VOC emissions from transfer operations may frequently vent to an add-on control device (*e.g.,* carbon adsorber or condenser), the most effective control options for storage tanks and equipment leaks are not add-on control equipment, but rather tank modifications (*e.g.,* seal systems) or work practices (*e.g.,* equipment leak detection and repair).

VOC category. As shown in Figure 2, projected expenditures are fairly evenly split among technologies for reducing aliphatic hydrocarbons, aromatic hydrocarbons, halogenated hydrocarbons, and alcohols, glycols, ethers, epoxides, and phenols. These categories of VOC constitute the majority of hazardous air pollutants (HAPs) that must be controlled under Title III of the CAAA of 1990.

Emission stream flow rate. Users and suppliers of VOC abatement technologies made significantly different estimates based on exhaust stream flow rate. These results are shown in Figure 3.

The users project that about 40% of their capital expenditures will be used to control VOC emissions from low-flow streams (<500 scfm) and that 80% of their outlays will be used to control low- to mid-flow streams

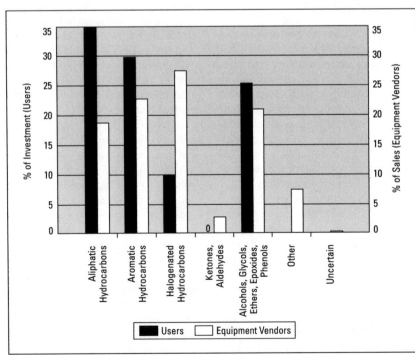

■ *Figure 2. Investments will be split among different VOC categories.*

AIChE's Center for Waste Reduction Technologies

The Center for Waste Reduction Technologies (CWRT) is a partnership among industry, academia, and government that serves as a focal point for research, education, and information exchange on the innovative waste reduction technologies needed for economically competitive processing and manufacturing facilities.

CWRT's unique program revolves around three components: targeted research, technology transfer, and enhanced education. The research program will be based on the identification of target waste streams and the development of a hierarchy of technological solutions to effect their elimination or reduction. Technology transfer and related information will be disseminated through tools like practical "how-to" publications, topical conferences, and continuing education courses for practicing engineers, as well as through cooperation with other organizations that have related interests. And, to ensure that environmentally compatible design becomes a permanent feature of industrial practice, CWRT will develop new course materials for undergraduate and graduate engineering curricula and promote student internship programs.

Established in 1989, the Center currently has 27 sponsoring members. Among them are operating companies (chemical, petroleum, and diversified manufacturing), the engineering and construction contractors that serve them, and firms specializing in environmental consulting services.

To obtain more information about the Center or to inquire about becoming a sponsor, contact Lawrence L. Ross, CWRT Director, at 212/705-7407; Fax: 212/752-3297.

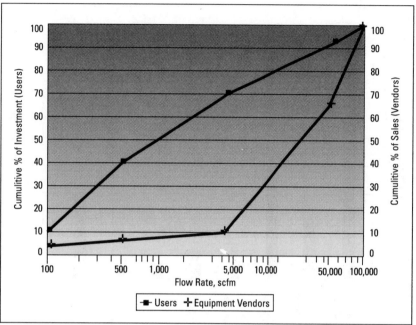

■ *Figure 3. Users plan more expenditures for low- to mid-flow streams, whereas suppliers project greater sales for high-flow streams.*

(<5,000 scfm). Low-flow streams, which have typically been exempt from regulations, will likely be subject to regulations and require a large share of the projected investment on VOC control.

Information from suppliers, however, does not support the notion of a greater need for low-flow VOC abatement technologies. Nearly 90% of the estimated domestic sales is projected to be for control of high-

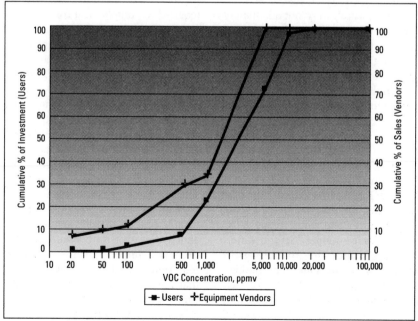

■ *Figure 4. Equipment suppliers and users differ on their projections regarding control of lean VOC streams.*

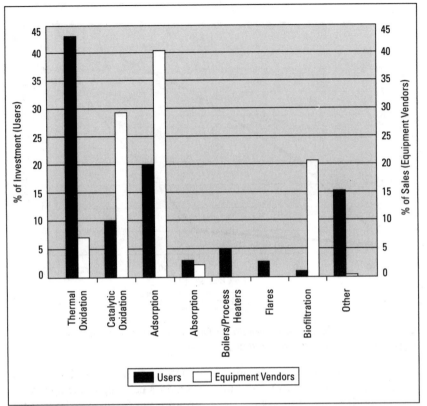

■ *Figure 5. The most striking differences between the estimates of users and equipment vendors deal with investments for various types of control equipment.*

flow streams (>5,000 scfm). Equipment suppliers may be basing their projections on historical sales profiles, which may not adequately describe the variety and number of sources that will be affected by forthcoming regulations. On the other hand, some of the investment to control high-flow streams can be attributable to the large number of currently unregulated dilute emission sources that will be subject to future regulation and to the replacement of existing controls with new or alternative technologies.

VOC concentration. Figure 4 illustrates the distribution of capital expenditures and sales among VOC concentration ranges. Several differences between the projections of users and suppliers are worth noting.

Users expect less than 8% of their expenditures to be for the control of lean VOC exhaust streams (<500 ppm) and nearly half to be spent on streams with VOC concentrations

ranging from 1,000 ppm to 5,000 ppm. There are two possible reasons for this. First, facilities that will be required to comply with forthcoming regulations may believe that lean VOC streams may be exempt from regulatory action. Alternatively, facilities may be considering alternative control strategies instead of end-of-pipe technologies for lean VOC streams, since the cost to control lean streams may not justify the use of expensive end-of-pipe controls.

Suppliers of VOC abatement technologies project that 99% of their sales will be attributable to VOC streams less than 5,000 ppm. In this case, it appears that the suppliers are anticipating that forthcoming regulations will impose stringent control requirements on lean VOC streams that have historically been exempt from regulatory action.

Control technology. Figure 5 illustrates the distribution of investment among types of VOC abate-

Literature Cited

1. **AIChE Center for Waste Reduction Technologies,** "Current and Potential Future Industrial Practices for Reducing and Controlling Volatile Organic Compounds," American Institute of Chemical Engineers, New York, NY (Sept. 1992).
2. **Bohn, H.,** "Consider Biofiltration for Decontaminating Gases," *Chem. Eng. Progress,* **88**(4), pp. 34–40 (Apr. 1992).
3. **Leson, G., and A. M. Winer,** "Biofiltration: An Innovative Air Pollution Control Technology for VOC Emissions," *J. Air and Waste Management Association.,* **41**(8), pp. 1045–1054 (Aug. 1991).
4. **Baker, R. W., et al.,** "Membrane Vapor Separation Systems for the Recovery of VOCs," presentation at the 85th Annual Air and Waste Management Association Meeting and Exhibition, Kansas City, MO (June 1992).
5. **Shugarman, L. E.,** "Ultraviolet/ Activated Oxygen: A New Air Pollution Control Technology Comes of Age," presentation at the 84th Annual Air and Waste Management Association Meeting and Exhibition, Vancouver, BC (June 1991).
6. **Nunez, C. M., et al.,** "Corona Destruction: An Innovative Control Technology for VOCs and Air Toxics," presentation at the 85th Annual Air and Waste Management Association Meeting and Exhibition, Kansas City, MO (June 1992).
7. **Thorpe, M.,** "Plasma Energy: The Ultimate in Heat Transfer," *Chem. Eng. Progress,* **85**(7), pp. 43–53 (July 1989).
8. **Ondrey, G., and K. Fouhy,** "Plasma Arcs Sputter New Waste Treatment," *Chem. Eng.,* **98**(12), pp. 32–35 (Dec. 1991).
9. **Chadha, N., and C. S. Parmele,** "Minimize Emissions of Air Toxics via Process Changes," *Chem. Eng. Progress,* **89**(1), pp. 37–42 (Jan. 1993).

Further Reading

U.S. Environmental Protection Agency, "Control Technologies for Hazardous Air Pollutants," EPA/625/6-91/014 (June 1991).

E. C. MORETTI is an engineer with Baker Environmental, Inc., Coraopolis, PA (412/269-6055; Fax: 412/269-6097), where his current responsibilities include air permitting and emissions inventory and estimation procedures for hazardous air pollutants. Prior experience includes regulatory support to the EPA in the development of several National Emission Standards for Hazardous Air Pollutants (NESHAPs) and New Source Performance Standards (NSPSs). He is the main author of an inspection manual prepared for EPA for enforcement of several equipment leak NSPSs and has extensive experience in equipment leak emission estimation and control. He is a 1984 graduate of the Univ. of Pittsburgh with a BS in chemical engineering and is a member of AIChE.

N. MUKHOPADHYAY is manager of process and environmental services with Baker Environmental, (412/269-6138), where he is primarily responsible for overseeing all of Baker's pollution prevention, waste minimization, and process engineering services. Prior to joining Baker, he served as director of environmental affairs for Ciba-Geigy Corp.; he also worked for Sherwin Williams Chemicals, CFBraun, and Tenneco Chemicals. He has 25 years of industrial and consulting experience in the fine chemicals, petroleum, and pharmaceuticals industries. He holds a master's degree in chemical engineering from the Univ. of Colorado, Boulder, a BSChE from Jadavpur Univ., Calcutta, India, and a BS in chemistry from Calcutta Univ.

Future research needs

The primary focus of the CWRT study was end-of-pipe technologies for controlling atmospheric VOC releases. The project also identified several additional VOC-related topics that invite further research as the imperative to reduce or eliminate VOCs becomes increasingly important. These include:

- Evaluation of source reduction, waste minimization, or pollution prevention initiatives for VOC abatement.
- In-depth study of the performance capabilities, as well as environmental and economic impacts, of combination VOC abatement technologies, such as catalytic-oxidation/carbon-adsorption systems.
- In-depth study of the performance capabilities, as well as environmental and economic impacts, of biofiltration, membrane separation, UV oxidation, corona destruction, and plasma technology.
- Development of a single computer software package for estimating capital and annualized costs of all the VOC abatement technologies described in the report.
- Evaluation of continuous emissions monitors and their role in forthcoming VOC regulations.
- In-depth study of "leakless" process equipment to identify the design and operating criteria that will meet maximum achievable control technology (MACT) requirements for equipment leak standards.
- Study of VOC removal from groundwater, wastewater, and soil.
- Study of VOC control technologies for indoor air, particularly the industrial workplace.

ment technologies. For nearly all types of equipment, significant differences exist between the projections of users and suppliers.

Users project that about 43% of their capital expenditures will be spent to purchase thermal oxidizers, approximately 10% to buy catalytic incinerators, and nearly 20% to purchase adsorption equipment. Absorbers, boilers and process heaters, flares, and biofiltration systems combined are projected to account for approximately 12% of capital investment. The remaining 15% of the expenditures will be spent on other technologies — respondents specifically identified condensers and membrane systems in this category.

Some of the reasons for these projections may be the traditionally conservative philosophy of many operating companies with regard to regulations. Many firms trust that proven, mature technologies will meet their performance needs. Newer technolo-

gies may be overlooked or avoided until their performance capabilities have been exhaustively evaluated. In addition, technologies such as boilers, process heaters, and flares are not widely considered to be dedicated VOC abatement technologies, so they are not considered as such during VOC equipment evaluation and selection.

Control equipment suppliers estimate that 7% of their sales will be attributable to thermal oxidizers, 29% to catalytic oxidizers, 40% to adsorbers, 2% to absorbers, 21% to biofiltration systems, and a small percentage to other technologies. Consistent with the information from users, sales of absorbers, boilers and process heaters, and flares are expected to be minimal. **CEP**

Select the Best VOC Control Strategy

Use this comparison of emission control technologies to identify those suitable for the application at hand.

**Edward N. Ruddy
and Leigh Ann Carroll,**
Burns & McDonnell
Engineering Co., Inc.

Volatile organic compounds (VOCs) are among the most common pollutants emitted by the chemical process industries (CPI). Accordingly, VOC emission control is a major portion of the CPI's environmental activities.

Reduction of VOC emissions in areas that exceed the current national ambient air quality standard for ozone of 0.12 ppm is mandated under Title I of the Clean Air Act Amendments of 1990. In addition, Title III of the Amendments requires reduction of the emission of 189 hazardous air pollutants, most of which are included under the definition of VOCs as well. Thus, some sources may be controlled under two separate sets of regulations with potentially differing requirements. This makes the selection of controls especially difficult — what may be the cost-effective choice today could prove inadequate in the near future and costly to upgrade. Therefore, intelligent planning is essential for defining a cost-effective emissions control strategy.

This article discusses VOC emission control from a process design standpoint. It first outlines the factors that must be considered when selecting a technology. Then it reviews the technologies available for VOC control and where each can and cannot be used.

The article focuses mainly on add-on (or "end-of-pipe") controls rather than process modifications, as the latter were covered recently in *(1)*. While process and equipment modification is generally the preferred alternative for reducing emissions, in most CPI plants some type

of end-of-pipe control will still be necessary. The technologies discussed are primarily applicable to point sources of emissions (that is, vents or stacks) rather than fugitive emissions (or equipment leaks), which were discussed extensively in *(2–5)*.

Know your emissions

VOCs are carbon compounds that react with nitrogen oxides and other airborne chemicals, in the presence of sunlight (photochemically), to form ozone, which is a primary component of smog. VOCs include most solvents, thinners, degreasers, cleaners, lubricants, and liquid fuels. Table 1 is a brief list of some common VOCs. The list is not complete. Note the last three items are basic categories of organic

Table 1. Some common VOCs.
Toluene
Xylene
Isopropyl alcohol
Glycol ethers
Petroleum distillates, naphthas, and mineral spirits
Methyl ethyl ketone (MEK)
Acetone
Paraffins
Olefins
Aromatics

chemicals encompassing thousands of individual compounds.

The U. S. Environmental Protection Agency (EPA) will not regulate, under the Clean Air Act's ozone provisions, organic compounds that do not contribute to ground-level ozone. Such compounds include methane, ethane, carbon monoxide, carbon dioxide, carbonic acid, metallic carbides or carbonates, ammonium carbonate, 1,1,1-trichloroethane, methylene chloride, and various chlorofluorocarbons and perfluorocarbons. [For a complete listing, see (6).]

The first task in evaluating VOC control techniques is to prepare a comprehensive emissions inventory. The emissions inventory provides the basis for planning, determining the applicability of regulations, permitting, and the selection of control options for further consideration.

The inventory should cover the entire facility source-by-source. Each piece of equipment within the facility should be characterized based on:

• pollutant(s) emitted;

• the individual chemical species within each vent stream (to identify any non-VOC materials that may have detrimental effects on particular types of control equipment);

• hourly, annual, average, and worst-case emissions rates;

• equipment status;

• existence and condition of current pollution control equipment; and

• regulatory status.

For further information on how to compile an emissions inventory, see (7–9).

Selection issues

The available emission reduction options depend upon the nature of the process. For example, many simple actions, such as the covering of vessels, can significantly reduce emissions with minimum expenditure. A comprehensive VOC emissions reduction strategy addresses the VOC content of the raw materials, process operating characteristics,

final product requirements, and the potential application of end-of-pipe control equipment. The options discussed here are generally applicable to emissions sources and should be considered in one form or another in the planning stages of any emissions reduction program.

Process modifications. Process modifications are usually the preferred alternative for reducing emissions. Modifications include the substitution of materials to reduce VOC input to the process, changes in operating conditions to minimize the formation or volatilization of VOCs, and the modification of equipment to reduce opportunities for escape into the environment. The first two mechanisms vary with the process being

---

When evaluating VOC abatement technologies, the first step is to compile a comprehensive emissions inventory.

---

addressed and are not further discussed here.

Equipment modification can take many forms, but the objective is always to prevent the escape of VOCs. VOCs can be emitted through open vessel tops, vents, or leaks at flanges or valves, or they may be the result of process conditions (such as spray painting). The starting point is the vessel or structure in which the operation takes place. Vessels can be capped or equipped with rupture disks or pressure/vacuum vent caps to contain vaporous emissions. Monitoring and repair programs can be instituted to consistently reduce emissions due to leaks from valves, pumps, and process piping flanges; the latter might even be replaced with welded piping connections in order to reduce emissions.

Similarly, process enclosures can be designed to reduce emissions. By enclosing the source, a positive means of collecting the emissions can be provided. However, simply providing an enclosure is not enough to reduce emissions. If the emissions are captured in the enclosure but no additional measures are taken, the pollutants will eventually escape into the environment as raw materials and products are transferred into and out of the enclosure. Typically this situation is handled by end-of-pipe solutions discussed later.

Add-on controls. In addition to, or sometimes instead of, process modifications, add-on controls are often used for emission control. The most common such technologies include: thermal oxidation, catalytic oxidation, condensation, carbon adsorption, and absorption.

End-of-pipe controls are often the more expensive options available. But those methods incorporating recycling of wastes can produce paybacks, significantly reducing overall costs. Because capital and operating costs are strongly dependent upon the airflow to be treated, some degree of process modification to reduce the volume of air needing treatment generally accompanies the installation of these units. In some cases the systems here may be used in parallel modules to accomodate large airflows.

Equipment selection considerations. Listed below are ten factors that are important in the selection of VOC control equipment. These general information items should be identified and considered early in the selection process and used to guide further evaluations.

1. Recycling potential. Recycling presents the opportunity to partially recover costs of control equipment. Before the project has progressed significantly, the viability of this option must be assessed. Can the recovered materials be reused in the process? In another process within the operation? Used as boiler fuel? Perhaps reused as an equipment wash solvent? Sold to another user in the area? If recycling is desirable and achievable, further consideration of direct destruction devices such as oxidation systems may be unnecessary.

2. Variability of loading. The nature of the process being controlled determines the variability of the loading to which the control device will be exposed. High flow and/or concentration variability, such as may exist with batch operations, can produce additional wear and tear on equipment, reduce thermal energy recovery efficiency, and possibly reduce the actual Destruction and Removal Efficiency (DRE) of the selected device.

3. Average loading. The average loading determines the applicability of various processes. Recovery options such as direct condensation require high inlet concentrations to operate successfully. Very low average inlet concentrations may require concentration prior to actual treatment.

4. Diversity of VOCs present. The mixture or mixtures of VOCs to be treated in the control equipment has an effect on system applicability. The greater the variability of substances to be controlled, the greater the limitation placed upon the selection process. The DRE is dependent upon the "worst player" (that is, the most difficult compound to remove or destroy) in the exhaust stream. Recovery systems may require additional separation equipment to recycle materials.

5. Lower explosive limit (LEL). VOC and air mixtures are characterized by explosive limits. Operation between the LEL and the upper explosive limit (UEL) is very dangerous, because the mixture can detonate upon exposure to a spark or other source of sufficient energy. Because the gas mixture in a duct may not be completely mixed at the time of introduction to a control unit, manufacturers of thermal destruction devices normally will not design for mixtures with the potential to exceed 25% of the LEL. The potential for

VOC	LEL, %*	UEL, %*
Toluene	1.2	7.1
Xylene	1.0	7.0
Isopropyl alcohol	2.0	12.7
Ethylene glycol monomethyl ether	1.8	14.9
Petroleum distillates (naphtha)	1.1	5.9
Methyl ethyl ketone	1.4	11.4
Acetone	2.5	13.0
Heptane	1.05	6.7
Naphthalene	0.90	5.9
Styrene	1.1	7.0

Table 2. Lower and upper explosive limits of some common VOCs.

*Percentages are given by volume, in air.

exceeding the LEL must be evaluated during normal process operating conditions and during startup, shutdown, and failure modes. Table 2 lists LEL and UEL values for some common VOCs.

6. Upper explosive limit. Certain operations may routinely produce mixtures that exceed the UEL. In these cases, the high concentrations normally make recovery by condensation a very viable option. However, as the recovery operation proceeds, the mixture will proceed through the explosive zone between the UEL and LEL. Process design must take this into account. Dilution with outside air also carries this danger. Dilution with an inert gas, such as nitrogen or carbon dioxide, which also reduces the oxygen content of the mixture, is a potential alternative.

7. High discharge temperature. Relatively high vent discharge temperatures may preclude the use of condensation or adsorption options because of the cost of cooling the gas stream. However, thermal or catalytic oxidation would benefit from such a preheated gas stream.

8. Non-VOCs. The presence of non-VOC contaminants in the vent stream can produce problems for VOC control equipment. Particulate matter in the exhaust stream can plug adsorption beds and heat recovery modules on thermal oxidizers. Halogenated

materials can be oxidized to form their acidic counterparts, possibly requiring additional pollution control equipment and costly corrosion resistant materials of construction.

9. Location. A central plant location may require that certain technology applications include long ductwork runs to reach peripheral locations. Roof mounted systems may require extensive reinforcement of the roof support system as part of the design.

10. Maintenance. The level of maintenance the facility is capable of providing is also a consideration. If current maintenance is not very sophisticated (for example, the source is neither large nor complex), the choice of a control system requiring continual monitoring and adjustment is not recommended.

Many of these items are contained in the emissions inventory report and should be included in any communication with consultants or equipment vendors to accurately convey the status of the operations to be controlled. With this information in hand, the engineer is ready to begin consideration of control alternatives.

Thermal oxidation

Thermal oxidation systems, also known as fume incinerators, are no longer simple flares or afterburners. The modern thermal oxidizer is designed to accomplish from 95% to 99+% destruction of virtually all VOCs. These systems can be designed to handle a capacity of 1,000 to 500,000 cfm and VOC concentration ranges from 100 to 2,000 ppm. Nominal residence times range from 0.5 s to 1.0 s. Available with thermal energy recovery options to reduce operating costs, thermal oxidizers are very popular.

Thermal oxidation systems combust VOCs at temperatures from 1,300°F–1,800°F. Actual operating temperature is a function of the type and concentration of materials in the vent stream and the desired DRE. Compounds that are difficult to combust or that are present at low inlet concentrations will require greater heat input (greater fuel costs) and retention time in the combustion zone to ensure that the desired DRE is accomplished. High DRE requirements will also require higher temperatures and longer retention times. Inlet concentrations in excess of 25% of the LEL are generally avoided by oxidizer manufacturers because of potential explosion hazards.

Operating temperatures near 1,800°F can produce elevated levels of nitrogen oxides (from nitrogen in the air), a secondary pollutant that may, in turn, require further treatment, such as selective catalytic reduction. Halogenated compounds in the vent stream are converted to their acidic counterpart. Sufficient quantities may necessitate the use of expensive corrosion resistant materials of construction and the use of additional acid gas controls, such as scrubbing, as follow-up treatment.

Two types of thermal energy recovery systems are in common use today, regenerative and recuperative. Both use the heat content of the combustion exhaust stream to heat the incoming gas stream prior to entering the combustion zone.

Regenerative systems, illustrated in Figure 1a, use ceramic (or other dense, inert material) beds to capture heat from gases exiting the combus-

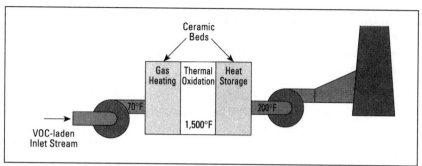
■ *Figure 1a. Regenerative thermal oxidations uses a ceramic bed to capture heat from exiting gases.*

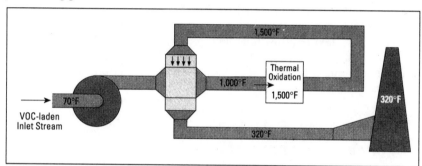
■ *Figure 1b. Recuperative thermal oxidation employs a heat exchanger to recover heat.*

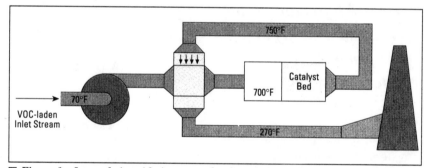
■ *Figure 1c. In catalytic oxidation, a catalyst lowers the combustion temperature.*

tion zone. As the bed approaches the combustion zone temperature, heat transfer becomes inefficient and the combustion exhaust gas stream is switched to a lower temperature bed. The incoming gas stream is then passed through the heated bed where it recovers the captured heat prior to entering the combustion zone. By using multiple beds, regenerative systems have achieved up to 95% recovery of the thermal energy input to the system as fuel and the heat content of the combusted VOCs. Where the incoming gas stream contains sufficient thermal energy potential from VOC combustion, regenerative systems can operate without

external fuel (excluding the need for a pilot light).

The efficiency of the thermal recovery system depends on process operating characteristics. A process where the flow rate and VOC content are relatively constant has a good potential for achieving virtual no-fuel operation. Cyclic processes generally are not as compatible with regenerative oxidation systems. The absorbed heat is lost to the environment during periods of low activity (that is, when air flow or VOC content is reduced). Operation with insufficient VOC content to supply thermal input requirements necessitates the use of external fuel sources.

Recuperative thermal oxidation systems, shown in Figure 1b, recapture thermal energy with a simple metallic heat exchanger, typically a shell-and-tube design. The maximum thermal energy recovery of a recuperative system is around 70% of the fuel and VOC combustion energy input to the system. The advantage over regenerative systems comes from the relatively short period required for the heat exchanger to reach operating conditions. The larger mass of the regenerative heat recovery system requires time and relatively large initial fuel inputs to reach operating conditions, while the recuperative heat exchanger reaches

Condensation produces a liquid that must be treated to remove water and possibly to separate various chemical components.

operating conditions within several minutes of startup. Recuperative systems are best suited to cyclic operations where the versatility of an oxidation system is required along with the ability to respond to cyclic operating conditions.

Catalytic oxidation

Catalytic oxidation systems directly combust VOCs in a manner similar to thermal oxidizers. The major difference is that the catalytic system operates at a lower temperature — typically about 700°F–900°F. This is made possible by the use of catalysts that reduce the combustion energy requirements. As depicted in Figure 1c, the incoming gas stream is heated, most often in a recuperative heat exchanger followed by additional input from burners if needed, and

passed through a honeycomb or monolithic support structure coated with catalyst.

Catalytic systems can be designed to handle a capacity of 1,000 to 100,000 cfm and VOC concentration ranges from 100 to 2,000 ppm. The catalytic system is well-suited to low concentration operations or those that operate in a cyclic manner. They are often used for vent controls where flow rates and VOC content are variable. Destruction efficiencies in excess of 90% are common with a maximum DRE of 95%. High concentration vent streams can also be treated with catalytic technology; however, as with thermal oxidation, vendors are generally unwilling to touch concentrations in excess of 25% of the LEL. Lower operating temperatures, combined with a recuperative heat exchanger, reduce the startup fuel requirement. Large catalytic systems have been installed, but are not as popular as direct thermal oxidation systems at this time due to the high costs of catalyst replacement.

Catalytic systems, like thermal oxidizers, can produce secondary combustion wastes. Halogens and sulfur compounds are converted to acidic species by the catalytic combustion process; these are treated by using acid-gas scrubbers. Also, the spent catalyst materials can require disposal as a hazardous waste if they are not recyclable. However, the lower operating temperature precludes the formation of significant quantities of nitrogen oxides.

Catalyst materials can be sensitive to poisoning by non-VOC materials such as sulfur, chlorides, and silicon. Many catalyst manufacturers have overcome sensitivity to one or more of these substances, but every catalyst has susceptibilities that must be considered in the process

selection stage. For example, some catalysts are sensitive to deactivation by high-molecular-weight hydrocarbons or polymerizing materials. Also, the catalyst support may become deformed in high-temperature, high-concentration situations. Researching these issues should be part of the process selection activity if catalytic oxidation is under consideration.

Table 3. Use this chart to

Control Technology	Applicable Concentration Range, ppm	Capacity Range, cfm	Removal Efficiency
Thermal Oxidation	100–2,000	1,000–500,000	95–99+%
Catalytic Oxidation	100–2,000	1,000–100,000	90–95%
Condensation	>5,000	100–20,000	50–90%
Carbon Adsorption	20–5,000	100–60,000	90–98%
Absorption	500–5,000	2,000–100,000	95–98%

Legend:
Recup = Recuperative
Regen = Regenerative
Fixed = Fixed-Bed
Fluid = Fluidized-Bed

Condensation

The driving force for condensation is over-saturation, which is achieved by chilling or pressurization (or both) of the waste gas stream. Condensation is most efficient for VOCs with boiling points above 100°F at relatively high concentrations — above 5,000 ppm. Low-boiling VOCs can require extensive cooling or pressurization, which sharply increases operating costs.

Exceeding the 25% LEL threshold is more common with condensation systems. In fact, some systems begin operation above the UEL. This is dangerous, however, because the concentration will likely fall through the explosive range during the condensation process. Inert gas blanketing of vessels or unmanned process enclosures can avoid the explosion hazard associated with high VOC concentrations, but cause additional operating costs. Polymerizing materials should also be avoided in condensation systems due to the potential for fouling heat-transfer surfaces.

Best suited to monosolvent systems, condensation produces a liquid product that must be treated to remove condensed water and possibly to separate various chemical species. Recovered VOCs can be reused within the process, used as wash solvents during equipment cleanup, burned as an alternative boiler fuel, shipped off-site for disposal (which is expensive and difficult), or resold for reuse by others. Recovered water should be sent to a wastewater treatment plant prior to discharge if exposed to miscible VOCs.

Carbon adsorption

Carbon adsorption is a very common method of VOC emission control. VOCs are removed from the inlet air by physical adsorption onto the surface of the carbon.

Variable flow rates and VOC concentrations are not disruptive to carbon adsorbers. These systems can be designed to handle a capacity of 100 to 60,000 cfm and VOC concentrations ranging from 20 to 5,000 ppm. They can easily handle VOC concentrations in excess of the 25% LEL threshold mentioned earlier. The system is sized according to the maximum flow and concentrations expected, and anything less usually improves efficiency. Carbon adsorption systems

reen VOC control technologies.

apital osts	Annual Operating Cost	Secondary Wastes	Advantages	Limitations and Contraindications
ecup: 0–200/cfm egen: 0–450/cfm	Recup: $15–90/cfm Regen: $20–150/cfm	Combustion products	Up to 95% energy recovery is possible	Halogenated compounds may require additional control equipment downstream. Not recommended for batch operations.
xed: 0–250/cfm uid: 5–220/cfm	Fixed: $10–75/cfm Fluid: $15–90/cfm	Combustion products	Up to 70% energy recovery is possible	Thermal efficiency suffers with swings in operating conditions. Halogenated compounds may require additional control equipment downstream. Certain compounds can poison the catalyst (lead, arsenic, phosphorus, chlorine, sulfur, particulate matter).
0–80/cfm	$20–120/cfm	Condensate	Product recovery can offset annual operating costs	Not recomended for materials with boiling points >100°F. Condensers are subject to scale buildup, which can cause fouling.
5–120/cfm	$10–35/cfm	Spent carbon; Collected organic	Product recovery can offset annual operating costs. Can be used as a concentrator in conjunction with another type of control device. Works well with cyclic processes.	Not recommended for streams with relative humidity >50%. Ketones, aldehydes, and esters clog the pores of the carbon, decreasing system efficiency.
5–70/cfm	$25–120/cfm	Wastewater; Captured particulate	Product recovery can offset annual operating costs	Might require exotic scrubbing media. Design could be difficult in the event of lack of equilibrium data. Packing is subject to plugging and fouling if particulates are in the gas stream. Scale formation from absorbent/absorber interaction can occur.

are flexible and inexpensive to operate. Installation costs are often lower than those of other systems.

Typically, a carbon adsorption system consists of two parallel adsorption trains. While one carbon bed is on-line, the other is being regenerated. The carbon supplier may retrieve the saturated carbon and either replace it with fresh carbon or regenerate it off-site and return it to the facility. Replacement and off-site regeneration are expensive and result in wastes being generated at another facility with the generator's name attached. The liability associated with this is an unseen and often unconsidered cost of this type of service.

Alternatively, regeneration can be done on-site with steam, hot air, or hot nitrogen. Which method is chosen depends upon process conditions and available local utilities. One advantage of nitrogen regeneration is that it provides an inert atmosphere for operation under otherwise potentially explosive conditions. All three regeneration methods create a concentrated form of the original pollutant(s), which must then be recycled or treated by condensation, thermal oxidation, catalytic oxidation, or some other method. Because the VOCs have been concentrated by the adsorption/desorption process, the cost of the second control unit has been significantly reduced.

Carbon attrition is a fact of life with adsorption systems. The carbon adsorbs VOCs onto active sites on the surface of the porous carbon granule. This adsorptive attraction is overcome by heating in the desorption (regeneration) process. Because the carbon is flammable, the temperature can only be raised to a few hundred degrees during regeneration. When these temperatures are insufficient to overcome the attraction between carbon and VOC, the bond becomes permanent and sites are no longer available for adsorption. Eventually, lost bed capacity results in decreased performance and the bed must be replaced. Waste carbon

must be disposed of in a responsible manner. Certain carbons, due to the type of service in which they were employed, will be classified as hazardous wastes and must be disposed of accordingly.

E. N. RUDDY is a project manager for industrial air-pollution control with Burns & McDonnell Engineering Co., Kansas City, MO (816/333-4375; Fax: 816/333-3690), where he is currently the environmental compliance manager for the expansion of Lambert Airport in St. Louis. Previously, he worked in radioactive waste disposal with the Univ. of Missouri and in air filtration systems with American Air Filter. A registered professional engineer in Kansas, he has both bachelor's and master's degrees in chemical engineering from the Univ. of Missouri, Columbia. He is a member of AIChE and the Air and Waste Management Association.

L. A. CARROLL is an air-pollution control engineer with Burns & McDonnell Engineering Co., where she is serving as environmental engineer for the design and construction of a three-bay aircraft maintenance hangar in Taiwan. She holds a bachelor's degree in chemical engineering from Kansas State Univ. and is a registered engineer-in-training in the state of Kansas. She is a member of AIChE and the Air and Waste Management Association.

Carbon bed performance is sensitive to the moisture content of the gas stream being treated. As the relative humidity of the gas stream exceeds 50%, the performance of the carbon (that is, its ability to adsorb VOCs on a per-pound basis) decreases. This tendency must be taken into account in design calculations to allow for local ambient conditions and the effects of winter weather on outdoor installations.

Carbon adsorption is not recommended for VOC streams containing ketones. The carbamyl carbon (the molecular structural unit that makes the VOC a ketone) can polymerize exothermically on the carbon surface. Continued exposure to ketones has been known to create exothermic conditions and result in bed fires. In order to avoid this problem, the exhaust streams to be treated must be ade-

quately analyzed to determine the potential for ketone presence.

Further details on carbon adsorption for VOC control can be found in the article by Ruhl following this one. The basics of carbon adsorption were covered in *(10)*.

Absorption

Absorption can be used to remove VOCs from gas streams by contacting the contaminated air with a liquid solvent. Any soluble VOCs will transfer to the liquid phase. In effect, the air stream is "scrubbed." This takes place in an absorber tower designed to provide the liquid-vapor contact area necessary to facilitate mass transfer. This contact can be provided by using tower packing or trays, as well as liquid atomization. Packed-bed and mist scrubbing are discussed here.

Absorption systems can be designed to handle a capacity of 2,000 to 100,000 cfm and VOC concentration ranges from 500 to 5,000 ppm. Absorbers can achieve VOC removal efficiencies of 95% to 98%. The design of absorption systems for VOC control is similar to design of absorbers for process applications, using vapor-liquid equilibrium (VLE) data, liquid and vapor flux rates, equipment liquid- and vapor-handling information, and material balances.

Packed-bed scrubbing uses packing material to improve liquid-vapor contact. Packings can either be randomly dumped or stacked in the tower. Packings vary widely in size, cost, contact surface area, pressure drop, and materials of construction, and each packing design has its own advantages under different conditions. Packed-bed scrubbers work well with low-solubility systems due to the higher liquid and vapor residence times (>10 s) associated with the entraining nature of packing. Packed-bed scrubbers should not, however, be used when liquid flow rates are low, which causes inadequate wetting of the packing material. Also, particulate in the incoming air stream or absorbent/absorbate reaction products can foul or plug the packing.

Mist scrubbers use spray nozzles to atomize the liquid stream into tiny droplets. These droplets provide surface area for liquid-vapor contact. Mist scrubbers require a very low pressure drop and are not fouled by particulate in the incoming gas stream. Residence times of the liquid and vapor are low (generally 1–10 s). Therefore, mist scrubbing should only be applied to highly soluble systems.

Absorption is not particularly suitable for cyclic operation due to start-up time constraints. It is, however, good for high humidity air streams (>50% R.H.).

Technology selection

Table 3 summarizes the control technologies discussed in this article. The capital and annual operating costs cited reflect the use of each system at the extremes of applicable concentration and capacity ranges. Price variations are due to the wide ranges of operating conditions and materials of construction.

Wrap-up

VOC controls can take many forms. The least costly is usually modification of the process to eliminate the use of VOC containing materials. However, many processes are not amenable to this approach. In this case, one must determine the options available for control and the regulatory requirements for the particular process, and then select an applicable technology.

The first step in selecting a VOC control method is to characterize the system operations and emissions via an emissions inventory. Then an equipment and operations review should be conducted to identify opportunities to reduce or eliminate emissions by materials substitution or the modification of process equipment.

When these options are exhausted, the emissions inventory and regulatory requirements should be reviewed to determine what additional end-of-pipe control measures should be taken.

Before a review of control devices is conducted, review the operations to identify any opportunities to reduce airflow. Remember, the price of most air-pollution control devices is directly tied to the volume of air processed. Reducing the airflow will result in direct savings of both capital and operating funds. Also, determine if several sources might be combined prior to treatment to reduce the number of control devices at the facility. Once these measures have been completed, the selection of VOC control equipment can begin in earnest. **CEP**

Literature Cited

1. Chadha, N., and C. S. Parmele, "Minimize Emissions of Air Toxics via Process Change," *Chem. Eng. Progress*, **89**(1), pp. 37–42 (Jan. 1993).
2. Colyer, R. S., and J. Meyer, "Understand the Regulations Governing Equipment Leaks," *Chem. Eng. Progress*, **87**(8), pp. 22–30 (Aug. 1991).
3. Schaich, J. R., "Estimate Fugitive Emissions from Process Equipment," *Chem. Eng. Progress*, **87**(8), pp. 31–35 (Aug. 1991).
4. Adams, W. V., "Control Fugitive Emissions from Mechanical Seals," *Chem. Eng. Progress*, **87**(8), pp. 36–41 (Aug. 1991).
5. Brestel, R., *et al.*, "Minimize Fugitive Emissions With a New Approach to Valve Packing," *Chem. Eng. Progress*, **87**(8), pp. 42–47 (Aug. 1991).
6. *Federal Register*, **57**(22), pp. 3945–3946 (Feb. 3, 1992) and Code of Federal Regulations, Part 51 — Requirements for Preparation, Adoption, and Submittal of Implementation Plans.
7. Kuhn, L. A., and E. N. Ruddy, "Comprehensive Emissions Inventories for Industrial Facilities," Proceedings of the Air and Waste Management Association's 85th Annual Meeting and Exhibition, Kansas City, MO, Paper No. 92-139.12 (June 21–26, 1992).
8. Walther, E. G., *et al.*, "How to Develop Your Toxic Emissions Inventory: Approaches, Problems and Solutions," Proceedings of the National Research and Development Conference on the Control of Hazardous Materials, Anaheim, CA (Feb. 20–22, 1991).
9. U.S. Environmental Protection Agency, "Compiling Air Toxics Emissions Inventories," U.S. EPA, Research Triangle Park, NC, EPA/450/4-86-010, NTIS No. PB86-238086 (July 1986).
10. Stenzel, M. H., "Remove Organics by Activated Carbon Adsorption," *Chem. Eng. Progress*, **89**(4), pp. 36–43 (Apr. 1993).

Further Reading

AIChE Center for Waste Reduction Technologies, "Current and Potential Future Industrial Practices for Reducing and Controlling Volatile Organic Compounds," American Institute of Chemical Engineers, New York, NY (Sept. 1992).

Buonicore, A. J., *et al.*, "Air Pollution Engineering Manual," Air and Waste Management Association, Van Nostrand Reinhold, New York, NY (1992).

Marks, J. R., and T. Rhoads, "Planning Saves Time and Money When Installing VOC Controls," *Chem. Proc.*, pp. 42–48 (Mar. 1991)

Patkar, A. N., and J. Laznow, "Hazardous Air Pollutant Control Technologies," *Hazmat World*, pp. 78–83 (Feb. 1992).

PEER Consultants, P.C., Research Triangle Institute, "Organic Air Emissions from Waste Management Facilities," CERI 90-124a, U.S. Environmental Protection Agency, Research Triangle Park, NC, p. 5–4 (Dec. 1990).

Perry, R. H., and D. W. Green, "Perry's Chemical Engineers' Handbook," 6th ed., Chapter 14: Mass Transfer and Gas Absorption, McGraw-Hill, New York, NY (1984).

U.S. Environmental Protection Agency, "EPA Handbook: Control Technologies for Hazardous Air Pollutants," U.S. EPA, Office of Research and Development, Washington, DC, EPA/625/6-91/014 (June 1991).

Minimize Emissions of Air Toxics via Process Changes

Follow these generic strategies to reduce air emissions from CPI plants.

Nick Chadha and Charles S. Parmele,
IT Corp.

Numerous articles have discussed the necessary steps for setting up a pollution prevention program and some of the classical pollution prevention strategies, such as raw materials substitution and waste recycle and reuse. An exhaustive review of industrial pollution prevention programs, procedures, and incentives has recently been prepared by the Risk Reduction Engineering Laboratory at the U.S. Environmental Protection Agency (EPA) *(1)*.

Pollution prevention requires a change in philosophy so that the generation of emissions and wastes is viewed as an inefficiency in the production process rather than as an inevitable environmental problem. Just as with effective safety and quality programs, management commitment is critical for implementing a successful pollution prevention program.

This article discusses opportunities for minimizing emissions of hazardous air pollutants that can be uncovered by looking at production processes very closely and asking questions relating to generation of emissions. Many of the commonsense approaches described here are simple, do not require the development of breakthrough process technologies, and could easily be applied to any size chemical process industries (CPI) facility. The technical strategies discussed are based on experience with the synthetic organic chemical manufacturing industry (SOCMI), the pharmaceutical industry, and other CPI segments that use volatile organic compounds (VOCs) as raw materials or process solvents. Although successful applications from two specific industries are highlighted, the approaches outlined are applicable to most facilities that use or manufacture chemicals.

Incentives for minimizing emissions of air toxics

The term "air toxics" refers to the 189 hazardous air pollutants regulated by the Clean Air Act Amendments (CAAA) of 1990. Nearly half of the chemicals on this list are VOCs. This diverse list includes common solvents, such as methylene chloride and methyl ethyl ketone, and SOCMI products, such as acrylonitrile and ethylene oxide. The incentives for minimizing emissions of air toxics are related to economic and regulatory benefits.

A major advantage of minimizing air toxic emissions is reduced capital and operating costs for air-pollution control equipment. Other benefits include cleaner and safer working conditions, as well as the reduction of potential environmental liabilities and risks. Where it is not possible to completely eliminate emissions via process changes, capture of emissions and reuse in a process can be beneficial, especially for process solvents. This allows recovery of the chemical and/or energy value from streams that did not get converted to products or useful byproducts.

Although prudent management of chemicals makes good business sense and several companies have been focusing on minimizing emissions since the 1970s, recent regulatory developments have heightened industry awareness in this regard. One of these is the requirement for annual reporting of emissions and other releases under the community right-to-know provisions of the Superfund Amendments and Reauthorization Act of 1986, known as SARA Title III. Others include the Pollution Prevention Act of 1990, the Early Reduction Program under Title III of the CAAA, and the EPA's 33/50 Program, which are discussed by Doerr in another article in this issue, pp. 24–29.

Emission sources

SOCMI plants produce a large number of intermediate and finished organic products starting from a small number of basic feedstock materials which, in turn, are derived from crude oil, natural gas, or coal. Major operations involve conversion processes, such as air oxidation, and separation processes, such as distillation.

In SOCMI air oxidation processes, one or more chemicals are reacted with oxygen, which is supplied as air or oxygen-enriched air. Vents from these processes exhaust large volumes of inert gases containing VOCs because the nitrogen in the oxidation air passes through unreacted and carries VOCs out with it. Emissions of VOCs from distillation process vents depend on the operating conditions and various other factors.

Because of the diversity of operations, emissions from SOCMI plants vary widely in the types and concentrations of VOCs emitted.

Pharmaceutical synthesis operations are typically conducted in batch processes and involve one or more chemical reactions followed by a series of product concentration and purification steps. Physically, a pharmaceutical plant consists of reactors, process vessels, centrifuges, dryers,

crystallizers, distillation columns, and other process equipment. Different products are made in the same equipment at different times, or "campaigns," and product purity is critical.

Nitrogen is used abundantly in pharmaceutical plants, for example, for pressure transfer of vessel contents, purging of lines and vessels, direct-contact cooling of vessel contents, and other uses. Volatile organic solvents, including many that are flammable, are commonly used. The nitrogen and solvents typically do not take part in the synthesis reactions.

The liquid nitrogen and organic solvents are usually much cheaper than the high-value pharmaceutical products and byproducts manufactured. Thus, in many cases, there is limited economic incentive for solvent recovery (especially if multiple solvents are involved in the same process), so the waste solvents are often sent to an on-site hazardous waste incinerator for disposal. Solvent losses represent the predominant air toxic emissions from a pharmaceutical plant.

Although each SOCMI and pharmaceutical plant is unique in terms of its process operations and the types of products manufactured, there are similarities when these processes are closely examined from the viewpoint of environmental discharges. As shown in Figure 1, emissions and wastes from any CPI facility represent materials that did not get converted to products and/or useful byproducts. These emissions and wastes are either raw materials, impurities in raw materials, solvents, process catalysts, products, byproducts (useable or nonusable), or products of combustion. Emissions from combustion sources typically contain more inorganic pollutants than organic pollutants, and are not discussed further.

Generic emission categories

Emissions from any CPI facility can be classified into four major categories to provide a structured approach to minimizing emissions. The generic categories used by the EPA to establish New Source Performance Standards, National Emission Standards for Hazardous Air Pollutants, and other air permitting regulations (2) are:

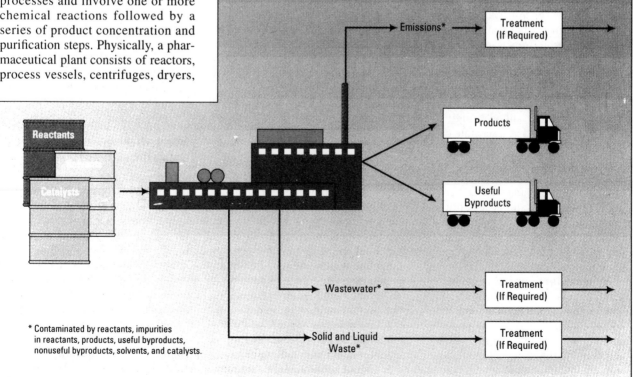

* Contaminated by reactants, impurities in reactants, products, useful byproducts, nonuseful byproducts, solvents, and catalysts.

■ *Figure 1. Materials that are not converted to products or useful byproducts become wastes or emissions.*

• **Storage and handling emissions.** These emissions are a function of the construction and size of the storage tank, the vapor pressure of the stored organic liquid, and ambient conditions at the tank location. For fixed-roof tanks, emissions are the sum of working losses (due to filling and emptying of tanks) and breathing losses (due to changes in ambient temperature or pressure). Handling losses occur from railcar, tank truck, or marine stations used for loading/unloading, and transfer of volatile organic liquids.

• **Process emissions.** Sources for process emissions are stacks and vents from process reactors and recovery and control equipment, such as vent condensers, carbon adsorbers, and absorbers or scrubbers. Process emissions also include losses from downstream product separation and purification equipment such as filters, centrifuges, and distillation columns. For example, if distillation is used for product fractionation, potential emission sources include condensers, product receiver vessels, vacuum pumps, and steam ejectors associated with the distillation unit.

• **Fugitive emissions.** These emissions occur from "leaking" pumps, valves, flanges, open-ended lines, agitator seals, instrument connections, sample connections, and so on. Fugitive emissions from leaking equipment depend on the nature of the chemical process, the level of preventive maintenance, and other site-specific variables. Although fugitive emissions are listed as a separate category, they can occur from many plant sources, including storage tanks and process operations. Examples of process fugitive sources include covers for manholes and access ports, drains from process vessels, and hotwells (troughs used for collecting discharges from steam ejectors and barometric condensers).

• **Secondary emissions.** These refer to organic emissions from wastewater collection and treatment sources, such as trenches, sumps, lift stations, surface impoundments, and aeration basins. This category also includes organic emissions from solvent recovery operations, liquid waste

incineration, accidental spills and leaks, and other miscellaneous sources.

Emissions from storage and handling, fugitive, and secondary sources have been addressed by numerous technical publications and EPA guidance documents. This article focuses specifically on emissions from process sources, which represent the major component of a plant's total air toxic emissions.

Generic ways of creating process emissions

To make the identification of process emissions sources easier, it is helpful to review the generic ways in which emissions are created. In general, emissions are created when a noncondensible material such as air or nitrogen is introduced into a process containing a volatile organic or when any uncondensed material leaves a process.

The relative importance of the different ways of creating emissions can be predicted somewhat — for example, the introduction of air into a vessel will likely create more emissions than displacement due to liquid transfer. However, the exact emission-generating potential will depend on many operating variables and will need to be estimated on a site-specific basis.

Some common generic ways in which process emissions are created are as follows.

1. Introduction of air (oxygen) into reactors for oxidation of volatile organic reactants during chemical synthesis.

2. Introduction of air into reactors when manhole covers are opened for charging solid powders or other raw materials — that is, deliberate ventilation designed to provide an adequate face velocity at the reactor opening.

3. Leakage of air into any process equipment containing volatile organics and operating under vacuum — that is, incidental ventilation due to poor equipment maintenance.

4. Use of liquid nitrogen or dry ice to provide direct-contact cooling in vessels containing volatile organics.

5. Use of nitrogen (or other noncondensible materials) for pressure

transfer of volatile organics from one vessel to another or for blowing lines to clear residual liquids.

6. Use of nitrogen for breaking vacuum or providing an inert atmosphere (via constant purging) in process equipment containing volatile organics.

7. Generation of noncondensible gases (such as carbon dioxide or hydrogen) as a product or byproduct of the reaction.

8. Generation of VOCs with vapor pressures greater than atmospheric pressure at process temperatures (such as ethane or isobutylene) as a product or byproduct of the reaction.

9. Evacuation of vessels containing volatile organics — that is, venting a vessel to vacuum equipment to reduce its operating pressure.

10. Heating of vessels containing volatile organics so as to cause expansion and increase in organic vapor pressures.

11. Stripping of volatile organics from reaction mixtures during vacuum distillations.

12. Boiling of pure solvents in vessels to clean them between batches or different product campaigns.

13. Charging of liquids into a vessel containing volatile organics, or level equalization between two vessels, so as to cause volume displacement.

14. Exhaustion of uncondensed organic vapors from vent condensers and other recovery or control equipment that handles VOCs (this could be due to several reasons).

Start with an accurate emissions inventory

An accurate baseline emissions inventory is required to demonstrate emissions reductions to EPA and to help identify potential opportunities for minimizing emissions. Experience indicates that the "80/20 rule" will apply in many plants — that is, 80% of the emissions will be generated by 20% of the sources. Therefore, it is important to focus on identifying and quantifying the major contributors.

Typically, engineering calculations are used to quantify emissions from process sources. However, this approach has some pitfalls related to the various simplifying assumptions that must be made to estimate process emissions.

In fact, engineering estimates can, in some cases, hide potential opportunities for minimizing emissions. For example, in one pharmaceutical plant, centrifuge exhausts were a major (unidentified) source of solvent emissions because air in-leakage rates were several orders of magnitude higher than what had been assumed in the engineering calculations. This, in turn, was due to high equipment vibration, missing gaskets, age of the machines, and other maintenance factors.

Unidentified, or fugitive, emissions. The total fugitive emissions from leaking process equipment and other sources typically comprise 20% to 50% of the total emissions from a plant. The balance is contributed by stack emissions. However, it is often difficult to close the material balance adequately for batch processes even though the emission estimates developed by engineering calculations are usually conservative. All "unidentified emissions" are then incorrectly accounted as "fugitive emissions" in the baseline emissions inventory.

Many of these emissions, however, are actually unidentified point-source emissions from some not-so-obvious process operations and sources. These sources must be identified by addressing the generic emission-generation mechanisms that were previously discussed.

Review the data. This systematic approach to minimizing emissions begins with a pollution prevention audit to review available emission data and identify data gaps. During this audit, the generic emissions categories and the generic ways of creating emissions are addressed so that any previously unknown point-source emission sources can be uncovered. During problem definition, one needs to go beyond the traditional development of a baseline emissions inventory to ask not only "what and how much," but also "why, how, and when" for emissions from each source.

The generation of emissions fol-

lows repeating patterns that are independent of the specific industry. This is because the true source of air toxic emissions is either the process chemistry, or the way in which the process is engineered, operated, or maintained. For example, if emissions are created due to generation of a volatile organic product or byproduct during a chemical reaction, then the true source of emissions is the process chemistry, and changing the engineering design will not solve the problem.

Identifying the true source of emissions in this way helps identify correct strategies for treating the problem and not just the symptom. The following sections outline some

---

The generation of process emissions follows repeating patterns that are independent of the specific industry.

---

generic strategies for minimizing air toxic emissions and provide specific examples to illustrate those strategies.

Process chemistry modifications

In some cases, emissions are generated by reasons related to the process chemistry, such as the reaction stoichiometry, the kinetics, or the conversion or yield. Emission generation may be minimized by strategies varying from simply adjusting the order in which reactants are added to major changes that require significant process development work and capital expenditures.

Change the order of reactant additions. A pharmaceutical plant made process chemistry modifications to minimize the emissions of an undesirable byproduct, isobutylene, from a mature synthesis process (3). As shown in Figure 2, the process consisted of four batch operations. Emissions of isobutylene were reduced by identifying the process conditions that led to its formation in the third step of the process.

In the first reaction, tertiary butyl

alcohol (TBA) was used to temporarily block a reactive site on the primary molecule. After the second reaction was complete, TBA was removed as tertiary butyl chloride (TBC) by hydrolysis with hydrochloric acid. To improve process economics, the final step involved recovery of TBA by reacting TBC with sodium hydroxide. TBA recovery was incomplete because isobutylene (a chemical related to both TBA and TBC) was inadvertently formed during the TBA recovery step.

Investigation of the true source of these emissions indicated that isobutylene formation was not inevitable. The addition of excess NaOH caused alkaline conditions in the reactor, accidentally creating conditions that favored the formation of isobutylene over TBA. When the order of addition of the NaOH and TBC reactants was reversed and the NaOH addition rate was controlled so as to maintain the pH between 1 and 2, isobutylene formation was almost completely eliminated. As a result of these process changes, it was unnecessary to install add-on emission controls, and the only capital expense was for the installation of a pH control loop.

Change the chemistry. In one plant, odorous emissions had been observed for several years near a drum dryer line used for volatilizing an organic solvent from a reaction mixture. Although there were two dryer/product lines, the odors were noticed in the vicinity of only one line.

Analysis and field testing indicated that the chemical compounds causing the odors were actually produced in upstream unit operations due to the hydrolysis of a chemical additive used in the process. The hydrolysis products were stripped out of solution by the process solvent and appeared as odorous fumes at the dryer. Conditions for hydrolysis were favorable at upstream locations because of temperature and acidity conditions and the residence time available in the process, and the water for hydrolysis was provided by another water-based chemical additive used in the dryer line that had the odor problem.

Because the true cause of the odor-

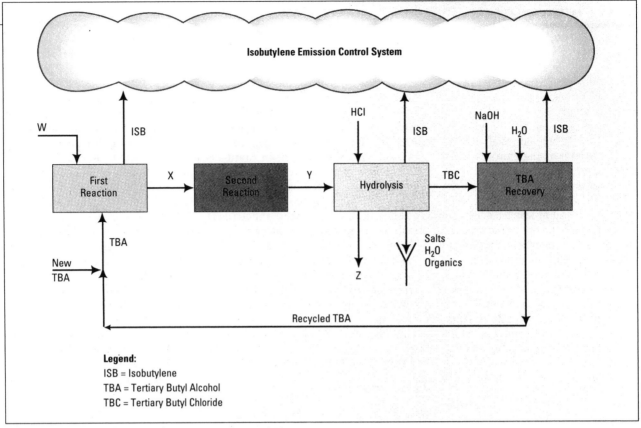

Figure 2. Process chemistry changes can reduce emissions.

ous emissions was the process chemistry, the plant needed to evaluate ways to minimize hydrolysis and the resulting formation of undesirable odorous byproducts. Ventilation modifications to mitigate the odor levels would have served only to transfer the fumes from the drum dryers to locations outside the process building and not provided a long-term solution to the odor problem.

Engineering design modifications

Emissions may be created due to such factors as equipment operating above its design capacity, pressure and temperature conditions, improper process controls, or faulty instrumentation. Strategies to minimize emissions can vary from troubleshooting and debottlenecking existing equipment to designing and installing new hardware.

Vent condensers. Vent condensers are commonly used in SOCMI and pharmaceutical processes for the recovery and/or control of organic pollutants. In several plants, vent condensers were significant emission sources because of one or more of the following conditions:

• Field modifications bypassed the vent condenser but the associated piping changes were not documented in the engineering drawings.

• The vent stream was too dilute to condense because of changes in process conditions.

• The condenser was overloaded (that is, the heat-transfer area was not adequate) due to gradual increases in production capacity over the years.

• The overall heat-transfer coefficient was much lower than design, due either to gradual fouling by dirty components or to condenser flooding with large quantities of noncondensible nitrogen gas.

• The condenser's cooling capacity was limited by improper control schemes. In one case, the coolant supply was unnecessarily throttled because the coolant supply valve was adjusted by a flow controller that was equipped only with a proportional-band control (that is, no reset control). In another case, only the coolant return temperatures were controlled (that is, process stream outlet temperatures were not monitored).

In each case, design modifications were needed to reduce emissions.

Nitrogen usage. In some large pharmaceutical plants nitrogen con-

sumption can be several billion cubic feet per year. In many plants, nitrogen flows for purging of reactors and other process equipment are not monitored or controlled. Pressure transfer and line blowing operations are conducted manually and the duration of the transfer operation depends on the operator. Often, the operator may not be present to shut off the nitrogen supply valve when the operation is complete.

Identifying ways to reduce nitrogen usage will help minimize solvent emissions from a process. For example, every 1,000 ft^3 of nitrogen (if it is assumed to be completely saturated) will vent approximately 970 lb of methylene chloride with it at 20°C and 132 lb of methylene chloride with it at −10°C. The problem is aggravated if fine mists or aerosols are created due to pressure transfer or entrainment and the nitrogen becomes supersaturated with the solvent.

Some plants, for instance, could monitor and reduce nitrogen consumption by installing flow meters in the nitrogen supply lines to each process building. Within each building, simple engineering changes such as installation of flow rotameters, programmable timers, and automatic shutoff valves could help minimize solvent emissions.

Operational modifications

Operational factors that can impact emissions include the operating rate, scheduling of product campaigns, and the plant's standard operating procedures. Implementation of operational modifications often requires the least capital compared to the other strategies discussed in this article.

One SOCMI plant wanted to reduce emissions of cyclohexane solvent from storage and loading/unloading operations. The tank farm had organic liquid storage tanks with both fixed-roof and floating-roof construction. The major source of cyclohexane emissions was found to be liquid displacement due to periodic filling of fixed-roof storage tanks. Standard operating procedures for cyclohexane loading were modified so that the fixed-roof tanks were always kept full and the cyclohexane liquid volume varied only in the floating-roof tanks. This relatively simple operational modification reduced cyclohexane emissions from the tank farm by more than 20 tons/yr.

A pharmaceutical manufacturer wanted to reduce emissions of methylene chloride solvent from a process consisting of a batch reaction step followed by vacuum distillation to strip off the solvent. The batch distillation was conducted by piping the reactor to a receiver vessel evacuated via a vacuum pump. The following changes were made in existing operating procedures to minimize emissions:

1. The initial methylene chloride charge was added at a reactor temperature of −10°C rather than at room temperature. Providing cooling on the reactor jacket lowered the methylene chloride vapor pressure and minimized its losses when the reactor hatch was opened for charging solid reactants later in the batch cycle.

2. The nitrogen purge to the reactor vessel was shut off during the vacuum distillation step. The continuous purge had been overloading the

N. CHADHA is manager, air engineering, with IT Corp., Knoxville, TN (615/690-3211; Fax: 615/693-4944). He has more than 14 years of experience in the design, management, and implementation of engineering solutions to air-pollution control and other environmental problems. He has conducted many multimedia pollution prevention audits and assisted clients in a variety of industries to minimize emissions and wastes via process changes. He previously worked as a process/project engineer for Davy McKee Corp. and Sherex Chemical Co. He has an MS in chemical engineering from the Univ. of Kentucky, is a registered professional engineer in four states, and has been elected a Senior Technical Associate for IT Corp.

C. S. PARMELE is an engineering manager with IT Corp., Knoxville, TN (615/690-3211; Fax: 615/690-3626). He has more than 22 years of experience in environmentally-related technology development and application activities. He is responsible for developing and implementing pollution prevention and/or treatment services by applying chemical engineering technologies, and for the evaluation and application of physical-chemical systems for treating chemical process discharges. He has a BS in chemical engineering from Northwestern Univ. He, too, has been elected a Senior Technical Associate for IT Corp.

downstream vacuum pump system, and it was unnecessary because methylene chloride is not flammable. This change reduced losses due to stripping of methylene chloride from the reaction mix.

3. The temperature of the evacuated receiving vessel was lowered during the vacuum distillation step. Providing maximum cooling on the receiving vessel helped to minimize methylene chloride losses due to revaporization at the lower pressure of the receiving vessel.

Maintenance modifications

Strategies to minimize emissions in any industrial facility should also be linked to preventive maintenance practices. Emissions are generated by leaking equipment (inadequate sealing or corrosion), infrequent preventive maintenance, lack of instrument calibration, and the maintenance chemicals and procedures used. Maintenance modifications to minimize emissions are often obvious once the specific source and cause are identified.

For example, if a pump seal leaks, fixing the packing material or mechanical seal is the first strategy that is usually addressed. For many

SOCMI and pharmaceutical plants, simply expanding the existing preventive maintenance program to include centrifuges, dryers, and other process equipment that use volatile organics has resulted in significant reductions in emissions.

Lessons learned

In summary, our experience in successfully minimizing emissions of air toxics via process changes indicates that:

• Generation of emissions can be minimized even for mature production processes.

• Generation of emissions follows repeating patterns that are independent of the industry.

• Defining the true source of emissions is critical for quickly focusing on long-term solutions for minimizing emissions.

• An accurate emissions inventory aids in uncovering potential opportunities for minimizing emissions.

• Unidentified emissions in a plant's emissions inventory may not always be fugitive emissions.

• Evaluating generic ways of creating emissions from a process can help identify unexpected opportunities for minimizing emissions.

• Not all pollution prevention strategies require process development and/or significant capital investments. **CEP**

LITERATURE CITED

1. **Freeman, H.,** *et al.,* U.S. Environmental Protection Agency, Risk Reduction Engineering Laboratory, "Industrial Pollution Prevention: A Critical Review," *J. Air Waste Manage. Assoc.,* **42**(5), pp. 618-656 (May 1992).

2. **U.S. Environmental Protection Agency,** Office of Research and Development, "Handbook — Control Technologies for Hazardous Air Pollutants," EPA/625/6-91/014 (June 1991).

3. **Parmele, C. S.,** *et al.,* "Waste Minimization through Effective Chemical Engineering," presented at the Air Pollution Control Association Conference on Performance and Costs of Alternatives to Land Disposal of Hazardous Waste (December 1986).

Explore New Options For Hazardous Air Pollutant Control

By keeping abreast of advances in emission control technologies, engineers will be able to select the alternatives that best meet their needs.

Robert G. McInnes, P.E.,
ENSR Consulting and Engineering

The control of hazardous air pollutant (HAP) emissions, also known as air toxics, from industrial facilities is becoming increasingly important and expensive to the chemical process industries. One estimate suggests that control of volatile organic compounds (VOCs) alone from all mandates under the Clean Air Act Amendments of 1990 will cost industry $1.2 billion/yr by 1996 *(1)*.

Under Title III of the Clean Air Act, 174 industry-specific maximum achievable control technology (MACT) standards will be issued on a phased schedule. Both existing and new major emitters of the 189 listed HAPs will be required to comply with these standards. Major source (for this rule) is defined as a source that emits or has the potential to emit more than 10 tons per year (tpy) of any single HAP or 25 tpy of all HAPs combined. In addition, each MACT standard specifies applicability criteria in terms of emission, design, and/or operating characteristics.

As of mid-September 1995, 14 MACT standards have been finalized and another seven are in draft form. Each standard is unique in the types of emission controls and/or practices required to limit HAP emissions. Several standards require techniques such as product substitution (for example, replacing chromium-based agents in cooling towers with nonchromium substitutes), while others require changes in design or operating practices, such as the installation of enclosures to minimize releases. Yet others rely on conventional technologies to establish a level of emission control that must be achieved by affected sources.

Although the source does not need to apply the specific control technique cited in the MACT standard, it must achieve a comparable level of control. This flexibility to use the most appropriate technique, as long as the level of emission reduction is achieved, provides an opportunity for each facility to select the most cost-effective and directly applicable technology for its specific set of emissions and conditions.

So what choices does a facility have for HAP emission control? There are many conventional technologies available, ranging from thermal and catalytic incineration and carbon adsorption for VOCs to fabric filtration and electrostatic precipitation for heavy metals and other particulates. These technologies are provided by numerous equipment vendors, and each has demonstrated advantages and disadvantages. These techniques are expected to be widely used in future HAP reduction efforts.

However, some new emission control techniques show promise for providing cost-effective HAP reduction. Many of these techniques (some of which are commercially available) are based on approaches that have been used for decades. In addition, there have been recent enhancements to conventional technologies to improve their performance, thus making them more cost-effective or increasing the number or types of compounds they can control.

This article examines these new and enhanced emission control technologies for organic (*i.e.,* VOC) emissions. It will help the chemical engineer understand the range of choices currently available — as well as possibly available in the future — for the most efficient and cost-effective reduction of HAP emissions.

New and novel techniques

A discussion of new and novel HAP emission control techniques must necessarily begin with the statement that most of the technologies that are discussed here are not new at all. They are existing techniques that have been used in other environmental disciplines or in other uses for other purposes, such as odor control. Only now, with the increasing focus on HAP emission control, are they being seriously considered for use in emission control in the U.S.

The extent to which one or more of these techniques is viable for a specific emission control problem depends (as with conventional technologies) on the specific details of the emission stream, site-specific constraints that may inhibit or prohibit use at a given location, regulatory concerns that may dictate specific control levels, and economic considerations, which always play an important, if not crucial, role in control device selection. The selection process, whether for conventional or novel technology, will be the same, beginning with a full characterization of the nature and variability of the HAP emission streams.

Biofiltration

Biofiltration uses microorganisms to biologically degrade both organic and inorganic air contaminants in a solid-phase reactor. The concept of employing biodegradation to destroy environmental contaminants has been widely used in the area of wastewater treatment for several decades. Moreover, this technique has been successfully applied in Germany and the Netherlands for the control of odors from a wide range of sources (2).

The basic design of a biofilter is shown in Figure 1 (2). The system consists of a blower or fan to bring the HAP-contaminated gas to the biofilter, a humidification section to condition the gas stream, and one or more biofil-

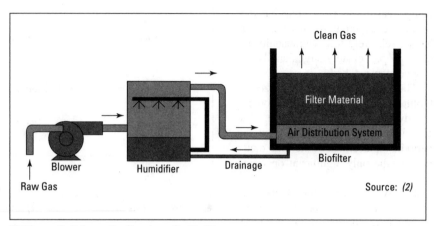

■ *Figure 1. Schematic diagram of a biofilter system.*

ter beds. The filter beds are typically 2–4 ft deep. The biologically active material is usually based on compost, peat, or soil.

Given sufficient residence time in the bed, the air contaminants will diffuse into the wet, biologically active layer, which coats the filter particles. Aerobic degradation of the contaminant then occurs on this active layer. Details of the key design and operating principles of biofiltration are found in the literature (2–6).

Biofilters can be used for a variety of organic HAPs. Table 1 (4) summarizes the relative biological degradability of these compounds. The concentration of the HAP in the treated gas will impact treatability. Generally, total VOC concentrations should not exceed 3,000–5,000 mg/m³ (3). Compound-specific limits can be much lower and depend on the pollutant in question.

The end products of the biofiltration process include carbon dioxide, water, mineral salts, and microbial biomass. The biomass can build up and compact the bed over time, so bed replacement every 3–5 yr may be required.

Biofilters appear to be most applicable to high-volume/low-concentration (volumes over 10,000 cfm, concentrations below 1,000 ppm) gas streams that do not have extensive pretreatment requirements and where there is sufficient space for the filter beds. Due to the site-specific characteristics of each gas stream, it is advisable to conduct a pilot test of the gas stream in question.

The capital costs for biofilters are typically lower than competing emission control technologies if large volumes (over 10,000 cfm) of dilute (50–500 ppm) gases require treatment. However, the capital costs are directly related to the amount of pollutant

Table 1. Relative biodegradability of HAPs.

High	Medium	Low
Acetaldehyde	Acetonitrile	Dioxane
Butadiene	Benzene	1,1,1-Trichloroethane
Cresols	Carbon disulfide	Trichloroethylene
Ethylbenzene	Hexane	Perchloroethylene
Formaldehyde	Methylene chloride	
Methanol	Methyl ethyl ketone	
Styrene	Phenols	
	Toluene	
	Xylene	

Source: (4)

being treated and its degradation rate — high concentrations of compounds having relatively low degradability require proportionately larger, and hence more expensive, beds.

The operating costs of all sizes of biofilters tend to be the major economic advantage of this technique. The electricity to operate the fan and the periodic replacement of the filter material are the only significant operating costs.

Finally, biofiltration offers the added advantage of being a benign treatment technology. It is characterized by low overall energy requirements, no secondary pollutant discharge, and no cross-media contamination (except for some water discharge associated with the humidification of the gas stream). For these reasons, the use of this technology is expected grow during the 1990s.

Activated carbon biofilters. A new biofiltration system, called Biocarb, consists of an activated carbon bed (which serves as the support medium) inoculated with microorganisms. VOCs are adsorbed onto the carbon, and are then slowly released for degradation in the surrounding water film.

The advantages of an activated carbon biofilter are that it can handle spike loads, it does not suffer from the effects of aging that conventional media (such as compost and soil) do, it can be used with a shorter residence time, and it has lower overall maintenance requirements. Based on pilot tests conducted by system developer Wheelabrator Clean Air Systems, the performance of this process compares favorably with that of conventional compost-based biofilters (7).

The main disadvantage of an activated carbon biofilter is cost — on a per-pound basis, carbon costs about five times more than compost or peat. This is offset, though, by reduced operating costs (the pressure drop across the bed is minimal) and maintenance costs (the carbon does not need to be replaced periodically as compost and soil do) (7).

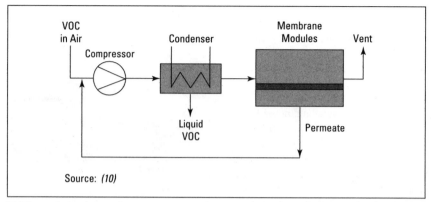

Source: (10)

■ *Figure 2. Basic membrane vapor separation process for VOC removal.*

Membrane separation

Another recent development for VOC and organic HAP control is membrane separation. Membrane separation is a viable emission control approach when the gas stream contains a relatively high concentration (over 10,000 ppm) of vapors. In this regard, it is expected to actively compete with and/or supplement refrigerated condensation as a control method. Several emerging applications for membrane technologies in both air- and liquid-based environmental cleanups have been reported in the literature (8–11).

As shown in Figure 2 (9), the basic membrane system for VOC control consists of a compression/condensation step followed by a membrane separation step. The VOC-contaminated vapor stream is first compressed to 45–200 psig. The compressed mixture is then sent to a condenser, where the organic vapor condenses and is recovered for reuse. In some applications, a dryer precedes the condenser to remove moisture from the gas stream prior to condensation.

The noncondensed organic stream, typically containing up to 1% organics, is then directed through the membrane. The membranes, which are of proprietary designs, are 10 to 100 times more permeable to organic vapor than to air.

To induce the selective permeability, a pressure difference is created across the membrane, typically using a vacuum pump on the permeate side of the membrane or by compressing the feed stream. Air and organic vapors permeate through the membrane at rates determined by their relative permeabilities and the pressure difference across the membrane. Since the membrane is significantly more permeable to the organic vapors, there is an enrichment of the vapor concentration on the permeate side of the membrane. Depending on the system design, fivefold to fiftyfold enrichments can be realized, with resulting HAP control efficiencies in excess of 90% (9).

Membrane separation can be used to treat a wide range of organic vapor streams, including those summarized in Table 2 (10). The most promising

Table 2. HAPs recovered by membrane separation.

Benzene	Toluene
Methanol	1,1,1-Trichloroethane
Methyl bromide	Trichloroethylene
Methylene chloride	Vinyl chloride
Methyl ethyl ketone	Xylene
Perchloroethylene	

Source: (10)

applications appear to be those involving compounds that are difficult to remove by condensation and carbon adsorption, such as compounds with low boiling points and/or the chlorinated organics. In addition, the more expensive solvents (those in the $1–5/lb cost range) are also suitable for control by membrane separation.

According to membrane manufacturer Membrane Technology and Research, the advantages of membrane separation over carbon adsorption are the elimination of the desorption step and the elimination of all secondary waste streams. Membrane technology also avoids the main problems associated with low-temperature condensation, including poor recovery efficiency due to low concentration and/or low boiling point of the condensible component, the need to regularly defrost the condenser, and the high cost of maintaining low condensing temperatures (10).

The initial applications of this technology have focused on air streams containing more than 1% VOC and flow rates of less than 200 cfm. Future applications are expected to include a much wider range of both VOC concentrations and flow rates.

Ultraviolet oxidation

Ultraviolet oxidation technologies for the gas-phase destruction of organics have been studied under laboratory conditions for several years (12). At least one of these technologies has been commercialized, by Terr-Aqua Enviro Systems (13), and evaluated for paint spray emission control (14).

This process uses ultraviolet (UV) light and "activated oxygen" (AO) to destroy VOCs in the vapor phase. Specifically, the photocatalytic effects of tuned-frequency UV light on volatile or reactive organic compounds combine with an oxidizer such as ozone or peroxide to destroy VOCs (13). In certain applications a wet scrubber and/or carbon adsorber are also used.

Figure 3 shows one UV/AO system (13). VOC-contaminated gas from a process is first directed to a filter for re-

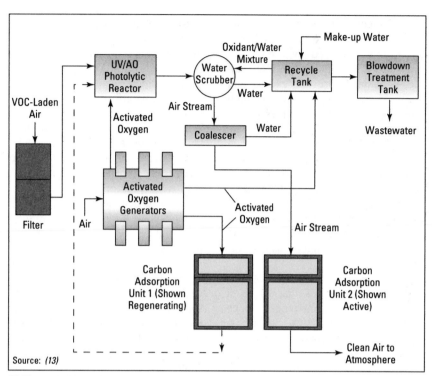

■ *Figure 3. Typical ultraviolet/activated oxygen system for VOC destruction*

moval of particulate matter. The gas then passes through a photolytic reactor, where the tuned UV light and injected oxidant begin to destroy the VOCs. The photolytic reactor uses a process similar to that which occurs in nature, where ultraviolet light from the sun works in combination with naturally occurring oxidants to oxidize and reduce VOCs. The photolytic reactor utilizes a specially selected ultraviolet light frequency range [200–280 nm (12)] to catalyze or enhance the oxidation reaction, thereby increasing the effective destruction of VOCs (13).

Next, the gas flows through a water scrubber, where ozonated or oxidant-containing water passes countercurrent to the upflowing VOC-laden gas. Through a water scrubbing and oxidation process, most of the VOCs are captured or ultimately reduced to carbon dioxide and water.

As the air stream passes through the photolytic reactor and water scrubber, some organic compounds are partially oxidized, others are completely reduced, and some are simply removed by the water scrubbing action. The re-

sult is an air stream leaving the water reactor with lower concentrations of some of the original organic compounds, plus new intermediate compounds from the partial or incomplete oxidation processes. The reduction or removal efficiency of each active stage of this technology depends on the organic compounds involved, the relative concentrations, and the total organic loading (13).

Finally, for installations with a carbon adsorber, the gas is ducted to one of two beds, where the residual VOCs are removed. These beds are regenerated with the oxidant (not with steam as in conventional installations), which converts the adsorbed VOCs into CO_2 and water. Thus, there is no VOC byproduct for disposal. In addition, because the VOCs are not incinerated, no combustion byproducts (e.g., NO_x and CO) are produced.

The design principle behind the UV/AO reactor is the photocatalytic effect of UV light. To control this process to obtain beneficial results, the UV frequency must be tuned to the type of VOCs present and there must be sufficient residence time for the VOC/UV exposure. Some compounds are readily destroyed by certain frequencies of UV light; other, more stable, compounds are just excited and can be destroyed only with very high intensity UV light in the presence of oxidants. In general, most VOCs are only excited by the UV light used in this process — the destruction is typically achieved with the oxidants.

Several full-scale UV/AO systems, ranging in size from 20,000 to 90,000 cfm, have been installed to control organic emissions from spray painting operations and a furniture manufacturing plant.

An important advantage of the UV/AO technology is that this is a highly energy efficient process, since only the contaminant VOCs are affected (and not the surrounding air mass). The only energy requirements are the electricity associated with the fan and UV light source. In addition, no pollutant byproducts are generated, the process operates at low temperatures, and removal efficiencies are high (typically 90–95%).

On the down side, the technology may require wastewater treatment. And, the range of HAPs potentially treatable by this technology has not yet been fully defined.

Nevertheless, this technique is potentially applicable to streams with high gas flow rates and low VOC concentrations, such as those from paint spray booths, that will be newly subject to MACT control requirements in the future.

Modifications to technologies

Other VOC/HAP control technologies are associated with the more conventional control methods, including carbon adsorption and thermal and catalytic incineration. While each of these methods has been widely used to control organic or particulate emissions in the past, recent refinements and improvements in their operation promise to increase their control efficiency, extend their range of applicability, or reduce their costs.

Carbon adsorption

Activated carbon adsorption controls VOCs and organic HAPs in a variety of industries. The design and operating principles of this technology have been described extensively in the literature (15–17).

Carbon-based systems do have limitations, particularly where the VOC-laden stream requiring control is hot (above 100°F) or saturated with moisture. As a result of these inherent disadvantages, new and novel adsorption methods have been developed to overcome these problems. These new methods can generally be classified as relating to new adsorbents or relating to the desorption portion of the process. In addition, carbon and other adsorbents are being used in new systems to concentrate the VOCs from industrial sources, thereby providing an emission stream that can subsequently be controlled more cost-effectively.

New adsorbents

New adsorbent compounds that have been developed to replace granular activated carbon (GAC) in otherwise conventional adsorption systems include zeolites, polymeric adsorbents, and carbon fibers.

Zeolites. Zeolites are hydrated alkali-aluminosilicates. There are about 40 natural zeolites, each with a different chemical composition and crystalline structure (18). In addition, silicate research has allowed the manufacture of entirely new synthetic forms. As a result, there are approximately 70 known zeolite structures.

All zeolite compounds share a common tetrahedral structure, in which the tetrahedra are connected to each other at the corners they share with an oxygen atom. The structure is solid, but also very open, which makes zeolites well-suited to VOC control.

Zeolites have been widely used as dehumidifying agents, as ion-exchange resins, and as catalysts in the petroleum refining industry. Recent developments in the processing technology have enabled manufacturers to produce hydrophobic (water-repelling) ze-

olites that, when their pores are not filled with water, can attract and hold organic molecules. This property is now being exploited in zeolite-based adsorption systems.

The size of the pores in each zeolite structure will determine what molecules the material can adsorb — the more tetrahedra in the structure, the larger the pore size and the larger the compounds that can be adsorbed. Hydrophobic zeolites can adsorb organic molecules up to about 8 Å in diameter; larger molecules will not be adsorbed and will bypass the zeolite filter. Thus, the diameter of the organic compound of interest determines which type of zeolite can be used.

In addition, because of the zeolite's strictly regular crystalline structure, all pores are essentially the same size. Thus, a bed of zeolites functions like a molecular sieve — the material captures the smaller molecules that fit into its pores while the larger molecules pass through the bed with the air.

The polarity of the organic molecule also influences the adsorption reaction. The smaller the molecule's dipole moment, the better it adsorbs onto the zeolite. Compounds that have a low solubility in water, such as toluene, benzene, xylene, and styrene, adsorb very well, whereas substances that are highly water soluble, such as alcohols, do not adsorb as well on any hydrophobic adsorbents, including zeolites (18).

Although zeolites generally have a lower adsorption capacity than carbon, they have demonstrated higher removal efficiencies for certain compounds (e.g., xylene) in the early minutes of the adsorption cycle (19), particularly with low (less than 100 ppm) inlet concentrations. This efficiency drops off with increasing cycle time, suggesting that the best overall efficiency can be achieved with a short adsorption cycle.

Finally, hydrophobic-zeolite-based systems adsorb very little water until the relative humidity exceeds 90%. This contrasts to carbon systems, where inlet humidity levels above 60% reduce organic adsorptive capacity.

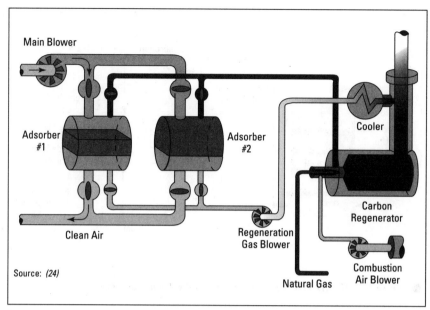

Figure 4. CADRE adsorption system uses a high-temperature regeneration gas.

Labels in figure: Main Blower; Adsorber #1; Adsorber #2; Cooler; Clean Air; Regeneration Gas Blower; Carbon Regenerator; Combustion Air Blower; Natural Gas; Source: (24)

Thus, zeolites may be applicable for gas streams with high humidity levels.

Polymeric adsorbents. Several adsorbent vendors offer products that seek to avoid the humidity and adsorptive capacity limitations of coal- and coconut-based GAC. These products are based on styrene/divinylbenzene polymers.

One such synthetic adsorbent (Bonopore 1120, from Chematur Engineering AB) is particularly effective in controlling contaminants with high boiling points and low polarity, including most common VOCs (20). As a synthetic adsorbent, its properties can be customized to optomize performance for specific emissions situations.

Other adsorbents (Ambersorb carbonaceaous adsorbents, from Rohm and Haas Co.) are manufactured from a highly sulfonated styrene/divinylbenzene macroreticular ion-exchange resin that is subsequently pyrolyzed in a patented process. The pyrolysis significantly increases the resin's microporosity (pore diameter < 20 Å), provides the ability to control the resin's surface properties (degree of hydrophobicity), and imparts excellent mechanical stability (21).

Thus, these adsorbents are extremely reproducible, have excellent capacity, and are hydrophobic. This latter characteristic allows them to be used under high humidity conditions while retaining a relatively high adsorptive capacity, a key advantage of these adsorbents over conventional GAC.

A third line of synthetic adsorbents (these from Dow Chemical) are based on methylene-bridged styrene/divinylbenzene copolymers. These products can be used to control contaminants with low boiling points and high polarity that are not geneally well controlled by other adsorbents, such as acetone, ethanol, methanol, and methyl chloride (20).

Carbon fibers. Another alternative to GAC is carbon fiber. Because of its shape, carbon fiber has more surface micropores available for adsorption and desorption (22). A a result, the desorption cycle of a typical carbon fiber system may take 5–10 min, compared to 30–60 min for a typical GAC system.

In addition, carbon fiber contains fewer transition metals (50–90% less) than GAC (22); with fewer catalytically active metals available to promote hydrolysis, the formation of corrosives is significantly reduced. This is particularly important when chlorinated solvents are being recovered. The desorption of such chlorinated compounds with steam in conventional GAC sys-

tems can result in the generation of corrosive acid gases as the chlorinated compounds react with the metals in the carbon to release hydrochloric acid, which creates maintenance problems with the carbon adsorption vessel.

Another conventional GAC problem addressed by the carbon fiber system are the bed fires associated with the recovery of ketones, particularly cyclohexanone and methyl ethyl ketone. These fires are caused by the reaction of the ketones on the carbon surfaces — the ketones break down on the carbon in the presence of air and/or steam and the transition metals of the carbon to form adipic acid and biacetyl, respectively (23). The exothermic heat of reaction is retained by the carbon granule, and if the reaction continues for an extended period, a smoldering fire can result.

Carbon-fiber-based systems can minimize this problem because: the low metals content of the carbon fiber results in lower rates of reaction between the solvent heel and the air or steam; the frequent, but short, regeneration cycles, plus the thoroughness of regeneration, leaves only a small heel of solvent on the bed to react and build up heat in the carbon bed; and the small fiber diameter, low total weight, and low fiber density allow quick removal of exothermic reaction heat from the carbon bed in the exhaust stream.

New system configurations

Several new adsorption system designs seek to eliminate the generation of a recovered solvent or VOC-contaminated wastewater associated with conventional steam-regenerated carbon adsorption systems. Some of these new system configurations use conventional carbon, while others use a different material as the adsorbent.

Integral hot gas regeneration. The CADRE adsorption-regeneration process (developed by Calgon Carbon Corp.), shown in Figure 4 (24), employs a conventional GAC for the adsorption mode. However, for regeneration, CADRE uses its own high-temperature source of on-demand regeneration gas. The hot regeneration gas, typically at 250°F, quickly raises the temperature of the carbon bed, desorbing the highly concentrated VOCs. The desorbed organics are then destroyed in a controlled oxidation process *(24)*.

This process offers several advantages. Because the desorbed organics are so highly concentrated in the regeneration cycle, they may serve as auxiliary fuel to reduce energy requirements in the carbon regenerator. Compared to steam regeneration, use of this high-temperature gas provides improved VOC desorption from the carbon pore structure. Thus, the CADRE system avoids the use of steam, does not require a steam supply, and does not generate liquid organics or condensed steam that must be treated or disposed of. Moreover, the desorbed organics will be completely destroyed in a thermal oxidizer.

Fluidized-bed adsorption. Another new system design is the POLYAD process (developed by the Swedish company Chematur Engineering AB and available in the U.S. from Weatherly, Inc.), illustrated in Figure 5. This system uses macroporous polymer spheres in a fluidized-bed configuration to adsorb VOCs from contaminated air streams. The VOCs are subsequently desorbed from the polymer in a thermal desorption unit and can either be recovered or destroyed.

The adsorbent is the Bonopore 1120 styrene/divinylbenzene adsorbent described above. The polymer spheres have a diameter of approximately 0.02 in., an average pore diameter of 80 Å, and a high specific surface area of approximately 8,600 ft²/g *(25)*.

The adsorption unit consists of one or more fluidized beds of polymer spheres. Solvent-laden air enters at the bottom and fluidizes the bed. The solvent is adsorbed onto the spheres as it flows upward. Adsorbent, saturated with solvent, flows from the bottom of the adsorption vessel through the desorption unit, where it is regenerated, and is then vented back into the top of the adsorption unit. Thus, a continuous flow of adsorbent is maintained.

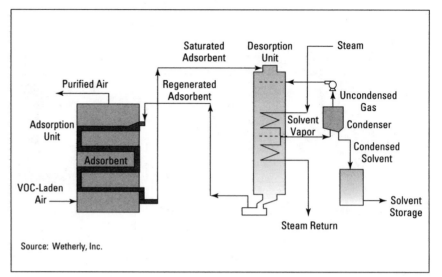

Source: Wetherly, Inc.

■ *Figure 5. POLYAD VOC emission control and solvent recovery process.*

In the regeneration process, the saturated adsorbent is pneumatically transported to the top of the desorption unit. As the adsorbent descends through the desorption unit, it is heated to a temperature at which the solvent is released from the polymer. The heat is provided by steam traveling through a heat exchanger within the desorber. The released solvent is drawn out of the desorption column and into a condenser. Condensed solvent is collected into a storage tank, while uncondensed gas is fed back to the desorption column. The regenerated adsorbent is pneumatically conveyed back to the top of the adsorption unit, completing the cycle.

POLYAD systems can be used to purify air containing aromatics, aliphatics, alcohols, aldehydes, ketones, and some chlorinated compounds, and are recommended for solvent concentrations between 0.1 and 10 g/m³. They are not recommended for solvents with low boiling points (below 60°F) or solvents that are very polar, such as methanol or methylene chloride. In addition, it may be impractical or uneconomical to use this system for control of air streams that contain several different organic constituents at low concentrations. While a control efficiency in excess of 99% can be achieved, most systems are de-

signed to remove between 90% and 95% of the incoming solvent, as the cost of achieving in excess of 95% is disproportionately high *(25)*.

Regeneration via refrigeration. Another alternative to steam regeneration of conventional GAC-based systems uses the refrigeration process known as the reverse Brayton (or Joule) cycle for solvent recovery *(26)*. The process, trademarked Braysorb, is patented by 3M Corp. and licensed to Nucon.

The system reportedly offers several advantages for both carbon regeneration and solvent recovery: it produces dry (less than 1% moisture), relatively pure recovered solvent; it allows the uncomplicated recovery of solvents that are water-miscible; and it enables the recovery of chlorinated solvents that are water-miscible *(26)*.

The system, shown in Figure 6 *(26)*, uses two carbon beds. To completely saturate the beds, solvent-laden air is first directed through only one of the beds. The outlet gas stream from the bed is monitored until the solvent concentration reaches a predetermined level. At this point, the outlet gas from the first carbon adsorber is redirected through the second adsorber. The first bed continues to adsorb solvents until the bed is saturated. Because carbon preferentially adsorbs solvents, any water in the air stream that may have been adsorbed by the first bed is likely to be desorbed and replaced by solvent as the bed nears saturation. This desorption of water helps the system achieve less than 1% water content in the recovered solvents.

When the solvent concentration of the outlet air from the first carbon bed is equal to the inlet air concentration, the bed is taken off-line for desorption and the inlet air is redirected to the second bed for adsorption. Desorption of the first bed begins with the evacuation of the saturated carbon by a dedicated vacuum pump followed by the introduction of nitrogen until the oxygen level is less than 3–5%. Reduction in the oxygen level is a fire prevention measure, to keep the bed from exceeding the lower explosive limits of the contaminants.

The bed is then ready for desorption. Clean, hot nitrogen at 310°F and 15 psia is introduced into the adsorber and liberates solvent vapors from the bed. Solvent-laden nitrogen leaving the bed enters a glycol-cooled heat exchanger, which reduces the gas stream temperature to 80°F. The gas then enters a regenerative heat exchanger, where it is further cooled to 10°F. At this temperature, over 90% of the solvents in the gas stream condense and are removed in a liquid separator.

Next, nitrogen with trace solvents passes to the turbocompressor, where it is cooled to −48°F and its pressure is reduced to 8 psia. A second liquid separator removes the remaining solvents, which are now condensed. The cooled nitrogen enters the other side of the heat exchanger, where its temperature is increased to 60°F. It then passes through a vacuum pump and the compressor side of the turbocompressor and exits at 225°F and 15 psia. A steam heat exchanger raises the nitrogen temperature back to 310°F, and the hot nitrogen is returned to the carbon bed for desorption. When the carbon bed is completely desorbed, the nitrogen is cooled to 80°F (instead of heated to 310°F) and is used to cool the bed before adsorption is resumed *(26)*.

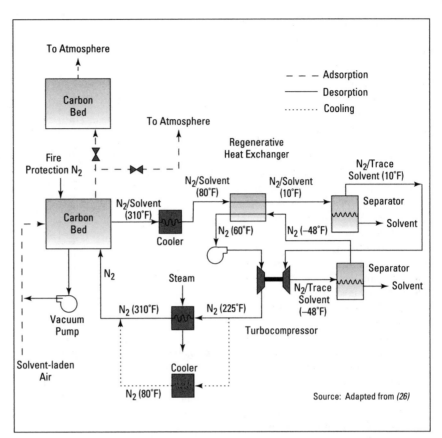

■ *Figure 6. Reverse Brayton cycle solvent recovery system.*

The Braysorb system is currently in use at a plant where solvent recovery is projected at greater than 99.5% for a system in the 8,000–10,000 cfm size range *(26)*.

Regeneration using solvents. Another carbon regeneration technique that is not really new but that has received increased attention lately is solvent regeneration. It uses a solvent to first dissolve adsorbed material out of the pores of the carbon, then the solvent is removed by steam. The key to the process is choosing a solvent that efficiently solubilizes the contaminants out of the carbon pores yet can be subsequently removed by steaming. Typical solvents include acetone, methanol, toluene, and mixed aliphatic hydrocarbons. Solvent regeneration is reportedly only a few cents per pound more expensive than steam regeneration, and less than one-tenth as expensive as thermal regeneration *(27)*.

Steam regeneration is generally preferred for vapor-phase applications where a relatively volatile, very clean organic vapor is adsorbed. However, as the adsorbed contaminant becomes less volatile (atmospheric boiling points above about 120°C), or if less-volatile compounds are also present, steam regeneration may be a poor choice. If the contaminants will dissolve in an organic solvent, then solvent regeneration may be a cost-effective alternative *(27)*.

Each of the novel adsorption systems described above seeks to avoid the perceived disadvantages associated with desorbing conventional carbon-based systems with steam and thereby creating a mixed liquid waste of organics and water that may need to be further pretreated prior to ultimate disposal. At the same time, these systems seem to broaden the applicability of adsorption to additional organic compounds that currently may not be controllable with conventional GAC systems.

Concentrating wheels

An outgrowth of the research into new uses of carbon and other adsorbent media has been the development of concentrating wheels fabricated with adsorbent materials *(28)*. These wheels continuously adsorb VOCs from dilute gas streams and subsequently continuously desorb these compounds from the adsorbent with a second, heated air stream that is 5–10% of the original exhaust flow. This action lowers the size of the gas stream requiring ultimate treatment while increasing the VOC concentration in this stream by a factor of 10 to 20. This smaller, more-concentrated gas stream can then be treated in a much more cost-effective manner, such as by conventional adsorption or incineration.

Several equipment vendors provide concentrating wheels that use a variety of commercially available adsorbents, including carbon fiber *(22)*, zeolites *(18)*, and conventional carbon *(17)*. The units typically consist of a honeycomb support structure in the shape of a rotating wheel. The major portion of the wheel, the adsorption side, is exposed to the high-volume, low-concentration exhaust gas, while the smaller part of the wheel, the desorption side, is exposed to a heated low-volume gas stream that strips the organics and becomes enriched with VOCs. This desorption stream is fed directly to the ultimate control or VOC-recovery device.

Concentrating wheels are useful in applications where the VOCs must be controlled yet the concentrations are too low (less than about 1,000 ppm) to provide cost-effective emissions reductions (for example, in automotive and aerospace paint spray shops).

While the concentrating wheel is an additional piece of hardware in the VOC control system, its one-time cost can be offset by reductions in the operating costs of the final control system, particularly if the latter is an incinerator with significant auxiliary fuel costs. An added advantage is that a concentrating wheel has a very low pressure

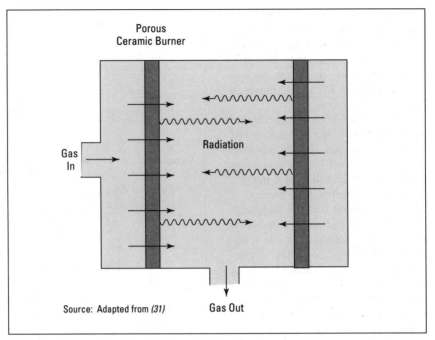

■ Figure 7. Inward-fired, adiabatic radiant combustor.

drop (on the order of 0.2–0.4 kPa [about 1 in. of water]), contributing to the low operating costs *(28)*.

Incineration

Thermal and catalytic incinerators have been used for VOC and organic HAP control for decades. They are a proven destruction technology, offered by multiple vendors, and result in the destruction of the organic molecule at levels in excess of 98%. Incinerators rely on high temperatures, long residence times, and good turbulence between the organic-laden gas stream and oxygen in the combustion air to provide for the permanent destruction of the organics.

Over the past 10 to 15 years, the primary development in this technology, particularly thermal incineration, has been the improvement of the thermal efficiency of these units using ceramic or stoneware heat-transfer media. These regenerative incinerators, with their 85–95% heat recovery, minimize the need for auxiliary fuel, particularly with gas streams that have relatively low VOC content. While costing more initially, they can result in lower life-cycle costs than recuper-

ative incinerators, which use metallic shell-and-tube heat exchangers with heat-recovery levels in the 55–70% range. [Ref. *(29)* provides a detailed comparison between recuperative and regenerative heat recovery.]

However, both thermal and catalytic incinerators have certain perceived disadvantages. These include the generation of nitrogen oxides (NO_x) during high-temperature combustion, slow thermal response, and the general inability to handle chlorinated materials satisfactorily. In response to these concerns, equipment vendors and catalyst suppliers have developed new combustion-based and catalyst-based techniques to enhance incinerator performance.

Combustion modifications

New burner designs. A variety of new burners that generate lower levels of NO_x emissions have been developed recently *(30)*. One particularly interesting design from a VOC control standpoint is the inward-fired, adiabatic radiant burner technology developed by Alzeta Corp. *(31)*. VOC-laden gases are mixed with fuel, and the mixture is forced through a porous ceramic burn-

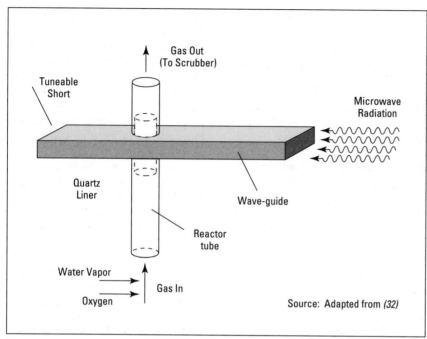

■ Figure 8. Microwave-discharge plasma reactor.

er. Because the individual molecules of the gaseous mixture pass through the combustion zone as they exit the burner, complete combustion can be achieved at lower temperatures and shorter residence times. Because turbulence is essentially nonexistent, NO_x emissions are minimized. And, the burner can be heated to operating temperature in a matter of seconds, which makes it useful in batch operations.

According to the manufacturer, the adiabatic radiant burner offers several advantages for VOC control: greater than 99.9% control of both chlorinated and nonchlorinated hydrocarbons; NO_x and CO levels below 10 ppm (@3% O_2); heating to an acceptable operating temperature from a cold start in approximately 2 s; relatively small sizes; low fuel requirements; and automated operations.

The key to the system design is the combustor itself, which is shown schematically in Figure 7 (31). The gas mixture enters an annular space surrounding the cylindrical burner element. The burner element is made of a porous ceramic material, the inner surface of which serves as the combustion zone through which all of the gas mix-

ture must pass. The burner operates at 1,600–1,800°F and can handle approximately 15–20 scfm per square foot of burner surface area.

With the inward flow design, radiant heat is exchanged between the inner surfaces of the burner. This allows the burner to maintain the required operating temperatures even with 100% or more excess air. The hot, combusted gas exits the burner at a temperature just below the combustion zone temperature. The exhaust gas then passes through a recuperator, thereby providing heat for the incoming gas stream.

Several phenomena are believed to contribute to the low NO_x emissions achieved by this unit. First, since the inside of the burner glows incandescently but no flame is produced, the temperature is fairly uniform across the surface. This eliminates localized hot spots that can result in NO_x emissions. In addition, the mixture can include a large quantity of air, thereby ensuring that the natural gas burns quickly and at a relatively low temperature.

A 100-cfm prototype unit has been tested on a gasoline and hexane mixture and several field units are now opera-

tional treating the off-gas from gasoline-contaminated soil remediation, and other units are being planned. Currently available systems can handle flow rates of 100–6,000 cfm and concentrations up to the lower explosive limit.

Plasma technology. A different type of combustion device, a microwave-discharge plasma reactor, has been developed by Argonne National Laboratories (32). Research on this technology has been spurred by the problem of incinerating low-concentration VOC emissions, particularly halogenated hydrocarbons, which require higher operating temperatures, longer residence times, and more auxiliary fuel for combustion than nonhalogenated organics.

Plasma is essentially a high-temperature ionized gas state that is extremely reactive. Two methods of plasma generation have been used for chemical destruction applications — arc discharge and high-frequency discharge. In an arc-discharge unit, an electric field is used to produce a discharge, or arc, between two electrodes, and the intense energy from the arc generates and sustains the plasma.

High-frequency-discharge units use microwave or other radio frequency radiation to create and sustain the plasma state. One advantage of this approach is that no electrodes, which are possible sources of contamination, are used.

As illustrated in Figure 8 (32), contaminated gas is mixed with oxygen, water vapor, or a combination of the two and pumped into a quartz-lined reactor tube. The reactor tube passes through a wave-guide, which serves to direct entering microwave radiation to one portion of the tube. A tuneable "short" on one end of the wave-guide is used to couple the microwave energy with the plasma. Thus, by adjusting the tuneable short, plasma generation can be optimized.

As the contaminated gas flowing in the reactor tube reaches the wave-guide, it enters a plasma zone created within the tube. This plasma zone does not extend beyond the wave-guide. In this zone, the gas reacts with the water and/or oxygen to form carbon dioxide, carbon monoxide, hydrogen, water, and hydrochloric acid. Gas exiting the reactor passes through a caustic scrubber to remove the hydrogen chloride. The reactor is cooled by blowing air down the wave-guide from the same end that the microwave radiation enters.

Laboratory-scale testing has demonstrated a destruction efficiency of over 99% for trichlorethane and trichlorethylene. While this system has only been evaluated on a laboratory scale, there has been some commercial interest in it, and it does appear suitable for VOC control in certain applications, specifically for the treatment of off-gas from vapor extraction, air stripping, and incineration systems (32).

Another plasma-based technology that has been evaluated at the bench scale is a cold plasma technology called the silent discharge plasma (SDP) device (33). This technique uses a cold plasma process in which gaseous emissions are combusted at relatively low temperatures (approximately 932°F).

The advantage of using a cold plasma technique relates to the creation of the free radicals needed to complete the destruction of organic compounds. During conventional incineration, organic compounds are destroyed by chemical reactions that depend on the presence of free radicals, such as $O^{\bullet}$, $OH^{\bullet}$, and $H^{\bullet}$, which are very effective at breaking the carbon-carbon and carbon-halogen bonds in the waste. The efficiency at which these free radicals are produced in an incinerator depends on the temperature of the combustion process, with more free radicals produced at higher temperatures.

These same free radicals can be generated at low temperatures using nonthermal or cold plasma. Such a plasma can be created by electrical discharges in gas. The plasmas produce energetic electrons that effectively create these free radicals at or near ambient temperatures, so high temperatures are not required (33).

The SDP process is currently used in industrial applications to generate large quantities of ozone. By modifying the conventional design to include a dielectric barrier and applying alternating high voltages ranging from 50–60 Hz to several kHz, large quantities of plasma are generated in the gas along the electrodes. These discharges are transient in nature and develop in only a few nanoseconds in oxygen. Due to the short duration of the discharges and the low ion mobilities, the electrical energy in these discharges is coupled into electron channels. This means that only the electrons are "hot," while the other species are "cold." As a result, electrical energy is used very efficiently to generate the radicals, which destroy the organic molecules in the plasma (34).

In the SDP reactor, water is needed to generate $OH^{\bullet}$ free radicals, and it is provided by bubbling the contaminated gas stream through water to saturate it. The free radicals generated by the plasma then react with organic compounds in the gas, producing CO_2, H_2O, and, if operating conditions are not optimized, low levels of carbonyl halides; other combustion products may also be produced.

Results of bench-scale tests indicate that the estimated energy consumption for 90% and 99% destruction of trichloroethylene is 12 kW•h/kg and 84 kW•h/kg, respectively, and 270 kW•h/kg for 90% destruction of carbon tetrachloride. Based on these tests, the Los Alamos National Laboratory, where the research is being conducted, plans to design and construct a pilot-scale SDP treatment unit (33).

The modifications to thermal incineration have all been conducted on bench- or pilot-scale units. Depending on the success of ongoing research and development programs, it is conceivable that a full-scale commercial size unit of one or more of these technologies may be built in the coming years.

Catalyst-based modifications

Catalytic incinerators employ a bed of active material (catalyst) to facilitate the overall combustion reaction. The catalyst has the effect of increasing the reaction rate, enabling conversion of the organic molecule to carbon dioxide and water vapor at lower temperatures than required for thermal incinerators. This results in lower overall operating temperatures for the system and the need to use less auxiliary fuel to support combustion. Thus, when applicable, catalytic incinerators can be less expensive to operate.

However, to be widely applicable, catalysts must successfully address potential flue-gas-related problems that can render them inappropriate for a specific emission control problem or cause them to fail prematurely. These potential problems include: fouling with particulates, silicon, polymers, and ash; catalyst suppression when exposed to halogens, sulfur compounds, or NO_2; and poisoning due to exposure to phosphorus, lead, zinc, arsenic, mercury, or other heavy metals.

In addition, the ideal catalyst for VOC control should have a small physical volume relative to its VOC control efficiency (a small space velocity, where space velocity is defined as the volumetric air flow rate divided by the catalyst bed volume), it should result in a small pressure drop, it should resist attrition (that is, it should not break down and wear away), and it should not become brittle when exposed to high (800°F and higher) operating temperatures (35, 36).

The major catalyst suppliers are continually improving the range of acceptable operating conditions of their products. Progress, though, is evolutionary rather than revolutionary. Some of the advances in this area were discussed in an earlier *CEP* article (36).

Catalyst beds. Traditionally, catalytic incinerators have used a noble metal (platinum or palladium) catalyst supported on a fixed bed. A common bed design is the catalyst monolith, where the catalyst is deposited on a solid block containing parallel, nonintersecting channels aligned in the direction of the gas flow. Monoliths offer the advantages of minimal attrition due to thermal expansion and contraction

Literature Cited

1. **ICF Resources Inc., and Smith Barney, Harris Upham and Co.,** "Business Opportunities of the New Clean Air Act: The Impact of the CAAA of 1990 on the Air Pollution Control Industry," Draft Report prepared for the U.S. EPA, Office of Air and Radiation, Research Triangle Park, NC (Jan. 1992).

2. **Leson, G., and A. M. Winer,** "Biofiltration: An Innovative Air Pollution Control Technology for VOC Emissions," *Journal of the Air and Waste Management Association,* **41**(8), pp. 1045–1054 (Aug. 1991).

3. **Bohn, H.,** "Consider Biofiltration for Decontaminating Gases," *Chem. Eng. Progress,* **88**(4), pp. 34–40 (Apr. 1992).

4. **Leson, G., et al.,** "Control of Hazardous and Toxic Air Emissions by Biofiltration," Paper 92-116.03, presented at the 85th Annual Meeting of the Air and Waste Management Association, Kansas City, MO (June 1992).

5. **Allen, E. R., and Y. Yung,** "Operational Parameters for the Control of Hydrogen Sulfide Emissions Using Biofiltration," Paper 92-116.06, presented at the 85th Annual Meeting of the Air and Waste Management Association, Kansas City, MO (June 1992).

6. **Togna, A. P., and B. R. Folsom,** "Removal of Styrene from Air Using Bench Scale Biofilter and Biotrickling Filter Reactors," Paper 92-116.04, presented at the 85th Annual Meeting of the Air and Waste Management Association, Kansas City, MO (June 1992).

7. **Pisotti, D. A., and R. Newton,** "An Innovative Use of Activated Carbon," Paper No. 45C, presented at the AIChE Summer National Meeting, Boston, MA (July-Aug., 1995).

8. **Tyagi, R. K., et al.,** "A Novel VOC Removal Technology Using Pervaporation Membranes," Paper 95-WA77A.03, presented at the 88th Annual Meeting of the Air and Waste Management Association, San Antonio, TX (June 1995).

9. **Baker, R. W., et al.,** "Membrane Vapor Separation Systems for the Recovery of VOCs," Paper 92-114.03, presented at the 85th Annual Meeting of the Air and Waste Management Association, Kansas City, MO (June 1992).

10. **Baker, R.W., et al.,** "Membrane Systems for VOC Recovery from Air Streams," presented at the AIChE Summer National Meeting, Seattle, WA (Aug. 1993).

11. **Koros, W. J.,** "Membranes: Learning a Lesson from Nature," *Chem. Eng. Progress,* **91**(10), pp. 68–81 (Oct. 1995).

12. **Moralejo, C., and R. J. Cody,** "Degradation of BTEX Contaminants in Air by Excimer Laser Irradiation," Paper 95-WP77B.03, presented at the 88th Annual Meeting of the Air and Waste Management Association, San Antonio, TX (June 1995).

13. **Shugarman, L.E.,** "Ultraviolet/Activated Oxygen — A New Air Pollution Control Technology Comes of Age," Paper 91-104.9, presented at the 84th Annual Meeting of the Air and Waste Management Association, Vancouver, BC (June 1991).

14. **Ayer, J., and C. H. Darvin,** "Cost-Effective VOC Emission Control Strategies for Military, Aerospace, and Industrial Paint Spray Booth Operations: Combining Improved Ventilation Systems with Innovative, Low Cost Emission Control Technologies," Paper 95-WA77A.02, presented at the 88th Annual Meeting of the Air and Waste Management Association, San Antonio, TX (June 1995).

15. **Stenzel, M. H.,** "Remove Organics by Activated Carbon Adsorption," *Chem. Eng. Progress,* **89**(4), pp. 36–43 (Apr. 1993).

16. **Ruhl, M. J.,** "Recover VOCs via Adsorption on Activated Carbon," *Chem. Eng. Progress,* **89**(7), pp. 37–41 (July 1993).

17. **Graham, J. R., et al.,** "The Application of Carbon Adsorption Systems for VOC Recovery," presented at the Eighth Annual Hazardous Materials and Environmental Management Conference West/Fall, Long Beach, CA (Nov. 1992).

18. **Falth, L.,** "Hydrophobic Zeolites for VOC Abatement," Munters Zeol, Amesbury, MA (1987).

19. **Crompton, D., and A. Gupta,** "Removal of Air Toxics: A Comparison of the Adsorption Characteristics of Activated Carbon and Zeolites," Paper 93-TP-31B.06, presented at the 86th Annual Meeting of the Air and Waste Management Association, Denver, CO (June 1993).

20. "Newly Improved Synthetic Absorbents and Adsorbents Increase Emission Control Options," *The Air Pollution Consultant,* **5**(3), pp. 1.6–1.12 (May/June 1995).

21. **Rohm and Haas Co.,** "Ambersorb Carbonaceous Adsorbents — Specialty Purifications," Technical Notes published by Rohm and Haas Co., Philadelphia, PA (Oct. 1990).

22. **Kenson, R. E.,** "Recovery and Recycling of Chlorinated Solvents from Industrial Air Emissions," Paper 92-114.05, presented at the 85th Annual Meeting of the Air and Waste Management Association, Kansas City, MO (June 1992).

23. **Kenson, R. E.,** "Recovery and Reuse of Solvents from VOC Air Emissions," *Env. Progress,* **41**(3), pp 161–164 (Aug. 1985).

24. "CADRE VOC Control Process," Calgon Carbon Corp., Pittsburgh, PA (1987).

25. "Polymer Adsorption Process Used for VOC Control," *The Air Pollution Consultant,* **3**(3), pp. 1.19–1.21 (May/June 1993).

26. **Jain, N. K.,** "Brayton Cycle Solvent Recovery," Paper 92-114.09, presented at the 85th Annual Meeting of the Air and Waste Management Association, Kansas City, MO (June 1992).

27. **McLaughlin, H. S.,** "Regenerate Activated Carbon Using Organic Solvents," *Chem. Eng. Progress,* **91**(7), pp. 45–53 (July 1995).

28. **Keller, G. E.,** "Adsorption: Building upon a Solid Foundation," *Chem. Eng. Progress,* **91**(10), pp. 56–67 (Oct. 1995).

29. **Klobucar, J. M.,** "Choose the Best Heat-Recovery Method For Thermal Oxidizers," *Chem. Eng. Progress,* **91**(4), pp. 57–63 (Apr. 1995).

33. **Garg, A.,** "Specify Better Low-NO$_x$ Burners For Furnaces," *Chem. Eng. Progress,* **90**(1), pp. 46–49 (Jan. 1994).

31. **Bartz, D. F., et al.,** "Ultra-High VOC Destruction with Low NO$_x$ and CO in Adiabatic Radiant Combustors," presented at HAZMACON '92, The Ninth Annual Hazardous Materials Management Conference, Long Beach, CA (Mar.-Apr. 1992).

32. **Krause, T. R., et al.,** "Chemical Detoxification of Trichloroethylene and 1,1,1-Trichloroethane in a Microwave Discharge Plasma Reactor at Atmospheric Pressure," presented at Emerging Technologies for Hazardous Waste Treatment, American Chemical Society, Atlanta, GA (Oct. 1991).

33. **Rosocha, L. A., et al.,** "Nonthermal Plasma Alternative to the Incineration of Hazardous Organic Wastes," presented at the 11th International Incineration Conference, Albuquerque, NM (May 1992).

34. **Hung, S. L., et al.,** "Methyl Chloride and Methylene Chloride Incineration in a Catalytically Stabilized Thermal Combustor," *Environmental Science and Technology,* **23**(9), pp. 1085–1091 (Sept. 1989).

35. **Dowd, E. J., et al.,** "A Historical Perspective on the Future of Cataltytic Oxidation of VOCs," Paper 92-109.03, presented at the 85th Annual Meeting of the Air and Waste Management Association, Kansas City, MO (June 1992).

36. **Bell, A. T., et al.,** "Protecting the Environment Through Catalysis," *Chem. Eng. Progress,* **91**(2), pp.26–34. (Feb. 1995).

37. **Kittrell, J. R., et al.,** "Direct Catalytic Oxidation of Halogenated Hydrocarbons," *Journal of the Air and Waste Management Association,* **41**(8), pp. 1129–1133 (Aug. 1991).

38. **Lester, G. R., et al.,** "Activity, Stability, and Commercial Applications of the Allied-Signal Halohydrocarbon Destruction Catalyst," Paper 92-109.05, presented at the 85th Annual Meeting of the Air and Waste Management Association, Kansas City, MO (June 1992).

39. **Patkar, A. N., and J. M. Reinhold,** "Evaluation of Control Systems for Reduction of Air Toxic Emissions," presented at the Air and Waste Management Association International Conference on Tropospheric Ozone, Boston, MA (Oct. 1992).

during startup and shutdown, as well as a low overall pressure drop.

This fixed-bed design has evolved to a honeycomb-like support, typically made of stainless steel. These honeycomb supports offer the resistance to thermal and mechanical shock required, while providing higher surface area per unit volume (low space velocity), better thermal conductivity, and the ability to accommodate a wide variety of base metal and platinum group catalysts (37).

Another design is the simple packed-bed reactor, in which the pelletized catalyst particles are supported in shallow trays through which the gases pass. Pelletized catalyst offers the advantage of easy replacement, which is beneficial for applications where phosphorus or silicon compounds are present in large quantities, as these compounds may lead to attrition and/or deactivation (35).

Both fixed-bed designs can suffer from the inability to readily transfer heat released by the combustion process. Consequently, care must be taken to limit the temperature rise across the catalyst bed.

A third reactor design uses a fluidized bed to provide the catalyst/gas stream contact. Due to the rapid solids mixing in such units, they have much higher particle-to-particle heat-transfer rates, and the formation of hot spots, which can form in fixed-bed designs, is reduced. Therefore, gases with higher heating values can be processed with these reactors. Fluidized-bed systems are also more tolerant of high inlet particulate loadings. At the same time, care must be taken in the preparation of the catalyst to prevent attrition due to abrasion between catalyst particles.

New catalysts. New catalysts have been developed that address many of the fouling and/or poisoning problems of the past. For example, the halohydrocarbon destruction catalyst (HDC) developed by Allied Signal Corp. is much more resistant to phosphorus and related materials (38). When tested on the exhaust from a lithographic web offset press that used phosphorus-containing additives in the ink solutions, this catalyst demonstrated hydrocarbon destruction efficiencies in excess of 97% over more than 40 months of operation.

Other new catalysts have been developed that can treat gases containing halogens, and several equipment vendors now offer catalytic incineration systems that successfully treat halogenated streams (37, 39). These catalysts are generally chromia-alumina-based materials (typically 12–25% chromium calculated as Cr_2O_3). The systems are available in both fixed- and fluidized-bed designs, and the catalysts can be extruded or made into granules or pellets.

With these advances in catalyst development, catalytic incinerators can now be used under gas stream conditions that previously would have poisoned or fouled the catalyst, thereby leading to premature system failure. The availability of catalysts to handle both phosphorus- and halogen-containing streams is an important development for the chemical process industries.

Closing thoughts

The chemical process industries are facing increased regulation of hazardous air pollutants for the rest of this decade and beyond. These regulations will require that sources previously exempt from control or where controls have not been cost-effective in the past must now act to reduce emissions. The new regulations will impact sources ranging from paint spray booths, with high-flow, low-concentration emissions, to processes emitting chlorinated organic species, which have always been a special emission control problem.

In response to the increasing regulatory requirements and the market that these control demands represent, equipment vendors, EPA, national research laboratories, and universities have responded with new and novel control measures. Some of these have been used successfully for years in limited applications, but are now finding more universal application. Others are recent developments that will fill a need in certain air emission situations. Still other new technologies consist of modifications and enhancements to existing, conventional HAP-control methods designed to improve the performance of the equipment or to overcome shortfalls or operating limits.

To take advantage of these new emission control options, the user must (more than ever) be able to fully characterize the emission stream of concern. Not only is the total amount of pollutants important, but now the number and types of species in the stream, the variation in the emissions rate and number of species over time, and the specific emission limits to be achieved for each compound must also be well documented. Such characterization of the emission source will play a critical role in the selection of an emission control technology that achieves regulatory limits while also being cost-effective.

The 1990s will be a time of increasing development and refinement for HAP emission control technology. Potential users of such controls must keep up with these changing developments if they expect to select the best control option for their emission control needs and maintain continuous compliance with all regulatory requirements. **CEP**

R. G. McINNES is principal air quality engineer with ENSR Consulting and Engineering, Acton, MA (508/635-9500; Fax: 508/635-9180; e-mail: RMcInnes@ENSR.com), where he directs programs involving the evaluation, selection, and costing of air-pollution control equipment. He has over 20 years of air quality engineering experience in such areas as emissions inventory development, preparation of control technology studies, Title III MACT compliance evaluations, Title V permit application preparation, and control system design. He has authored a number of technical papers relating to the quantification and control of emissions from industrial facilities. He received his BS in mechanical engineering from the City College of New York and his MS in environmental engineering from Northeastern Univ.

Select the Right NO$_x$ Control Technology

Consider the degree of emission reduction needed, the type of fuel, combustion device design, and operational factors.

Stephen C. Wood,
Energy Technology Consultants, Inc.

Most major industrialized urban areas in the U.S. are unable to meet the National Ambient Air Quality Standards (NAAQS) for ozone. Atmospheric studies have shown that ozone formation is the result of a complex set of chemical reactions involving volatile organic compounds (VOCs) and nitrogen oxides (NO$_x$). These studies indicate that many urban areas with VOC/NO$_x$ ratios greater than 15:1 can reduce ambient ozone levels only by reducing NO$_x$ emissions. Many states, therefore, are implementing NO$_x$ control regulations for combustion devices in order to achieve compliance with the NAAQS ozone standard.

This article discusses the characterization of NO$_x$ emissions from industrial combustion devices. It then provides guidance on how to evaluate the applicable NO$_x$ control technologies and select an appropriate control method.

Characterizing emissions

Most industrial combustion devices have not been tested to establish their baseline NO$_x$ emission levels. Rather, the NO$_x$ emissions from these units have been simply estimated using various factors. In light of recent regulations, however, it is mandatory that the NO$_x$ emissions from affected units now be known with certainty. This will establish each unit's present compliance status and allow definition of the applicable control technologies for those units that will require modification to achieve compliance.

Many of the emissions inventories supplied by industry were derived from the U.S. Environmental Protection Agency's "AP-42 Emission Factors" document *(1)*. NO$_x$ emissions from combustion devices are quite variable and depend on a number of design, operational, and fuel conditions. Thus, the AP-42 factors, which were derived from the results of a number of test programs sponsored by the EPA during the 1970s, may be inaccurate when applied to a particular combustion device.

It is, therefore, important to test each combustion device to verify its NO$_x$ emissions characteristics. The testing process should be streamlined to provide timely and necessary information for making decisions regarding the applicability of NO$_x$ control technologies.

The basic approach is to select one device from a class of units (that is, of same design and size) for characterization testing (NO$_x$, CO, and O$_2$). Testing is conducted at three load points that represent the normal operating range of the unit, with excess oxygen variation testing conducted at each load point. Figure 1 illustrates the typical characterization test results. The remaining units in the class are tested at only one load point, at or near full load.

The operational data obtained during testing, in conjunction with the NO$_x$ and CO data, are used to define the compliance status of each unit, as well as the applicable NO$_x$ control technologies for those devices that must be modified. In most instances, this approach will allow multiple units to be tested in one day and provide the necessary operational data the engineer needs to properly evaluate the potential NO_x control technologies. Table 1 lists the operational data that should be obtained during the characterization testing.

Fundamental concepts

Reasonably available control technology (RACT) standards for NO_x emissions are defined in terms of an emission limit, such as 0.2 lb NO_x/MMBtu, rather than mandating specific NO_x control technologies. Depending on the fuel fired and the design of the combustion device, a myriad of control technologies may be viable options. Before selecting RACT for a particular combustion device, it is necessary to understand how NO_x emissions are formed so that the appropriate control strategy may be formulated.

NO_x emissions formed during the combustion process are a function of the fuel composition, the operating mode, and the basic design of the boiler and combustion equipment. Each of these parameters can play a significant role in the final level of NO_x emissions.

NO_x formation is attributed to three distinct mechanisms:

• thermal NO_x formation;

• prompt (*i.e.*, rapidly forming) NO formation; and

• fuel NO_x formation.

Each of these mechanisms is driven by three basic parameters — temperature of combustion, time above threshold temperatures in an oxidizing or reducing atmosphere, and turbulence during initial combustion.

Thermal NO_x formation in gas-, oil-, and coal-fired devices results from thermal fixation of atmospheric nitrogen in the combustion air. Early investigations of NO_x formation were based upon kinetic analyses for gaseous fuel combustion *(2)*. These analyses by Zeldovich yielded an Arrhenius-type equation showing the relative importance of time, temperature, and oxygen and nitrogen concentrations on NO_x formation in a premixed flame (that is, the reactants are thoroughly mixed before combustion).

While thermal NO_x formation in combustion devices cannot actually be deter-

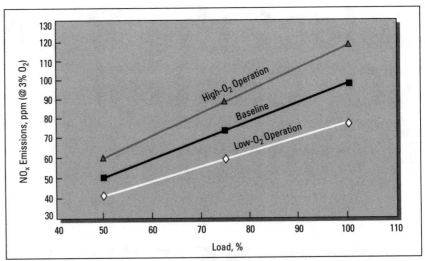

■ *Figure 1. Characterization testing gathers NO_x emission data as a function of oxygen level and operating load.*

Table 1. Collect the following operational and emissions data for each device during characterization testing.

Flow Rates	Pressures, psig	Emissions
Fuel, MMBtu	Steam	Plant Oxygen, %
Steam, klb/h	Drum	Test Equipment
Feed Water, klb/h	Natural Gas Supply	Oxygen, %
Natural Gas, kscfh	Burner	Carbon Monoxide, ppm
Draft, in. H_2O	Temperatures, °F	NO_x, ppm
Forced-draft Fan	Feed Water/Economizer Exit	NO_x Corrected to 3% O_2
Windbox	Steam	ppm
Furnace	Ambient, Wet Bulb	lb/h
Flame	Ambient, Dry Bulb	lb/MM Btu
Width, ft	Air Preheater	
Length, ft	Air In	
Furnace Vibration, mils	Air Out (combustion)	
Ultimate Fuel Analysis	Gas In	
(specifics depend	Gas Out	
on type of fuel)	Stack	
	Firebox	

mined using the Zeldovich relationship, it does illustrate the importance of the major factors that influence thermal NO_x formation, and that NO_x formation increases exponentially with combustion temperatures above 2,800°F.

Experimentally measured NO_x formation rates near the flame zone are higher than those predicted by the Zeldovich relationship. This rapidly forming NO is referred to as **prompt NO**. The discrepancy between the predicted and measured thermal NO_x values is attributed to the simplifying assumptions used in the derivation of the Zeldovich equation,

Table 2. Use this chart to screen potential NO$_x$ control technologies.

Technique	Description	Advantages	Disadvantages	Impacts To Consider	Applicability	NO$_x$ Reduction
Low Excess Air (LEA)	Reduces oxygen availability	Easy operational modification	Low NO$_x$ reduction potential	High carbon monoxide emissions; Flame length; Flame stability	All fuels	1–15%
Off-Stoichiometric Combustion (OS) a. Burners-out-of-service (BOOS) b. Overfire Air (OFA) c. Air Lances	Staged combustion, creating fuel-rich and fuel-lean zones	Low operating cost; No capital equipment required for BOOS	a. Typically requires higher air flow to control carbon monoxide b. Relatively high capital cost c. Moderate capital cost	Flame length; Forced-draft fan capacity; Burner header pressure	All fuels; Multiple-burner devices	30–60%
Low-NO$_x$ Burners (LNB)	Provides internal staged combustion, thus reducing peak flame temperatures and oxygen availability	Low operating cost; Compatible with FGR as a combination technology to maximize NO$_x$ reduction	Moderately high capital cost; Applicability depends on combustion device and fuels, design characteristics, waste streams, etc.	Forced-draft fan capacity; Flame length; Design compatibility; Turndown flame stability	All fuels	30–50%
Flue Gas Recirculation (FGR)	Up to 20–30% of the flue gas is recirculated and mixed with the combustion air, thus decreasing peak flame temperatures	High NO$_x$ reduction potential for natural gas and low-nitrogen fuels	Moderately high capital cost; Moderately high operating cost; Affects heat transfer and system pressures	Forced-draft fan capacity; Furnace pressure; Burner pressure drop; Turndown flame stability	Gas fuels and low-nitrogen fuels	40–80%
Water/Steam Injection (WSI)	Injection of steam or water at the burner, which decreases flame temperature	Moderate capital cost; NO$_x$ reductions similar to FGR	Efficiency penalty due to additional water vapor loss and fan power requirements for increased mass flow	Flame stability; Efficiency penalty	Gas fuels and low-nitrogen fuels	40–70%
Reduced Air Preheat (RAPH)	Air preheater modification to reduce preheat, thereby reducing flame temperature	High NO$_x$ reduction potential	Significant efficiency loss (1% per 40°F)	Forced-draft fan capacity; Efficiency penalty	Gas fuels and low-nitorgen fuels	25–65%
Selective Catalytic Reduction (SCR)	Catalyst located in flue gas stream (usually upstream of air heater) promotes reaction of ammonia with NO$_x$	High NO$_x$ removal	Very high capital cost; High operating cost; Extensive ductwork to and from reactor required; Large volume reactor must be sited; Increased pressure drop may require induced-draft fan or larger forced-draft fan; Reduced efficiency; Ammonia sulfate removal equipment for air heater required; Water treatment of air heater wash required	Space requirements; Ammonia slip; Hazardous-waste disposal	Gas fuels and low-sulfur liquid and solid fuels	70–90%
Selective Noncatalytic Reduction (SNCR) — Urea Injection	Injection of urea into furnace to react with NO$_x$ to form nitrogen and water	Low capital cost; Relatively simple system; Moderate NO$_x$ removal; Nontoxic chemical; Typically low energy injection sufficient	Temperature dependent; Design must consider boiler operating conditions and design; NO$_x$ reduction may decrease at lower loads	Furnace geometry and residence time; Temperature profile	All fuels	25–50%
Selective Noncatalytic Reduction (SNCR) — Ammonia Injection (Thermal DeNO$_x$)	Injection of ammonia into furnace to react with NO$_x$ to form nitrogen and water	Low operating cost; Moderate NO$_x$ removal	Moderately high capital cost; Ammonia handling, storage, vaporization, and injection systems	Furnace geometry and residence time; Temperature profile	All fuels	25–50%

such as the equilibrium assumption that $O = \frac{1}{2}O_2$. Near the hydrocarbon-air flame zone, the concentration of the formed radicals, such as O and OH, can exceed the equilibrium values, which enhances the rate of NO_x formation. However, the importance of prompt NO in NO_x emissions is negligible in comparison to thermal and fuel NO_x.

When nitrogen is introduced with the fuel, completely different characteristics are observed. The NO_x formed from the reaction of the fuel nitrogen with oxygen is termed **fuel NO_x**. The most common form of fuel nitrogen is organically bound nitrogen present in liquid or solid fuels where individual nitrogen atoms are bonded to carbon or other atoms. These bonds break more easily than the diatomic N_2 bonds so that fuel NO_x formation rates can be much higher than those of thermal NO_x. In addition, any nitrogen compounds (*e.g.*, ammonia) introduced into the furnace react in much the same way.

Fuel NO_x is much more sensitive to stoichiometry than to thermal conditions. For this reason, traditional thermal treatments, such as flue gas recirculation and water injection, do not effectively reduce NO_x emissions from liquid and solid fuel combustion.

NO_x emissions can be controlled either during the combustion process or after combustion is complete. Combustion control technologies rely on air or fuel staging techniques to take advantage of the kinetics of NO_x formation or introducing inerts that inhibit the formation of NO_x during combustion, or both. Post-combustion control technologies rely on introducing reactants in specified temperature regimes that destroy NO_x either with or without the use of catalyst to promote the destruction. Table 2 summarizes the commercially available NO_x control technologies, as well as their relative efficiencies, advantages and disadvantages, applicability, and impacts.

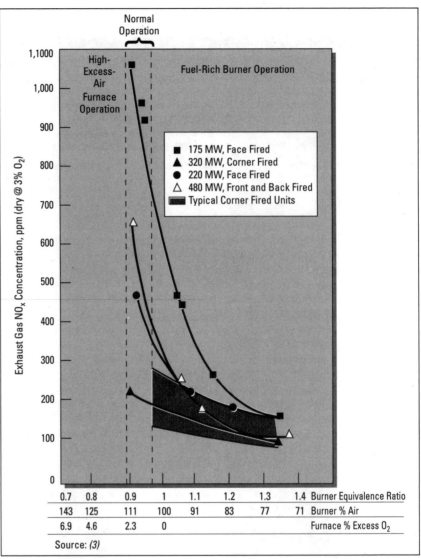

■ *Figure 2. Burners-out-of-service can be an effective combustion staging technique.*

Combustion control

The simplest of the combustion control technologies is **low-excess-air operation** — that is, reducing the excess air level to the point of some constraint, such as carbon monoxide formation, flame length, flame stability, and so on. Unfortunately, low-excess-air operation has proven to yield only moderate NO_x reductions, if any.

Three technologies that have demonstrated their effectiveness in controlling NO_x emissions are off-stoichiometric combustion, low-NO_x burners, and combustion temperature reduction. The first two are applicable to all fuels, while the third is applicable only to natural gas and low-nitrogen-content fuel oils.

Off-stoichiometric, or staged, combustion is achieved by modifying the primary combustion zone stoichiometry — that is, the air/fuel ratio. This may be accomplished operationally or by equipment modifications.

An operational technique known as burners-out-of-service (BOOS) involves terminating the fuel flow to selected burners while leaving the air registers open. The remaining burners operate fuel-rich, thereby limiting oxygen availability, lowering peak flame temperatures, and reducing NO_x formation. The unreacted products combine with the air from the terminated-fuel burners to complete burnout before exiting the furnace. Figure 2 illustrates the effectiveness of this technique applied to electric utility boilers.

Staged combustion can also be achieved by installing air-only ports, referred to as overfire air (OFA) ports, above the burner zone, redirecting a portion of the air from the burners to the OFA ports. A variation of this concept, lance air, consists of installing air tubes around the periphery of each burner to supply staged air.

BOOS, overfire air, and lance air achieve similar results. These techniques are generally applicable only to larger, multiple-burner, combustion devices.

Low-NO_x burners are designed to achieve the staging effect internally. The air and fuel flow fields are partitioned and controlled to achieve the desired air/fuel ratio, which reduces NO_x formation and results in complete burnout within the furnace. Low-NO_x burners are applicable to practically all combustion devices with circular burner designs. [Low-NO_x burners are covered in more detail in the article by Garg, pp. 46–49. — Editor]

Combustion temperature reduction is effective at reducing thermal NO_x, but not fuel NO_x. One way to reduce the combustion temperature is to introduce a diluent. **Flue gas recirculation** (FGR) is one such technique.

FGR recirculates a portion of the flue gas leaving the combustion process back into the windbox. The recirculated flue gas, usually on the order of 10–20% of the combustion air, provides sufficient dilution to decrease NO_x emission. Figure 3 correlates the degree of emission reduction with the amount of flue gas recirculated.

On gas-fired units, emissions are reduced well beyond the levels normally achievable with staged combustion control. In fact, FGR is prob-

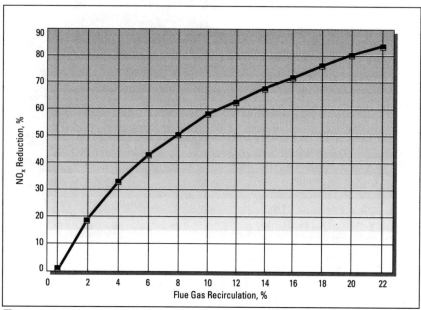

■ *Figure 3. NO_x reduction is a function of the amount of flue gas recirculated.*

ably the most effective and least troublesome system for NO_x reduction for gas-fired combustors.

An advantage of FGR is that it can be used with most other combustion control methods. Many industrial low-NO_x burner systems on the market today incorporate induced FGR. In these designs, a duct is installed between the stack and forced-draft

inlet (suction). Flue gas products are recirculated through the forced-draft fan, thus eliminating the need for a separate fan.

Water injection is another method that works on the principle of combustion dilution, very similar to FGR. In addition to dilution, it reduces the combustion air temperature by absorbing the latent heat of vaporiza-

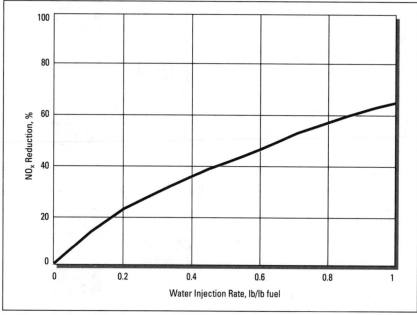

■ *Figure 4. Water injection can reduce NO_x emissions.*

tion of the water before the combustion air reaches the primary combustion zone.

Few full-scale retrofit or test trials of water injection have been performed. What little data exist suggest that the NO_x reductions follow the relationship shown in Figure 4.

Until recently, water injection has not been used as a primary NO_x control method on any combustion devices other than gas turbines because of the efficiency penalty resulting from the absorption of usable energy to evaporate the water. In some cases, water injection represents a viable option to consider when moderate NO_x reductions are required to achieve compliance.

Reduction of the air preheat temperature is another viable technique for cutting NO_x emissions, as shown in Figure 5. This lowers peak flame temperatures, thereby reducing NO_x formation. The efficiency penalty, however, may be substantial. A rule of thumb is a 1% efficiency loss for each 40°F reduction in preheat. In some cases this may be offset by adding or enlarging the existing economizer.

Post-combustion control

There are two technologies for controlling NO_x emissions after formation in the combustion process — selective catalytic reduction (SCR) and selective noncatalytic reduction (SNCR). Both of these processes have seen very limited application in the U.S. for external combustion devices. In **selective catalytic reduction**, a gas mixture of ammonia with a carrier gas (typically compressed air) is injected upstream of a catalytic reactor operating at temperatures between 450°F and 750°F. NO_x control efficiencies are typically in the 70–90% percent range, depending on the type of catalyst, the amount of ammonia injected, the initial NO_x level, and the age of the catalyst.

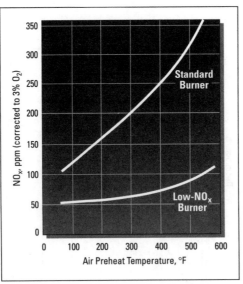

■ *Figure 5. Air preheat temperature affects boiler NO_x emissions.*

The retrofit of SCR on existing combustion devices can be complex and costly. Apart from the ammonia storage, preparation, and control monitoring requirements, significant modifications to the convective pass ducts may be necessary. [SCR is discussed by Cho in the article following this one, pp. 39–45. — Editor]

In **selective noncatalytic reduction**, ammonia- or urea-based reagents are injected into the furnace exit region, where the flue gas is in the range of 1,700–2,000°F. The efficiency of this process depends on the temperature of the gas, the reagent mixing with the gas, the residence time within the temperature window, and the amount of reagent injected relative to the concentration of NO_x present. The optimum gas temperature for the reaction is about 1,750°F; deviations from this temperature result in a lower NO_x reduction efficiency. Application of SNCR, therefore, must be carefully assessed, as its effectiveness is very dependent on combustion device design and operation.

Technology selection

As noted previously, selection of applicable NO_x control technologies depends on a number of fuel, design, and operational factors. Figure 6 sum-

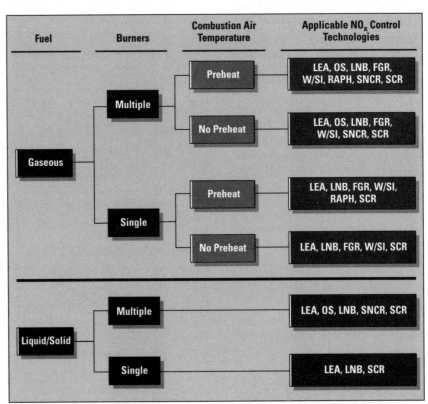

■ *Figure 6. Use these guidelines to identify potential NO_x control technologies.*

Table 3. NO$_x$ control efficiencies are multiplicative.

Control Technology	Efficiency	Final NO$_x$ Level, ppm
Low-Excess-Air	10%	100(1–0.10) = 90
Low-NO$_x$ Burner	40%	90(1–0.40) = 54
Flue Gas Recirculation	60%	54(1 – 0.6) = 21.6

marizes the choices for various scenarios. Figure 6 and Table 2 can be used to identify the potential control technologies for boilers and process heaters. After identifying the applicable control technologies, an economic evaluation must be conducted to rank the technologies according to their cost effectiveness. Management can then select the optimum NO$_x$ control technology for the specific unit.

It should be noted that the efficiencies of NO$_x$ control technologies are not additive, but rather multiplicative. Efficiencies for existing combustion devices have been demonstrated in terms of percent reduction from baseline emissions level. This must be taken into account when considering combinations of technology.

Consider, for example, the following hypothetical case. Assume a baseline NO$_x$ emissions level of 100 ppmv and control technology efficiencies as follows: low-excess-air operation (LEA), 10%; low-NO$_x$ burners (LNB), 40%; and flue gas recirculation (FGR), 60%. The three controls are installed in the progressive order of LEA-LNB-FGR. The stage-wise emission reductions are shown in Table 3.

It should also be noted that combining same-principle technologies (for example, two types of staged combustion) would not provide a fur-ther significant NO$_x$ reduction than either of the combination, since they operate on the same principle.

It must be emphasized that virtually all of the available control technologies have the potential for adversely affecting the performance and/or operation of the unit. The operational data obtained during the NO$_x$ characterization testing, therefore, must be carefully evaluated in light of such potential impacts before selecting applicable control technologies. Operational limitations such as flame envelope, furnace pressure, forced-draft fan capacity, and the like must be identified for each potential technology and their corresponding impacts quantified. (Reference (4), for example, discusses these items in detail.)

As anyone familiar with combustion processes knows, one technology does not fit all. Careful consideration must be used to select the appropriate, compatible control technology or technologies to ensure compliance at least cost with minimal impact on performance, operation, and capacity. **CEP**

Literature Cited

1. U.S. Environmental Protection Agency, "Compilation of Air Pollution Emission Factors," 4th ed., Document No. AP-42, EPA, Research Triangle Park, NC (Sept. 1985).
2. Zeldovich, J., "The Oxidation of Nitrogen in Combustion and Explosion," *Acta Physiochim, U.S.S.R. (Moscow)*, **21**(4), pp. 577–628 (1946).
3. Mormile, D., *et. al.*, "Analysis of Minimum Cost Control Approach to Achieve Varying Levels of NO$_x$ Emission Reduction from the Consolidated Edison Co. of NY Power Generation System," 1991 Joint Symposium on Stationary Combustion NO$_x$ Control, sponsored by U.S. Environmental Protection Agency and Electric Power Research Institute, Washington, DC (Mar. 25–28, 1991).
4. U.S. Environmental Protection Agency, "Alternative Control Techniques Document — NO$_x$ Emissions from Process Heaters," EPA-453/R-93-015, EPA, Research Triangle Park, NC (Feb. 1993).

Further Reading

American Petroleum Institute, "NO$_x$ Emissions from Petroleum Industry Operations," Publication No. 4311, API, Washington, DC (Oct. 1979).

Bartok, W., and A. F. Sarofim, "Fossil Fuel Combustion, A Source Book," John Wiley & Sons, New York (1991).

National Research Council, "Rethinking the Ozone Problem in Urban and Regional Air Pollution," National Academy Press, Washington, DC (1991).

Smith, L. L., and S. C. Wood, "Advances in NO$_x$ Control from Utility Power Plants," Second Conference on Air Quality Management in the Electric Power Industry, Austin, TX (Jan. 22–25, 1980).

Texas Air Control Board, "Proposed NO$_x$ RACT Rules," *Texas Register*, **17**(87), pp. 8136–8160 (Nov. 20, 1992).

U.S. Environmental Protection Agency, "Field Testing: Application of Combustion Modifications to Control Pollutant Emissions from Industrial Boilers — Phase II," EPA-600/2-76-086a, EPA, Research Triangle Park, NC (Apr. 1976).

Vatavuk, W. M., "Estimating Costs of Air Pollution Control," Lewis Publishers, Chelsea, MI (1990).

S. C. WOOD, southern regional manager for Energy Technology Consultants, a division of Woodward-Clyde, Inc., Houston, TX (713/690-0700; Fax: 713/744-9053), has 25 years of combustion process engineering experience. He has been responsible for analytical studies, performance, availability, and reliability optimization, as well as combustion-generated air-pollution reduction programs, on several hundred utility and industrial combustion devices throughout the U.S. and Europe. He is currently working with utility and industrial clients to achieve compliance with NO$_x$ RACT rules. He has authored a number of technical articles on NO$_x$ control technology and combustion process optimization. He received a BS in mathematics from the Univ. of Houston and completed graduate studies in mechanical engineering at the Univ. of Houston, and he is a member of the American Society of Mechanical Engineers and the Air and Waste Management Association.

Properly Apply Selective Catalytic Reduction for NO$_x$ Removal

Follow this design and operating guidance to maximize denitrification effectiveness.

Soung M. Cho,
Foster Wheeler Energy Corp.

Selective catalytic reduction (SCR) is a process in which nitrogen oxides (NO$_x$) are removed by the injection of ammonia (NH$_3$) into the flue gas. Chemical reactions in the presence of a catalyst produce nitrogen and water vapor, as indicated in Table 1. Table 2 highlights the wide range of applications in which SCR has been used.

This article discusses the design and operating experience of SCR systems for denitrification of flue gas in chemical process industries (CPI) and petroleum refinery heaters and boilers, gas turbine systems, and coal-fired steam plants. It provides an overview of the general SCR design approaches, the effects on the denitrification efficiency of major design and operating parameters, various ammonia vaporization and injection methods, and catalyst management strategies. Special precautions and design features associated with coal-fired units are also covered. Although the technical information here is primarily based on the experience of the author's company, it is considered to be typical of SCR industrial practice.

Table 1. SCR converts nitrogen oxides to nitrogen and water by these reactions.

$$4NO + 4NH_3 + O_2 \rightarrow 4N_2 + 6H_2O \qquad (1)$$

$$6NO + 4NH_3 \rightarrow 5N_2 + 6H_2O \qquad (2)$$

$$2NO_2 + 4NH_3 + O_2 \rightarrow 3N_2 + 6H_2O \qquad (3)$$

$$6NO_2 + 8NH_3 \rightarrow 7N_2 + 12H_2O \qquad (4)$$

$$NO + NO_2 + 2NH_3 \rightarrow 2N_2 + 3H_2O \qquad (5)$$

The results of SCR performance tests in actual applications indicate that the reaction represented by Eq. 1 is the dominant reaction.

Table 2. SCR has been used in a variety of applications such as these.

Process Types:	CPI plant and refinery heaters and boilers, gas turbines, and coal-fired cogeneration plants
Fuels:	Natural gas, industrial gas, crude oil, light or heavy oil, and pulverized coal
Gas Flow Rates:	10,000 to 1,200,000 Nm³/h
Gas Temperatures:	280 to 400°C
Gas Flow Directions:	Horizontal, vertical upward, and vertical downward
Ammonia Systems:	Anyhdrous and aqueous
Inlet NO$_x$ Levels:	25 to 270 ppmvd
NO$_x$ Removal Efficiencies:	42% to 90%

Design approaches

An SCR system (Figure 1) consists mainly of an ammonia injection grid, an SCR reactor, and associated ducting. The SCR reactor is the core of the system. Basically a container that houses the catalyst, it is an assembly of catalyst baskets, each consisting of square elements made of homogeneous ceramic honeycomb catalyst cells. (Nonhomogeneous catalysts such as a plate type are also available.) Cell sizes vary typically from 3.7 mm to 7.5 mm. The smaller cells are normally used for clean gas applications and larger cells for "dirty" gas applications.

A variety of catalysts are available for various applications. The major components of the catalysts are titanium dioxide (TiO_2), tungsten trioxide (WO_3), vanadium pentoxide (V_2O_5), and molydbenum trioxide (MoO_3). The actual compositions of the catalysts are, in general, proprietary to the catalyst supplier.

For given gas conditions, the performance of an SCR system depends upon:
• the type of catalyst;
• the area of catalyst surface exposed to the flue gas;
• the residence time of the gas in the reactor;
• the amount of ammonia injected upstream of the reactor;
• the degree and effectiveness of the mixing of ammonia and gas;
• the fuel's sulfur content; and
• the extent of any dust loading.

The major factors that determine the type of catalyst are gas temperature, sulfur content, and dust loading.

Once the catalyst is chosen, the performance of the SCR reactor depends significantly upon the diffusion surface area of the catalyst, known as the specific surface area. The larger the specific surface area, the better the SCR performance. Since the specific area is the gas diffusion area of the catalyst including the surface of all pores, it is characteristic of the catalyst, but not a convenient design parameter.

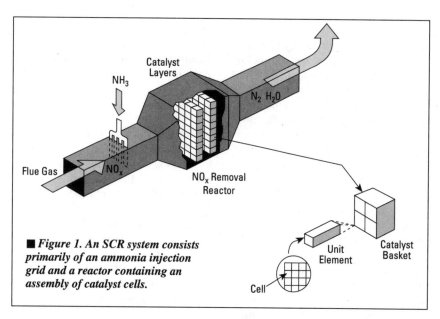

■ Figure 1. An SCR system consists primarily of an ammonia injection grid and a reactor containing an assembly of catalyst cells.

More practical design or performance parameters are those that combine the specific surface area with the residence time of the gas through the catalyst layer. These include the area velocity, the space velocity, and the linear velocity, all of which are related to one another. The most popular and convenient velocity parameter for SCR reactor design and performance evaluation is the space velocity (*SV*).

Denitrification of flue gas via SCR requires the supply of ammonia. According to the most dominant reaction process, Eq. 1 (Table 1), one mole of ammonia is required for each mole of nitrogen oxide (NO) for stoichiometric reaction. Therefore, the larger the ammonia injection quantity, the greater the probability of denitrification reaction and consequently the better the SCR performance. However, excessive ammonia injection may result in deleterious side effects, as will be discussed later.

The denitrification, or NO_x removal, efficiency (η) of an SCR reactor is defined as the quantity of NO_x removed divided by the quantity of NO_x in the inlet gas stream. This efficiency depends upon various parameters, as mentioned previously. Based on reaction kinetics and laboratory and field test data, one may express the NO_x removal efficiency in terms of the space velocity and the molar ratio of ammonia to NO_x:

$$\eta = m\left(1 - e^{-k/SV}\right) \qquad (6)$$

where *k* is an activity constant that is a function of many factors, including the catalyst composition, diffusion characteristics of ammonia and NO_x in the gas stream and catalyst layer, oxygen concentration, moisture concentration, gas temperature, gas velocity, and catalyst aging.

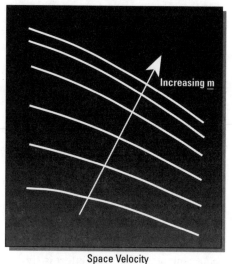

■ Figure 2. NOx removal efficiency is a function of space velocity and the ratio of ammonia to NO_x.

Figure 2 shows typical curves for the NO_x removal efficiency in terms of the space velocity and the molar ratio of ammonia to NO_x. In general, the lower the space velocity and the higher the molar ratio of ammonia to NO_x, the greater the denitrification efficiency. The characteristic curves depicted in Figure 2 are corrected for the effects of various operating parameters, the most significant being flue gas temperature, oxygen concentration, water vapor concentration, aging effect, and excess quantity of ammonia (also known as "ammonia slip"). There are a series of charts of this kind for various catalyst types and they are, in general, proprietary to each catalyst supplier.

From design curves such as Figure 2, the SCR reactor can be sized as follows. First, the NO_x removal efficiency is determined based on the inlet and outlet NO_x levels. Next, the molar ratio of ammonia to NO_x is determined based on the limit on the residual ammonia or ammonia slip, using the following definition of S:

$$S = (m - \eta)(NO_{x,in}) \qquad (7)$$

or

$$m = \eta + S/NO_{x,in} \qquad (8)$$

These expressions are valid when the gas temperature is below approximately 400°C. At higher temperatures, ammonia tends to be oxidized.

Knowing the values of η and m for design, one can read the required space velocity value from the curves of Figure 2. The superficial volume of the catalyst is then determined by the volumetric gas flow divided by the required space velocity.

Care should be exercised in preparing the physical design of the SCR reactor, particularly for the gas stream pressure drop. The linear velocity and the length of the catalyst layer should be sized so that the most economical system is obtained. A typical SCR system pressure drop is on the order of 50 mm to 100 mm water. Because of relatively small catalyst cell dimensions, the gas flow inside the reactor is typically laminar. The Fanning friction factor for laminar flow may be obtained from (1):

$$f = C/Re \qquad (9)$$

where the value of C depends upon the geometry of the flow cross-section as follows: $C = 14.2$ for square cross-section; 18.3 for rectangular cross-section with an aspect ratio of 4; 20.5 for rectangular cross-section with an aspect ratio of 8; and 24.0 for rectangular cross-section with an aspect ratio of infinity.

If the flow is turbulent, one may use the Moody diagram to determine the friction factor (2).

Table 3. Sulfur trioxide can react with excess ammonia and water vapor to form ammonium sulfates.

$SO_3 + 2NH_3 + H_2O \rightarrow (NH_4)_2SO_4$	(10)
$SO_3 + NH_3 + H_2O \rightarrow NH_4HSO_4$	(11)

Effects of operating parameters

The curves in Figure 2 are corrected for the effects of various operating parameters. The most significant parameters are discussed in this section.

Flue gas temperature. This is perhaps the most important operating parameter that influences the choice of the catalyst. The NO_x removal efficiency peaks at a certain temperature that is specific to each catalyst. Thus, the choice of catalyst must consider the range of operating temperatures in the plant.

Oxygen concentration. As indicated in Eqs. 1 and 3, oxygen is needed in the flue gas for dentrification. As oxygen concentration increases, catalyst performance improves until it reaches an asymptotic value. This effect is significant only when the oxygen content is less than about 2–3%.

Water vapor concentration. The water vapor content in the flue gas has an adverse effect on the NO_x removal efficiency. The higher the water vapor content, the lower the catalyst performance.

Aging effect. The performance of catalyst tends to deteriorate over time. The rate of deterioration is large in the beginning of operation and becomes rather gradual after initial settlement.

Ammonia slip. Theoretically, the amount of ammonia to be injected should be based on a molar ratio of ammonia to NO_x, which is numerically the same as the NO_x removal efficiency (based on the dominant denitrification reaction process of Eq. 1). However, since ammonia is not completely and uniformly mixed with NO_x, more than the theoretical quantity is normally injected. The excess residual ammonia in the downstream flue gas, known as the ammonia slip and typically specified in parts per million, is defined by Eq. 7. The NO_x removal efficiency increases with increasing ammonia slip and reaches an asymptotic value at a certain level of excess ammonia. This means that there is a limit to the advantageous effect of excess ammonia in removing NO_x.

Furthermore, excess ammonia is environmentally harmful when discharged to the atmosphere through the stack, and therefore its quantity should be minimized. The other important reason for limiting the ammonia slip to a low value is to reduce the chances of forming ammonium sulfates in the presence of SO_3. Sulfur-containing fuels produce SO_2 and a small quantity of SO_3. A small fraction of SO_2 in the flue gas is also converted to SO_3 by the SCR catalyst. When combined with excess ammonia and water vapor, SO_3 may form ammonium sulfates by the reactions shown in Table 3.

Ammonium sulfate, $(NH_4)_2SO_4$, is powdery and contributes to the quantity of particulates in the flue gas. Ammonium bisulfate, NH_4HSO_4, is a sticky substance that can deposit on the catalyst layers and/or downstream equipment, causing flow blockage and equipment performance deterioration.

Temperature is an extremely

important factor in the formation of sulfates. The lower the temperature, the higher the probability of sulfate formation. When natural or industrial gas or low-sulfur oil is used as the combustion fuel, the deteriorating effects discussed above are not likely to occur if the ammonia slip is limited to less than 10 ppm and the SO_3 concentration to less than 5 ppm (unless the gas temperature is very low).

SCR systems for coal-fired plants

The SCR units installed in coal-fired facilities require special design considerations, mainly due to generally high dust loading. The example described here is typical of pulverized-coal-fired plants. Further details of the installation can be found in Reference (3).

This cogeneration plant, nominally rated 285 MWe, burns a 2%-sulfur, Eastern bituminous coal. The NO_x removal efficiency is 63%, with an ammonia slip limit of 5 ppm (by volume, dry) at 7% O_2.

Nomenclature

AV	= area velocity = volumetric gas flow rate divided by the geometric catalyst contact surface area, m/h
C	= constant to account for the geometry of the flow cross-sectional area
D_H	= hydraulic or equivalent diameter, m
f	= Fanning friction factor
k	= activity constant
LV	= linear velocity = volumetric gas flow rate divided by the flow area of the gas through the catalyst, m/h
m	= molar ratio of ammonia to NO_x
$NO_{x,in}$	= inlet NO_x concentration, ppm
$NO_{x,out}$	= outlet NO_x concentration, ppm
Re	= Reynolds number = $\rho V D_H/\mu$
S	= ammonia slip, ppm
SV	= space velocity = volumetric gas flow rate divided by the superficial volume of the catalyst, h^{-1}
V	= gas velocity, m/s

Greek Letters

η	= NO_x removal efficiency, dimensionless
μ	= gas viscosity, N•s/m²
ρ	= gas density, kg/m³

Table 4. Flue gas conditions vary with plant load.

Load, % of Maximum Continuous Rating	100	70	50	35
Gas Flow Rate, thousand kg/h	610.8	442.4	322.1	225.4
Gas Temperature at SCR Inlet, °C	390	350	320*	320*

*With partial economizer bypass to maintain this temperature.

The plant, illustrated in Figure 3, has two 50% pulverized-coal-fired steam generators. The SCR unit is placed between the economizer and the air heater because this location gives the most optimum flue gas temperature range for commercially available catalysts. The SCR system consists of a partial economizer bypass duct, an ammonia injection grid, turning vanes, a flow rectifier, an SCR reactor with several catalyst layers, soot blowers, and associated ducting.

The principal design features of this coal plant, compared to plants burning other fuels, are the economizer bypass duct, turning vanes, flow rectifier, and soot blowers:

Economizer bypass duct. In general, the plant flue gas conditions vary with load requirements. Table 4 shows the range of flue gas flow and temperature for the example plant. For 35–50% load operation, part of the flue gas is bypassed around a portion of the economizer to maintain the flue gas temperature above 320°C to avoid the formation of ammonium sulfate and bisulfate.

Turning vanes. Turning vanes provide uniform gas flow distribution and, at the same time, minimize gas-side pressure drop. The exact geometric shapes of these vanes are determined based on the results of scale-model flow testing that simulates the flow configuration from the economizer exit to the air heater inlet.

Flow rectifier. Because of high dust loading associated with coal firing, the linear velocity of the gas stream is limited to approximately 6 m/s to prevent catalyst erosion. The gas flow entering the reactor is straightened out by a flow rectifier made of square pipes. The flow rectifier provides an inlet velocity vector that is in line with the catalyst honeycomb flow path, which tends to decrease erosion potential.

Soot blowers. Soot blowers are placed upstream of the catalyst layers for occasional "dust off" of the catalyst surface.

It should be noted that an "optimum" SCR design requires proper distributions of flue gas and ammonia gas at the reactor inlet. The ammonia injection grid, the turning vanes, the flow rectifier, the catalyst layers, and the associated ducting are all designed to achieve a uniform, balanced mix of ammonia gas with flue gas at the SCR reactor inlet.

Extreme care should be exercised in the design of an SCR unit for a plant burning sulfur-containing coal. The coal burned at this plant has a high sulfur content (2%). Thus, the following three features have been incorporated to minimize the chances of ammonium sulfate formation:

• The largest catalyst cell size, 7.5 mm pitch, is selected. Note that natural gas applications normally utilize a 3.7 mm catalyst cell size.

• Flue gas temperature is maintained above the critical ammonium sulfate formation temperature of 320°C throughout the load operating range. As stated previously, this involves bypassing part of the flue gas around a portion of the economizer at loads of 35–50%.

• Ammonia slip is kept below 5 ppm at all times.

Ammonia injection systems

The traditional method of injecting ammonia into the flue gas stream (Figure 4) utilizes "pure" anhydrous ammonia. Liquid ammonia is normally stored in a pressurized tank. From there it is piped to a heater (typically an elec-

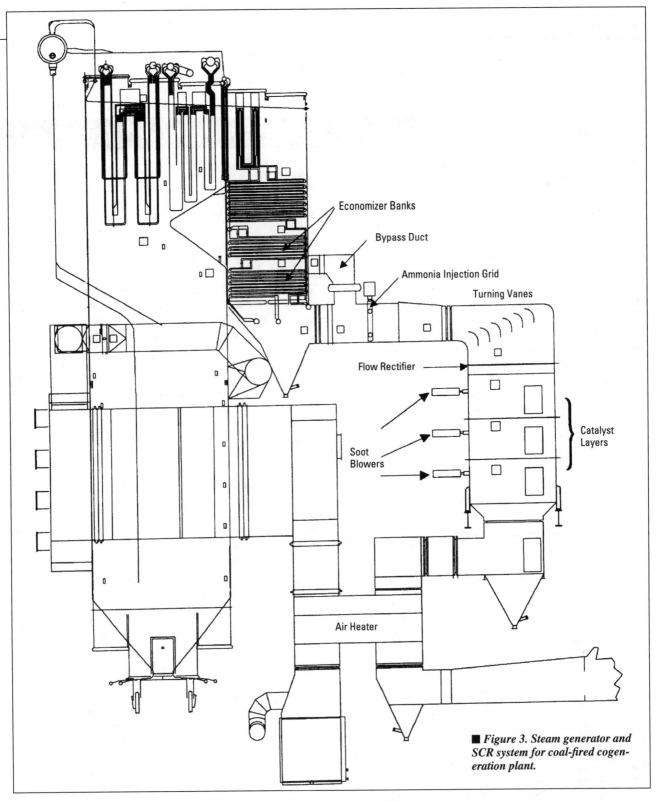

Economizer Banks

Bypass Duct

Ammonia Injection Grid

Turning Vanes

Flow Rectifier

Soot
Blowers

Catalyst
Layers

Air Heater

■ *Figure 3. Steam generator and SCR system for coal-fired cogeneration plant.*

tric heater), where the liquid vaporizes. The ammonia vapor is next routed into a mixing chamber, where it mixes with ambient air supplied by a fan or blower in a predetermined ratio. The ammonia-air mixture is then directed to the distribution grid system for subsequent injection into the flue gas stream at a location upstream of the SCR reactor.

However, anhydrous ammonia is toxic and hazardous. It has a high vapor pressure at ordinary temperature, and thus requires thick-shell storage tanks, piping, and vessels. Its release to the atmosphere may present a hazard, which makes transportation of pure anhydrous ammonia less desirable from a safety standpoint.

An alternative approach for ammonia injection is the use of aqueous ammonia, $NH_3 \cdot H_2O$, which is less hazardous. A typical industrial-grade aqueous ammonia contains approximately 27% ammonia and 73% water by weight. This ammonia-water mixture has a nearly atmospheric vapor pressure at ordinary temperature, and it can be safely transported on U.S. highways.

Figure 5 is a schematic of an aqueous ammonia injection system. It contains two trains of piping, one for aqueous ammonia and the other for carrier air, that feed into a vaporizer vessel. Aqueous ammonia is stored at ordinary temperature in a tank; from there it is pumped, metered, and sprayed through atomizing nozzles (using air to atomize it) into the vaporizer. Ambient air is drawn by a blower and heated, then enters the vaporizer via nozzles located at the top of the vessel. The vaporizer may be packed with metallic rings to promote the mixing and vaporizing processes. (Some vaporizers do not use packing, and those tend to be larger.)

The mixture of ammonia vapor, water vapor, and air is routed from the vaporizer to the injection grid network via a balancing skid for subsequent injection into the flue gas stream. The injection grid network consists of a number of pipes connected in parallel, each of which contains several orifice holes. The pipes and holes are sized to provide balanced ammonia flow distribution throughout the flue gas.

There are several different ways to heat the carrier air. One is to use an electric heater, as shown in Figure 5. In the second method, part of the flue gas is routed out of the flue duct and through a heat exchanger to heat the carrier air, then it is returned to the duct. However, this method requires a hot gas recirculation fan, which is generally not highly reliable, and possibly frequent maintenance for the heat exchanger due to "not-so-clean" gas conditions.

In the third scheme, ambient air drawn by a fan is heated in a heating coil placed directly in the flue gas stream. This method of in-duct air heating is attractive for several reasons: a fan which draws clean air from outside is more reliable than a "dirty gas" fan; the temperature drop of the flue gas across the air heater is negligible (on the order of 1°C for most applications), so it has no effect on the flue gas system; and the energy consumption is less than for the other two methods.

An alternative method of vaporizing the aqueous ammonia is the direct use of the thermal energy of the hot flue gas. A slip stream of flue gas is drawn into a vaporizer, where it mixes

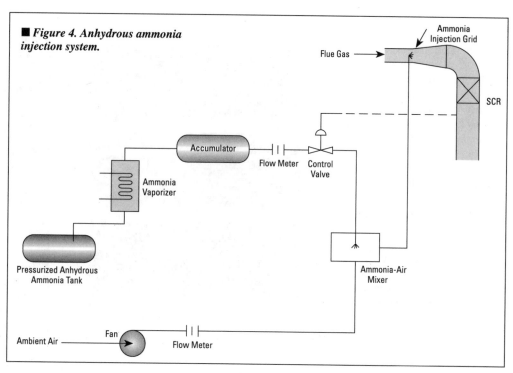

■ *Figure 4. Anhydrous ammonia injection system.*

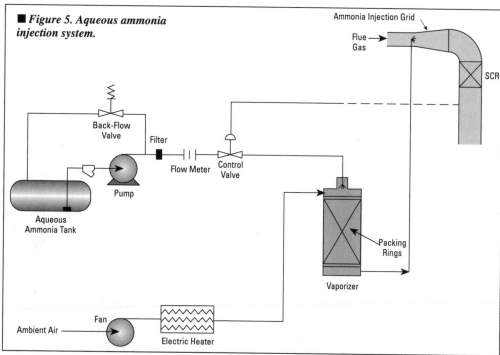

■ *Figure 5. Aqueous ammonia injection system.*

with and consequently vaporizes aqueous ammonia. The major drawbacks of this approach are the lower reliability of a hot gas recirculation fan, and the possible contamination and resulting frequent maintenance for the vaporizer and the injection pipes and nozzles.

Catalyst management strategies

As noted earlier, the NO_x removal efficiency of the catalyst decreases over time. Catalyst suppliers guarantee their catalyst performance typically for two to four years, while plants are designed for much longer periods (30 to 40 years). Loading a large quantity of catalyst initially is, in general, not attractive from economic and gas-side pressure-drop standpoints, so a catalyst management plan for the life of the plant is needed.

Let's consider again the coal-fired cogeneration plant example discussed earlier. Its SCR system is designed to limit the stack NO_x emission rate to 0.1 lb/MMBtu for 63% NO_x removal efficiency with an ammonia slip limit of 5 ppmvd for a ten-year period, with up to 50% of the initial catalyst charge to be added and/or replaced during each of the first and second five-year periods.

One of the catalyst management plans considered to meet the above requirements is schematically shown in Figure 6. The SCR reactor consists of three layers. Initially, the top two layers of the reactor are charged with catalyst. Figure 6 shows the potential catalyst NO_x removal efficiency vs. time (assuming that only coal is fired for the entire ten-year period). The initial two layers of catalyst maintain the NO_x removal efficiency above the design value of 63% or the outlet NO_x concentration below 0.1 lb/MMBtu for the first seven-year period. At the end of the seventh year, fresh catalyst

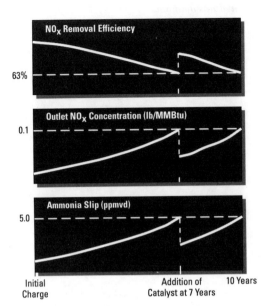

Figure 6. Adding fresh catalyst at varying intervals can be an effective catalyst management strategy.

is added in the third layer of the reactor. This will ensure that the NO_x removal efficiency remains above 63% or the outlet NO_x concentration below 0.1 lb/MMBtu for the remaining three-year period.

Note that the NO_x removal efficiency and the outlet NO_x concentration curves shown in Figure 6 are theoretical curves. In reality, ammonia injection is controlled to maintain the NO_x removal efficiency at the design value (63% for coal firing) and the outlet NO_x concentration at 0.1 lb/MMBtu. The true variable of the operation is the ammonia slip, whose performance characteristic for the planned catalyst management scenario is also shown in Figure 6. The ammonia slip is kept below 5 ppmvd (referenced at 7% O_2 level) at all times.

In general, catalyst management strategies should consider the addition and/or replacement of catalyst of varying quantities at different time intervals. The main goal of the strategy is minimum lifetime cost to the plant.

Operating experience

The author's firm has design and operating experience with over 30 SCR units since 1982 (Table 2). All of the units have been operating free of

Literature Cited

1. **Kays, W. M., and A. L. London,** "Compact Heat Exchangers," McGraw-Hill, New York (1964).
2. **Streeter, V. L., and E. B. Wylie,** "Fluid Mechanics," McGraw-Hill, New York (1975).
3. **Cho, S. M., and S. Z. Dubow,** "Design of a Selective Catalytic Reduction System for NO_x Abatement in a Coal-Fired Cogeneration Plant," Proceedings of the American Power Conference, sponsored by Illinois Institute of Technology, Vol. 54, pp. 717–722 (1992).

Further Reading

Cho, S. M., et al., "Design and Operating Experience of Selective Catalytic Reduction Systems for NO_x Control in Gas Turbine Systems," presented at the International Gas Turbine and Aeroengine Congress and Exposition, Orlando, FL, ASME Paper No. 91-GT-26 (1991).

S. M. CHO is director of engineering in the energy applications div. of Foster Wheeler Energy Corp., Clinton, NJ (908/236-1420; Fax: 908/236-1425). He has over 25 years of engineering experience in the design and analysis of heat transport systems and components in power and process plant systems, and has been with Foster Wheeler since 1973. He is also an adjunct professor of mechanical engineering at Stevens Institute of Technology. He received a BS from Seoul National Univ. and MS and PhD degrees from the Univ. of California, Berkeley, all in mechanical engineering. A Fellow of the American Society of Mechanical Engineers, he is a member of the ASME Heat-Transfer Div. Executive Committee and past chair of the ASME Nuclear Engineering Div.; he is also a member of the American Nuclear Society. He is an associate editor of an ASME transaction journal and has published over 55 technical papers.

trouble since their initial startup. In fact, most of them are overperforming, so that the NO_x contents at the SCR outlets are much lower than originally designed for and the original catalysts have not been replaced. CEP

Specify Better Low-NO$_x$ Burners For Furnaces

Air staging, fuel staging, and internal flue gas recirculation are among the design features that help reduce NO$_x$ emissions.

Ashutosh Garg,
Kinetics Technology International Corp.

Specifying the right requirements for low-NO$_x$ burners can significantly reduce nitrogen oxides (NO$_x$) emissions from a furnace. Ultra-low-NO$_x$ burners that can meet even the most stringent emission control limits imposed by some states are now available and offer a very attractive route to NO$_x$ reduction. However, burner selection and specification should be done very carefully, because burner operation has a direct effect on furnace performance. This article describes the various types of low-NO$_x$ burners and outlines the main design parameters that must be considered when selecting a burner system.

Low-NO$_x$ burners generally modify the means of introducing air and fuel to delay the mixing, reduce the availability of oxygen, and reduce the peak flame temperature. Whether for a new furnace or a retrofit application, these burners must meet five major requirements:

1. operation with lower NO$_x$ formation;
2. a flame pattern compatible with furnace geometry;
3. easy maintenance and accessibility;
4. a stable flame at turndown conditions; and
5. the ability to handle a wide range of fuels.

Burner types

Table 1 lists the types of burners currently in use in chemical process industries (CPI) plants and petroleum refineries. Figure 1 compares staged-combustion burners with standard gas and oil burners.

Staged-air burners. Combustion air is split and directed into primary and secondary zones, thus creating fuel-rich and fuel-lean zones.

These burners are most suitable for forced-draft liquid-fuel-fired applications. The combustion air pressure energy lends itself to better control of the staging air flows. It ensures a high enough air velocity to produce good air-fuel mixing and a good flame.

Staged-air burners lend themselves very well to external flue gas recirculation (FGR). In such designs, flue gas is generally introduced into the primary combustion zone.

Staged-fuel burners. The fuel gas is injected into the combustion zone in two stages, thus creating a fuel-lean zone and delaying completion of the combustion process. The fuel supply is divided into primary fuel and secondary fuel in a ratio that depends on the NO$_x$ level required. The flame length of this type of burner is about 50% longer than that of a standard gas burner.

Staged-fuel burners are ideal for fuel-gas-fired natural-draft applications.

Table 1. Different low-NO$_x$ burners have different NO$_x$ reduction capabilities.

Burner Type	NO$_x$ Reductions*
Staged-Air Burner	25–35%
Staged-Fuel Burner	40–50%
Low-Excess-Air Burner	20–25%
Burner with External FGR	50–60%
Burner with Internal FGR	40–50%
Air or Fuel-Gas Staging With Internal FGR	55–75%
Air or Fuel-Gas Staging With External FGR	60–80%

*Reductions compared to a standard gas burner.

Low-excess-air burners. These burners reduce NO_x emissions by completing combustion with the lowest amount of excess air possible, usually no more than 5–8%. Increases in excess air result in increases in NO_x formation (Figure 2a).

Most forced-draft burners have the ability to operate at very low levels of excess air. In a multiple-burner installation, it is essential that all burners receive equal amounts of air. This can be achieved by simulating the flow profiles in the ducts and burners. Flow deficiencies and other irregularities can then be detected and corrected using splitters and vanes, ensuring equal air distribution within ± 1%.

Flue gas recirculation burners. In these burners, 15–25% of the hot (300–500°F) flue gas is recirculated along with the combustion air. The flue gas acts as a diluent, reducing flame temperature and suppressing the partial pressure of oxygen, thus reducing NO_x formation. Flue gas can be injected into burners through a separate scroll into the primary zone or mixed with incoming air. External FGR can be used with natural-draft burners, although it is mostly used with forced-draft preheated air burners.

In some new burner designs, flue gas is internally recirculated using the pressure energy of fuel gas, combustion air, or steam. This makes the operation of burners simple and eliminates the FGR fan and its controls, although burner size becomes large.

Ultra-low-NO_x burners. Several designs are available today that combine two NO_x reduction steps into one burner without any external equipment. These burners typically incorporate staged air with internal FGR or staged fuel with internal FGR. In the former design, fuel is mixed with part of the combustion air, creating a fuel-rich zone. High-pressure atomization of liquid fuel or fuel gas creates FGR. The secondary air is routed by means of pipes or ports in the burner block to complete combustion and optimize the flame profile.

In staged-fuel gas burners with internal FGR, fuel gas pressure induces recirculation of flue gas, creating a fuel-

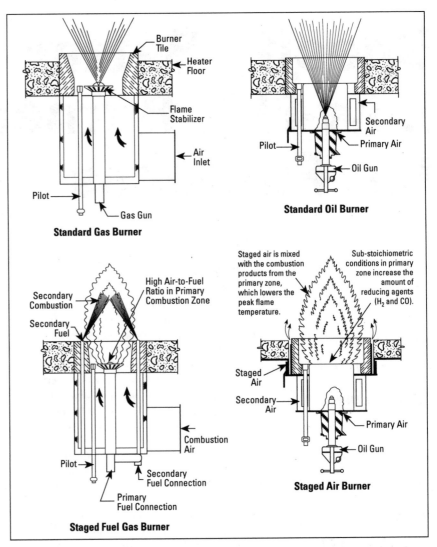

Figure 1. Off-stoichiometric combustion can be achieved by air staging or fuel staging.
Courtesy of John Zink Co.

lean zone and a reduction in oxygen partial pressure.

The former design can be used with the liquid fuels, whereas the latter design is used mostly for fuel gas applications.

The following parameters require attention during system design.

Fuel specifications. Correct and accurate fuel specifications are essential for predicting NO_x emissions.

For gaseous fuels, the complete analysis listing all the constituents is required, as well as any possible variations in gas composition. Major components affecting NO_x emissions are hydrogen and hydrocarbons in the C_3–C_4 range. Other physical properties, such as pressure, temperature, and heating value, are required for burner design.

For liquid fuels, the most important parameter is the fuel's nitrogen content — about 40–90% of the fuel nitrogen shows up as NO_x in the flue gas. Other liquid fuel parameters required by burner vendors are pressure, temperature, viscosity, and heating value.

Atomization medium. For low-NO_x burners, steam is preferred as the atomization medium over compressed air, because higher quantities of steam decrease the amount of NO_x in the flue gas. Increases in steam temperature increase NO_x emissions.

Fuel filters. Staged-fuel gas burners have more gas tips and risers than standard burners, and the fuel gas flow per tip is reduced to as low as one-fourth. It is important that these burners be used with clean fuel gases. To accomplish this, installation of fuel gas filters and knockout pots to remove particulates and condensate is recommended.

Some plants have opted for low fuel-gas pressures or double orifice designs for the gas tips to keep the tip size large enough to avoid plugging when firing dirty gases. These options generally do not give good results, and they also produce longer flames (flame length is discussed later).

Heat release and turndown. Plant engineers have typically specified a margin on the heat release rate as high as 30–50% over the design heat release. Furthermore, on most standard burners, the turndown for gas fuel is generally 5:1 and for oil it is 3:1. These two parameters offered virtually unlimited flexibility to overfire and underfire the furnaces.

However, to ensure optimum performance of low-NO_x burners, it is important to limit the overdesign margin to only 10%. And in most cases, the turndown should be limited to 3:1 for gas and 2:1 for oil. This will require

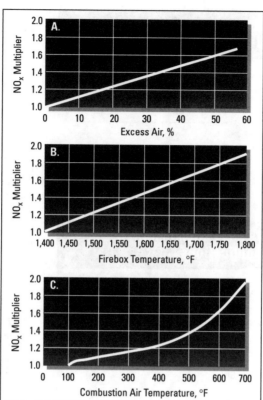

■ *Figure 2. NO$_x$ emissions are a function of various furnace parameters.*

more attention from the operators and minimization of burner outages. The result, though, will be better performance from the low-NO_x burners.

Heater draft. The available heater draft is a very important design parameter, especially for natural-draft burners, because it directly indicates the air pressure energy available for air/fuel mixing. It is, therefore, important that available draft be specified correctly.

In some cases, it may be advantageous to increase the available draft by increasing the stack height or diameter. Increased draft availability can reduce the size of the burners. However, increased draft at the hearth also increases the likelihood of air leakage into the furnace, so the furnace should be made leak-tight to prevent such air infiltration.

Firebox temperature. In the past, standard burners were specified independent of the furnace design parameters. However, the performance of

low-NO_x burners is closely linked with furnace design and firing arrangements. NO_x formation is dependent on firing density and firebox temperature. The burner vendor needs the firebox temperature and geometry to predict NO_x emissions correctly. Higher firebox temperature leads to higher NO_x formation, as depicted in Figure 2b.

Combustion air temperature. Combustion air temperature has a direct bearing on flame temperature, and the higher the flame temperature, the more thermal NO_x is formed, as shown in Figure 2c. If the heater is already equipped with an air preheater, then burners utilizing flue gas recirculation offer a good degree of NO_x reduction. In new heaters, alternative methods of waste heat recovery should be investigated.

Flame length. This parameter has the most important effect on the operation of the furnace.

Traditionally, furnace operators are accustomed to short, crisp flames, which prevent flame impingement damage to the furnace tubes. The key to getting a short flame has been to increase excess air until the flames are blue and short. This practice has been curbed to some extent by the installation of oxygen analyzers.

The basic design principle of low-NO_x burners calls for staged combustion and cooler flames. This is in direct conflict with the good mixing of air and fuel required for efficient combustion. Thus, a balance needs to be struck between the two requirements so as to achieve acceptable NO_x levels and flame dimensions.

A typical low-NO_x burner has a flame that is about 50–100% longer than the flame in a standard burner (when operated at design conditions). Any variation in operating conditions tends to increase the flame length in low-NO_x burners, thereby increasing the chance of flame impingement.

The expected flame length must be kept in mind when specifying the heat release rate and the total number of

Table 2. Keep the distance from the burner to the tubes within these limits.

Maximum Heat Release per Burner, MMBtu/h	Vertical Clearance from Burner to Roof Tubes or Refractory, ft		Horizontal Clearance from Burner Centerline to Tubes, ft	
	Oil	Gas	Oil	Gas
2	12	8	2.5	2.25
4	16	13	3.0	2.75
6	24	18	3.5	3.25
8	30	24	4.0	3.75
10	36	28	4.5	4.25

Table 3. Low-NO_x burners have been installed in a variety of applications.

Heater Application	Burner Type	NO_x Emission Level, ppm
Crude, vacuum, and coker heaters (cabin)	Forced-draft, staged-fuel, preheated air	60
Vertical cylindrical refinery heaters	Natural-draft, staged-fuel, with internal FGR	25
Down-fired hydrogen reformer	Induced-draft, staged-fuel-gas	60*
Vertical cylindrical refinery heaters	Forced-draft, staged-fuel, with internal FGR, preheated air	60
Upfired ethane cracker	Natural-draft, staged-fuel	85

*With steam injection.

Acknowledgments

The author is thankful to KTI management for permission to publish this article. Thanks are also due to Rose Williams for repeatedly typing the manuscript.

Further Reading

Bell, C. T., and S. Warren, "Experience with Burner NO_x Reduction," *Hydrocarbon Processing,* **62**(9), pp. 145–147 (Sept. 1983).

Garg, A., "Trimming NO_x from Furnaces," *Chem. Eng.,* **99**(11), pp. 122–130 (Nov. 1992).

Johnson, W. M., and R. R. Martin, "Staged Fuel Burners for NO_x Control in Fired Heaters," presented at the 1984 Winter National Meeting of AIChE, Atlanta, GA (Mar. 1984).

Kunz, R. G., et al., "Control NO_x from Gas Fired Hydrogen Reformer Furnaces," presented at the National Petroleum Refiners Association Annual Meeting, New Orleans, LA (Mar. 1992).

Waibel, R., et al., "Fuel Staging Burners for NO_x Control," presented at the 1986 Symposium on Industrial Combustion Technologies, sponsored by Gas Research Institute, U.S. Dept. of Energy, American Flame Research Committee, and American Society for Metals (now ASM International), Chicago, IL (Apr. 1986).

A. GARG is manager of thermal engineering at Kinetics Technology International Corp. (KTI), Houston, TX (713/974-5581; 713/974-6691). He has more than 19 years of experience in process design, sales, and commissioning of fired heaters and combustion systems. Previously, he worked for Engineers India, Ltd., and for KTI in India. He received a B.Tech. in chemical engineering from the Indian Institute of Technology. He is a registered professional engineer in California and a member of AIChE.

burners. It is also recommended that the maximum heat release rate per burner be limited to 10 MMBtu/h. Furthermore, the burner flame length should be kept to a third of the firebox height for long vertical cylindrical heaters and no more than two-thirds of firebox height for low-roof cabin heaters. Typical clearances for low-NO_x burners are recommended in Table 2.

Burner size. Today's low-NO_x burners are much larger than standard burners for several reasons:

1. Air staging has led to the use of secondary and tertiary air controls.

2. Fuel staging has led to the segregation of gas tips and, thus, larger-diameter burners. It also requires more gas piping and separate gas controls.

3. Recirculation of flue gases requires separate gas tubes. And the increased volumes of gas and air require larger burner throats.

4. Internal flue gas recirculation calls for larger burner tiles and recirculation flue gas ports.

Thus, it is becoming very difficult to fit the new low-NO_x burners in an existing heater floor without sacrificing some degree of operational and maintenance flexibility. It is essential that the engineering contractor be given drawings of the general arrangement of the heater and the steelwork to work out the installation details.

Burner testing

Burner design is mostly empirical and predicted design and operating conditions can only be verified through performance tests. Thus, burner testing is strongly recom-

mended for all new low-NO_x burners.

But testing of these burners should be handled with care. The flue gas flow and the expected temperature profile in the furnace usually cannot be reproduced exactly in the test furnace. For this reason, emission test results should be considered estimates, and actual emission calculations should incorporate a margin to account for this.

Low-NO_x burners have been installed in a variety of applications, in both new facilities and in revamped plants. Table 3 summarizes several installations. **CEP**

Stationary Source Testing: The Fundamentals

Use this review of the basics of air emissions testing to evaluate your needs and select a testing contractor.

Helen Yoest,
Entropy, Inc.

Robert W. Fitzgerald,
3M Corp.

Air quality is an important concern for the chemical process industries (CPI). As air quality issues have become more complex, so, too, have the methods for determining air quality.

This article provides a brief overview of one type of air emissions testing — stationary source testing. It first covers the common reasons for conducting testing and how the Code of Federal Regulations (CFR) specifies the test method to be used. Then it offers guidance on how to solicit a bid from a testing company and select from among the proposals that are received.

Types of test programs

First, it is important to distinguish between the four different types of air emissions testing: ambient, fugitive, indoor, and source.

Ambient. Ambient testing usually involves fenceline testing. For example, a neighbor is complaining about an odor coming from your plant. In response, you hire a firm that specializes in ambient testing to monitor the air blowing across your property.

Typically, the air is monitored at the four fenceline compass points. At least one additional measuring station is placed either in the predominant upwind (or downwind) location or in a direct line between your plant and your neighbor. The resulting data should yield information concerning the concentration of ambient emissions leaving your property (minus the emissions from adjacent facilities).

Fenceline disputes are very common. In many areas of the country, several facilities may share property boundaries delineated by a fenceline. Since each facility is regulated according to total emissions, it is critical that your neighbor's "drifting" emissions be qualified and quantified. Depending on your neighbor's production rate, the atmospheric conditions, and seasonal climate, their emissions could impact the operation of your facility or endanger the health of your employees.

For example, many facilities in the U.S. must continuously monitor downwind fenceline emission of hydrocarbons. If a neighbor's emissions of hydrocarbons or adjacent freeway hydrocarbon emissions drift across your fenceline and combine with your own hydrocarbon emissions, your total facility hydrocarbon emission limit could be violated. This scenario seems extreme, but it has occurred more than once in recent years.

Fugitive. Fugitive testing is a hybrid of ambient and source testing. It usually involves the monitoring of either particulate or gaseous emissions from sources open to the atmosphere. It may involve testing sources such as valves, flanges, pumps, and similar hardware for leaks, and it may include quantifying emissions from open drums, open vats, and surface impoundments such as lagoons, pits, water basins, and ponds. Sources of particulate matter fugitive emissions include construction activities, stone crushing, sand blasting, agricultural activities, drying equipment, coal piles, and vehicular traffic from unpaved roads.

Typically, fugitive testing is conducted using one or more of the four most common techniques: use of a handheld organic analyzer, "bagging" suspect sources for subsequent analysis, capturing and scrubbing fugitive emissions using a floating flux chamber/summa canister, or measuring particulate matter greater than or less than 10 microns in diameter (PM10) following promulgated Environmental Protection Agency (EPA) test methods.

Selection of the test method depends on factors such as type of emissions, source type, temperature, pressure, concentration, and so on. For example, a plant operator who suspects a leaking valve may use a handheld organic analyzer to verify the presence or absence of a leak. If the analyzer is not able to quantify the concentration of the leaking gas, then the bagging technique may be employed.

To determine the amount and type of organic emissions escaping from a settling pond or wastewater treatment tank, a floating flux chamber/summa canister may be preferred. This is a box that isolates a portion of the pond to determine volumetric flow. The box acts as a floating stack in which emissions are captured into a canister for analysis.

For stone crushing or material transfer operations, it is obviously not practical to use a handheld analyzer or to "bag" the source (especially something as large as a coal pile). In such cases of particulate matter fugitive emissions, a high-volume ambient PM10 sampling system is used, or the emissions are ducted through a temporary stack for direct measurement using an isokinetic sampling train such as EPA Method 201A.

Correct method selection is both scientific and subjective. Knowing when to utilize the appropriate method for a given circumstance is very important, since incorrect or inaccurate measurement, compounded by sampling dozens or even hundreds of sources, can lead to severe regulatory and subsequently financial punishments.

Indoor air. Indoor air testing, or indoor air quality investigation, is another common type of testing that has been receiving a considerable amount of attention in recent years. Indoor air testing may involve monitoring the air in the general vicinity of a workstation, a mobile worker, or localized areas within a plant, office building, or residence.

Understanding the problems associated with poor indoor air quality is difficult because relatively little research attention has been given to this issue. A tremendous amount of effort is currently underway to establish efficient, systematic protocols to ascertain the source of poor indoor air quality. For now, the best rule of thumb for approaching indoor air quality issues centers around careful investigations and interviews before any attempts are made to qualify (or quantify) the source of the problem.

Stationary sources. Stationary source emissions testing involves the direct sampling of an air stream in a duct, stack, or pipe that is the end source of an emission release point. Stationary source testing differs from other types of air emissions testing in that it is more labor-intensive in the field and in preparing the test report. This type of testing is commonly referred to as "stack testing," and is the focus of this article.

Why conduct source testing?

The three major reasons for which stationary source testing is required are to show compliance with (or applicability of) regulations or permitting authorities, to provide engineering data, and to support performance guarantees.

Compliance testing. A source subject to a regulation must demonstrate that it complies with the standard(s) outlined in the applicable regulation(s). These regulations include standards of performance for new stationary sources (NSPS), national emission standards for hazardous air pollutants (NESHAP), the hazardous organ-

■ *Stack testing is being conducted on the roof of a building at a chemical plant to collect engineering data.*

ic NESHAP (HON), solid waste thermal treatment requirements such as those in the Resource Conservation and Recovery Act and the Toxic Substance Control Act (RCRA and TSCA), and the standards that are now being written as mandated by the Clean Air Act Amendments of 1990.

For a facility to comply with a permitting authority, it must comply with the operating permit conditions written under the authority of the air-pollution control agency. This agency will enforce the minimum federal guidelines, and may also impose more-stringent compliance requirements written into the operating permit.

For example, an affected sulfuric acid plant is required to comply with a federal standard for opacity of visible emissions according to EPA Method 9. If the facility is located in New Jersey, the plant may be required to measure the opacity on a continuous basis. States (including New Jersey) may go one step further and require that the continuous measurement of opacity be relayed via telemetry to a central office.

In addition to demonstrating compliance with an emission limit, the state (or local) agency may also require air-pollution control equipment to operate at a minimum efficiency. For instance, if a facility operates a scrubber, the state regulatory agency may write into the permit that the scrubber is to be operated at 98% efficiency. As you can imagine, two chemical plants in different states may be regulated equally at the federal level, but they may operate under different permit conditions because of the control of each state.

Engineering testing. Engineering testing can be performed to:
• characterize the emissions for air-pollution control equipment installations;
• evaluate the performance of air-pollution control equipment;
• study emissions from a new process line or product;

• optimize a process, boiler, incinerator, or other such device; or
• conduct emissions testing before compliance testing to ensure the system is operating within design or permit parameters.

Engineering testing can be done using routine manual EPA test methods, instrumental methods such as on-line continuous emissions monitoring, Fourier transform infrared (FTIR) spectrometry, or gas chromatography.

Guarantee testing. Guarantee testing is similar to compliance and engineering testing. When a product is sold, in this case air-pollution control equipment, the product is usually

The test methods that must be used are typicallly specified by applicable regulations.

guaranteed to meet some degree of performance. Guarantee testing is typically performed to ensure that the process, control equipment, or system as a whole is operating within design limits, efficiency limits, or regulatory limits. After installation of the equipment or a hardware upgrade or retrofit, the unit must be tested to determine whether all contract guarantees are being met. In other words, did you get what you paid for, and what margin of operating error do you have?

Guarantee testing can be procured by either the buyer or the vendor, depending on the arrangements agreed upon by both parties. The vendor may provide testing to demonstrate that the unit is meeting the design specifications, and will procure the testing contractor. The buyer may approve the selected contractor, or may prefer an independent verification and will hire another emissions testing contractor. In some instances, the buyer disputes the results of the vendor's selected contractor and hires an independent contractor to provide additional data

for comparison. Sometimes, this testing is combined with compliance testing to minimize the costs to the buyer and the vendor.

CFR and method selection

The type of facility will determine the applicable federal standard. The local air-pollution control agency may enhance this regulation with more-stringent requirements. In general, most EPA test methods applicable to a facility will be contained in the Code of Federal Regulations *(1)*. Other test methods may be specified by the EPA Office of Solid Waste *(2)* or the National Institute for Occupational Safety and Health (NIOSH) *(3)*. Additionally, some states (such as New Jersey, California, Georgia, Florida, and Pennsylvania) have their own test methods that differ from EPA methods, the use of which they may require in lieu of the EPA methods.

The CFR is a codification of the rules published in the *Federal Register* by the executive departments and agencies of the federal government. The Code is divided into 50 titles covering broad areas subject to federal regulation. Each title is divided into chapters, which usually bear the name of the agency issuing the regulation. Each chapter is subdivided into subchapters, and also into parts, subparts, and sections, which cover specific topics.

Title 40, Protection of the Environment, consists of 16 volumes. It is revised once each calendar year, in July, and is kept up-to-date by the individual (daily) issues of the *Federal Register*. These two publications must be used together to determine the latest version of any given rule.

To determine whether a Code volume has been amended since its revision date, consult the "List of CFR Sections Affected," which is issued monthly, as well as the "Cumulative List of Parts Affected," which appears in the Reader Aids section of the daily *Federal Register*. These two lists will identify the *Federal Register* page number of the latest amendment of any given rule. These amendments will be incorporated into the next annual update of Title 40.

The CFR specifies test methods for testing for numerous compounds and various parameters necessary for determining concentrations and emission rates. The EPA test methods contained in Title 40 for compliance testing are far too numerous to list here. Table 1 lists some of the methods most relevant to the CPI.

Leafing through the CFR can be overwhelming at first glance. However, we will briefly and gently navigate you through Title 40, Part 60, using a nitric acid plant as an example. (If you can get your hands on a copy of Title 40, Part 60, follow along.)

The standards of performance for nitric acid plants can be found in Subpart G. To begin a search for this topic, turn to the beginning of Part 60, where you will find a heading entitled "Standards of Performance for New Stationary Sources." Glance down through the bold subheadings called subparts until you reach Subpart G, "Standards of Performance for Nitric Acid Plants." Make a note of the numerical address of this subpart (60.70) just below the bold heading.

As you leaf through the pages that follow, you will see that the top of each page has a numerical address. Follow the increasing numbers until you reach 60.70. (In the July 1, 1993 issue, 60.70 is found on p. 136.) You have now found Subpart G, "Standards of Performance for Nitric Acid Plants."

This standard regulates the entire nitric acid industry. Keep in mind that this subpart only regulates the actual nitric acid unit, not the entire facility. Other emission sources at a nitric acid plant may also be regulated under different documents. Subpart G stipulates this in the very first sentence: "The provisions of this subpart are applicable to each nitric acid production unit, which is the affected facility." If you have a nitric acid production unit that was constructed or modified after Au-

gust 17, 1971, your facility qualifies as an affected facility.

In Subpart G, we learn that no facility affected by this subpart shall emit more than 3 lb of NO_2 per ton of acid produced or exhibit 10% or more opacity from the production unit. Reading further, we learn that each production unit must be equipped with a nitrogen oxides (NO_x) continuous emissions monitor (CEM) calibrated according to Performance Specification 2 (PS 2).

PS 2 is found immediately following the air emission test methods in Appendix B. Before going any further, let us flip to that section. (In the July 1, 1993 issue, PS 2 is found on p. 906.) In PS 2, you will find everything you need to know about installation of the CEM, performance specifications, test procedures, calculations, and alternative procedures.

Now, let us flip back to Subpart G. Continuing on, we learn how to establish a conversion factor for converting the CEM output in terms of production, reporting requirements, and the test methods used to demonstrate compliance.

Note that Subpart 60.74, "Test Methods and Procedures," stipulates that Method 7 shall be used to verify compliance with emission limits, so,

now let us take a quick glance at Method 7. Turn to the appendix at the end of Part 60 (p. 481). Scan down the list until you reach Method 7. Notice that this method has several alternative techniques: 7A, 7B, 7C, 7D, and 7E. The confusion here is in guessing which method applies to your situation.

As mentioned earlier, some states may require the use of other methods to demonstrate compliance. If so, this is stipulated in your permit. If not, you have two choices — EPA Method 7E, the instrumental analyzer procedure, or the old wet chemistry technique of Method 7 or 7A. As a rule of thumb, most states prefer instrumental testing. If you are trying to conduct a PS 2 test on your nitrogen oxide CEM, EPA Method 7E is the preferred method because you can determine the pass/fail status while the test team is on-site. The wet chemistry techniques require subsequent analysis in a laboratory, so you cannot make quick process adjustments to ensure compliance before the stack testing contractor leaves the plant site.

Congratulations! You have successfully navigated through the CFR. You now know what regulations apply to your nitric acid production unit, monitoring and recordkeeping requirements, and testing procedures.

Table 1. EPA test methods are available for a variety of compounds and parameters.

- Sample and Velocity Traverses for Stationary Sources
- Stack Gas Velocity and Volumetric Flow Rate
- Gas Analysis for Carbon Dioxide, Excess Air, and Dry Molecular Weight
- Moisture Content in Stack Gases
- Particulate Matter
- Sulfur Dioxide
- Nitrogen Oxide
- Visible Emissions
- Carbon Monoxide
- Hydrogen Sulfide
- Inorganic Lead
- Total Fluoride
- Gaseous Organic Compounds
- Volatile Organic Compound (VOC) Leaks
- Polychlorinated Dibenzo-*p*-dioxins and Polychlorinated Dibenzofurans
- Hydrogen Chloride

Table 2. A matrix such as this can be used to evaluate contractors' bids.

Company	Experience	Correct Scope?	Price	Hours/day and Overtime Rate		Report Delivery Schedule	Availability
ABC	Good	Yes	$38,995	8 h/d	$80/h after 8 h	35 Business Days	3 Weeks
XYZ	Excellent	Yes (offered on-site data)	$43,000	12 h/d	$80/h after 12 h	25 Business Days	4 Weeks
Stacks-R-Us	Fair	No (price did not include inlet testing)	$30,250, plus $6,900 for inlet = $37,150 total	8 h/d	$85/h after 8 h	30 Calendar Days	2 Weeks

However, new regulations are being developed for many compounds that, as yet, have no promulgated test methods. Air emission testing specialists or consultants can often determine appropriate test methods for most of these compounds. Many such companies have had previous experience testing these compounds. Usually, the testing involves adapting an existing method to the pollutant of interest.

Remember, it is much easier to convince the state and federal agencies to accept an existing method than a newly developed test method. If it "looks like, tastes like, and smells like" an existing EPA method, you have a good chance of success. But if using an existing method is impractical, you have only two remaining choices. Either prevent the pollutant from being discharged to the atmosphere, or develop a test method particular to that pollutant. Whichever route you take, make sure you do your homework and seek the advice or assistance of a professional.

Soliciting bids

Before contacting a stack testing company, it is best to prepare a bid request package. The following information should be included in the bid request:
• when and where the testing will take place;
• the reason for conducting the test;
• the regulatory requirements that apply to the source;

• information on each source that must be tested, such as process type (batch, continuous, or semicyclic);
• the type of each test location (duct, stack, tank, vent);
• the test location configuration (horizontal, vertical, or remote);
• the temperature, concentration, and moisture content at the test location;
• the number and size of available test ports;
• all safety considerations (such as whether hazardous air streams or hazardous materials are present);
• the test report delivery schedule;
• any unusual requirements (such as the maximum number of hours per day that a test team can work);
• contractual or safety information (terms of sale, plant safety);
• the date the bid is due; and
• the technical, contractual, and safety contacts, including telephone and fax numbers.

This information must be provided to ensure that all bidders completely understand the scope of work and other requirements for the testing. By doing your homework now, you can save yourself and your company substantial cost overruns associated with additional charges from the stack testing contractor.

If you need assistance in developing your bid package, contact someone in your department who contracted a stack testing company previously. If no one is available, your corporate

environmental or safety personnel may be an excellent resource. If all else fails, call an experienced stack testing company for help.

After you have completed a bid package, you can solicit bids from air emission testing contractors. Now you should ask, "Who should bid on this project?" and "What selection criteria should I use in selecting a source testing contractor?"

If your company has no internal list of bidders, you may want to contact your industry peers, industry association, or local or state regulatory agency. Your company may have an environmental staff that tracks this kind of information. Or, someone else in your department may have experience with one or several vendors. In addition, your purchasing department may have a strict format to follow that has been successful during previous procurements. Although their process may seem cumbersome and complex, experience has taught the company that one cannot be too careful once the bidding process begins. Contractual, legal, and financial considerations must be addressed when inviting vendors to provide a bid for services. By discussing your concerns within your own organization, you will gain valuable information that can be used in the decision-making process.

Once the bid request has been mailed to all potential bidders, make sure someone is available to answer any questions the bidders may have.

No matter how clear your package is, there will, and should, be questions from the vendors. After you receive all of the bids, the review process begins.

Choosing a testing contractor

There are four main decision-making techniques companies use in selecting emissions testing contractors: the corporate review, the facility team review, the engineering review, or a combination of these three. In the past, someone from a company's engineering staff hired stack testing contractors. Today, it is very common for corporate/plant teams to make the decision jointly.

Regardless of your company's review process, the matrix approach is most often used to evaluate bidders. Each bid is broken down into six categories that are listed in the order of their importance, starting with the most important criterion. Table 2 illustrates this approach. A related decision-making technique, the use of utility functions, is explained in *(4)*.

Literature Cited

1. Code of Federal Regulations, Title 40 — Protection of the Environment, Parts 60, 61, 63, and 51.

2. **U.S. Environmental Protection Agency,** "Test Methods for Evaluating Solid Waste," 3rd ed., U.S. EPA Office of Solid Waste and Emergency Response, Washington, DC, Document No. SW-846 (Nov. 1986).

3. **National Institute for Occupational Safety and Health,** "Manual of Analytical Methods," 3rd ed., NIOSH, U.S. Department of Health and Human Services, Washington, D.C. (1989).

4. **Christian, J. C.,** "Use Utility Functions to Select Equipment," *Chem. Eng. Progress,* **91**(3), pp. 92–94 (Mar. 1995).

Further Reading

U.S. Environmental Protection Agency, "Quality Assurance Handbook for Air Pollution Measurement Systems," Vol. III, U.S. EPA, Office of Research and Development, Environmental Monitoring and Support Laboratory, Research Triangle Park, NC (1982).

Remember, you can fine-tune the information in the matrix by contacting each bidder and negotiating items, such as report delivery schedule and price revisions. In this example, it appears that Company XYZ is the highest bidder and Stacks-R-Us is the lowest. Let us now consider one probable scenario.

We know, for example, that the process is interrupted every 4–6 h for an average of 1 h. Therefore, it is not practical to assume the testing can be completed in less than 12 h each day. If a test crew of four people is on-site for 4 days, and we add overtime to each bidder's price for 12-h work days, what is the resulting cost for each company? Company ABC's price would rise to $44,115, XYZ's price would remain $43,000, and Stacks-R-Us' price goes to $42,590. Now, Company XYZ's price is in the middle, and we see very little difference in price among all three bidders. Because experience is our most important consideration, and on-site data are offered as a bonus, our choice would be Company XYZ.

The importance of each category will vary from project to project. For example, if availability is critical, then this column may come first or second. A simple spreadsheet is a very versatile and easy way to present this information. As the project's priorities change, the columns can be easily rearranged. If the matrix approach yields two or more companies as acceptable, then ask each company to attend a "best and final" meeting, at which you should be able to clarify which company is best prepared to conduct the project.

Final thoughts

The complexity of air emissions testing is due, in part, to changing regulations, updates, or additions of new test methods. Adding to this complexity are the choices for testing compounds in the absence of a promulgated test method. It is difficult to fully understand all of the requirements and test methods associated with your facility, unless source testing is your profession. If it is not, you cannot be expected to know all of the facts. Even when your source is only affected by a few parameters, frequent changes in regulations and test methods, process operations, and air-pollution control technology can pose daunting challenges.

Although the cost of hiring a stack testing company may seem unreasonable, you must consider the long-term effects of taking shortcuts. Working closely with a reputable contractor from start to finish of a test program, and on a continual basis, can save your company valuable time and minimize your effort. In addition, the contractor can share the responsibility both from a financial and legal standpoint. Finally, the contractor can apply knowledge gained from similar circumstances which can make a critical difference in agency negotiations.

Remember, hiring a stack testing company is no different than purchasing chemical processing equipment — do your homework! Beyond that, good luck! **CEP**

H. YOEST is vice president of sales and marketing at Entropy, Inc., Research Triangle Park, NC (919/781-3550; Fax: 919/787-8442). In her 12 years in the environmental field, she has performed and directed numerous air emissions testing programs, with a particular emphasis on primary and secondary metals industry facilities, and sources that burn refuse-derived fuels and hazardous wastes, such as incinerators, boilers, and industrial furnaces. She has also directed test programs for maximum achievable control technology (MACT) for clients in the primary and secondary aluminum industries, pulp and paper industry, and coke ovens. And, she has developed and instructed training programs in source testing. She has presented papers at many association meetings and is widely published. She holds a BS in environmental health from Old Dominion Univ. and an MS in environmental engineering from Brunel Univ., London. She is a member of AIChE, the Air and Waste Management Association, and the Carolina Air Pollution Control Association.

R. W. FITZGERALD, formerly sales manager with Entropy, Inc., is now sales manager with 3M Corp. He has eight years of experience in the air emissions testing industry, in the areas of method development, data handling and reporting, field testing, project management, and sales. He received a BS in mechanical engineering from North Carolina State Univ.

Select a Continuous Emissions Monitoring System

Continuous emissions monitoring systems offer many advantages for demonstrating ongoing regulatory compliance. Here's how to choose a CEMS.

Karen Walker,
Horiba Instruments, Inc.

The Environmental Protection Agency (EPA) is incorporating flexibility in the new regulations it is writing to implement the monitoring requirements of Title VII of the Clean Air Act Amendments of 1990. These requirements are commonly known as compliance assurance monitoring (CAM), which is the new name for "enhanced monitoring." CAM requirements are scheduled to be promulgated in the Code of Federal Regulations as 40 CFR Part 64 in July 1996.

The new flexibility being written into the CAM regulations is likely to result in reduced costs for industry and additional headaches for the engineer or manager responsible for implementing CAM. Continuous emissions monitoring systems (CEMS) were once the only choice available for continuous compliance monitoring. The primary monitoring strategies expected to be allowed under the CAM rules include not only CEMS, but also predictive emissions monitoring systems, parametric monitoring, and operations and maintenance recordkeeping.

A CEMS provides direct measurement of the pollutants exiting the stack on a continuous basis (at least one sample every fifteen minutes). Because it provides a direct measurement offering high accuracy, a CEMS has been the preferred method for quantifying emissions for the past two decades. A CEMS can be costly to purchase, install, and maintain, although recent advances and competitive pressures have brought costs down considerably. Besides the cost, another drawback of a CEMS is that because it is a point monitoring system it does not measure fugitive emissions, which the operating permit may require the facility to monitor.

A predictive monitoring system (PEMS) is a software-based system that estimates emissions indirectly by measuring operating parameters that affect emissions, such as fuel flow, operating temperature, and raw materials content. A simple PEMS may use just a few operating parameters, and this may be sufficient to accurately estimate emissions if the correlation between the chosen parameters and emissions is high. For instance, just two parameters, fuel sulfur content and fuel flow, predict sulfur dioxide emissions with high accuracy for certain applications. However, many applications require a complex PEMS that uses mathematical models to estimate emissions based on a large number of operating parameters. These PEMS can cost more to purchase and install than a conventional CEMS. The advantages of a PEMS are potentially lower maintenance costs and the collection of process information that can help the plant run more efficiently. Also, it may be possible to include fugitive emissions in the estimation of pollutants.

Parametric monitoring, sometimes referred to as demonstrated compliance parameter limits (DCPL), is similar to a simple PEMS. The difference is that DCPL does not attempt to estimate emissions on a continuous basis. Instead, one or more parameters are chosen that correlate with emissions, and a limit is set for the parameter(s). For instance, if a plant had emissions less than the permitted levels whenever the operating temperature was over 280°F, the

■ *You don't want to have to climb a stack like this to maintain in situ equipment.*

permit would be written so that the plant was in compliance when it operated above that temperature. The drawback of DCPL is that a plant may sometimes be out of compliance with respect to its permit even though its emissions are actually less than regulations allow. In the example above, the plant will be considered out of compliance if it is operating at 275° F, but the emissions may not be exceeding the allowed limits. The most attractive feature of parametric monitoring is its low price, since a plant may have to install little or no equipment to put a parametric monitoring program in place.

An operations and maintenance (O&M) recordkeeping program is the newest strategy being suggested for CAM, and it may initially be the least expensive strategy a plant can implement, since it requires no capital expenditures for equipment. This strategy involves putting in place a program to operate and maintain the pollution control equipment and keeping detailed records on the maintenance and quality assurance procedures. O&M recordkeeping has hidden costs in terms of the labor required and possible risks due to the inability to prove compliance. This approach is also the most vulnerable to human error — for ex-

ample, a plant could be fined if a technician forgot to enter information into the maintenance log.

Any of these choices may be valid for a particular plant. Thus, a thorough evaluation should be made to determine the best solution.

Advantages of CEMS

Even with all these choices, many plants will still opt for a CEMS, even though it initially may appear to be the most expensive option. There are several factors that make CEMS attractive:

1. A CEMS can provide important information to improve operating efficiency (a benefit also provided by a PEMS). This can result in decreased operating costs that offset the cost of the CEMS (or PEMS).

2. A CEMS monitors what is coming out of the stack with exceptional accuracy. Regulations require relative accuracy of 20% or better, sometimes providing incentives for relative accuracies better than 7.5%, and most CEMS exceed these accuracy requirements. Because of this high accuracy, a CEMS may be the best choice to prove compliance and protect against fines.

3. A CEMS may allow a plant to operate closer to its emissions limits than if it uses DCPL or O&M recordkeeping.

This will provide more flexibility and may allow the plant to forego the purchase of expensive pollution control equipment. (A sophisticated PEMS that has an accuracy equivalent to that of a CEMS may also provide this benefit.)

4. A CEMS will allow maximum flexibility in operations. A PEMS may require reprogramming or recertification if operating parameters change outside allowable ranges, and parametric monitoring requires that certain parameters remain within a set range. O&M recordkeeping would seem to offer flexibility, but since it requires that operating procedures be incorporated into the operating permit, it may actually offer the least flexibility of all the options.

5. A CEMS may be the only option that will allow a plant to keep its operating procedures confidential. PEMS and parametric monitoring may require certain operating information to be incorporated into the operating permit, which is available for public review. O&M recordkeeping will require detailed operating information to be included in the permit.

6. No one strategy can be considered *the* low-cost option. In many circumstances, an examination of the true overall costs of various options will determine that a CEMS is the most cost-effective solution for compliance monitoring. In other cases, a PEMS, parametric monitoring, or O&M recordkeeping may be the lowest cost. In evaluating these options, it is important to take into account the labor requirements, risk of fines, and potential savings from improvements in operations.

7. A properly designed and configured CEMS requires minimal operator input, allowing easier reporting with less chance of errors. (A well-designed PEMS will also have this advantage over parametric monitoring or O&M recordkeeping.)

CEMS technology

Once a plant has decided to install a CEMS, it must choose the type of technology to use and a vendor to supply the equipment and services. CEMS

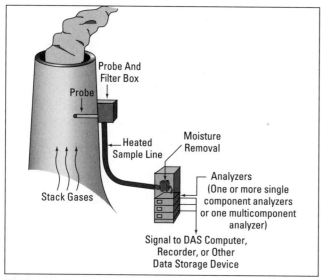

■ *Figure 1. Fully extractive CEMS.*

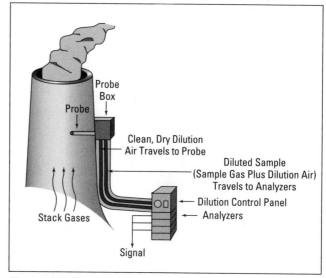

■ *Figure 2. Dilution-extractive CEMS.*

technologies can be divided into two broad categories: extractive and in situ.

Extractive systems are located anywhere from a few feet to several hundred feet from the sampling location. As the name implies, they extract a sample that is then conditioned and transported to the analyzer(s). The two main types of extractive systems are fully extractive (sometimes called simply "extractive") and dilution-extractive (also known as "dilution").

In situ systems are mounted at the sampling location and monitor the sample without removing it from the source.

Fully extractive systems

Fully extractive systems have been built for decades, and a number of them are still in use after ten or more years of operation, attesting to their durability. A typical fully extractive system (Figure 1) has a stainless steel probe, with a mesh filter to remove gross particulate, installed in the stack or duct. A heated sample umbilical containing sample tubing delivers the sample gas unchanged to the system enclosure (which may be a NEMA cabinet or environmentally controlled shelter). A second tube delivers calibration gas from the enclosure to the probe and back through the sample tubing to calibrate the system. To help keep the probe filter from plugging, air

can be forced back through the probe via the sample tube or a third tube.

Most fully extractive systems remove moisture from the sample gas with a sample conditioner located in the system enclosure. The simplest sampling conditioning method is cooling the sample (using a thermoelectric cooler or other method) and allowing the moisture to condense and drain out of the system. This must be done quickly, and exposure to liquid water must be minimized to avoid losing pollutants of interest in the condensate.

The design of the sample conditioning system is of special importance when measuring low levels of water-soluble gases such as sulfur dioxide. For example, when measuring 10 ppm of SO_2, a 10% relative accuracy requirement allows only 1 ppm error.

The sample conditioning system may include acid mist filters, particulate filters, or other components to prepare the sample gas for analysis.

The analyzer(s) may consist of one or more single component analyzers or one multicomponent analyzer. Infrared analyzers are widely used for measuring CO, CO_2, NO_x, and SO_2, and paramagnetic or zirconium oxide analyzers are generally used for oxygen measurement. For low levels of NO_x or SO_2, chemiluminescence and ultraviolet methods, respectively, are also commonly used.

An advantage of fully extractive systems, and one reason for their wide acceptance for many applications, is their ability to be configured to meet various conditions. Many CEMS vendors offer standard options to meet most requirements, such as a high-nickel alloy probe and probe filter blowback for sample gas containing high levels of particulate matter.

Fully extractive systems are sometimes used without moisture removal. One example of such an application is sludge incinerators that must measure total hydrocarbons (THC) — the regulations (in 40 CFR Part 503) stipulate that THC must be measured on a wet basis with the sample gas maintained at 150°C.

Fully extractive CEMS are recommended for:

• most applications measuring NO_x, SO_2, CO, CO_2, and/or O_2.

• plants that need to report measurements on a dry basis, such as is required under most EPA regulations.

• time-share systems (that is, where one set of analyzers is used to measure sample gas extracted from two or more locations).

Fully extractive CEMS may not be the best choice for:

• reporting pollutants on a wet basis, if required by the operating permit or regulation. (Wet-basis measurement is mandated for utilities required

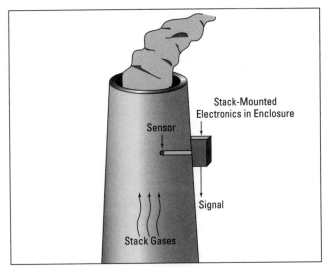

■ *Figure 3a. Point-type in situ CEMS with its sensor mounted at the end of the probe.*

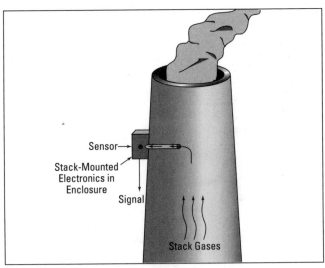

■ *Figure 3b. Point-type in situ CEMS with its sensor mounted in the box with the sensor electronics.*

to install CEMS to comply with the acid rain requirements of the Clean Air Act Amendments (promulgated in 40 CFR Part 75).)

• very wet (over 20% moisture), corrosive (containing Cl, SO_3, or other corrosive gases), or high-particulate samples (greater than 0.1 g/Nm³), especially those with a combination of these.

Fully extractive CEMS without moisture removal are recommended for:

• measuring highly reactive gases that could be lost during moisture removal, such as HCl, NH_3, and THC.

Dilution-extractive systems

Dilution-extractive systems (Figure 2) add clean, dry air to the sample gas at or adjacent to the sample probe, analyze the diluted sample, and multiply the measured concentration by the dilution ratio to determine the actual concentrations. Because the ratio of clean dilution air to sample is usually large (from 50:1 to 250:1 is common), sample handling is greatly simplified.

The addition of clean, dry air also substantially reduces the dewpoint of the sample gas so that, in theory, sample lines can be unheated. In practice, freeze-protected sample lines (which are maintained at temperatures above 10°C) are often used to ensure against any moisture condensing in the lines.

The diluted sample is similar to ambient air. In fact, ambient analyzers are used to measure the low levels of gases.

Most dilution-extractive systems are affected by the changes in temperature and barometric pressure that are normally encountered in stack gases. For this reason, it is recommended that temperature and pressure sensors be installed at the sampling location so that the data acquisition system can compensate for these effects.

Dilution-extractive systems have been used by many utilities to comply with the acid rain regulations. Utilities, especially coal-fired plants, represent the type of application that lends itself to dilution — high particulates, high concentrations of measured gases, and reporting of pollutants required on a wet basis. Another application that often utilizes dilution is the measurement of total reduced sulfur (TRS) in pulp and paper plants. In this case, diluting the sample increases the life of SO_2 scrubbers, decreasing the operating costs.

Dilution systems are recommended for:

• plants that must report pollutant concentrations on a wet basis according to the regulation or operating permit.

• applications with high levels of particulates (greater than 0.1 g/m³) or corrosive substances (*e.g.*, HCl or SO_3).

Dilution systems may not be the best choice for:

• reporting on a dry basis (additional equipment is required, which will increase the system cost).

• measuring pollutant concentrations less than 100 ppm.

• time-share applications (since response time is usually slower than for a fully extractive system due to the slower response time of ambient analyzers).

• plants required to measure O_2 (additional equipment is required).

In situ CEMS

In situ CEMS are mounted on the stack or duct and do not require sample conditioning or transport of the sample gas. There are two types of in situ systems: point and path.

The primary consideration in determining whether an in situ CEMS is acceptable for a certain installation is a review of the sampling location. In situ CEMS generally require little maintenance, but most maintenance that is needed must be done at the location where the CEMS is installed (unlike an extractive system, which will usually be on the ground out of the sun and wind). If the sampling location is easily accessible and not exposed to extreme weather conditions, an in situ CEMS may be the most cost-effective solution, especially if the monitor will be measuring only one or two gases.

The original in situ systems were developed for process applications and were not designed to be calibrated on a regular basis. Many of these early in situ systems required intricate modifications to meet the EPA's daily calibration requirements (contained in 40 CFR Part 60). These modifications increased the initial cost and maintenance costs of the systems.

Point monitoring systems. Point monitoring systems (Figures 3a and 3b) have their sample probe and analyzer installed right in the stack (they are also called in-stack monitors) and measure gas at a single point. It is important to choose a location that is representative in terms of the components of interest.

The most successful applications for point monitors have been in measuring SO_2 and O_2 in combustion sources, since point monitors are very cost-effective for measuring only one or two components. In addition, the calibration, accuracy, and other requirements are not as stringent for process monitors as they are for compliance monitors, allowing very inexpensive monitors to be used for process monitoring.

In situ point monitoring systems are recommended for:
• measuring a few pollutants, primarily SO_2 and O_2.
• locations with easy stack access.
• reporting on a wet basis.
In situ point monitoring systems may not be the best choice for:
• sample locations without easy access for maintenance.
• measuring low concentrations of gases (below 100 ppm).
• locations subject to stack or duct vibration.
• applications where particulate concentrations are high (greater than 0.1 g/m³).
• reporting on a dry basis (additional equipment is required).

Path monitoring systems. Path monitors (Figures 4a and 4b) measure gas concentration along a path, usually across the diameter of the stack or duct.

The most common method of measurement is differential absorption, in which a light source is mounted on one side of the stack and a beam is passed through to the other side. Since gas molecules absorb light energy at wavelengths that correspond to their absorption spectra, an analysis of the light waves reaching the receiver can accurately determine concentrations for certain gases. A single-pass system measures and analyzes the light waves that reach the other side of the stack, whereas a double-pass system uses a reflector and passes the light back across the stack before performing the measurement.

Because a single instrument can monitor a number of gases, the cost of an in situ path monitoring system can be very attractive. Since there are fewer components, less maintenance may be required.

In situ path monitoring systems are recommended for:
• locations with easy stack access.
• reporting on a wet basis.
In situ path monitoring systems may not be the best choice for:
• sample locations without easy access for maintenance.
• locations with high stack or duct vibration.
• high-particulate (opacity greater than 30%) applications (the ability of the light to pass through the stack gas is limited).

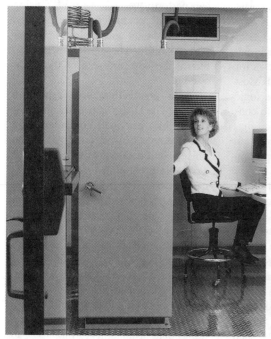

■ *The analyzer cabinet and DARS (left photo) can be installed in an environmentally controlled shelter, such as the enclosure in the photo on the right, which houses a fully extractive CEMS.*

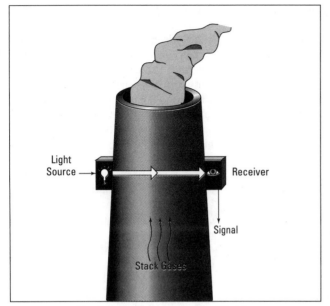

■ *Figure 4a. Single-pass path-type in situ CEMS.*

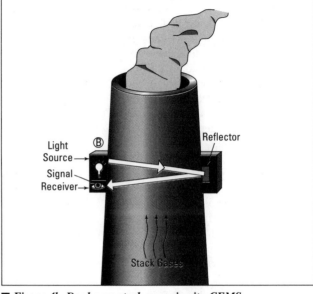

■ *Figure 4b. Dual-pass stack gases in situ CEMS.*

• measuring pollutant concentrations less than 100 ppm (interference effects from other gases may affect measurement accuracy).

• applications with high stack temperatures (the electronics required for this type of CEMS typically cannot tolerate temperatures above 150°F).

• reporting on a dry basis (additional equipment is required).

• measuring O_2 or other diatomic molecules that do not absorb infrared energy (additional instruments will be required, generally a point monitoring analyzer).

Review costs

Since there may be more than one type of measurement that is satisfactory for a particular application, the choice will often be determined by the total cost — initial cost plus maintenance costs — of the system.

Initial cost is relatively easy to quantify by obtaining bids from various vendors. Ongoing maintenance costs need to include the consumable supplies and replacement parts requirements, hours of maintenance required, and training of personnel. Quotations for service and maintenance contracts can also be obtained, giving an indication of the time and expense the ven-

dors feel their equipment requires. Be aware, however, that their numbers may not accurately reflect the costs of maintaining the system yourself, since their estimate will include travel time and expenses for service personnel. Their quotation is also an estimate based on experience with similar systems, and may differ from actual costs.

Data handling

After the type of CEM technology has been selected, the requirements for data handling must be addressed. Most systems utilize a programmable logic controller (PLC) or data logger to control system functions and collect analyzer and system data, which are then sent to a data acquisition and reporting system (DARS).

PLC or data logger. The data from the analyzers must be collected, the concentrations must be averaged or corrected, and the data must be stored in a form from which reports can be generated. A PLC or data logger is often used for this interim data-handling function. The PLC can accept analog signals from the analyzers and other equipment, along with digital signals. Digital signals from plant inputs, such as a boiler on/off signal, can be used to validate data or provide alarms.

After performing some calculations, such as one-minute averages, the PLC sends the data to the DARS. The PLC can also provide backup data storage in the event that the link between the PLC and the DARS is temporarily interrupted.

A recent advance has been the development of "smart" analyzers that can perform the tasks of a PLC using an internal computer chip, memory, and both digital and analog input/output capability. These instruments can reduce the cost and size of the CEM system.

Data acquisition and reporting system. The DARS is usually an IBM-compatible personal computer with software that can be as simple as an off-the-shelf database or as complex as a custom-programmed data management system. The primary requirement of a DARS is that it meet the relevant federal, state, and local regulatory reporting requirements.

Some regulations, such as EPA's acid rain regulations for utilities and the South Coast Air Quality Management District's RECLAIM program, require electronic data reporting via a modem. In addition, regulations require data availability from 75% to 100%. One-hundred percent data availability, which utilities must meet for the acid rain regulations, requires a DARS that can provide data substitution, whereby estimates are put into the database in place of any missing measurements. There are many DARS systems that are preprogrammed to meet regulatory requirements, and these systems may be easily configured to meet a plant's needs.

The second most important requirement for a DARS is arguably that it be easy to use. The popularity of Windows or other graphical user interfaces has resulted in DARS programs that can be used by individuals with minimal computer experience or training.

The DARS may even be able to display information from several CEMS in a central location, such as corporate headquarters. This requires sophisticated networking that may substantially increase the cost and complexity of the DARS.

CEMS vendors

Vendors supplying CEMS include manufacturers of analyzers and/or systems, and systems integrators who simply package other vendors' equipment, sometimes with their own DARS.

Analyzer/system manufacturers. These manufacturers produce their own hardware, and may also provide their own DARS software. When buying from a CEMS vendor in this category, the purchaser benefits from the vendor's excellent understanding of its own equipment. Also, the entire CEMS will have been designed around one

Table 1. Two ways to measure emissions are continuous emissions monitoring and stack sampling.	
Continuous Emissions Monitoring	**Stack Sampling**
• One sample at least every 15 minutes	• One sample one or more times per year
• Limited number of gases can be measured	• A larger range of gases can be measured
• Provides a reliable indication of pollutants being released into the atmosphere on an ongoing basis	• Can provide an estimate of pollutants being released into the atmosphere on an ongoing basis

analyzer line. Systems supplied by manufacturers are often very competitively priced.

Systems integrators. Systems integrators purchase analyzers from one or more manufacturers and package them into a CEMS. The integrator is not tied to using a single type of analyzer, and so can be more flexible when choosing manufacturers and technologies. However, to maintain this flexibility, its system is not specially designed around the analyzer. In this case, it is especially important that the integrator have the expertise and experience to put together a well-designed system for the application.

Choosing a vendor

The most important factors in choosing a CEMS vendor are the company's experience, level of expertise, and ability and willingness to stand behind the system it offers.

The best way to assure getting a vendor who meets these qualifications is by prequalifying the bidders list. An initial list can be obtained by calling other companies that have installed a CEMS, from buyer's guides, or from CEM seminars.

Once an initial list is prepared, a request for qualification information can be made to each vendor. This information should include an installation list, reference list with at least two comparable installations, and an annual report. Also, a description of additional services supplied by the vendor should be included. Such services may include startup and installation assistance, monitoring plans, certification, and

maintenance contracts. Brochures can also be requested to help the purchaser become familiar with the equipment the vendor typically supplies.

After reviewing this information and calling all references supplied, it should be an easy task to narrow the list to four or five vendors. From these qualified vendors, a firm price proposal can be requested in response to your specification or permit requirements.

CEMS vs. stack sampling

An alternative to continuous monitoring is periodic sampling by a stack testing company. In many cases, periodic sampling will meet the regulatory requirements. For monitoring some pollutants, especially the hazardous air pollutants (HAPs) for which regulations are now being written, periodic stack sampling is the only method currently in use.

Table 1 compares the two approaches. The following article discusses stack testing in more detail. **CEP**

Further Reading

Jahnke, J. A., "Continuous Emissions Monitoring," Van Nostrand Reinhold, New York (1993) (also available through Source Technology Associates, 919/929-4447).

Novello, D. P., "Clean Air Operating Permits: A Practical Guide," Air & Waste Management Association, Pittsburgh, PA (1995).

White, J. R., "Technologies for Enhanced Monitoring," *Pollution Engineering*, 27(6), pp. 46–50 (June 1995).

K. WALKER is a market planning specialist for Horiba Instruments, Inc., Irvine, CA (800/446-7422; Fax: 714/250-0924), a manufacturer of analytical instruments. She has specialized in continuous emissions monitoring systems since 1992. In this field, she has managed CEM projects for major utilities, provided regulatory support and technical sales assistance, and managed the introduction of a new line of CEM analyzers and systems. She is a member of the Air and Waste Management Association and the Institute of Clean Air Companies' Enhanced Monitoring Rule Industry Task Force. She holds a BA in economics from the Univ. of California at Irvine.

Map an Effective Odor Control Strategy

Use this quantitative method to identify and prioritize odor sources.

**Sharon L. Nolen,
Stacey E. Swisher,
Joseph N. Bays, and
Elaine G. Zoeller,**
Eastman Chemical Co.

I n recent years, increasing attention has been directed toward an obvious indicator of contaminants in the air — odor. Odors that were once tolerated as a sign of industrial prosperity are now perceived as evidence of environmental problems. Chemical process industries (CPI) facilities often experience odor problems associated with their various production and waste treatment processes.

Pinpointing the source of odors and quantifying them remains a difficult technical problem that must be overcome before odor reduction can be addressed. This is particularly difficult for large facilities that may have hundreds or even thousands of emissions sources — these sources are likely grouped in different geographical areas, and problems may cut across organizational lines, making it difficult to prioritize the sources without a quantitative method.

Another complication is that some chemicals have very low odor thresholds, which means that they can be detected by the human nose at very low concentrations — even lower than parts per billion. Any control technologies used must have extremely high removal efficiencies to reduce noticeable odors. In addition, some different chemicals may have unsuspected mixing or masking effects, so that the combination or removal of one or more of them is unpredictable and may result in a more (rather than less) offensive odor.

This article outlines a scientific, quantitative approach to odor management. The program has been used successfully at one large CPI facility, and can serve as a model for other plants.

Assemble the team

A multidisciplinary team is required for tackling odor investigations. Technical training, experience, and personal style are important in the team makeup.

Most team members will probably be engineers, their specific disciplines depending on the type of processes and equipment that will be included in the study. However, other types of expertise are also needed. Dispersion modeling is a key part of the study, and knowledge of the factors that affect dispersion of odors and the tools available to predict this phenomenon will be very helpful. Familiarity with environmental regulations is also an asset, since existing and anticipated regulations need to be considered in developing proposed control projects. While odor itself is generally not regulated, many of the chemicals that cause odor are (or will be) subject to regulations.

The experience of the team members is also important. Previous or current work in the areas where odor is suspected to be a problem is advantageous. In addition, any type of operations experience will be helpful in balancing the demands of production with the desire to reduce odors from the processes.

Finally, the personal style of the team members is a factor in the team's ultimate success. Participants should be chosen who are willing to actively seek information in the plant and solicit input from all levels of personnel within the manufacturing areas. In many cases, process operators may already be aware of the sources of odors and have useful ideas for changes. The ability to work across organizational boundaries is also important — while it is likely that the project will be supported by upper management, the team must also gain support from line management.

Our project team had the following makeup:
- Project manager;
- Three chemical engineers;
- Mechanical engineer;
- Environmental engineer;
- Sampling coordinator; and
- Operations representative.

The operations representative slot was rotated among the particular areas being studied. One of the chemical engineer positions was also rotated based on expertise in the processes under investigation. However, a core group was maintained to provide continuity throughout the study. This project was the major commitment for the core group members for the duration of the study.

In addition, an odor consultant provided independent review to the team throughout the study.

Define the problem

The first step in defining the problem is to develop a logical division of the plant. This may be done by organizational or physical divisions. Since support of the local management is essential, it may often be most efficient to approach the problem along organizational boundaries. However, you may want to consider a physical subdivision if the facility is not all on a common site.

We chose to segment the plant by organizational divisions. As a result, some areas that were not geographically connected were studied simulta-

Table 1. Average odor threshold concentrations for some common chemicals.

	Average OTC, ppm
Butyric Acid	0.0001
Hydrogen Sulfide	0.1
Acetone	10
Methanol	100
n-Butane	1,000

neously. However, since recommended projects would compete for funds within a division, the organizational approach was thought to be the best choice.

Once the plant has been divided into smaller areas for study, the initial work can begin.

An informal poll can be taken to help compile a preliminary list of the offensive sources. The team should develop a list of personnel to survey and generate a list of standard questions to be asked.

Operations personnel from operators to line management may be included, as well as senior company management. Expect to get very specific discussions from those closely involved with the processes. However, it is also important to talk to company executives to ensure that their concerns are addressed and to get additional information and direction.

Questions should include whether the odor is constant or intermittent, a description of the odor, and any comments on when the odor occurs. While the team can start with a rough list of sources, it should also ask about any other sources known to the person being interviewed.

The team may also want to survey the community, either through a specific poll to learn about where the odors are impacting the neighbors or by including a question about odor control performance in an existing community poll. In either case, improving the community's perception of the plant is an integral part of the study, and communicating with the community about these efforts may be a first step toward this goal.

In addition to the informal poll, the team should further define the study by gathering any available information on the emissions from the odor sources. This may include reports required to comply with various regulations, air permits, heat and material balances, and other sources.

Emissions data should be evaluated with respect to odor threshold concentration (OTC) data published in the literature (1). OTC can be defined several ways, the most typical being the 50% recognition threshold. This is the concentration at which 50% of an odor panel defines the odor as representative of the odorant. Table 1 summarizes the average OTC values for some common chemicals and illustrates the wide variability that exists.

Once the OTCs have been identified for the compounds emitted from the facility, the odor potential of each source is defined by a simple equation:

$$\sum \frac{Emissions}{OTC} = Odor\ Potential \quad (1)$$

The sources can be ranked based on their odor potential to develop an initial prioritized list. This list can then be used to identify the areas within the plant that warrant further study and any areas that may be eliminated at this stage. This step (indeed, the entire study approach) is geared toward cost minimization, that is, reducing the scope of the project at each stage. (In our study, emission data sheets were completed on approximately 200 sources, about 100 were sampled, and 50 were selected for modeling. Final plantwide project recommendations were made for the top 10 sources, and many other smaller recommendations were made to individual divisions.)

Overall approach

In determining the overall technical approach, the team should set goals for the project. These may deal with budget, schedule, and various other items of importance for the particular plant site.

Our goals included:
• complete the project within allocated budget;
• complete the project in one year;
• provide quantitative measures of odor sources;
• provide a prioritized list of odor sources; and
• develop recommendations for reducing odors from each priority source.

In setting these goals, the team recognized that the study would have some limitations. We specifically decided not to design a statistically based sampling program because of the potentially large costs and time involved. There is also variability within the processes studied, which impacts emissions and odors. The techniques used in the study are not exact. For example, the analytical technique chosen, an odor panel, is known to have an error of ±25%. Likewise, the ambient modeling results, while the best available, rely on an uncalibrated model. And, the meteorological data were collected during only one year, and while collected on-site, they were taken at a higher elevation than many of the emission points.

The major steps in an odor management strategy (each of which will be discussed in more detail later) are:

1. Select and sample sources.

2. Analyze samples.

3. Model to determine community impact.

4. Prioritize sources.

5. Develop recommendations for odor reduction.

The study proceeds sequentially, and the team executes each phase of the program for one division at a time. Overlap between divisions should be minimized — studying three or fewer divisions at one time, as shown in Figure 1, works well. This allows the team to maintain a steady project work load while keeping the technical information manageable.

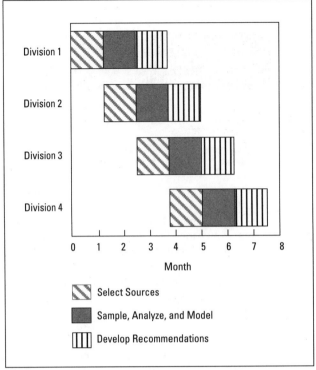

■ *Figure 1. Milestone schedule for odor control project.*

Focusing one of the chemical engineers and the operations representative on only one division assists in keeping the team organized. And, maintaining a core team throughout the project helps ensure continuity.

Identify sources and collect data

It is important for the team to agree on standard information to be collected for each source. A sample emission data sheet is shown in Table 2. This form includes information to help the team select the sources that will be sampled and also serves to collect information that will be used later for modeling the impact of the source. For consistent documentation, air permit numbers can be used (where possible) for the odor ID number.

Odors are first characterized using descriptive adjectives. Then hedonic tone, which represents a judgment of the relative pleasantness or unpleasantness of the odor, is ranked on a scale of –10 to +10, with –10 being very unpleasant and +10 being very pleasing. Perception of hedonic tone outside the laboratory is influenced by such factors as subjective experience,

frequency of occurrence, odor character, odor intensity, and duration (2).

Space is available for sketching the location of the source and the connected process.

Finally, questions to assist in assessing controls and any current work on the process or equipment complete the survey.

Once the standard format for collecting data has been established, the team is ready to identify the initial sources of concern within the selected plant area. This assessment may take several forms.

Initially, team members may want to simply walk through the area of study to become familiar with the equipment layout and any noticeable odors. Each engineer can be assigned to a particular area, with responsibility for collecting all necessary information and coordinating sampling and development of recommendations for sources within that area.

As sources come under investigation, operations personnel should be consulted to discuss known causes of the odors or to determine if the occurrence coincides with particular steps or fluctuations in the process. Area personnel should also be consulted for ideas on reducing the odors.

Before selecting the sources for further evaluation, the engineers should concurrently review the chemistry and any other available information on the process or on the delivery of a utility service.

When the sources of concern have been selected, an emission data sheet should be completed for each source.

Select sources for sampling

Once emission data sheets have been completed for each potential odor source, the team must determine which of the sources will actually be sampled. Since budget will certainly be a major factor, it is prudent to continue studying only those sources with the largest expected impacts. Key criteria to consider are location, hedonic tone, and flow rate. Knowledge of dispersion modeling will assist in determining critical locations, taking into account predominant wind directions and surrounding buildings. In terms of community impact, flow rate is usually a very important factor.

One technique to reduce costs is to sample one of a group of process-similar sources that are located close together. The remaining sources in the group can be incorporated in the modeling to estimate a combined effect. For example, at an installation where several scrubbers located close together are treating the air from a large room, a sample may be taken from one scrubber and the result multiplied by the number of scrubbers.

The analytical technique

A quantitative measure for odor sources can be developed using either a sampling and analytical approach or a calculation-based approach. We chose the former for several reasons. First, the estimated cost of a calculation-based study was higher. We also had concerns about the complexity of the calculations, the time required to perform them, and whether consistent and needed data would be available throughout the plant. And, because any calculation or instrumental analysis approach assumes pure chemicals and cannot deal with mixing and potential masking of odors, it is impossible to predict whether combinations of chemicals would produce an odor that is less or more noticeable.

Once the choice of a sampling and analytical approach was made, a specific analytical technique was needed. While many analytical techniques are available, we selected an odor panel consisting of trained personnel.

There are three common options for staffing the odor panel: employees, community members, or a commercial panel.

Using employees has several dis-

About the project

The odor control strategy described here was developed by the Tennessee Eastman Div. of Eastman Chemical Co. The plant, located in Kingsport, TN, was established in 1920 and now covers approximately 1,000 acres. Chemicals, fibers, and plastics are manufactured at the facility.

Because of concern about odors in the community, company management decided to make a concerted effort to reduce odors at the site and initiated a year-long study in the fall of 1991. A team was formed to identify and prioritize odors and develop plans for their reduction, focusing first on those odors that were detected in the community. Management set the direction for the team with a publicly stated vision of an "odor-free" plant.

advantages. Serving on the odor panel requires about two hours per day for analysis of the samples, and it is often difficult to find personnel who could be available on a consistent basis for that period of time without a detrimental effect on other work. In addition, employees may have developed some tolerance for odors associated with the plant's processes and may not provide the objective analysis required. The location of the analysis can also be a problem. A room with a high air exchange rate is required to avoid interference from other samples, and the intake air must be odor-free. There may be no location in the plant that can meet these criteria all of the time.

One of the potential benefits of using a community odor panel is that additional information regarding the company's efforts will be in the public domain. However, nearby residents' tolerance of some of the odors and the location of an appropriate facility can be concerns.

A commercial odor panel offers several advantages. First, screening, selection, and training of panel members has already been done by a third-party consultant. Facilities are available at a remote location, so tolerance for the odors to be tested is not a problem. Commercial odor panels are

administered by a third party and have no preconceived notions or knowledge of the sources sampled. Finally, this option can often have the lowest cost. Expect to pay about $150 per sample for analysis by a trained panel.

When using a commercial panel, the shipping procedure is critical to the testing. Analysis should take place within 24 hours of sample collection. Thus, once the samples are collected, there will probably be only a few hours available to get the samples prepared and submitted for next-day airfreight delivery (unless the panel is local). It is helpful for the shipping coordinator to fax each sample shipment's air bill number to the odor panel manager; if the sample's arrival the next day is delayed, this number can be a key factor in locating the package. To ensure that the process is error-free, it may also be worthwhile to have a trial run and collect samples, prepare them for shipping, and ship a package prior to the actual initiation of the sampling effort.

Sampling equipment and procedures

Sampling is done in the morning so that the samples are ready for afternoon shipping. Sampling personnel should work in teams of two for safety. They should certainly wear respirators when the odors are strong, and, to be conservative, should consider wearing them at all times. (The need for respirators will also depend on the chemicals being emitted from the facility.)

The basic sampling equipment consists of an air pump and charger, polyvinylfluoride sample bags with stainless steel fittings and polytetrafluoroethylene (PTFE) tubing, and freeze-safe shipping containers. The sample bag is attached to the air pump through a needle and septum. The fabric filter and septum should be changed when dirty or wet (usually before each sampling period).

Each day begins with the air pump fully charged. With the pump operating at about 0.3 cfm, a sample is collected until the bag is about three-quarters full. If the bags are to be shipped via air freight, it is important that they not be completely filled prior to their exposure to the lower pressures experienced in-flight.

Each bag is labeled with an identification number. Particularly strong samples are noted so the odor panel can test these last to avoid olfactory fatigue.

While this type of study is not statistically based, it is possible to test the precision of the odor panel and the variability of the sources by including duplicate and replicate samples for testing. For example, each day, include one duplicate (a second sample collected at the same odor source, immediately after the first) and one replicate (a second sample taken from an odor source that had been sampled on a previous day).

In our study, the odor panel was generally within the ±25% error expected for duplicates. (There was one exception, which was attributed to sampling difficulties associated with low-flow-rate sources.) Results from replicate sources never varied by more than an order of magnitude and were typically within ±25% as expected (except where known process variability was significant). For sources with known process variability, the replicate samples were used to assess the range of variability.

The analysis

The odor panel evaluates the samples using an olfactometer. The standard olfactometer is the IITRI Dynamic Dilution Forced-Choice Triangle Olfactometer, developed by the Illinois Institute of Technology Research Institute. It conforms to ANSI/ASTM E679-79, Standard Practice for Determination of Odor and Taste Thresholds by a Forced-Choice Ascending Concentration Series Method of Limits (3). Odor panelists are recruited, screened, and trained according to "Guidelines for the Selection and Training of Sensory Panel Members," (4).

The olfactometer is used to dilute

the sample with nonodorous air. The diluted sample is fed to one of three ports while the odor-free dilution air is fed to the other two ports. The dilution rate is decreased until the panelists can distinguish the sample from the nonodorous air.

Results. The results received from the odor panel are expressed in terms of dilution to threshold (D/T). The numbers are an indication only of odor strength. They do not take into account whether the odor is offensive or how often it occurs. The odor strengths in our study varied over four orders of magnitude.

An issue that should be addressed is whether any of the chemicals to be examined in the odor study will diffuse through the sample bag. A chemist can identify which chemicals in the plant have a high potential for diffusion. Using process knowledge, the team can identify sources that contain those chemicals and arrange

for special handling of those samples. The odor panel can analyze samples from the suspect sources daily for several days, and a decay curve of D/T vs. time can be plotted. If D/T varies with time, a value for the original sample can be extrapolated from the graph.

Modeling

The next step is to model the estimated community impact from each source. A quick screening tool can be used to help select sources for modeling. Since flow rate is a critical factor, the flow rate measured at each source is multipled by the D/T determined by the odor panel. The sources are then ranked by this product. A lower limit is established, and those sources above the limit are modeled, along with any other sources that are added due to particular concerns.

Odor impacts in the community are evaluated using an air dispersion

Table 2. Data sheet for collecting information on odorous emissions.

Date _____
Odor Source ID Number _____   Fugitive or Vent _____
Description _____   Diameter _____ in.
Discharge Direction _____   Elevation _____ ft
Descriptors (Check all that apply)

Quality

❑ Sweet	❑ Fruity
❑ Sour	❑ Spicy
❑ Fishy	❑ Medicinal/Chemical
❑ Rotten Eggs	❑ Ammonia
❑ Pungent	❑ Floral

Strength

❑ Very Faint
❑ Faint
❑ Distinct
❑ Strong
❑ Very Strong

Hedonic Tone (–10 to +10) _____
Notes _____

Plan View of Building Showing Source	Sketch of Connected Process

Other Nearby Related Sources for Combination _____

Existing Nearby Control Device/Destination _____

Solve by Simple Modifications or Operational Changes? Describe_____

Is Source a Workplace Problem? Where?_____
Projects in Progess _____

Variability of Process _____
Next Scheduled Run_____

model of the type commonly used to predict the impact of pollutant emissions. One such model is supplied by the U.S. Environmental Protection Agency (EPA) to the regulated community to assure uniformity in air impact analysis throughout the country *(5)*. Additional information on air modeling can be found in *(6)*.

In this case, the input to the model is the *D/T* value and flow rate for each source rather than the typical mass emission rate. Likewise, the ambient levels predicted are in terms of *D/T* rather than ground-level concentrations.

Three model outputs are useful for analyzing odor sources. These results relate to magnitude, frequency of occurrence, and area impacted for each source.

First, a maximum value for *D/T* outside the plant perimeter is calculated based on the worst-case one-hour average.

Second, frequency of occurrence is determined as the maximum number of hours per year that any receptor location outside the fence experiences a *D/T* value greater than a predetermined cutoff value. The frequency for each intermittent source is multiplied by the fraction of the year that the source typically emits, to correct for the fact that the source is not operating during some of the hours modeled. This is especially important for comparing batch and continuous processes on an equal basis.

Before modeling begins, an appropriate *D/T* cutoff must be selected. We had initially planned to use *D/T* = 1.0 as the cutoff. However, modeling results indicated that only a few sources would exceed this value. Since *D/T* = 1.0 is the level at which an odor can just be detected, this means that only a limited number of sources from the plant can be detected individually in the community.

However, because multiple sources emit simultaneously, the 1.0 cutoff was considered unrealistic. Rather, a lower value of *D/T* = 0.01 was selected. This resulted in more sources falling above the cutoff, and the area

impacted by these sources better corresponded to the area that we were concerned about. It also enabled comparative data to be developed.

The cutoff value will be unique to each study, and its selection will be somewhat trial-and-error, based on the Pareto Principle (The 80/20 Rule). In choosing a *D/T* cutoff, it is helpful to consider the area impacted by the sources that fall above alternative cutoff points as well as specific odor-causing sources. A *D/T* cutoff can then be selected so that the particularly objectionable sources fall above the cutoff and the impacted area corresponds with the area of concern (for example, within the fenceline or at a certain distance from the plant). Note that as the area of concern is increased, the *D/T* cutoff value should be decreased.

We chose *D/T* = 0.01 during the first division study and maintained it throughout the project for consistency to provide a means for comparison. A problem did arise later in the study when the area impacted by one source was too large to quantify at the 0.01 level. For comparison at the end of the study, the impacts of the largest sources were recalculated at the 0.1 level.

It is often valuable to conduct example model runs prior to completing the first batch of samples. This will assist the team in formatting the results as they want to see them and help in selecting an appropriate cutoff.

The three results for all sources can be normalized to calculate one number that represents the impact of each source on the community. Each result is divided by the maximum value in that category, so that the largest source has a value of one. The three categories could be assigned different weights to reflect varying degrees of importance, if appropriate. The three normalized figures for each source are summed to give one number, which has a possible value between zero and three. This provides a comparative score for the community impact of each source. These final odor scores can be used to prioritize the sources in

each division and the plant as a whole.

While the quantitative technique described in this article is very helpful in setting priorities, other factors may also be used. For example, the mathematical method can be used to select the top sources in the plant, and then other criteria can be applied to fine-tune the rankings of projects that will reduce the odors from those sources. Such criteria can include:

• The Clean Air Act. Some odor sources are processes that will require further control under the Clean Air Act Amendments of 1990. The year that such controls are expected to be required can help in prioritizing sources.

• The Superfund Amendments and Reauthorization Act (SARA). Many major chemical companies have set voluntary goals of reducing SARA emissions. Greater weight can be given to those projects that would contribute toward this goal.

• Hedonic tone. As described, results of the odor panel's analysis and the modeling only indicated odor strength. Team members will be knowledgeable of hedonic tone, a more subjective characteristic, and can include it in the final rankings.

• Economics. Some proposed odor reduction projects may have a positive return on investment due to capture and recovery of chemicals that are currently lost to the atmosphere. This should be taken into account in the final recommendations.

General results

Overall, the odor scores in our study were less than expected. Initially, we believed that more sources would have a *D/T* value greater than one outside the plant perimeter. This would indicate that a single source could be detected by the average resident of the community. Of course, this situation is not very realistic at a large facility with hundreds of sources emitting together rather than individually. It should also be noted that these values depend on the particular olfactometer used by the commercial odor panel and even the flow rate used during the panel's analysis. Therefore, the scores determined should be viewed as comparative rather than absolute values.

Flow rate has a significant impact on modeling results and, thus, community impact. Tanks containing pure chemicals with very low odor thresholds (such as butyric acid) ranked low due to low flow rates, whereas sources with fairly low D/Ts but high flow rates ranked quite high. This led us to conclude that point sources are a more significant odor problem than fugitive sources. It also led to a recommendation that high flow sources with the potential to cause odor problems need to be operated under tighter controls to avoid any deviation in operation that may result in a significant odor release.

Finally, an interesting observation relates to the worst-case hourly odor impacts. Since each odor source is modeled for the entire year and then adjusted for frequency, the results in terms of the worst-case hour are determined solely by weather conditions. The worst predicted odors occurred during evening and early morning hours. This may be of concern to the surrounding community since more people are home in the evening, and this may contribute to a negative perception.

Control technologies

Odor control technologies can be categorized as follows:

Odor minimization involves modifications to existing conditions that result in odor formation. Such modifications can include process or piping changes, addition of physical barriers, or temperature adjustments.

Chemical or physical treatment can be used to separate the odorous

Literature Cited

1. **Verschveren, K.,** "Handbook of Environmental Data on Organic Chemicals." 2nd ed., Van Nostrand Reinhold Co., New York (1983).
2. **U.S. Environmental Protection Agency,** "Reference Guide to Odor Thresholds for Hazardous Air Pollutants Listed in the Clean Air Act Amendments of 1990," Document No. EPA/600/R-92/047, EPA Office of Research and Development, Washington, DC (March, 1992).
3. **American Society for Testing and Materials,** "1992 Annual Book of ASTM Standards," Publication No. ANSI/ASTM E679-79, ASTM, Philadelphia, PA (1992).
4. **American Society for Testing and Materials,** "Guidelines for the Selection and Training of Sensory Panel Members," Special Technical Publication 758, ASTM, Philadelphia, PA (1981).
5. **U.S. Environmental Protection Agency,** "Industrial Source Complex (ISC) Dispersion Model User's Guide," Vols. I and II, Documents EPA-450/4-88-002A and EPA-450/4-88-002B, EPA Office of Air Quality, Research Triangle Park, NC (1987).
6. **Seigneur, C.,** "Understand the Basics of Air-Quality Modeling," *Chem. Eng. Progress.* **88**(3), pp. 68–74 (March 1992).
7. **Martin, A. M.,** *et al.,* "Control Odors from CPI Facilities," *Chem. Eng. Progress.* **88**(12), pp.53–61 (Dec. 1992).

S. L. NOLEN is a group leader with Eastman Chemical Co., Kingsport, TN (615/229-8265; Fax: 615/229-6099). She has specialized in environmental strategic planning and air pollution during her five years at Eastman and concentrated on air-pollution control during her previous four years at the U.S. Environmental Protection Agency's Air and Energy Engineering Research Laboratory. She received her BS in chemical engineering from Tennessee Technological Univ., is a registered professional engineer (chemical) in Tennessee, and is a member of Tau Beta Pi and the Tennessee Society of Professional Engineers.

S. E. SWISHER is a mechanical engineer with Eastman Chemical Co. (615/229-1108). For the last three years, she has served in the Environmental, Facilities, and Utilities Technology Group of Eastman's Engineering and Construction Division. She received her BS in mechanical engineering from Tennessee Technological Univ. and her MS from Virginia Polytechnic Institute and State Univ. She is a member of Tau Beta Pi and the American Society of Mechanical Engineers.

J. N. BAYS is an advanced chemical engineer with Eastman Chemical Co. (615/229-5854). He has worked in the Process Engineering Dept. at Eastman for six years, specializing in air emissions recovery and vacuum technology. He received his BS in chemical engineering from Virginia Polytechnic Institute and State Univ., is a registered professional engineer (chemical) in Tennessee, and is a member of AIChE and Omega Chi Epsilon.

E. G. ZOELLER is an environmental representative with Eastman Chemical Co. (615/229-3983). Her work is directed toward air quality evaluations and air dispersion modeling. She received her BS from the Univ. of Tennessee and her MS and PhD in environmental engineering from Vanderbilt Univ. She is a member of Sigma Xi and the Air and Waste Management Association.

compound from the air stream being emitted or to destroy the stream's organic components with oxidation. Examples of such treatments are thermal or catalytic oxidation, adsorption, absorption, condensation, and advanced oxidation.

Biological treatment with biofilters or biological scrubbers can be used to biologically degrade the odor-causing compound.

Various other technologies involve combinations of technologies or innovative, proprietary techniques.

Details on odor control technologies are available in (7).

Ongoing progress

Our facility's management's stated goal of an "odor-free plant" has given team members and line management a vision that has resulted in increased awareness throughout the plant. Most of the progress that has been made to date has involved process changes that could be made quickly for low cost. Meanwhile, pilot studies and detailed engineering are underway to more fully develop other options that required additional study or investigation.

Information on odors has also been added to seminars for new process engineering employees. This enhances their understanding of these concerns and encourages them to incorporate these considerations in the design of new plants and processes.

While the stated goal is recognized as long-term, incremental steps have already been made and will continue. The ultimate success of this effort will continue to be evaluated as recommendations are implemented and community perception is assessed.

Finally, while an odor study focuses as much as possible on a quantitative, engineering approach, perception in the community must also be a major consideration. Frequent communications with the community through the plant's community advisory panel, the media, and community groups will enhance the team's technical efforts. **CEP**

Solve Plant Odor Problems

This four-step approach to identifying, quantifying, and controlling odors can help plants direct their efforts toward the most cost-effective solutions.

Jeffrey H. Siegell,
Exxon Research and Engineering Co.

Most complaints that facilities receive from their surrounding communities concern odors. In many cases, the public's perception is that an unpleasant odor indicates the presence of harmful chemical substances. Local regulatory authorities are more likely to be notified and asked to act on odor problems than on most other environmental issues. Thus, companies need to rapidly address community concerns about objectionable odors to minimize the potential impact on plant operations and maintain good community relations.

Odor problems are subjective. Different people detect odors at different concentrations and may disagree on whether an odor is pleasant or objectionable. In some cases, even a pleasant odor may produce complaints if it is persistent or of high intensity. If someone in the community perceives that an odor is a nuisance, it is usually advisable to accept this, look for the cause, and reduce the problem. Since the resolution of an odor problem is often political as well as technical, the plant's public affairs specialist should be included in technical meetings aimed at solving the odor problem. This can often improve interactions with the surrounding community and reduce the likelihood of any additional regulatory agency actions.

This article presents a step-by-step approach for addressing odor problems. These steps include establishing the need for further action, identifying the source of the odors, quantifying the ambient concentration impact of the emission, and reducing or eliminating the problem. This structured approach focuses emission reduction and control efforts on those sources that are the more likely causes of odors, which helps minimize the costs of solving the odor problem.

For each step, issues are first listed in the form of questions that should be considered in determining what actions may be required. (Note that some of the issues listed may not apply in certain situations.) Additional information and guidance for conducting each of the steps is provided, as are details and references for additional material on the techniques and methods described. Alternative approaches may be satisfactory, provided the substantive issues that are discussed are adequately addressed.

Each of the steps requires the collection of site-specific information. Close coordination among the public affairs, environmental, operations, and engineering functions is recommended in order to obtain a fully workable solution that can be implemented at the lowest cost and that will satisfy the local community and regulatory agencies.

Step 1: Establish the need for further action

• Is the action required because of a community complaint? How frequent are the complaints, and what portion of the community is complaining? Are the complaints mostly from a single geographic location or are they widespread?

• *Has worker concern about safety or health initiated the action? Has an industrial hygienist been included in the problem solving group? Are the potential odor-causing chemicals a safety or health problem?*

• *Are there local nuisance or odor regulations? Have there been any complaints from the community to the local regulatory agencies? Have the local regulatory authorities contacted the facility, and have they established time limits or other requirements for solving the odor problem? Has the facility previously had discussions with the local regulatory agencies about other odor problems?*

• *Is the control of odors driven by an internal initiative for safety, health, or general housekeeping? If so, identify the initiative.*

• *Are there any health concerns associated with the chemicals involved, for either the community or plant personnel? Are these based on immediate or long-term effects of exposure to the chemical?*

In this step, the urgency and potential resources that will be required to solve the problem are determined. The approach to solving an odor problem will depend on the source of the complaint, as well as the requirements of local regulations that govern odors or other nuisances.

Odor problems that result in frequent complaints from a large part of the community or that may present health problems for the community or plant personnel will require a much more rapid and comprehensive response than will other odor problems. Likewise, where local regulations restrict the generation of odors, a quick response is essential.

Most odor problems that require quick action are the result of complaints from the surrounding community. Many plants have well-established liaisons within a supportive community, and these may serve as reliable indicators of the severity of

the problem. In all cases, it is important to maintain open and frequent communication with the surrounding community as the odor problem is being resolved.

In some cases, it may be possible to eliminate the need for additional follow-up by showing that the odor is benign, that it is unlikely to continue, or that the source is not at your facility. These possibilities will be very site- and situation-specific. Care should be taken, however, not to give the appearance of minimizing the importance

> *The approach to solving an odor problem will depend on the source of the complaint, as well as the requirements of local regulations.*

that the facility places on solving the odor problem.

The sudden presence of an odor may indicate an operations problem that needs attention. When odor problems are identified by plant personnel, they can often be resolved before the surrounding community is affected and before complaints are received. Sometimes the presence of an odor indicates that the concentration of a particular chemical is above acceptable levels, so it may be appropriate to consult the plant industrial hygienist for all odor problems.

Specific local regulations may exist for controlling odors or eliminating nuisances. In some locations, regulations require that any offensive conditions be eliminated. Surrounding communities may be active in pressing for more-stringent regulations should severe odor problems persist. A recent practice is for those in the affected community to initiate legal action to seek "damages" for the perceived reduction in the quality of their environment. Good community relations and a rapid plant response are the best approaches to minimize

the likelihood of a more costly regulatory-agency-enforced response to an odor problem.

Step 2: Identify the cause of the odor

• *Can the type of odor be identified? Is it generally pleasant or objectionable? Is it being detected inside as well as outside the facility? In what locations is it being detected most frequently?*

• *When and how frequently does the odor occur? Does the occurrence correlate with any specific plant operation (e.g., startups, tank or vessel loading, etc.) or with any specific meteorological condition (e.g., wind speed or direction)?*

• *What are the likely chemicals that are causing the odor? Are those chemicals normally handled in the facility? Where in the facility? Are they also handled in adjacent or nearby facilities that could be causing or contributing to the odor problem?*

• *What are the potential plant sources for the chemicals causing the odor problem? Are they specific units in a small area or are they widespread? Do the location, timing of the complaints, wind direction, or other factors suggest a specific plant area that may be the source of the odorous chemicals?*

• *If sources outside the facility are contributing to the odor problem, can this be well-documented based on their typical emissions of the suspect chemicals? Are these outside facilities likely to be cooperative in solving the odor problem? Have they been contacted previously?*

In this step, the emission sources that are causing the odor problem are identified. Sometimes the sources of the odorous chemicals are obvious, and sometimes they are nearly impossible to determine.

The first step in locating the sources is to make an attempt at determining the chemicals present from their characteristic smell. Many classes of chemicals have distinctive odors that are easily recognized (Table 1). Some plant personnel have become quite adept at recognizing odors from particular process units, and they may be a good resource in locating both the source of the odor and the chemicals involved.

Once the chemicals causing the odor problem are identified, the specific sources of the emissions need to be determined. Typical sources include the wastewater treating system (including sewers, junctions, and access holes), fugitive equipment leaks (valves, pumps, compressors, and flanges), tanks (working and storage losses), vents, loading operations, land farms and other solid waste treatment, storage or disposal units, and intermittent operations (vessel degassing, exchanger cleaning, painting, etc.). In some cases, the odor may have been caused by a small spill that is easily located and treated. When the odor is only detected outside the facility boundary, an elevated source (i.e., stack or vent) may be the cause, with the odorous plume reaching ground level some distance from the plant.

Emission sources outside the plant should also be considered as the potential cause of the odor problem. Sometimes the surrounding community wrongly attributes an objectionable odor to a refinery or chemical plant just because it is a large (and perhaps historically likely) nearby source. Small adjacent facilities may be responsible for or may contribute significantly to the odor problem. If an adjacent facility is suspected, it is usually necessary to fully quantify the emissions and resultant ambient concentrations so that contributions to the odor problem can be attributed to the appropriate internal and external emission sources.

In some cases, the specific source of the odorous emission cannot be clearly identified. Application of low-cost emission reduction options (discussed later) should be considered for the more likely sources of the odor. Tracer studies, using a nontoxic identi-

Table 1. Detection thresholds, health exposure limits, and odor descriptors for common odorous compounds.

Compound	Odor Threshold, ppm	Health Limits, ppm		Descriptor
		TWA*	STEL†	
Acetaldehyde	0.067	25‡		Pungent/Fruity
Acrylic Acid	0.09	2		Rancid/Acrid
Acrylonitrile	1.6	2	10	Onion/Garlic
Ammonia	17	25	35	Pungent/Irritating
Amyl Mercaptan	0.0003			Unpleasant/Putrid
Benzyl Chloride	0.041	1‡		Pungent
Benzyl Mercaptan	0.0002			Unpleasant/Strong
Butadiene	0.45	10‡		Aromatic/Rubber
n-Butyl Amine	0.080	5‡		Sour/Ammonia
Carbon Disulfide	0.0081	1	10	Medicinal/Sulfur
Carbon Tetrachloride	250		2	Sweet
Carbonyl Sulfide	0.1			Sulfide
Chlorine	0.01	0.5	1	Pungent/Bleach
Chlorobenzene	1.3	10		Almond/Shoe Polish
Chloroform	192		2	Sweet
Cumene	0.032	50		Sharp
Dibutyl Amine	0.016			Fishy
Diisopropyl Amine	0.13	5		Fishy
Dimethyl Amine	0.34	5	15	Fishy/Putrid
Dimethyl Sulfide	0.001			Decayed Cabbage
Diphenyl Sulfide	0.0001			Unpleasant
Ethyl Amine	0.27	5	15	Ammoniacal
Ethyl Mercaptan	0.0003	0.5‡		Decayed Cabbage
Ethylene Dichloride	26	1	2	Chloroform
Ethylene Oxide	257	1		Sweet
Hydrogen Sulfide	0.0005	10	15	Rotten Eggs
Methanol	100	200	250	Sour/Sweet
Methyl Amine	4.7	5	15	Putrid/Fishy
Methyl Chloroform	385	350	450	Sweet
Methyl Ethyl Ketone	10	200	300	Sweet/Sharp
Methyl Isobutyl Ketone	0.47	50	75	Sweet/Sharp
Methylene Chloride	144	50		Sweet
Methyl Mercaptan	0.0005	0.5‡		Rotten Cabbage
Naphthalene	0.027	10	15	Tar/Creosote
Nitrobenzene	0.0047	1		Almond/Shoe Polish
Phenol	0.047	5		Medicinal/Acid
Phenyl Mercaptan	0.0003	0.5		Putrid/Garlic
Propylene Oxide	35	20		Sweet
Propyl Mercaptan	0.0005			Unpleasant
Styrene	0.047	50	100	Sweet/Aromatic
Styrene Oxide	0.06			Sweet/Pleasant
Sulfur Dioxide	2.7	2	5	Pungent/Irritating
Toulene	2.8	100	150	Sour/Burnt
Trichloroethylene	82	50	100	Ether/Solvent
Xylene	0.73-5.4	100	150	Sweet

Sources: (14–19)

* TWA = Threshold Limit Value — Time-Weighted Average. This is the average concentration for a normal 8-h/d, 40-h/wk workweek, to which nearly all workers may be repeatedly exposed, day after day, without adverse effect. Where noted (by ‡), the ceiling exposure level is listed instead.

† STEL = Threshold Limit Value — Short-Term Exposure Limit. This is the concentration to which workers can be exposed for a short period of time without suffering from adverse effects. Exposure at levels between TWA and STEL should not exceed 15 min and should not occur more than four times per day. There should be at least 60 min between successive exposures above the TWA.

‡ Ceiling Exposure Limit.

fier incorporated into other releases, may help to identify source and receptor location relationships and thereby reduce the number of potential emission sources that need to be considered for further action.

Step 3: Quantify the ambient concentration

• Has ambient monitoring been conducted to confirm the presence of the odorous compound? Is the detection threshold concentration below the instrument detection limit? Have monitors been placed so as to capture the likely path of the concentration plume based on the potential emission sources and local meteorology?

• What are the emissions rates from the likely sources of the odorous compounds? Have all potential sources been accounted for, including all fugitive, tank, loading, vent, and wastewater air emissions? Are combustion sources a likely contributor to the odor problem and should they be included in the analysis? Can the emission rates be monitored? What is the variability and uncertainty in the estimated emissions, both because of operations changes and because of assumptions used for the input data and calculation method?

• Have predicted concentration isopleths (lines of constant concentration) been plotted based on the estimated emissions of the odorous chemical? Have the sources been modeled appropriately and do they realistically represent the physical situation?

• Do the predicted ambient concentrations from emissions and dispersion modeling exceed the odor detection threshold? How conservative are the predicted concentrations? At the highest likely emission rate and worst dispersion conditions (i.e., the highest ambient concentration), are the predicted concentrations below the odor detection threshold? How close to the detection thresholds are the predicted ambient concentrations?

• Are the predicted concentrations near the short- or long-term health exposure limits for any of the compounds being modeled? Has an industrial hygienist reviewed the results? Is any immediate action required?

• Can some of the sources or chemicals be eliminated as potential contributors to the odor problem based on the monitoring or modeling? Can this be fully and defensibly documented?

• Are sources outside the facility the most likely cause or are they major contributors to the odor problem

> *Quantify the ambient concentrations of the odor-causing compounds to determine their contributions to the odor problem.*

(based on the emissions and dispersion modeling)? Can perimeter monitoring be used to establish a high upwind concentration of the odorous compound? Have these sources outside the facility been the cause of previous odor problems?

In this step, the ambient concentrations of the odor-causing compounds are quantified to help in determining their contributions to the problem and to focus odor reduction efforts. This can involve ambient air monitoring as well as emissions estimating and dispersion modeling. In cases where the odor source is not clearly identified, some combination of both approaches may be required to fully evaluate the situation. If working with a local regulatory agency, it is prudent to gain their acceptance of the chosen quantification approach before proceeding with extensive and costly monitoring or modeling efforts. The modeling impact analysis will also allow an industrial hygienist to compare the predicted concentrations of specific chemicals to the recommended health-based standards.

Ambient air monitoring is generally more costly than emissions and disper-

sion modeling and has some additional limitations. These limitations include the potential for missing the plume location, and therefore the maximum concentration, by having too few, or too widely spaced, monitors. There may also be some compounds whose odor detection threshold concentrations are lower than can be monitored with available equipment. If the odorous compound is detected at concentrations above threshold levels, however, monitoring provides good evidence of the problem and subsequent monitoring, after mitigation steps are implemented, can demonstrate progress toward solving the odor problem.

The modeling approach involves estimating the emissions of odorous compounds from the suspect process units and using dispersion models to predict downwind ambient concentrations. The basics of air quality modeling are outlined in (1). Techniques for estimating routine plant emissions are provided in (2–3) and other documents. These are likely emission sources to consider if the odor problem is continuous or occurs frequently. In some cases, odor complaints can be correlated with tank filling or other loading operations.

Transient emissions, such as from vents, safety valves, spills, and maintenance operations, require either a means to measure the released quantity or a basis to determine, through engineering analysis, how much material has been released. In these cases, a first approach is to assume that the largest quantity of material available is released, and the predicted ambient concentration is compared to the odor threshold. If this conservative approach results in predicted ambient concentrations well below the odor threshold, then the source is not likely to be the cause of the odor problem.

For dispersion modeling, sources must be described as either points, lines, areas, or volumes. The selection of source type has a significant impact on the magnitude of the predicted ambient concentration. Tanks, vents, and loading emissions sources are usually treated as point sources. Wastewater treating, land farms, and spills are area sources. Fugitive emissions are the most difficult to describe since, although they are discrete point sources, their large number precludes modeling each release point individually. Fugitives have been modeled as a single combined near-ground-level point source that includes all the fugitives for an entire unit, as an area source equivalent to the footprint of the unit, and as a volume source. If modeling is required by a regulatory agency, details of the modeling procedures should be reviewed with agency personnel before proceeding with the analysis to assure acceptance of the results.

Care must be taken in interpreting the results from standard dispersion models when predicting downwind ambient concentrations of the suspected odorous compound. Most of these models predict ambient concentrations that are time averages over about 15 minutes to one hour. If the odor problem is not constant, one may need to consider much shorter concentration averaging times. The technology to predict these very short average time concentrations, however, is still developing. A short-averaging-time peak concentration can be approximated by multiplying the concentration predicted by the standard models by a factor of three to six.

When conducting the impact analysis, it may be useful to include nearby emission sources that are not within the plant boundary. Although the plant sources may result in a predicted concentration that is well below the odor threshold, the combined sources may be sufficient at times of specific wind direction or plant operations to result in odor complaints from the surrounding community.

Application of the modeling approach is limited to cases where sufficient information to quantify the source emissions and atmospheric

Table 2. Exponents for Stevens' Law equation.

Compound	Exponent*
Acetaldehyde	0.26
Allylamine	0.58
Benzene	0.55[†]
Butyl Mercaptan	0.29
Chloroform	0.58
Cyclohexane	0.66
Ethyl Mercaptan	0.25
Ethanol	0.54
Heptane	0.43
Methyl Mercaptan	0.34
Nitrobenzene	0.74
Octane	0.36
Propyl Mercaptan	0.33
Styrene	0.48
Toluene	0.53

*Source (except as otherwise noted): (20)
[†]Source: (21)

dispersion is available. In some cases, there may not be a clear indication of either the chemical causing the odor or the potential sources. Modeling in these situations is inappropriate and can result in the use of unrealistic assumptions, leading to a poor choice of emission reduction and control.

Step 4: Reduce or eliminate the odor problem

• *How much emission reduction is required to lower the perceived odor concentration to acceptable levels (based on Stevens' Law, discussed later)? Is the reduction requirement being set by a regulatory agency or to meet some ambient concentration related to the odor threshold? Is it being set by a health-based standard?*

• *Have all source reduction opportunities been considered? Are there some low-cost equipment retrofits, or operational, housekeeping, or materials changes, that can be implemented to reduce emissions? Can ongoing emission reduction efforts be focused on the areas most likely to emit odorous compounds?*

• *Are emission source control actions required? What are the most applicable recovery or destruction options? Is there a regulatory requirement for a particular type of control device or efficiency?*

This step includes a determination of the amount of emission reduction required and the choice of reduction or control alternatives. In some cases, it can be beneficial to prioritize the emission sources based on the contribution of each source to the odor, the cost to control each source, or other factors.

A preliminary step in determining acceptable control options is to determine the amount of emission reduction required to eliminate or reduce the odor problem. A 50% decrease in emissions and, therefore, equivalent reduction in ambient concentration will usually result in less than a 50% perceived reduction in the odor intensity. The perceived odor intensity is related to the ambient concentration of the odorous compound by Stevens' Law:

$$I = kC^n$$

where C = ambient concentration, I = perceived intensity of the odor, and k and n are constants.

The value of the exponent in Stevens' Law (Table 2) ranges from about 0.2 to 0.6, with most sulfur compounds in the 0.3 range. Thus, for example, a 50% reduction in the ambient concentration of a compound with a Stevens' Law exponent of 0.3 will result in a perceived odor reduction of less than 20%. This means that much more source reduction is required than if the perceived odor intensity was proportional to the ambient concentration.

Stevens' Law exponents are available for only a limited number of chemicals. These exponents have been developed through the use of odor panels. Since there have been different protocols and panels used in the various studies, the actual reported values of the exponent tend to vary. However, a reasonable approach is to use a value of 0.3 if data are not available.

Table 3. Suggested source reduction opportunities.

Fugitive Emissions
- Initiate a leak detection and repair program.
- Install new packing sets in block and control valves.
- Upgrade pump seals. For example, use low-emission single seals with contacting back-up (possibly with a gas buffer) or a buffered dual-seal design (gas or liquid).
- Consider canned or magnetic-drive pumps.

Tanks

Fixed Roof
- Install a vapor balance system.
- Install a vapor recovery or destruction system
- Install an internal floating roof.

Floating Roof
- Check condition of existing seals and replace or upgrade if necessary.
- Install secondary rim seals.
- Replace vapor-mounted primary rim seal with liquid-mounted seal.
- Control losses from roof fittings (for example, add gasket and float to guidepole, add gasket to access hatch and bolt hatch closed, add gasket to gage hatch).
- Install a vapor recovery or destruction system (convert an external floating roof tank to an internal floating roof tank first).
- Install a geodesic dome roof over an external floating roof tank to convert it to an internal floating roof tank.

Wastewater Collection and Treating
- Decrease wastewater volume and organic concentrations by improved housekeeping and/or by employing low-cost control technologies.
- Install sewer system emission suppression (Seal drains: P-leg, seal pot, closed system. Junction boxes: seal pot, gas trap manhole).
- Reduce air/water contact. For example, install covers on separators and induced-air flotation units, use floating blocks or balls on API separators to reduce surface area, and replace basins and impoundments with covered tanks.
- Replace API separators with covered corrugated plate interceptors.
- Consider using "masking chemical agents" that can be sprayed over the equipment or added to wastewater to neutralize acid and alkali gases.

Loading
- Use "submerged" loading in place of "splash" loading.
- Use vapor balancing during loading.
- Install a vapor recovery or destruction system.

In choosing a mitigation approach, consideration must focus on what will satisfy the local community and regulatory authorities. It may be important to balance reductions in odor intensity with reductions in the frequency of occurrence. In some cases there is an advantage to reducing the frequency rather than the magnitude of an odor problem, since the former may involve operations and housekeeping while the latter is likely to involve some investment.

Suggestions on source reduction opportunities to reduce fugitive, tank, wastewater treating, and loading emissions are provided in Table 3 and in (4–9).

Fugitive emissions can be reduced by a combination of monitoring, maintenance, and equipment modifications; in extreme cases, leakless equipment (*e.g.,* bellows valves, canned pumps, etc.) may be required. Tank emission controls for floating roof tanks involve improved rim seals and controls on roof fittings; vapor control on the vent of an internal floating roof tank (as used for fixed roof tanks) may be required in extreme cases. Air emissions from wastewater systems can be reduced most cost-effectively by reducing the quantity of material that enters the system; additional controls include covers and "masking" chemicals. Masking

products may be especially useful as an interim measure until permanent control options can be installed.

Source control options involve either recovery or destruction of the chemical emission causing the odor problem. The available technologies include absorption, adsorption, condensation, flaring, thermal oxidation, catalytic oxidation, and biofiltration. [Refs. *(10–13)* provide more detailed information on these technologies.] The choice will depend on various technical, economic, and regulatory considerations.

For small and intermittent sources, such as tank vents and some wastewater

Literature Cited

1. **Seigneur, C.,** "Understand the Basics of Air-Quality Modeling," *Chem. Eng. Progress,* **88**(3), pp. 68–74 (Mar. 1992).

2. **U.S. Environmental Protection Agency,** "Compilation of Air Pollutant Emission Factors," US EPA, Research Triangle Park, NC, Document No. AP-42, 5th ed. (Jan. 1995).

3. **Li, R.,** *et al.,* "Develop a Plantwide Air Emissions Inventory," *Chem. Eng. Progress,* **91**(3), pp. 96–103 (Mar. 1995).

4. **Chadha, N.,** "Develop Multimedia Pollution Prevention Strategies," *Chem. Eng. Progress,* **90**(11), pp. 32–39 (Nov. 1994).

5. **Siegell, J. H.,** "Control VOC Emissions," *Hydrocarbon Processing,* **74**(8), pp. 77–80 (Aug. 1995).

6. **Chadha, N., and C. S. Parmele,** "Minimize Emissions of Air Toxics via Process Changes," *Chem. Eng. Progress,* **89**(1), pp. 37–42 (Jan. 1993).

7. **Lipton, S.,** "Reduce Fugitive Emissions Through Improved Process Equipment," *Chem. Eng. Progress,* **88**(10), pp. 61–68 (Oct. 1992).

8. **Adams, W. V.,** "Control Fugitive Emissions from Mechanical Seals," *Chem. Eng. Progress,* **87**(8), pp. 36–41 (Aug. 1991).

9. **Brestel, R.,** *et al.,* "Minimize Fugitive Emissions With a New Approach to Valve Packing," *Chem. Eng. Progress,* **87**(8), pp. 42–47 (Aug. 1991).

10. **Martin, A. M.,** *et al.,* "Control Odors from CPI Facilities," *Chem. Eng. Progress,* **88**(12), pp. 53–62 (Dec. 1992).

11. **Mukhopadhyay, N., and E. C. Moretti,** "VOC Control: Current Practices and Future Trends," *Chem. Eng. Progress,* **89**(7), pp. 20–26 (July 1993).

12. **Mukhopadhyay, N., and E. C. Moretti,** "Current and Potential Future Industrial Practices for Reducing and Controlling Volatile Organic Compounds," AIChE, New York, NY (1993).

13. **Ruddy, E. N., and L. A. Carroll,** "Select the Best VOC Control Strategy," *Chem. Eng. Progress,* **89**(7), pp. 28–35 (July 1993).

14. **U. S. Environmental Protection Agency,** "Reference Guide to Odor Thresholds for Hazardous Air Pollutants Listed in the Clean Air Act Amendments of 1990," U.S. EPA, Research Triangle Park, NC, Document No. EPA/600/R-92/047 (March 1992).

15. **Prokop, W. H.,** "Odors," in "Air Pollution Engineering Manual," A. J. Buonicore and W. T. Davis, eds., Air and Waste Management Association, Van Nostrand Reinhold, New York, NY, pp. 147–154 (1992).

16. **Sober, R. F., and D. Paul,** "Less-Subjective Odor Assessment," *Chem. Eng.,* **99**(9), pp. 130–136 (Sept. 1992).

17. **Robinson, S.,** "Develop a Nose for Odor Control," *Chem. Eng.,* **100**(10), pp. 20–27 (Oct. 1993).

18. **American Conference of Governmental Industrial Hygienists,** "1993–1994 Threshold Limit Values for Chemical Substances and Physical Agents and Biological Exposure Indices," ACGIH, Cincinnati, OH (1993).

19. **National Institute of Occupational Safety and Health,** "NIOSH Pocket Guide to Chemical Hazards," U.S. Dept. of Health and Human Services, National Institute for Occupational Safety and Health, Cincinnati, OH, Publication No. 90-117 (June 1990).

20. **Laffort, P.,** *et al.,* "Olfactory Coding on the Basis of Physicochemical Properties," *Ann. N. Y. Acad. Sci.,* **237,** pp. 193–208 (1974).

21. **Hyman, A. M.,** "Factors Influencing the Psychophysical Function for Odor Intensity," *Sensory Processes,* **1,** pp. 273–291 (1977).

Further Reading

Cain, W. S., "Olfaction," in "Stevens' Handbook of Experimental Psychology," R. C. Atkinson, *et al.,* eds., Vol. 1, Wiley, New York, pp. 409–459 (1988).

Cain., W. S., "The Odoriferous Environment and the Application of Olfactory Research," in "Handbook of Perception," E. C. Carterette and M. P. Friedman, eds., Academic Press, New York, pp. 277–304 (1978).

Devos, M., *et al.,* "Standardized Human Olfactory Thresholds," Oxford University Press, New York (1990).

Dravnieks, A., "Odor Perception and Odorous Air Pollution," *TAPPI,* **55**(5), pp. 737–742 (May 1972).

Hellman, T. M., and F. H. Small, "Characterization of Petrochemical Odors," *Chem. Eng. Progress,* **69**(9), pp. 75–77 (Sept. 1973).

Nolen, S. L., *et al.,* "Map an Effective Odor Control Strategy," *Chem. Eng. Progress,* **90**(9), pp. 47–53 (Sept. 1994).

Turk, A., and A. M. Hyman, "Odor Measurement and Control," in "Patty's Industrial Hygiene and Toxicology," G. D. Clayton, and E. Florence, eds., 3rd revised edition, Vol. 1, pp. 665–707 (1978).

treating units, adsorption (carbon canisters) will often be the lowest-cost option. It is not applicable, however, to high-temperature, high-humidity streams, or to those that contain contaminants that can poison or plug the adsorbent bed. For potentially high-flow-rate intermittent streams, routing to a flare may be the most economic alternative. Low-concentration streams may be controllable using a relatively low-cost biofilter. For continuous high-flow-rate and high-concentration emissions, the most appropriate controls include absorption, condensation, and oxidation.

Since these controls may represent a significant capital (and in some cases operating) expense, the ability of the control technology to eliminate the odor problem should be confirmed. **CEP**

Acknowledgment

This article is based on a paper that is part of the Petroleum Environmental Research Forum (PERF) Cooperative Air Program (CAP), which is aimed at developing approaches, data, and technologies that lead to cost-effective compliance with the Clean Air Act Amendments.

J. H. SIEGELL is a senior staff engineer with Exxon Research and Engineering Co., Florham Park, NJ, where he heads activities on estimating and controlling air toxic, hydrocarbon, and odorous emissions. He has over 20 years of industrial experience, including fundamental and applied research and technical support to refineries and chemical plants. He holds a BE(ChE) and an ME(ChE) from the City College of New York and a PhD(ChE) from the City University of New York.

Control Odors from CPI Facilities

Selecting an odor control technology depends on the compounds causing the odors and their concentrations, as well as the air stream flow rate, moisture content, and variability. Follow these guidelines to narrow the choices.

Alison M. Martin,
James M. Montgomery
Consulting Engineers

Sharon L. Nolen,
Eastman Chemical Co.

**Paula S. Gess and
Todd A. Baesen,**
James M. Montgomery
Consulting Engineers

Various production and waste treatment operations at chemical process industries (CPI) facilities produce odor emissions. Public awareness of CPI plants has heightened in recent years, and odors that were once tolerated are now a cause for concern. Thus, many plants are now implementing odor reduction programs.

Installation of odor control systems may also help a facility meet the Title III air toxics requirements of the Clean Air Act Amendments (CAAA) of 1990. If a facility's odorant compounds are on the CAAA's list of 189 toxic compounds, then appropriate odor control devices will aid in compliance with air toxics regulations.

CPI facilities are difficult to characterize in terms of odors, since sources have different odorous compounds, flow rates, and variability. Because each source is unique, much consideration will be required to reduce odor emissions at CPI plants.

Most likely, many different technologies will be required to reduce odors at a specific facility. Site constraints may eliminate some choices, and energy requirements may eliminate others. Cost-effective odor treatments, however, should be viable for most sources.

Odor control technologies fall into four categories:
- odor minimization;
- chemical or physical treatment, such as incineration (or oxidation), adsorption, absorption, condensation, and advanced oxidation;
- biological treatment; and
- other technologies.

This article discusses both conventional and innovative odor control technologies. Many of these are the basic techniques for controlling emissions of volatile organic compounds (VOCs). The article also provides guidance on how to select from among the technologies — a choice that will depend heavily on the air stream flow rate, compounds present, compound concentrations, moisture content, and stream variability.

Odor minimization

The best way to control odors is to prevent their release. Odor minimization techniques are very site-specific and depend on such factors as the odor compound(s) in question, concentration, flow rate, parameter variability, and the emission source. No one method will reduce or eliminate odors at every emission point.

Odor minimization basically consists of modifying the existing conditions that result in odor formation. Such modifications can include process changes, piping changes, the addition of physical barriers, temperature adjustments, or the use of chemical additives.

Uncovered liquids. One can minimize odors from uncovered liquid sources, such as process sumps, wastewater collection lines, or liquid storage/treatment basins, in a variety of ways:

1. Minimize surface area. A reduction in the surface area of the liquid or solid surfaces that emit odors will decrease the mass-transfer area available. This can be done by reconfiguring the existing equipment or in the design of new equipment. For example, a tall tank with a small diameter can be used for material storage, or deep, narrow sumps can be designed.

2. Increase freeboard depth. An increase in freeboard depth can decrease wind or wave action on liquid surfaces, which reduces turbulence. A depth increase of about 2 ft for a large liquid surface is typical. The ratio of freeboard depth to the diameter of the liquid surface is the critical parameter for the effectiveness of this type of control measure. In general, a minimum of 0.02 ft freeboard depth per foot of liquid surface diameter is a good rule of thumb.

3. Increase submergence of inlet and out-

let piping. Inflow and outflow piping from sumps and other liquid-containing areas should be positioned well below the surface of the liquid. This will result in minimal disturbance of the liquid surface.

4. Adjust temperature. Increasing liquid temperature increases vapor pressure. Therefore, a reduction in liquid temperature will reduce emissions. Additionally, liquids discharging into a large basin should be maintained at a temperature as close as possible to that of the basin. This is because when liquids of two different temperatures are mixed, convective currents are induced, which cause mixing and increased turbulence, resulting in greater emissions.

5. Use chemical additives. The addition of chemicals to liquids can be extremely effective at reducing or eliminating odors. For example, oxidizing agents are frequently added to sewage collection systems and treatment plants to minimize the formation of hydrogen sulfide. Common additives for this application include chlorine, hydrogen peroxide, pure oxygen, iron salts, sodium hydroxide, and sodium nitrate, as well as several proprietary compounds that are available commercially. The additive is chosen based on the specific odorants involved.

6. Install physical barriers. Popular physical barriers include synthetic covers, floating spheres, and surfactant/foam layers. Synthetic covers reduce wind over the emitting surface and can contain emissions for further treatment. Covers are generally used for small surfaces (less than 20 ft in diameter or 300 ft^2), such as process sumps, and are frequently equipped with a vent for sending the emissions to an adjacent treatment area.

Hollow spheres are generally used over large surface areas (greater than 50–60 ft in diameter or 1,000 ft^2) to serve as a barrier to evaporative losses and emissions. Normally, the spheres are designed with an anti-rotation collar that locks them in place. About 91% of the exposed surface can be covered, resulting in emission reductions of 90%. However, the spheres are typically effective only during quiescent conditions.

Table 1. Strategies for reducing or eliminating odorous air emissions.

I. Eliminate the source of the pollutant.
   1. Seal the system to prevent interchanges between it and the atmosphere.
     a. Use pressure vessels.
     b. Interconnect vents on receiving and discharging containers.
     c. Provide seals on rotating shafts and other necessary openings.
   2. Change raw materials and/or fuels.
   3. Change the manner of process operation to prevent or reduce the formation or air entrainment of a pollutant.
   4. Change the type of process step.
   5. Use a recycled gas, or recycle the pollutants, rather than using fresh air for venting.
II. Reduce the quantity of pollutant released or of carrier gas to be treated.
   1. Minimize entrainment of pollutants into a gas stream.
   2. Reduce the number of points in the system in which materials can become airborne.
   3. Recycle a portion of process gas.
   4. Design hoods to exhaust the minimum quantity of air necessary to ensure pollutant capture.
III. Use equipment for dual purposes, such as a fuel combustion furnace as a pollutant incinerator.

Source: *(1)*

Surfactant and foam layers have been used to reduce evaporation from water reservoirs and, more recently, to retard emissions from organic-laden water and soils. These products must be reapplied periodically to ensure continued effectiveness.

Process areas. In general, elimination or reduction of odors in process areas can be accomplished by changing the equipment, process, materials, or operational practices. Process modifications can include changes in temperature, pressure, or gas volume. Table 1 *(1)* lists some ways to reduce odorous air emissions.

Incineration

Incineration is a common and effective method for the destruction of VOCs in air streams. Many different types of incineration devices exist.

Thermal incinerators can be used to treat a wide variety of continuously emitted streams containing VOCs. Thermal incineration is not as sensitive to fluctuations in VOC concentration and air stream characteristics as other incineration techniques. Thermal incinerators are not well suited to streams that exhibit a highly fluctuating flow rate due to the poor mixing and incomplete combustion that results. Under steady flows, however,

thermal incinerators usually exhibit destruction efficiencies up to 99%.

Very dilute air streams (less than 20 ppmv) may require supplemental fuel to maintain necessary combustion temperatures, which are usually between 1,000°F and 1,500°F. Fuel requirements can be reduced by heat recovery from the hot exhaust gas. This can be done in a variety of ways, such as those discussed by Shook *(2)*.

Catalytic incinerators operate much like thermal incinerators but they employ catalysts that permit combustion at lower temperatures — normally 600°F–900°F. Fuel usage with this type of unit is less than that of a thermal incinerator, although initial capital costs are higher and there are added costs associated with catalyst replacement.

Catalytic incinerators are, however, more sensitive to air stream characteristics than thermal incinerators. Certain substances in an air stream can poison the catalysts, which reduces treatment efficiency. Typical catalyst poisons include phosphorous, bismuth, lead, arsenic, antimony, mercury, iron oxide, tin, zinc, and sulfur.

Catalytic incinerators can usually achieve destruction efficiencies of 95%–99%. Continuous air flows of up to 100,000 scfm can be treated.

Flares are normally used at CPI

plants to treat waste gases during process upsets and emergencies. They are usually regarded as safety devices that provide destruction of waste emissions. Because they are direct combustion devices, flares require high operating temperatures — in the range of 2,500°F. Many different types of flares are used at industrial facilities, such as the steam-assisted flare depicted in Figure 1.

When the heat content of the air stream is greater than 300 Btu/scf, 98% destruction efficiencies can be achieved. Flares can be designed to handle very large flow rates — up to 100,000 lb/h for ground flares and up to 2,000,000 lb/h for elevated flares.

Boilers and process heaters are sometimes used for VOC control. If the heating value of the air stream is above 300 Btu/scf, then the stream can be used as supplemental fuel. Emission streams with lower heat contents can also be used as long as the flow rate of the emission stream is small compared to that of the air/fuel mixture.

Under the right circumstances, boilers or process heaters can provide destruction efficiencies up to 98%. Only a small capital cost and little or no operating cost is required. Additionally, almost complete heat recovery is possible.

However, the use of emission streams for this application is subject to many limitations. Because the devices are essential to the plant's operation, only streams that will not affect the unit's performance or reliability can be used. Variations in flow or VOC concentration are not well tolerated and could decrease unit performance. Corrosive impurities in an air stream will also render it unsuitable for this application.

Technical considerations. Incineration devices can be designed to successfully treat a variety of VOC-laden air streams. Unfortunately, incineration is not always the lowest cost alternative.

Additionally, partial combustion of an organic compound may result in formation of another with even more odorous characteristics. For example, butyl aldehyde or butyric acid may result from incomplete oxidation of butanol. Nitrogen in the air stream and combustion air form NO_x, and sulfur compounds form SO_x. It may, therefore, be necessary to provide a secondary control system such as a scrubber to remove SO_2, HCl, and so on.

Costs for incineration are highly dependent on the waste stream in question. Costs can range from small capital and operating costs to moderate capital and large operating costs.

Incineration has several advantages:

• It is a proven and effective technique for treatment of VOCs.

• Odorous streams can sometimes be vented to boilers or incinerators already on-site.

• Incineration units can be designed to treat streams with high or low flow, high or low VOC concentration, and high air stream variability.

• The organics are destroyed and not simply transferred to another phase.

• This technology is suitable for complex mixtures of VOCs.

But it also has some drawbacks and limitations:

• Catalytic units cannot tolerate poisons or high VOC concentrations (above approximately 10,000 ppmv).

• Most incineration processes operate best under stable air stream characteristics and flows.

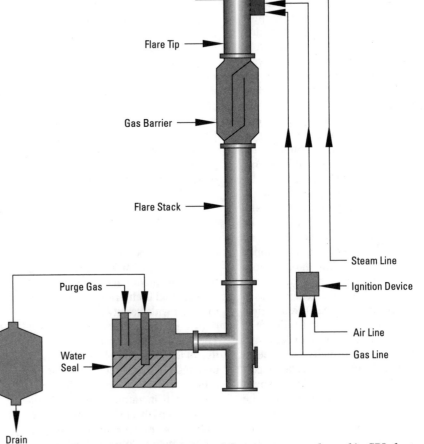

Figure 1. Steam-assisted elevated flares are commonly used in CPI plants.

• Secondary control devices may be required to treat combustion products.

• Combustion processes may not be suitable for streams that contain sulfur and nitrogen compounds due to SO_x and NO_x formation.

Adsorption

Gas adsorption involves passing a gas stream through a porous solid material. Gas molecules are removed from the stream by a combination of physical and chemical adsorption. Physical adsorption occurs when gaseous molecules are held to the adsorbent by van der Waals' forces; it is readily reversible by raising the temperature or lowering the pressure of the system. In chemical adsorption, or chemisorption, gas molecules form a chemical bond with the adsorbent and the gas is held to the adsorbent by strong valence forces; it is generally considered irreversible.

For odorous gases, activated carbon is the most commonly used adsorbent. Gases that may not be readily adsorbed onto activated carbon can often be treated using chemically impregnated activated carbon. Carbon impregnated with potassium hydroxide is recommended for use with organic acids, hydrogen sulfide, and sulfur dioxide; phosphoric acid impregnated carbon is recommended for use with ammonia streams.

There are several configurations for adsorption systems. The most common is the fixed-bed adsorber. Moving-bed adsorption is also available. In one such system, the adsorbent is contained in a slowly rotating drum and the effluent gas stream is moved by a fan into the rotating drum. The air is forced through the adsorbent material then the treated air leaves by ports at the end of the drum. In this configuration, a small portion of the housing is separated for use as a continuous desorption area. The carbon in the desorption zone is subjected to hot air at a lower flow rate than that of the influent dirty air. The VOCs are thereby desorbed as a higher concentration stream for further recovery.

Adsorption units can be designed to accommodate virtually any size stream that may be encountered. For streams with high VOC concentrations (above 10,000 ppmv), dilution prior to treatment may be required.

Fixed beds have typically achieved the highest removal efficiencies (usually greater than 90%) for streams with inlet concentrations in the 1,000–5,000 ppmv range. Tests of moving bed adsorption systems have shown adequate treatment (77%–99% efficiency) of inlet concentrations as low as 1 ppmv.

In general, activated carbon is suitable for the following types of VOC (3):

• those with molecular weights between 50 and 200 g/mol;

• compounds whose boiling points are between 68°F and 350°F;

• aliphatic and aromatic hydrocarbons between C_4 and C_{14};

• most common halogenated solvents (such as carbon tetrachloride, ethylene dichloride, methylene chloride, perchloroethylene, and trichloroethylene);

• most common ketones (acetone, methyl ethyl ketone) and some esters (butyl and ethyl acetate); and

• common alcohols.

Several types of organic compounds are not suitable for vapor-phase activated-carbon adsorption. These include species that react with carbon, react with the regeneration steam, polymerize on the carbon, or are difficult to remove by regeneration (basically high molecular weight compounds). These species include organic acids, aldehydes, some ketones, some easily hydrolyzed esters, some halogenated hydrocarbons, plasticizers, resins, hydrocarbons greater than C_{14}, phenols, glycols, and amines (3). There are other adsorbents on the market, however, that may be successfully used for these compounds. One such system is based on the use of hydrophobic zeolites, which are not sensitive to moisture content in the air stream and can withstand aggressive regeneration.

Adsorption has the following advantages:

• It is a well-established treatment technology, and long-term operating data are readily available.

• Operation is relatively simple.

• The process can treat a wide variety of organic compounds to meet desired effluent standards.

• Product recovery is possible.

• The system is not prone to upsets when used with mixed waste streams.

Its disadvantages include:

• Secondary wastes (spent carbon and condensed VOC if not reuseable) are generated and these may require off-site treatment or disposal.

• Costs may be prohibitively high for high concentration streams.

• Some adsorbents cannot be easily regenerated, resulting in high costs for replacement and disposal.

Absorption

Absorption, or scrubbing, is a process in which one or more components (solute) in a gas stream are selectively transferred, or absorbed, into a relatively nonvolatile liquid (solvent). The driving force for absorption is the difference between the actual solute concentration in the solvent and the equilibrium concentration. The mechanism can be either physical, where the solute simply dissolves in the solvent, or chemical, where the solute reacts with the solvent or compounds dissolved in it.

The rate of absorption depends on many factors, including the physical properties of the gaseous/liquid system and the absorber operating conditions. The absorption rate can be enhanced by using lower temperatures, higher gas/liquid contact area, and higher liquid-to-gas ratios; absorption is also favored by higher contaminant concentrations in the gas stream.

Absorption equipment includes packed towers, plate towers, spray towers, spray chambers, and venturi scrubbers. These devices are designed to allow the maximum liquid/gas contact to enhance the mass-transfer rates from one phase to the other.

As odor control devices, scrubbers are used much more frequently for the removal of inorganic vapors than for organic vapors. Absorption as a control technology for the removal of VOCs has several limitations. First, the VOC must be readily soluble in the solvent. Second, the absorption solvent must also be treated. A water solvent can be treated in a wastewater treatment plant, whereas used organic solvents are typically stripped. For VOC removal, these stripping requirements are often very expensive due to the very low residuals necessary for solvent reuse.

Third, for odor control, very low outlet VOC concentrations will be required. (Just how low depends on the compound in question. For example, the odor threshold for ammonia is 17 ppm, whereas the odor threshold for hydrogen sulfide is about 5 ppb.) This low concentration may lead to extremely tall towers, long contact times, and high liquid/gas ratios, which may make absorption an expensive treatment alternative. Absorbers for VOC control are, therefore, most effective when used in combination with other control technologies, such as incineration.

The cost of a scrubbing system is usually low to moderate for the equipment. Operating costs can be low but may increase if chemical additives are needed. Additives to a water solvent are common, to provide destruction of the absorbed species. Popular additives include potassium permanganate (especially for H_2S, SO_2, aldehydes, and unsaturated organics), hydrogen peroxide (for organic constituents), sodium bisulfite (for aldehydes), sodium hypochlorite (for H_2S), mild acid solution (for ammonia), surfactants (for organics), and chemical counteractants (for a variety of chemicals) (4).

The advantages of absorption are:
• It is relatively inexpensive.
• It can be very effective when the solvent is carefully chosen for the particular compounds of concern.
• Many industrial facilities already have scrubbers on-site, and personnel may already be familiar with the operation and maintenance of these devices.
• A water solvent can be discharged directly into an on-site wastewater treatment system if one is present.

Among the disadvantages:
• For best performance, the constituents of concern cannot vary once the column is designed.
• Some types of additives to the scrubbing water may be expensive.
• Scrubbing may not be effective on streams that contain a mixture of many dissimilar species.
• Product recovery is probably impractical.
• For certain streams, absorbers may be unable to reduce the odorant concentrations down to acceptable levels.

Condensation

Condensation separates one or more volatile components from a vapor mixture by saturation and subsequent phase change. The induced phase change can occur either by increasing the system pressure at a constant temperature or by reducing the temperature at a constant pressure.

When used to control emissions, condensers are usually operated at the same pressure as the emission source. A refrigeration unit is frequently required to reduce the temperature enough for VOC condensation.

The two most common types of condensers are indirect-contact (or surface) condensers and direct-contact condensers. Surface condensers are often shell-and-tube heat exchangers, with the coolant on the tube side and condensing vapors on the shell side. Condensed VOCs are drained to a tank for collection or disposal.

Direct-contact condensers operate by intimate mixing of the vapor-laden air stream with a cooled liquid. Upon mixing, the cooled liquid may either remain a liquid or evaporate. The VOCs are separated from the cooling stream either through phase separation (in the case of a liquid/liquid mixture) or by venting the coolant stream (in the case of a gas/liquid mixture).

Condensers are frequently used as preliminary control devices for VOC removal prior to other treatment, such as incinerators or absorbers. They can be used by themselves for high concentration streams.

Normally, condensation is not a feasible technology unless the VOC concentration is greater than 3,000 ppm. But at concentrations above 5,000 ppm, 95% VOC recovery can be realized. Condensers are usually sized to treat flow rates of 2,000 scfm or less.

The advantages of condensation include:
• Condensation leads to VOC recovery, so the organics can be either recycled back into the process or inexpensively treated.
• It does not generate process residuals with the exception of the liquefied VOCs.
• Condensers are standard processing units and do not require extensive knowledge for operation.

A. M. MARTIN is a supervising engineer with James M. Montgomery, Consulting Engineers, Inc., Metairie, LA (504/835-4252; Fax: 504/835-8059). For the last four years, she has specialized in waste minimization, emissions control, and industrial wastewater treatment. She received her BS from the Univ. of California at Berkeley and her ME and PhD from Tulane Univ., all in chemical engineering. She is a registered professional engineer (chemical) in Louisiana and is a member of Tau Beta Pi, Sigma Xi, Omega Chi Epsilon, and AIChE.

S. L. NOLEN is a group leader with Eastman Chemical Co., Kingsport, TN (615/229-8265; Fax: 615/229-6099). She has specialized in air pollution and waste minimization during her three years at Eastman and her previous four years at the U. S. Environmental Protection Agency's Air and Energy Engineering Research Laboratory. She received her BS in chemical engineering from Tennessee Technological Univ., is a registered professional engineer (chemical) in Tennessee, and is a member of Tau Beta Pi and the Tennessee Society of Professional Engineers.

P. S. GESS is an associate with James M. Montgomery, Consulting Engineers, Inc., Metairie, LA (504/835-4252; Fax: 504/835-8059). Since joining Montgomery, she has focused on air pollution control and compliance issues. She received her BS in civil engineering from Louisiana State Univ. in 1990 and is a registered engineer-in-training in Louisiana.

T. A. BAESEN is a professional engineer with James M. Montgomery, Consulting Engineers, Inc., Walnut Creek, CA (510/975-3516; Fax: 510/975-3412), where his primary responsibilities include performing environmental feasibility studies and remediation design for industrial clients. He holds a BS in chemical engineering from the Univ. of California at Santa Barbara and is a member of the Environmental Div. of AIChE.

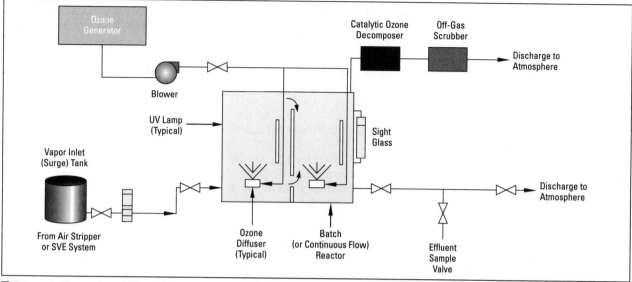

■ *Figure 2. Vapor-phase UV oxidation can treat a wide variety of VOCs.*

• Condensers can be designed to handle nearly any VOC constituent.

Its limitations are:

• If water vapor is present in the vapor stream, ice may form on the condenser tubes, so dehumidification of the air stream may be required.

• Condensation is not cost-effective for streams with low VOC concentrations (less than about 3,000 ppm).

• The stream needs to be fairly steady in flow, species types, and VOC concentrations.

• A condenser cannot lower the inlet VOC levels to below the saturation concentration at the coolant temperature.

Vapor-phase advanced oxidation

Vapor-phase advanced oxidation is an innovative technology that uses strong oxidizing agents and intense ultraviolet radiation for the photochemical stimulation of oxidation reactions. The UV light induces the oxidation of a wide range of organic compounds, leaving carbon dioxide and water as the major by-products. Figure 2 is a process flow diagram of a typical vapor-phase advanced oxidation unit.

UV oxidation processes generally involve mixing of the contaminated vapor with a strong oxidizing agent (typically ozone) and reacting it in the presence of ultraviolet radiation. A

$$3\,C_2H_4Cl_2 + 5\,O_3 \xrightarrow{\text{UV}} OH\cdot + \text{Intermediate Products}$$

$$\xrightarrow{\text{UV}} 6\,CO_2 + 3\,H_2O + 6\,HCl_{(g)}$$

■ *Equation 1.*

typical reaction using dichloroethane (DCA) is seen in Equation 1.

Reactions of this type leave the additional by-product of a halogen gas (HCl in the case of DCA). The reaction mechanism can be quite complex, but generally involves the formation of a hydroxyl radical (OH•) due to the intense UV bombardment. The hydroxyl radical is a strong oxidizer that initiates and propagates the subsequent reactions. The UV radiation also provides the activation energy necessary to drive this reaction to completion.

Batch and continuous flow reactors are available in sizes ranging from 10 ft³ to 800 ft³ and are capable of treating up to 200 cfm of vapor. These units are often mobile, and multiple modules may be placed in series (to increase system treatment capability) or in parallel (for increased capacity).

Residence time is the main operating parameter. It is a function of the influent stream matrix and the desired effluent characteristics, and it controls the size of the treatment unit. Residence times for vapor-phase UV oxidation may vary from a few sec-

onds to a few minutes. Residence time may be reduced by increasing the number of lamps (thus increasing capital costs) or by increasing lamp power (thus increasing operating costs).

UV oxidation has several advantages:

• The process is capable of treating a wide variety of organic compounds to meet any desired effluent standards.

• There are no secondary wastes generated that require off-site treatment or disposal.

• Since treatment costs are proportional to percent reduction (not concentration), UV oxidation will result in significant cost savings for many scenarios where high contaminant concentrations exist (such as where the carbon loading would result in prohibitively high costs for activated carbon).

Its disadvantages are:

• The process is less effective on highly chlorinated, saturated compounds such as chloroform and carbon tetrachloride, and where such compounds are present they will be the limiting factors (that is, controlling residence times, sizing, costs, and so on).

• The by-product of oxidation of chlorinated VOCs is gaseous HCl, and a fume scrubber may be needed to treat this gas.

• The capital cost of UV oxidation is prohibitively high for many small-scale (less than 10 cfm) treatment scenarios.

• Unlike the more established technologies, long-term operating data on UV oxidation are not available, so a treatability study is needed to determine the effectiveness of UV oxidation on a given vapor stream.

Biological treatment

In general, there are only two biological treatment methods that are fairly well established for odor control applications — biological scrubbers and biofilters. Occasionally, odors can successfully be removed by venting the air stream to a biologically active unit such as an activated sludge aeration basin. However, these applications are specific to the site and odorous compounds present; production of more odors or corrosive air stream properties are potential problems for this method. Biological treatment of odors using equipment specifically designed to treat the air stream is a much lower risk alternative than attempting to use an existing system designed for a different purpose.

Biological scrubbers. Biological scrubber systems are very similar in configuration to absorption scrubbers. A biological scrubber, as depicted in Figure 3, consists of a highly concentrated fixed growth attached to a plastic media with high specific area. To maintain the organisms in an active condition, primary or secondary wastewater effluent is sprayed on the packing material. The effluent provides the biomass with the moisture and nutrients required for survival.

In the biological tower, the influent air stream is forced upward through the tower while effluent water is sprayed down on top of the packed media. Prior to discharge from the tower, the air stream is passed through a mist eliminator section. In this layer, consisting of media similar to the main tower packing, water droplets are collected and removed from the effluent stream. The cleaned air is exhausted from the top of the tower.

Design criteria frequently used for biological towers include:

• minimum air retention time of 10 seconds;

• maintenance of a nitrifying environment of 10–25 lb of BOD_5 (biological oxygen demand) per 1,000 ft^3/d; and

• total hydraulic loading (feed, which is determined by waste strength, plus recirculation) of 1.5–4 gal/min per ft^2 of packing area.

Biological scrubbing has proven most suitable for compounds that can be oxidized relatively easily. Tests have indicated that about 90%–98% odor removal and up to 95% hydrogen sulfide removal can be achieved. This type of system is not recommended for use with organic sulfur compounds.

The advantages of biological scrubbing are:

• It is effective for the removal of low concentrations (less than 20 ppmv) of pollutant gases from air streams.

• A biological scrubbing tower is a relatively compact system.

• It has relatively low opera-

tion and maintenance requirements.

• Filter media replacement may be needed as infrequently as every five years.

• Treatment of residuals is minimized, as it generally consists of dewatering and disposal of the suspended biomass. This is usually a less expensive operation than the regeneration of carbon or scrubbing liquid.

Its disadvantages are:

• The system cannot handle compounds that are not oxidized relatively easily.

• If an excess of organic material is introduced into the system, either from the air stream or the wastewater, more odor problems may be created than are eradicated.

• Bioscrubbers have not been used widely in the U.S., so long-term operating data are scarce.

Biological filters. In biofiltration, off-gases containing biodegradable VOCs or oxidizable inorganic compounds are vented through a biologically active material, and microorganisms break down the pollutants into harmless products such as carbon dioxide, water, mineral salts, and microbial biomass. Typical filter media are primarily mixtures based on compost, peat, or biologically active soils.

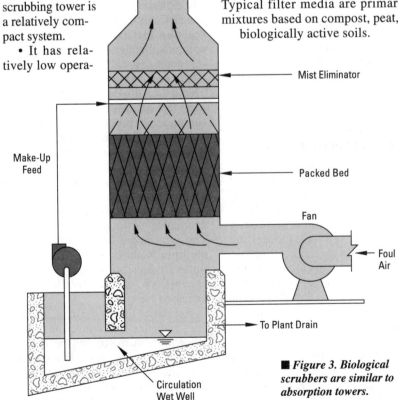

Make-Up Feed

Mist Eliminator

Packed Bed

Fan

Foul Air

To Plant Drain

Circulation Wet Well

■ *Figure 3. Biological scrubbers are similar to absorption towers.*

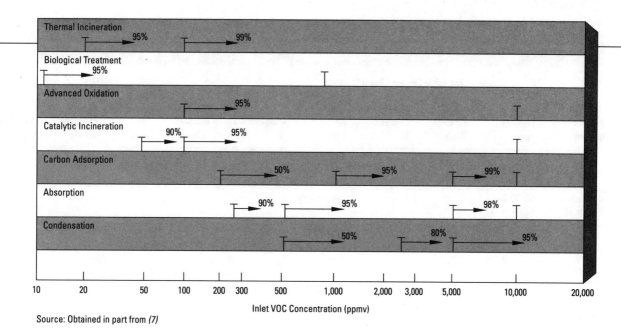

Source: Obtained in part from (7)

■ *Figure 4. Reduction efficiencies for add-on odor control technologies.*

Biofilters are effective for the treatment of streams with relatively low VOC concentrations. They are typically recommended for use with streams containing less than 1,000 ppm of easily biodegraded air pollutants. Alcohols, ethers, aldehydes, ketones, and several common monocyclic aromatics are generally well degraded. Highly chlorinated organics tend to degrade slowly, so biofiltration may not be suitable for streams containing them.

Biofilter beds may be either open single-bed units or multiple-layer systems. The former have the lowest capital costs, but the latter could be beneficial when space is limited. Enclosed systems can be used in areas where the climate may affect the operation of the filter system.

The advantages of biofiltration include:
• It achieves high removal efficiencies, between 90% and 99%.
• It has relatively low installation and operations/maintenance costs.
• It is environmentally safe in that it creates no secondary pollution.
• It is efficient in treating low-concentrations (less than 20 ppmv).
• It can handle variations in stream concentrations.

Its drawbacks:
• If channeling occurs in the bed, loss of treatment efficiency will be encountered.
• It may be difficult to maintain the pH of the filter media in the ideal range of 6–8.
• If proper moisture content in the bed is not maintained, for example by passing the air stream through a humidifier prior to introduction to the filter bed or by direct irrigation of the filter bed, microorganisms may be killed, thereby deactivating the filter material.
• Large amounts of space (approximately 1 ft^2 of filter area for every 0.15 cfm of air treated) may be required.

Further details on biofiltration are available in Reference (5).

Other technologies

Many other technologies exist for odor control. Some of these are proprietary, while others are combinations of technologies that have been discussed earlier. For example, some systems consist of a carbon bed to adsorb odorous compounds followed by an incinerator in which the concentrated desorbed stream is oxidized.

A fairly recent technology consists of an advanced oxidative treatment system utilizing ultraviolet light and the addition of an oxidant. The gas stream is then scrubbed using ozonated water. After scrubbing the air is passed through carbon beds for final polishing. Water used in the scrubbing step is continuously cleaned.

An interesting proprietary process that uses chemical and electrochemical technology to remove and destroy airborne toxic chemicals, odors, bacteria, and viruses has also recently been developed. The system is effective for the treatment of both low and high concentration streams — as low as 5 ppm up to as high as 5,000 ppm — but the destruction efficiency depends on the compounds being treated. The gas stream is fed through a scrubber employing a proprietary aqueous solution as the scrubbing fluid. The solution absorbs, dissolves, complexes, and partially or totally destroys the contaminants. The absorbent liquid is constantly regenerated in an electrolytic cell, where additional destruction of the pollutants takes place. By-products of treatment are water, carbon dioxide, nitrogen, oxygen, or some combination of these substances. [For more information, see (6).]

Finally, chemical counteractants are becoming increasingly popular at CPI facilities. Chemical counterac-

LITERATURE CITED

1. **Calvert, S., and H. M. Englund, eds.,** "Handbook of Air Pollution Technology," John Wiley, New York, p. 669 (1984).
2. **Shook, J. R.,** "Recover Heat from Flue Gas," *Chem. Eng. Progress,* **87**(6), pp. 49–54 (June 1991).
3. **Spivey, J. J.,** "Recovery of Volatile Organics from Small Industrial Sources," *Env. Progress,* **7**(1), p. 31 (1988).
4. **Cheremisinoff, P. N., et al.,** "Techniques for Industrial Odor Control," *Pollution. Engineering.,* **7**(10), pp. 24–31 (Oct. 1975).
5. **Bohn, H.,** "Consider Biofiltration for Decontaminating Gases," *Chem. Eng. Progress,* **88**(4), pp. 34–40 (April 1992).
6. **Genders, J. D., and N. L. Weinberg, eds.,** "Electrochemistry for a Cleaner Environment," published by Electrosynthesis Co., East Amherst, NY, pp. 323–330 (1992).
7. **Purcell, R. Y., and G. S. Shareef,** "Control Technologies for Hazardous Air Pollutants," U.S. Environmental Protection Agency, Research Triangle Park, NC, EPA-625/6-86-014, pp. 23–96 (1986).

Control Device	Emission Stream Characteristics					VOC Characteristics*			
	VOC or Organics Content[†], ppmv	Heat Content, Btu/scf	Moisture Content, %	Flow Rate, scfm	Temperature, °F	Molecular Weight, lb/lb-mole	Solubility	Vapor Pressure, mm Hg	Adsorptive Properties
Thermal Incinerator	>20; <25% of LEL[‡]			<100,000[§]					
Catalytic Incinerator	50–10,000; <25% of LEL[‡]			<100,000					
Flare		>300[‖]		<2 million[#]					
Boiler/ Process Heater**		>150[††]		Steady					
Carbon Adsorber	1,000–10,000; <25% of LEL[‡]		<50%[‡‡]	300– 100,000	100–200	45–130			Must be able to adsorb on and desorb from available adsorbents
Absorber	250–10,000			1,000– 100,000			Must be readily soluble in water or other solvents		
Condenser	>5,000			<2,000				>10 (at room temperature)	
Biological Treatment	<1,000		40–60%	<90,000[§§]	68–105				
Advanced Oxidation	100–10,000		>50%	<5,000	75–85				

Notes:
*Refers to the characteristics of the individual VOC if a single VOC is present and to that of the VOC mixture if a mixture of VOCs is present.
[†]Determined from VOC/hydrocarbon content.
[‡]For emission streams that are mixtures of air and VOC; in some cases the limit can be increased to 40% – 50% of LEL with proper monitoring and control.
[§]For packaged units; multiple-package or custom-made units can handle larger flows.
[‖]Based on EPA's guidelines for 98% destruction efficiency.

[#]Units are lb/h.
**Applicable if such a unit is already available on-site.
[††]Total heat content.
[‡‡]Relative humidity.
[§§]Less than 16 scfm/ft².

Source: Obtained in part from (7)

tants are generally mixtures of essential oils, such as eucalyptis or pine oil, and may contain other chemicals, depending on the intended application. These compounds are designed to deal with low-concentration (less than 30 ppmv) odorous streams. Chemical counteractants are generally either atomized directly into the vent stream or used in a wet scrubber system. The counteractants reduce odors by physically combining with the odorous molecules and removing the odorous properties of the molecules.

The advantages of chemical counteraction are

• It can treat low concentration odorous streams.

• The counteractants are generally products that are safe and easy to handle.

• It requires low capital and operation/maintenance costs.

• Its space requirements are low. Its disadvantages are:

• The odor causing molecules are not destroyed but rather their odorous properties are altered.

• Chemical counteractants should not be used on waste streams that contain toxic compounds, where the presence of the toxic substance may be disguised.

• In order to achieve effective odor control, a different counteractant may be required for each waste stream.

Selection

Figure 4 (7) shows the ranges of VOC reduction efficiencies that can be achieved over various VOC concentration ranges by the various technologies. Table 2 (7) shows the general air stream conditions required for specific technologies.

Once an emissions stream has been characterized in terms of flow rate, chemical content, and variability, Table 2 and Figure 4 can be used to determine the most applicable and most cost-effective technologies. More detailed cost evaluations between promising technologies can then be conducted on a case-specific basis. **CEP**

Recover VOCs via Adsorption on Activated Carbon

Adsorption is a proven and reliable pollution control technology that has the added benefit of recovering valuable materials for reuse.

M. John Ruhl,
Dedert Corp.

One of the most effective methods of controlling emissions of volatile organic compounds (VOCs) is also one of the most economical — adsorption, usually using activated carbon as the adsorbent. This process is cost-effective because it is typically able to recover many VOCs for reuse. A particularly common application of carbon adsorption for VOC control is solvent recovery.

In general, solvent recovery via carbon adsorption is a logical consideration for any industrial process exhausting sizable quantities of valuable solvent (subject, of course, to the solvent's suitability for adsorption by activated carbon, as discussed later). The most commonly recovered solvents include:

- toluene;
- heptane;
- hexane;
- carbon tetrachloride;
- acetone;
- ethyl acetate;
- methyl ethyl ketone (MEK);
- naphthalene; and
- methylene chloride.

Many other solvents are also suitable for recovery by carbon adsorption.

A recent *CEP* article *(1)* discussed the basics of carbon adsorption and application of the technology to water and wastewater treatment. This article takes an in-depth look at the use of activated carbon adsorption of the control of airborne VOC emissions and solvent recovery. It outlines how to decide if carbon adsorption is suitable for an application and then explains how to implement the technology.

How it works

VOC molecules are physically attracted and held to the surface of the carbon. Activated carbon is such a good adsorbent because of its large surface area, the result of its vast infrastructure of pores and micropores and micropores within micropores.

In a commercial activated-carbon solvent-recovery plant, solvent-laden air passes through a tank containing a bed of activated carbon. The solvent is adsorbed on the carbon surface and clean air is exhausted to atmosphere. When all of the available surfaces of the carbon pores are occupied, the carbon will not capture any additional solvent.

To recover the solvent for reuse, it must be released from the carbon surface. This is most commonly done by heating the carbon with steam. The hotter the carbon, the less solvent it can hold, so as the steam heats the carbon, solvent is released and flushed away by the steam. The mixture of steam and solvent is condensed by cooling and then separated, in the simplest case by gravity decanting. If the solvent is soluble in water, distillation is required instead of decanting. The carbon can then be reused as well.

The batch process of adsorption and desorption as described above can be made continuous by the use of multiple carbon beds so that one is off-line for desorption (steaming) while the others are adsorbing. Figure 1 illustrates a basic carbon adsorption process as applied commercially. Figure 2 is a photograph of an operating plant.

Is solvent recovery the right choice?

Early solvent-recovery plants were installed for economic reasons — the value of the reusable solvent justified the investment in very short order. This has continued to be a major driving force in installation of recovery plants. Today, as environmental regulations tighten and treatment of more and more dilute exhaust air streams becomes required, the economic decision becomes one of recovery vs. destruction methods of VOC control, such as incineration.

Recovery plants usually cost more than alternative methods of control. If the value of the recovered solvent is not adequate to offset the higher cost of the plant in a very few years, nonrecovery technologies are likely to be chosen.

The first step in evaluating solvent recovery is to determine the value of the solvent that will be recovered. If it is less than $50,000 per year, recovery probably should not even be considered unless there are other unusual reasons for employing it. One such reason might be the fact that recovery is considered a "politically correct" technology whereas incineration can carry with it (rightly or wrongly) the opposite connotation. The potential for arousing community opposition, resulting in delays or bad publicity, might be enough to overshadow a recovery project's marginal economics.

Once the potential recovery value is known, the best way to estimate the costs of various solvent recovery options is to call a few suppliers to get budget prices. For those suppliers to provide reasonably accurate estimates, they will need the following data:
- description of the process emitting the VOC;
- exhaust volume;
- temperature, pressure, and relative humidity of the exhaust air;

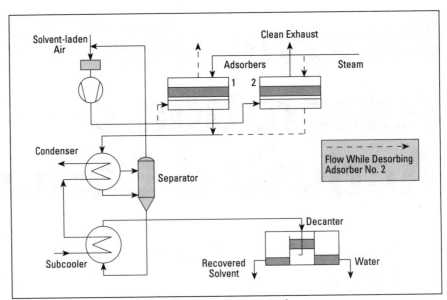

■ *Figure 1. A typical activated carbon solvent recovery plant.*

- composition of the VOCs and their concentrations;
- recovery (capture) efficiency required;
- quality of recovered solvent required;
- any contaminants present, such as dust, high-boiling components, resins, and so on;
- if future expansion of the plant is a possibility;
- whether cooling water is available, and if so, its temperature;
- whether steam is available, its pressure, and whether it is saturated; and

■ *Figure 2. A commercial solvent recovery plant.*

- steam costs (which will aid in determining steam recovery option economics).

Note that the data required are pretty much the same, regardless of the control technology being considered.

Other data would be required to complete the design of the plant. However, the above information should be adequate for a system supplier to estimate a budget price very closely.

Budget prices for various alternative technologies can then be compared using the company's economic evaluation criteria, such as payback period. At that point it may already be obvious which direction to take. If so, the next phase will be easier because the number and range of suppliers will be smaller. If the technology of choice is still not obvious, it would be wise to keep the options open throughout the formal bidding process.

Buying a solvent recovery plant

To obtain more precise costs for a solvent recovery plant, many of the smaller details of plant design must be evaluated. Many of these fine points can have a significant impact on cost and performance.

In addition to being as specific and accurate as possible on the data described above for budget inquiries, the following additional considerations must be made:

1. Pressure. What pressure needs to be maintained at the recovery plant inlet? Must the recovery plant provide suction to aid in exhausting?

2. Concentration control. Is monitoring and control of the exhaust solvent concentration with respect to the lower explosive limit (LEL), which is referred to simply as LEL control, desired? In general, VOC control systems that are designed to always operate below 25% of the LEL do not require continuous monitoring or control.

However, it is desirable to operate solvent recovery plants at higher inlet concentrations. This minimizes fan horsepower (by moving less air) and minimizes steam consumption (at higher inlet concentrations, the carbon has a higher capacity, thus decreasing the required frequency of steamings).

If continuous monitoring of the exhaust concentration is provided, continuous operation at up to 40% LEL, with automatic shutdown at 50% LEL, is typically allowable. Controlling at 40% LEL is done by limiting the flow of dilution air into the process, either by controlling the suction on the exhaust to induce more or less outside air (Figure 3), or by directly controlling the flow of dilution air with dampers.

If LEL control is feasible for a process, it should be included in the specification, at least as an option.

3. Fan control. If the process operates at constant rates, perhaps no control is required.

If the exhaust flow and VOC emission rates are variable, consideration should be given to control dampers or variable-speed drives. The dampers will cost less but will be a little less efficient overall. Variable-speed drives, commonly

variable-frequency drives with standard motors, will provide the most efficient operation and can usually be justified by the savings in operating cost. A few less horsepower per fan adds up to substantial annual electricity cost savings.

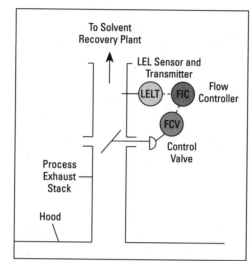

■ *Figure 3. LEL control allows operation at concentrations up to 40% of the lower explosive limit.*

4. Cooling water. Adsorption works best at low temperatures. Many solvent emitting operations are ovens or driers with exhaust air of 150°F–300°F and require cooling down to 100°F or lower. If adequate cooling water is not available, it will be necessary to add a cooling tower. This is an easy process addition, but it does add costs that must be included.

5. Particulates. Is there any possibility of dust or other particulates, which could clog the activated carbon should they reach it, getting into the adsorber? If so, it is prudent to install a filter upstream of the carbon beds. An estimate of the dust loading should be made to determine how frequently the filter medium will have to be cleaned or replaced. Many types of filters are available. The dust loading and type of dust are the major factors in making the selection.

6. Valve and damper actuators. If reliable, dry compressed air is available, air actuators are cost-effective. Most of the valves in a solvent

recovery plant are operated either fully open or fully closed, so simple double-acting cylinders or single-acting actuators with a spring return are adequate. If air quality or availability is a problem, electric actuators can be used, usually at a slightly higher cost. Limit switches should be included for positive indication of both open and closed positions of the valves.

7. Vessel and ductwork designs. These vary from light-gauge sheet metal enclosures to fabrications that meet American Society of Mechanical Engineers (ASME) codes. Be sure to understand the differences and how they pertain to the application at hand. If the recovery plant is the weak link in the operation, its being down will cause the rest of the plant to be down. Leaks in underdesigned flanges and manway covers can be dangerous and costly.

8. Adsorber designs. Adsorber designs also vary for process reasons. Manufacturers offer either upflow-air or downflow-air configurations, with steam flow typically in the opposite direction. Carbon bed support designs vary significantly, from exotic metal screens to specialty perforated plates.

Good arguments can be made for all design features, with each manufacturer claiming that its type is best. If a particular design feature is an important decision factor, be very sure that the reasons for favoring one or the other are pertinent to the particular situation. Otherwise a valid option might be excluded.

9. Type of carbon. Activated carbon is manufactured in many forms from a multitude of raw materials and within a wide price range. Vapor-phase solvent recovery plants tend to use pelletized, or in some cases granular, carbon manufactured from coal, wood, or coconut shells.

Specifications are readily available from manufacturers. One of the most useful specifications is the adsorption isotherm, in which adsorption capacity is plotted against inlet concentration of a specific component at a constant temperature. However, specification sheets only tell part of the story.

For a new application, pilot testing of the carbon, duplicating the process conditions as closely as possible, is recommended for verifying suitability and predicting commercial performance. The results of the pilot testing, however, will only be as good as the ability to match real process conditions in the pilot plant. The best type of pilot testing involves running the test on-site using a slip-stream from the actual exhaust requiring treatment. However, this is not always feasible.

10. Desorption method. The most commonly used method for removing adsorbed solvent from activated carbon is to directly heat the carbon with steam. The suitability of this method depends on several factors. If the solvent to be recovered is insoluble in water, the desorbate vapor can be condensed and the solvent decanted from the water. Some solvents and/or local regulations may make it necessary to further treat the decanted water by air or steam stripping. If air stripping is used, the stripper exhaust air is recycled to the front end of the solvent recovery plant, avoiding an emission to the atmosphere.

11. Fire suppression. Adsorption on carbon is an exothermic process, and the carbon and most solvents are combustible. Thus, it is not hard to imagine that a fire could happen at some time. Therefore, a fire suppression system is included in virtually all activated carbon solvent recovery systems.

The simplest and perhaps the most reliable (and safest) system senses high temperature and sounds

an alarm. An operator checks the system and upon verifying a probable fire, closes the adsorber valves, which causes the fire to suffocate itself due to lack of oxygen. The adsorber is then filled with water (from the bottom, not by sprays) to cool the carbon. Sprays must be avoided to prevent rapid overpressuring of the adsorber by steam generation as the water spray hits the hot carbon.

14. Future expansion. Most solvent recovery plants can be designed to accommodate future expansion. For example, the condensing/liquid-handling portion of the plant can be sized for ultimate capacity, thus requiring only additional adsorber trains to increase the plant capacity. If this is planned from the beginning, future expansion costs can be minimized.

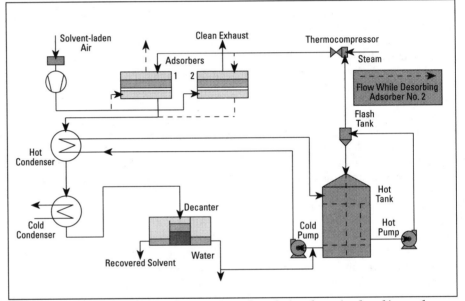

■ *Figure 4. A hot water storage tank can recover heat during the entire desorbing cycle.*

12. Materials of construction. Obviously, the solvent recovery system should be built from materials that will not corrode rapidly. Care should be taken to anticipate corrosion from trace byproducts that may result from steam regeneration or reaction with the carbon.

13. Control system. Modern recovery plants are typically controlled by programmable logic controllers (PLCs) with computer system interface capability either already built-in or optionally available. Operator interface can be through traditional push buttons and analog or digital displays or through a cathode ray tube (CRT).

Innovations

Adsorption on activated carbon is an old process that in itself has changed little over the years. However, a few innovations have been developed during the last decade that have increased operating efficiencies and expanded the suitability of adsorption to new applications.

Steam recovery. The largest operating expense in an activated carbon solvent recovery plant is the steam cost. A typical plant requires 3–5 lb of steam per pound of recovered solvent. Steam recovery techniques have made it possible to cut this in half.

One of the simplest steam recov-

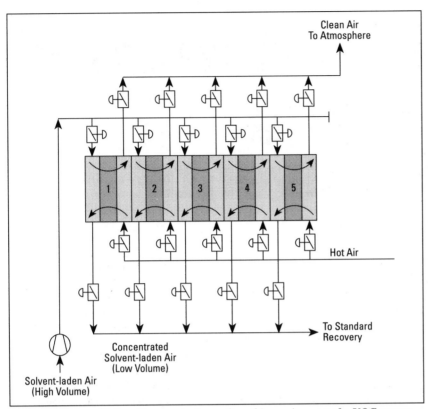

■ *Figure 5. A multiple-cell concentrator such as this can increase the VOC concentration in a stream to make recovery easier and more cost-effective.*

Literature Cited

1. **Stenzel, M. H.,** "Remove Organics by Activated Carbon Adsorption," *Chem. Eng. Progress,* **89**(4), pp. 36–43 (Apr. 1993).

ery schemes involves using a falling film evaporator in place of the condenser, thus using the latent heat of vaporization of the desorbate to generate steam from the aqueous condensate. This scheme can be used during about the second two-thirds of the desorbing cycle, after the carbon bed has heated up. The steam saving by this method is about one-third overall.

A more efficient method uses a hot water storage tank to store hot condensed water just below its boiling temperature, flashing it off under vacuum with a thermocompressor driven by plant steam, as shown in Figure 4. This method has the advantage of recovering heat during the entire desorbing cycle and is thus more efficient than the previously described method, resulting in about a 50% savings.

A combination of the two methods can achieve even higher efficiency. Both methods minimize cooling water requirements and wastewater effluent that might otherwise require treatment.

Alternative adsorbents. Activated carbon is not the only adsorbent medium available for solvent recovery. Hydrophobic zeolites, for example, can be manufactured with precise pore sizes, allowing selective adsorption of some compounds while excluding others. Since hydrophobic zeolite, as its name suggests, adsorbs little water, it can be used at higher humidities than carbon. Hydrophobic zeolite is also nonflammable, so it can be used for some compounds that might be a fire risk with activated carbon (cyclohexanone, for example). The cost of hydrophobic zeolite is still very high, so its use is economically limited to applications for which activated carbon is not well-suited.

Concentrators. Very low VOC concentrations in exhaust air historically have been expensive to treat.

For many low-concentration situations it is possible to use adsorption to increase the concentration to a level at which it is more feasible to clean up the air using a reasonably sized recovery plant. This is done by regenerating (desorbing) using a small volume of hot air. This smaller airflow then contains all of the VOCs at a higher concentration. The large volume of now clean exhaust air can be vented to atmosphere, and the smaller concentrated air stream goes to further treatment.

The process can be made continuous by desorbing in a small fraction of the carbon while adsorbing in the rest. This can be done by passing the solvent-laden air through a rotating wheel adsorbent bed while concurrently desorbing a pie-shaped segment of the wheel, or by using multiple cells of low-cost pelletized activated carbon (shown in Figure 5). The cells are individually valved to allow operation in either adsorption or desorption mode. **CEP**

M. J. RUHL is manager of the Solvent Recovery Div. of Dedert Corp., Olympia Falls, IL (708/747-7000; Fax: 708/755-8815). In addition to solvent recovery systems, he has experience with many types of VOC abatement technologies, including catalytic incineration and scrubbers, and also has an extensive background in fluidized-bed combustion of waste fuels. He has a BS in chemical engineering from Purdue Univ. and is a registered professional engineer in Illinois.

Regenerate Activated Carbon Using Organic Solvents

In applications where steam regeneration is not practical, solvent regeneration is a less-expensive alternative to thermal regeneration.

Hugh S. McLaughlin,
Waste Min Inc.

Adsorption by activated carbon is widely used to remove chemical species from waters and wastewaters *(1)* and volatile organic compounds (VOCs) from vapor streams *(2)*. In normal applications, the activated carbon gradually accumulates the chemical species removed from the liquid or vapor stream. This causes a progressive reduction in the carbon's ability to remove additional chemicals from the stream, and eventually the adsorption capacity is consumed and the carbon is spent. At this point, the carbon must either be replaced or be regenerated to restore its adsorptive capacity.

Many chemical engineers believe that the operating costs of carbon adsorption are usually high. This belief stems from the predominate industrial practice of using the carbon once and landfilling or thermally regenerating it when it is spent. However, other regeneration alternatives exist, and the engineer has to be able to recognize the appropriate applications for each technology.

Solvent regeneration is one such technology. It uses solvents to dissolve adsorbed material out of the pores of the activated carbon. Then the solvent is removed by steam (similar to steam regeneration of activated carbon).

The concept is not new — the original patent was issued in 1932 (U.S. Patent No. 1,866,417). But the technology is underappreciated by most chemical engineers. This article discusses where solvent regeneration fits into the overall scheme of activated carbon applications and regeneration options.

Cost-wise, solvent regeneration of activated carbon is only a few cents per pound more expensive than steam regeneration, and less than one-tenth as expensive as thermal regeneration. Because steam regeneration is appropriate for only certain industrial applications (discussed later), the choice is usually between solvent and thermal regeneration. Since solvent regeneration costs pennies per pound of carbon regenerated and thermal regeneration typically costs in the range of a dollar per pound, the savings can add up quickly.

What is activated carbon?

Activated carbon is a versatile and inexpensive adsorbent produced from a variety of carbon-containing raw materials, such as coal, wood, and coconut shells. Its unique properties relate to the carbon-based backbone, predominantly in the form of graphite planes or platelets, which are oxidized to varying degrees on the surface and dislocated in an irregular pattern to form adsorption sites that are referred to as "pores."

To understand how the spent adsorbent is regenerated, or reactivated, one should first be aware of how activated carbon is manufactured. Typically, the carbon-containing raw material is exposed to a series of three controlled temperatures, the final step also involving the use of steam. The first temperature regime, dehydration/devolatilization, occurs at 100–200°C and removes moisture and low-molecular-weight volatiles from the raw material. The next step is carbonization at 250–500°C, where the raw material decomposes into a graphitic backbone and emits the oxygen and hydrogen found in the raw material as gaseous compounds. The final step is activation at 800–1,000°C, where steam is reacted with the graphitic backbone to create the internal pores of the activated carbon.

The end result is a relatively inexpensive material with approximately 1,000 m²/g of internal surface area and capable of preferentially adsorbing organic compounds from both vapor and aqueous steams. Virgin activated carbon typically costs between $0.50 and $2.00 per pound, which corresponds to roughly 1¢ per acre of internal surface area.

Virgin activated carbon is manufactured with different pore-size distributions, depending on the intended application. Vapor-phase carbon has a substantially higher proportion of small pores, which are the principal active sites for vapor-phase adsorption (basically capillary condensation within the micropores). Liquid-phase carbon has a broader pore-size distribution to facilitate the diffusion of the adsorbates into the carbon and provide capacity for adsorption of larger molecules. Vapor-phase carbon has excellent capacity for small molecules in liquid-phase applications, but because the pores are small, more time is required to reach equilibrium, often rendering the process unacceptably slow. Thus, liquid-phase activated carbon is manufactured to facilitate faster adsorption, but it has a pore-size distribution that results in a lower overall capacity.

The adsorption and regeneration cycle

The carbon adsorption process can be viewed as two separate cycles — adsorption of contaminants from the waste stream onto the carbon, followed by desorption of the contaminants off the carbon.

In the adsorption mode, contaminants are transferred from a waste stream (either water or air) onto the surface of the carbon — specifically into the internal pores of the carbon and onto the surface area therein. For an application of activated carbon to be successful, the dynamics of the adsorption must be favorable.

The adsorption thermodynamics depend on many factors, but the most important are a compound's structure, as characterized by its molecular weight and functional groups, and its stability in the fluid phase (i.e., air or water). In general, compounds with higher molecular weights and lower volatilities tend to be better adsorbed onto the activated carbon in vapor applications. In liquid applications, the less soluble an organic compound is in water, the more likely it is to be adsorbed. The extent of adsorption is basically a balance between the contaminant's relative stability (or energy level) in the fluid phase and its stability adsorbed on the surface of the carbon.

The kinetics of adsorption are basically determined by the diffusivities of the contaminants in the carrier stream, which dictate how long it will take for the molecules to diffuse into the internal pores of the carbon. A typical vapor-phase adsorption process requires only one or two seconds of contact time, whereas a typical liquid-phase application may need 15 min or more of contact. The fundamental difference between these two applications is simply that the diffusion rates of the adsorbates in the liquid phase are slower than those in the vapor phase.

The desorption cycle consists of undoing what happened during adsorption — that is, removing the adsorbate from the carbon pores. The carbon regeneration process consists of altering the conditions of the carbon to effect the release of the contaminants.

Regeneration methods vary in the amount of adsorption capacity that is recovered, the fate of the adsorbates removed and unremoved, and the extent of deterioration of the residual activated carbon. Depending on the material adsorbed, the requirements to remove it and the impact of the removal conditions on the original activated carbon can vary widely.

For convenience, regeneration methods can be broken into two distinct types: methods that desorb the adsorbates into a liquid phase, and methods that heat the activated carbon and remove the adsorbates as vapors. The regeneration methods can be applied to any type of activated carbon, regardless of whether the loading cycle involved liquid-phase or vapor-phase adsorption.

Thermal regeneration

Thermal methods for regenerating spent carbon include steam regeneration, hot inert gas regeneration, thermal regeneration, and thermal reactivation.

Steam regeneration is routinely used for vapor-phase carbon. It uses direct-contact steam to strip the adsorbed organics away from the surface of the carbon, exploiting the phenomenon that the volatility of the adsorbed compounds increases with temperature. Thus, by increasing the temperature of the carbon, the adsorption equilibrium can be shifted to desorb some of the contaminants out of the carbon pores, resulting in the regeneration of some of the carbon's capacity.

Steam regeneration can successfully be utilized when the adsorbates are volatile organics with atmospheric boiling points up to about 120°C. It has the advantage that regeneration conditions are mild and the internal pore structure of the carbon is generally unaffected by the regeneration conditions. But it has the disadvantage that less volatile compounds, such as phenol and styrene, if present, are not effectively removed and can slowly poison the carbon.

Hot inert gas regeneration is similar to steam regeneration in that both regenerate carbon by heating it and volatizing contaminants from the surface of the carbon. The choice of steam vs. hot inert gas (often nitrogen) usually depends on how miscible the adsorbates are in water. Steam is generally used when the adsorbates are relatively immiscible in water, such as toluene and chlorinated solvents, as this allows the separation of the recovered organic from the condensed steam. If steam is used for water-miscible compounds, the recovery of the desorbing fluid (condensing the steam to water) requires distillation, which increases the equipment and energy requirements of the overall process.

Alternatively, hot inert gas can be used in cases where the adsorbates are miscible in water, as in the case of acetone and methanol. By using inert gas in these applications, the purification of the desorbing fluid is made easier. Instead of distillation, the vapor stream is simply cooled until the adsorbates condense out. Unfortunately, the substitution of hot inert gas for steam usually entails higher capital equipment and operating costs due to the requirement for low-temperature condensation.

Thermal regeneration involves heating the carbon to typically 300–500°C. Under those conditions, the adsorbates decompose into gaseous fractions, such as volatile hydrocarbons, water vapor, and oxides of carbon and nitrogen, and a carbonaceous residue or char.

The conditions encountered during carbonization regeneration do not appreciably deteriorate the original backbone of the activated carbon. Depending on the adsorbates, the relative fraction of char remaining in the carbon pores will vary. Whatever char remains consumes adsorption capacity and a slow poisoning of the carbon is observed when thermal regeneration alone is used.

Thermal reactivation relies on temperatures as high as 1,000°C, which oxidizes the adsorbed organics out of the pores of the carbon. Unfortunately, some of the internal pores of the carbon are also destroyed during thermal reactivation, leading to progressive deterioration and eventual loss of adsorption capacity and mechanical strength.

Another drawback of both thermal regeneration and thermal reactivation is economics. The capital equipment costs are high, and the high temperatures result in high energy and operating costs. In addition, some activated carbon and

Current solvent regeneration applications

Vapor-phase carbon

Styrene monomers generated during fiberglass manufacturing
VOCs from polymer compounding and polymer molding processes
Emissions from high-vacuum applications, such as distillation and drying
VOCs from point source abatement or fugitive emission capture
Organic contaminants in indoor air recirculation systems

Liquid-phase carbons

Textile and printing wastewater for recycle or industrial pretreatment
Phenol-containing wastewaters from coking and casting operations
Phenolic wastewaters from phenolic resin manufacturing and resin use
Rinsewaters from aqueous cleaning operations employing terpenes
Papermaking water containing soluble organics
Groundwater contaminated with hydrocarbons, such as gasoline and fuel oil

all adsorbed materials are destroyed; periodic replacement of the activated carbon and the inability to recover the adsorbates also increase the operating costs. Overall, there is nothing overtly wrong with the higher-temperature thermal options, except that more cost-effective and environmentally acceptable regeneration methods exist.

Liquid-phase desorption

Methods that desorb the adsorbates into a liquid phase include chemical regeneration, solvent regeneration, and supercritical extraction.

Chemical regeneration is encountered in liquid-phase carbon adsorption and typically involves altering the pH within the pores of the carbon. The pH change converts the organic adsorbate into an ionized form that has a less-favorable adsorption equilibrium or ion exchanges for an adsorbed inorganic cation. Typical applications include the removal of phenol from activated carbon by conversion to sodium phenate at high pH and the removal of soluble heavy metals by acid washing.

Solvent regeneration exchanges an organic solvent for the fluid in the pores of the carbon. The solvent dissolves the adsorbates and removes them from the carbon. The key to the process is choosing a solvent that effi-

ciently solubilizes the contaminants out of the carbon pores, yet can be subsequently removed by steaming. Of course, purification requirements for the desorbing fluids (in this case, the contaminated solvent and condensed steam) must also be considered. Factors affecting solvent choice will be discussed in greater detail later.

Supercritical extraction involves the use of a supercritical fluid (typically carbon dioxide, since its critical temperature [31.2°C] is near ambient) to extract adsorbates from the surface of the carbon. The technique is basically an exotic variation of solvent regeneration, with the desorbing solvent being a supercritical fluid. The difference lies in the recovery of the desorbing fluid, which is accomplished by reducing the pressure so the solvent reverts to subcritical conditions and, hence, the vapor phase. The adsorbates, being insoluble in the subcritical fluid state, are separated as a liquid from the subcritical vapor.

Unfortunately, even if supercritical extraction does prove the most successful regeneration method for a given application, economics usually eliminate it from further consideration. High capital and operating costs have kept this technology from being commercially viable.

How solvent regeneration works

Figure 1 depicts the three basic steps involved in the solvent regeneration of activated carbon. The first is the traditional adsorption step, where the contaminant is transferred from the liquid phase into the activated carbon. Step 1 is performed in standard adsorption equipment for both liquid- and vapor-phase adsorption applications.

Step 2 is the desorption step. The pores of the activated carbon are flooded with an organic solvent, which dissolves and removes the adsorbed material. The process is analogous to a solid/liquid leaching process, where the moving liquid phase extracts compounds from the stationary insoluble solid by dissolving them into the liquid phase and carrying them away.

At the end of the second step, the activated carbon has been stripped of the adsorbed material, but the pores of the activated carbon remain filled with solvent. The third step is the removal of this residual solvent by steaming, in a manner similar to steam regeneration of activated carbon.

Finally, the steamed carbon is flooded with water to cool the carbon and fill the internal pores with water. For vapor-phase applications, the regenerated and steamed carbon is cooled with dry air to remove the residual water from the internal pores.

Figure 1 shows the basic solvent regeneration cycle for a liquid-phase adsorption application being regenerated with a water-soluble solvent. The overall solvent regeneration cycle is slightly more complicated for the regeneration of vapor-phase carbon and when water-immiscible solvents are used.

One can view the solvent regeneration cycle as the steam regeneration method with a solvent extraction step inserted ahead of the steaming step. This additional step is an important distinction between solvent regeneration and steam regeneration. In steam regeneration, the steam removes the adsorbed material directly — by heating the carbon and vaporizing the adsorbed material. If the adsorbed material will not effectively vaporize at the temperatures provided by steaming, then efficient regeneration is not achieved. In contrast, in solvent regeneration, the adsorbed material is dissolved into the liquid phase by the solvent, avoiding the requirement of vaporizing the adsorbed material. Subsequently, the steam removes the solvent.

Martin and Ng (3–5) systematically studied the effects of a large number of variables on the desorption phenomenon. They observed a correlation between decreasing molecular weight of adsorbate and decreasing regeneration efficiency — the smaller the adsorbate, the further it could penetrate into the micropores of the carbon, thereby resisting displacement by the regenerant. Furthermore, in general, the smaller the organic regenerant, the further it could penetrate into the micropores of the carbon and displace the adsorbate. Thus, the success of the regeneration process is governed by the mechanism of physical displacement of the adsorbate molecule by the organic solvent molecule (4).

How well does it work?

The performance of solvent-regenerated activated carbon depends upon the adsorbates loaded on the spent carbon and the choice and quantity of solvent used to remove them. Typically there is an initial loss of adsorption capacity (relative to virgin activated carbon) for the first solvent regeneration cycle, followed by stable working capacity for subsequent adsorption/desorption cycles.

For example, an aqueous activated carbon adsorbing phenol and being regenerated with acetone had a reusable working capacity of roughly 75% of the adsorption capacity of virgin activated carbon.

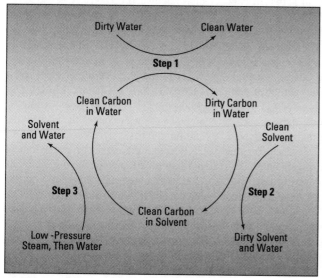

■ *Figure 1. Solvent regeneration cycle for liquid-phase applications.*

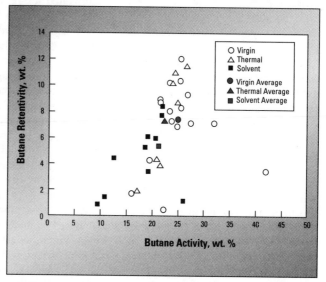

■ *Figure 2. Results of the butane working capacity assay for virgin, thermally reactivated, and solvent-reactivated carbons.*

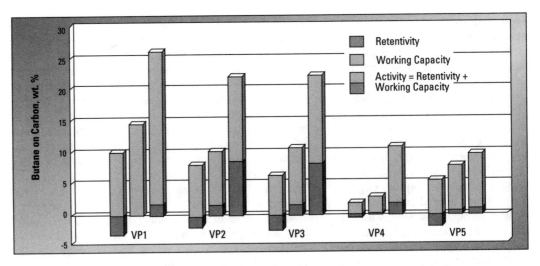

Figure 3. *Extent of solvent regeneration of five industrial vapor-phase activated carbons, for "as received" (left bars), dried (center bars), and solvent-regenerated (right bars) material.*

The carbon lost 25% of its capacity after the first loading cycle, after which the working capacity was stable for 50 cycles *(6)*.

The initial loss of capacity is attributed to adsorption into micropores with adsorption energies too high to be solvent-regenerated or pores that are structured such that the solvent cannot effectively access the adsorbates to dissolve them.

Testing activated carbon. Before we can characterize the effectiveness of solvent regeneration, we must establish how to measure the quality of an activated carbon, be it virgin or regenerated by any means. Adsorption quality of activated carbon is relatively difficult to assay because the extent of adsorption depends on many factors, including the types and concentrations of adsorbates. Various American Society for Testing and Materials (ASTM) tests have been used to measure carbon performance in adsorption applications.

One such test is ASTM D 5228, "Determination of the Butane Working Capacity of Activated Carbon." ASTM D 5228 first measures butane activity, which is the weight gain of a sample of activated carbon when equilibrated in a stream of pure butane vapor under controlled conditions. Then the test measures butane retentivity, which is the residual butane retention after purging the carbon for 40 min with dry air. The working capacity is defined as the activity minus the retentivity, expressed

as weight percent or weight per unit volume of carbon.

The value of the butane assay is that it gives insight into the micropore structure of the activated carbon. Butane activity is an indicator of micropore volume, and butane retentivity is an indicator of pore structure. These two measurements of micropore properties provide insight into how the activated carbon will perform in the real world. In general, higher butane activity and retentivity values correspond to superior performance in actual adsorption applications.

In Figure 2, the butane activity (x-axis) and the butane retentivity (y-axis) are plotted for 16 virgin activated carbons, seven thermally reactivated carbons, and ten solvent-regenerated carbons. Although the number of data points is too low for rigorous statistical accuracy, the average butane retentivities and average butane activities provide a good quick comparison: For the virgin activated carbons, these are 25.1 and 7.1; for the thermally reactivated carbons, 22.7 and 7.3; and for the seven highest solvent-regenerated activated carbons (three are excluded because they were poor candidates for solvent regeneration), 21.1 and 5.4.

Based on these limited butane assay data, virgin activated carbons are discernibly better than thermally reactivated and solvent regenerated carbons, but they are all more alike than dramatically different. However, the butane assay

is just one component of the overall evaluation of a regeneration option — the other big consideration is cost (discussed later).

Solvent regeneration of industrially loaded carbons. Figure 3 shows the butane assay results for five solvent-regenerated vapor-phase activated carbons, comparing "as received," dried, and solvent-regenerated samples. The extent of regeneration depends on several factors, including the adsorbates present and the degree of loading.

Note that the "as received" spent carbon can have a negative butane retentivity, which is common with vapor-phase activated carbons. A negative retentivity means that the sample weighed less after being loaded with butane and purged with dry air. This anomaly is due to the presence of condensed water vapor in the pores of the original spent carbon. During the butane loading and desorption, dry carrier gases evaporate some of the water out of the carbon, resulting in a net weight loss.

Because the extent of water loss is not controlled, neither the resulting butane retentivity measurement nor the butane activity measurement for the "as received" carbon is particularly meaningful. All one can conclude is that the carbon is wet and needs to be properly dried before a meaningful butane assay can be performed.

Upon solvent regeneration, three of the carbons (VP1, VP2, and VP3) re-

■ Figure 4. Extent of solvent regeneration of five industrial liquid-phase activated carbons, for dried (left bars) and solvent-regenerated (right bars) material.

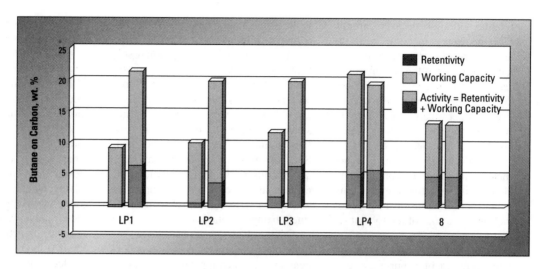

turned to near-virgin properties. These carbons were used to treat off-gases from a thermal desorption unit used to devolatilize contaminated soils, from a polymer compounding operation, and from a semiconductor manufacturing facility. VP1 exhibited low butane retentivity because the original virgin material was a wood-based carbon, which has a low butane retentivity even as initially manufactured.

The remaining two carbons (VP4 and VP5) responded less favorably to solvent regeneration. VP4 (from a polymer compounding application) was extremely heavily loaded and had been repeatedly thermally reactivated, and this degree of regeneration measured for VP4 is probably all the carbon had to offer. VP5 (which was used in polymer manufacturing) probably did not solvent-regenerate effectively due to the adsorbates present — unreacted prepolymer was adsorbed into the pores of the carbon, where it reacted to form a crosslinked polymer that was insoluble in the regenerating solvent (and thus not removable).

Figure 4 compares five solvent-regenerated liquid-phase activated carbons. Because these carbons were used in aqueous applications, they had to be dried prior to the base line butane assay. This can partially regenerate the carbon by stripping volatile compounds, but there is no realistic alternative for liquid-phase activated carbon.

The first three spent carbons (LP1, LP2, and LP3) returned to near-virgin properties. These carbons came from a terpene cleaner rinse-water recycle system and two groundwater treatment applications (gasoline and diesel fuel cleanup), respectively.

The performance of LP4 deteriorated upon solvent regeneration. The original spent carbon was lightly loaded, if at all. The decrease in butane activity upon solvent regeneration is attributed to adsorption of solvent into the very-high-adsorption-energy micropores (which do not normally solvent-regenerate well); in the absence of adsorbates to irreversibly consume these micropores, the solvent takes up permanent residence. This illustrates one reason why activated carbon should not be regenerated until it is significantly loaded or spent — which is sound advice for any regeneration method.

When LP5 (used to remove ppb-levels of chlorinated hydrocarbons from groundwater for several years) was solvent-regenerated using acetone, it exhibited only "middle of the road" performance. Such applications often slowly load humic acids and precipitate inorganic salts into the carbon's pore structure; acetone is not effective in removing humic acids, nor would it be expected to dissolve inorganic salts out of the carbon pore structure (5). However, other chemical regeneration methods are typically effective for such applications.

Factors influencing the choice of solvent

Many solvents can be used to regenerate activated carbon (3–5, 7), although solvent recovery and reuse concerns make solvents with certain properties better choices than others. The desirable properties of a successful solvent are:

• good solubility for the compounds adsorbed on the activated carbon;

• moderate adsorption energy onto activated carbon;

• the capability of being removed from activated carbon by low-pressure steam;

• easy recovery by decanting or distillation from water and the adsorbates; and

• affordable cost and environmental acceptability.

The first two properties ensure effective removal of the adsorbates out of the pores of the spent carbon. The third quality allows the solvent to be removed from the carbon at the end of the regeneration cycle. The fourth property addresses the recovery of the solvent for recycle.

These attributes point to acetone, methanol, toluene, and mixed aliphatic hydrocarbons (including gasoline cuts) as the preferred choices for solvent regeneration.

In general, methanol is not as good a solvent as acetone due to lower energy of adsorption onto activated carbon, but it is less expensive. Both solvents should be considered for most solvent regeneration applications.

Toluene and aliphatic hydrocarbons may be necessary in some cases to achieve complete extraction of the adsorbed material from the carbon pores. Toluene and gasoline cuts have the advantage that they are relatively immiscible with water, making them easy to recover for recycle. However, water-immiscible solvents have been less effective in solvent regeneration of aqueous-phase activated carbon due to problems with effectively exchanging the water for solvent inside the pores of the carbon at the start of the solvent regeneration process.

A similar problem exists for the solvent regeneration of vapor-phase carbons since the pores contain air. Simply flooding the carbon with solvent results in the carbon bed being "air bound." This is not only dangerous from a safety standpoint, but it also significantly slows the rate of flooding of the carbon pores with solvent.

Recent improvements (patent pending) in solvent regeneration technology have circumvented these solvent exchange obstacles, and the current state-of-the-art allows flexibility in choosing water-miscible and water-immiscible solvents for the regeneration of both aqueous-phase and vapor-phase activated carbons. These improvements should greatly expand the number of viable industrial applications for solvent regeneration.

A final consideration in choosing a solvent is the impact of trace residual solvent on regenerated carbon. Depending on the type of activated carbon, the organics previously adsorbed on the carbon, the choice of solvent, and the conditions used for steaming at the end of the regeneration cycle, some residual solvent will remain on the regenerated carbon. In general, the amount of residual solvent is negligible and does not represent a significant source when the activated carbon is returned to adsorption service. The steaming conditions at the end of the solvent regeneration cycle are tailored to produce an acceptable level of residual solvent. In air-pollution applications, for example, the level of solvent

in the effluent air is typically in the low parts-per-million range.

The expected level of residual solvent bleed after solvent regeneration varies in liquid-phase adsorption. Some solvents, such as methanol and ethanol, are removed to very low levels during steaming and have very poor adsorption equilibria on activated carbon. Such solvents rapidly desorb any trace solvent during the first bed volume of effluent after the carbon is returned to adsorption service, with subsequent effluent being solvent-free.

Other solvents, such as toluene, remain on the regenerated carbon at percent levels, which typically results in a consistent bleed of several parts per million. If such residual levels are not acceptable (although for many applications they are), then the residual toluene is removed by modifying the solvent regeneration process to include a methanol extraction step prior to steaming.

Acetone tends to bleed for the first 10 to 100 bed volumes of effluent at levels of about 10 ppm after the carbon is returned to adsorption service. The residual acetone level can be reduced by modifying the solvent regeneration process, but such low levels of residual acetone in the treated effluent are typically not objectionable.

In general, acetone and toluene emerge as the preferred solvents for new installations, depending on the adsorbates being removed from the carbon. Toluene is often considered because it can solubilize many unusual adsorbates, such as low-molecular-weight polymers. If an industrial site currently has the capability and capacity to recover a particular regeneration solvent by distillation, then that solvent should also be considered, due to the reduced capital investment for the solvent recovery facility.

The full-scale process

One of the attractions of the solvent regeneration technology is that it can be adapted to virtually any size industrial application. The major factors influencing the design are:

• regeneration in the adsorber vs. in a separate solvent regenerator; and
• new construction vs. retrofit to an existing industrial facility.

The overall solvent regeneration process consists of three separate operations, as shown in Figure 1.

The adsorption step and the regeneration process can be performed in the same vessel, or the carbon can be moved from the adsorber vessel to a solvent regeneration vessel. The integrated adsorber/desorber design is utilized only when relatively large amounts of activated carbon (at least 500,000 lb/yr) are being used at a single location, making the construction of dedicated solvent regeneration facilities economical.

Using a separate regeneration vessel has the advantage that one desorber can regenerate the carbon from a number of adsorbers throughout an industrial complex or even from many neighboring sites. In addition, since the solvents are flammable, a solvent regeneration facility has to be explosionproof and designed to the appropriate industrial standards for storing and handling bulk flammable liquids. With separate adsorption and regeneration facilities, the extra expense of explosionproof construction can be restricted to the regeneration facility alone and existing adsorption equipment can be utilized for the adsorption step.

Regardless of the configuration of the adsorbers and desorbers, every solvent regeneration facility will have a separate solvent recovery facility. Since the key to the process is the solubility of the adsorbate in the solvent, the separation of the adsorbates from the solvent involves fractionating a miscible solution. The technique of choice is multistage distillation. Most applications will use a small batch distillation column, since the total amounts of solvent to be recovered are too small for a continuous fractionation operation. As with the solvent desorber, the construction of the solvent recovery facility must be explosionproof.

Solvent regeneration can be integrated into an existing industrial operation. If steam and cooling water are available onsite, the steaming of the carbon at the end of the regeneration cycle and the solvent recovery operations simply draw from the available site utilities. In some instances, an appropriate solvent is already being used onsite and solvent recovery capabilities are available. Such systems end up being only sparingly more complicated than a carbon adsorption system with steam regeneration.

Full-scale industrial applications do exist, but they are typically integrated into chemical complexes and hidden from the public eye (such as Figure 5).

When to consider solvent regeneration

In terms of severity of regeneration conditions, solvent regeneration occupies the gap between steam regeneration and the higher-temperature thermal methods, thermal regeneration and thermal reactivation. Additionally, the capital investment and cost of operation increase as one goes from steam to solvent to thermal regeneration.

The first question, then, is: "Does steam regeneration work efficiently and provide stable working capacity?" If so, steam regeneration is the clear choice, since it is the least expensive technology to install and to operate.

In general, steam regeneration is efficient and stable basically for only vapor-phase applications where a relatively volatile, very clean organic vapor is adsorbed, such as acetone, methanol, or toluene. For those applications, the VOCs are uncontaminated and easily removed from the carbon by steaming.

As soon as the solvent becomes less volatile (atmospheric boiling points above about 120°C), as in the case of xylene or styrene, the amount of steam required to desorb the solvent increases and the economics of the steam regeneration process deteriorate. In addition, if less-volatile compounds are also present (for example, due to high-temperature or vacuum conditions, the presence of fumes and aerosols, or polymerization of monomers within the

carbon pores), then the carbon will slowly lose capacity and steam regeneration will prove to be a poor choice.

When steam or hot inert gas regeneration methods are not suitable, then the next logical alternatives are solvent regeneration, thermal regeneration, and thermal reactivation.

For both thermal methods, the carbon must be removed from the adsorber and regenerated in a large-scale operation similar to the original carbon

■ *Figure 5. Full-scale solvent regeneration activated carbon process.*

activation facility. The cost of transportation, energy to attain the elevated temperatures, and makeup carbon requirements result in high costs — typically $0.50 to $1.00 per pound of carbon regenerated.

In contrast, solvent regeneration is significantly more economical. The major energy inputs are the steam to remove the solvent at the end of the solvent desorption cycle and the energy to recover the solvent. Typically, these requirements amount to 1,000–2,000 Btu per pound of carbon regenerated. In addition, the energy is consumed as low-pressure steam, which is typically the cheapest form of energy at an industrial site. Additional costs for the solvent regeneration process are cooling water

and pumping costs, which are minor compared to the steam input. The total direct operating costs are $0.01 to $0.05 per pound of carbon regenerated.

Solvent regeneration will provide better regeneration of the activated carbon than the higher-temperature thermal methods in certain applications. A good example is the thermal reactivation of vapor-phase carbon, where elevated temperatures and oxidizing conditions gradually destroy the smaller micropores of the carbon. The carbon loses working capacity and becomes useless after as few as five regeneration cycles. This phenomenon does not occur during solvent regeneration.

Solvent regeneration is most appropriate for dissolving and desorbing organic chemicals or any other compounds readily soluble in the solvent used. Applications where trace heavy metals are adsorbed or the carbon is used for oxidation (typically removing residual chlorine), solvent regeneration (as described here) is not effective.

Of course, if steam regeneration is not appropriate and if the adsorbed contaminants will not dissolve in a solvent such as methanol, acetone, toluene, or gasoline, then thermal regeneration must be used.

Evaluating potential applications

The first step in the technology evaluation is to determine whether activated carbon will work in the particular adsorption application — that is, will activated carbon remove the contaminants from the liquid or vapor stream to an acceptable level? This can be determined through discussions with carbon suppliers, a literature search, or pilot testing.

The second step is evaluating whether the carbon can be regenerated by solvent desorption. The critical issue is whether the adsorbed compounds will dissolve in an organic solvent such as methanol, acetone, toluene, or gasoline. A second concern is whether there are any other adsorbed compounds that

Literature Cited

1. **Stenzel, M. H.,** "Remove Organics by Activated Carbon Adsorption," *Chem. Eng. Progress,* **89**(4), pp. 36–43 (Apr. 1993).
2. **Ruhl, M. J.,** "Recover VOCs via Adsorption on Activated Carbon," *Chem. Eng. Progress,* **89**(7), pp. 37–41 (July. 1993).
3. **Martin, R. J., and W. J. Ng,** "Chemical Regeneration of Exhausted Activated Carbon — I," *Water Research,* **18**(1), pp. 59-73 (Jan. 1984).
4. **Martin, R. J., and W. J. Ng,** "Chemical Regeneration of Exhausted Activated Carbon — II," *Water Research,* **19**(12), pp. 1527–1535 (Dec. 1985).
5. **Martin, R. J., and W. J. Ng,** "Repeated Exhaustion and Chemical Regeneration of Activated Carbon," *Water Research,* **21**(8), pp. 961–965 (Aug. 1987).
6. **Sutikno, T., and K. J. Himmelstein,** "Desorption of Phenol from Activated Carbon by Solvent Regeneration," *Ind. Eng. Chem. Fundam.,* **22**(4), pp. 420–425 (Nov. 1983).
7. **Cooney, D. O.,** *et al.,* "Solvent Regeneration of Activated Carbon," *Water Research,* **17**(4), pp. 403–410 (Apr. 1983).

will not dissolve and whether these nondissolving compounds can be removed prior to carbon regeneration, such as by filtration or precipitation, or removed by other chemical treatment methods, such as acid extraction.

If the answers to these questions are positive, then chances are good that solvent-regenerated activated carbon will perform successfully in the application. The next step is to demonstrate the adsorption and desorption cycles on the actual streams and contaminants.

The easiest way to demonstrate the adsorption cycle is to use a 55-gal adsorber designed for liquid-phase or vapor-phase service. Such units are large enough to accurately mimic the performance of larger-scale adsorbers, and (in liquid-phase applications) they yield enough treated water to allow for the evaluation of water recycle and reuse options. Yet they

are not excessively expensive, and many commercial suppliers offer this type of equipment.

A liter sample of the uncontaminated virgin activated carbon from the adsorber should be retained for later comparison with the solvent-regenerated carbon.

By the time the evaluation of the adsorption cycle is complete, the adsorbers have typically been on-line long enough to yield some saturated or spent activated carbon for evaluation of the desorption cycle. The most-saturated carbon is located at the inlet to the carbon bed, which may be at either the bottom or the top of the 55-gal canister. It is easiest to operate the pilot adsorption cycle downflow to ensure that the spent carbon is on the top of the carbon bed where it can be easily accessed.

A sample of spent carbon is obtained by opening the 55-gal canister and removing the top layer of carbon. Note whether any significant particulate deposits have developed on the surface of the carbon, since these will have to be filtered out in a full-scale application.

The solvent desorption cycle can be evaluated simply as follows. First prepare a slurry of the spent carbon with the solvent (liquid-phase carbon should first be drained but not dried). Allow the sample to equilibrate for at least 1 h, then filter the sample and analyze the filtrate by gas or liquid chromatography. If the contaminants removed by the activated carbon in the adsorption step are present in the filtrate, then the solvent regeneration step is working.

A more detailed test can provide additional insight into the extent of solvent regeneration taking place. First the spent carbon sample is extracted in a specialized laboratory apparatus called a soxhlet extractor, then the residual solvent is extracted with boiling water and the carbon is dried. The regenerated carbon is then tested by the butane working capacity assay and the recovered micropore structure compared to the virgin activated carbon sample taken prior to the adsorption studies.

The next step is to solvent-regenerate several pounds of spent carbon using excess solvent and excess hot water, then return the regenerated carbon to a small adsorber to confirm the performance of the solvent-regenerated activated carbon in the subsequent adsorption cycle. Multiple adsorption/desorption cycles are usually necessary to confirm that a stable adsorption working capacity has been reached for the solvent-regenerated carbon and that the effluent quality is acceptable.

Plant-wide integration. When evaluating solvent regeneration, it is important to consider *all* the potential applications of solvent-regenerated activated carbon for the entire manufacturing facility, from process water treatment for recycle to VOC abatement. Because one solvent regeneration facility can regenerate both liquid-phase and vapor-phase activated carbons using the same equipment, this can make carbon adsorption with solvent regeneration a versatile and cost-effective technology for addressing many of a plant's waste minimization and pollution abatement requirements. **CEP**

H. S. McLAUGHLIN, Ph.D., P.E., is vice president of Waste Min Inc., Groton, MA (508/448-6066; Fax: 508/448-6414), an engineering firm that specializes in waste minimization and process engineering. He has 16 years of experience with the solvent regeneration of activated carbon, performing fundamental research, process development, and commissioning of a full-scale industrial facility. He has recently applied for a patent titled "Process Improvements for Solvent Regeneration of Activated Carbon," and is currently licensing the technology and consulting on specific applications. He holds a BS in chemistry from Harvey Mudd College, an MS in chemical engineering from the Univ. of Southern California, and a PhD in chemical engineering from Rensselaer Polytechnic Institute. He is a registered professional engineer in Massachusetts, New York, and Puerto Rico.

Consider Biofiltration for Decontaminating Gases

Soil and compost beds effectively and inexpensively dispose of gaseous pollutants. Here's what you need to know to specify this as a pollution control option.

Hinrich Bohn,
Univ. of Arizona

Perhaps you have come across such terms as biofiltration, bioremediation, bioreclamation, activated sludge, trickle filters, fixed-film biological treatment, landfills, and landfarming of petroleum sludges. These all represent variations of the same process, wherein microorganisms in a moist, oxygen-rich environment oxidize organic compounds to carbon dioxide and water. The organics can be solid, liquid, or gaseous. The process can be carried out in water, packed columns, or beds of porous synthetic materials, compost, or soil. This article focuses on one type of biological treatment that chemical engineers may be less familiar with—the use of soil beds, or biofiltration, to treat contaminated gases.

Biofiltration is the removal and oxidation of organic gases (volatile organic compounds, or VOCs) from contaminated air by beds of compost or soil (1,2,3). Many chemical process industries (CPI) operations could utilize biofiltration to treat waste gases, just as they use biological treatment for liquid and solid wastes. Table 1 summarizes a few CPI applications.

Biofilters are beds of soil or compost, under which lies a distribution system of perforated pipe. Soil and compost are extensive networks of fine pores and have large surface areas. As contaminated air flows upward through the bed, VOCs sorb onto the organic surfaces of the soil or compost. The sorbed gases are then oxidized by microorganisms to CO_2. Biofilter beds also

■ *This biofilter bed controls odors from the treatment of wastewater from a brewery.*

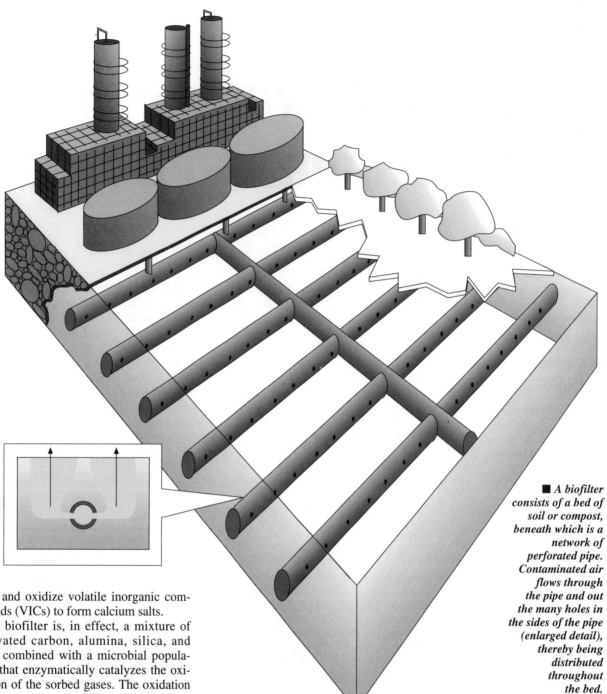

■ *A biofilter consists of a bed of soil or compost, beneath which is a network of perforated pipe. Contaminated air flows through the pipe and out the many holes in the sides of the pipe (enlarged detail), thereby being distributed throughout the bed.*

sorb and oxidize volatile inorganic compounds (VICs) to form calcium salts.

A biofilter is, in effect, a mixture of activated carbon, alumina, silica, and lime combined with a microbial population that enzymatically catalyzes the oxidation of the sorbed gases. The oxidation requires no fuel or chemicals. The sorption capacity of soil and compost beds for gases is relatively low, but the oxidation regenerates the sorption capacity. Thus, the beds are effective because of continuous VOC oxidation, rather than because of their sorption capacity.

The removal and oxidation rates depend on the biodegradability and reactivity of the gases. The half-lives of contaminants range from minutes for the most reactive gases to months for the least biodegradable. A general classification system is shown in Table 2. This list is much the same for all of the bioreclamation processes because they all rely on the same mechanism.

Why use biofiltration?

Biofiltration of gases has gained only slow acceptance, even though soil treatment of natural wastes, decomposition of solid and liquid organics in landfills, and biological wastewater treatment have long been accepted. However, biofiltration offers a number of advantages.

Gases are inherently more biodegradable than solids and liquids because they are molecularly dispersed. Biofiltration does not contaminate the soil because the loading rates are very low, the gases degrade rapidly, and oxygen is in excess. In contrast, soil contamination has resulted from adding liquid and solid organics at high loading rates without providing for microbial degradation.

Biofiltration of contaminated air is considered new and untested technology in North America. This is partly because incineration, water and chemical scrubbing, and activated carbon adsorption are entrenched as air-pollution control methods.

Actually, biofiltration is not "new." Rather, it is an adaptation of the process by which the atmosphere is cleaned naturally. VOCs exist in the atmosphere until plants and soils absorb and degrade them. The process has been going on for more than one billion years. However, it is inefficient due to the limited contact of soils and plants with the atmosphere and because the reactions are relatively slow. Biofiltration provides maximal contact and allows sufficient time for VOCs to react with soils.

Another reason for biofiltration's slow acceptance is the reluctance to put pollutants into the ground. In the case of gas treatment, however, the low loading rates, the rapid oxidation, the excess oxygen, and the upward mass flow prevent any possibility of soil or groundwater pollution. For example, a continuous input of 100 ppm of a two-carbon VOC to a soil bed corresponds to about 1 kg/m^2 of carbon annually. This compares with the approximately 10 kg/m^2 of natural plant carbon that is oxidized annually in soils and the 1–10 kg/m^2 per year of carbon that is safely added to soils as petroleum sludge in landfarming operations and as solid wastes in landfills.

A third reason may simply be the cynical perception that "if it's cheaper, it can't be any good." The low cost of biofiltration is due to its use of natural rather than synthetic sorbents and microbial rather than thermal or chemical oxidation.

Biofiltration has had more industrial success in Europe and Japan. Over 500 biofilters are operating in

Table 1. Biofiltration has a variety of CPI applications.

Company	Location	Application
S.C. Johnson & Son, Inc.	Racine, WI	Propane and butane removal from room air; 90% removal efficiency, about 3,000 cfm
Monsanto Chemical Co.	Springfield, MA	Ethanol and butyraldehyde removal from dryer air; 99% removal, 28,000 cfm; Styrene removal from production gases
Dow Chemical Co.	Midland, MI	Chemical process gases
Hoechst Celanese Corp.	Coventry, RI	Process gases
Sandoz	Basel, Switzerland	Chemical process gases
Esso of Canada	Sarnia, Ont.	Hydrocarbon vapors from fuel storage tanks (proposed)
Mobil Chemical Co.	Canandaigua, NY	Pentane from polystyrene–foam molding (proposed)
Upjohn Co.	Kalamazoo, MI	Pharmaceutical production odors; 60,000 cfm (proposed)

Table 2. Gases can be classified according to their degradability.

Rapidly Degradable VOCs	Rapidly Reactive VICs	Slowly Degradable VOCs	Very Slowly Degradable VOCs
Alcohols	H_2S	Hydrocarbons*	Halogenated
Aldehydes	NO_x	Phenols	Hydrocarbons†
Ketones	(but not N_2O)	Methylene chloride	Polyaromatic
Ethers	SO_2		Hydrocarbons
Esters	HCl		CS_2
Organic Acids	NH_3		
Amines	PH_3		
Thiols	SiH_4		
Other molecules containing O, N, or S functional groups	HF		

*Aliphatics degrade faster than aromatics such as xylene, toluene, benzene, and styrene

†Such as trichloroethylene, trichloroethane, carbon tetrachloride, and pentachlorophenol

Europe and many have been installed in Japan (3).

Still, in North America, more than 21 million biofilters are operating continuously—but are unrecognized. Biofilters are the part of household and industrial wastewater leach fields (septic systems) that completely and reliably remove the VOCs emitted by the wastewater.

Comparing air-pollution control methods

Organic compounds in air, water, and soils are removed by oxidation. The oxidation rates of VOCs from industrial waste air vary from fractions of a second in incinerators to seconds in chemical scrubbers to minutes or days in biofilters. The rapid reactions require fuel, chemicals, and maintenance. The slower, inexpensive microbial reactions in biofilters require large reactor volumes and bed areas.

Incineration at high temperatures is fast and 99+% complete. However, it requires large amounts of fuel and produces NO_x. Catalysis and heat recovery reduce fuel costs, but at the expense of higher installation and maintenance costs.

Chemical scrubbing with chlorine, ozone, hypochlorite, or permanganate is rapid, although ozone may require a reaction time of up to 30 seconds. Chemical scrubbing removes 95% of rapidly biodegradable or reactive VOCs and VICs, but it is ineffective for hydrocarbons and other slowly reactive compounds. In addition, the chemical oxidants are corrosive and hazardous. Furthermore, an expense is incurred for purchasing the scrubbing agents.

Water washing and activated carbon adsorption transfer the VOCs to water or carbon, and the VOCs must be disposed of later. Water washing is rarely effective for VOC treatment because it removes only water-soluble gases. Water usage is high and sewerage costs are considerable. Adding acids or bases is ineffective for VOCs, but may enhance the removal of VIC.

Activated carbon adsorbs 90–95% of incoming VOCs if the carbon is regenerated or replaced regularly. The VOCs are unchanged by the sorption and in some cases can be recovered economically. Removal effectiveness declines as the carbon becomes saturated, and any water vapor present interferes with VOC sorption. Also, under the dry conditions of gas sorption, activated carbon is flammable.

Biofiltration combines the sorption of activated carbon, the washing effect of water, and oxidation. It removes 80% to over 99% of a very wide range of VOCs and VICs depending on their chemical and biological reactivity.

Table 3 compares overall costs of air-pollution control by several methods. Biofiltration has a considerable advantage. Although the actual costs vary with each case, a situation in which biofiltration is more expensive than the other methods would be rare.

Biofiltration requires a longer reaction time, so the volume of the

Biofiltration is not new— it is an adaption of a natural atmosphere-cleaning process.

soil or compost bed must be relatively large compared to the space required for the other methods. Some industries in urban areas may have insufficient area for single-layer beds. Multilayer biofilter beds can overcome this problem, although the installation and maintenance costs will be higher.

Mechanisms of biofiltration

Soil and compost are porous. The pores are organic or organically coated to sorb VOCs from the air flowing through the pores. Soils have porosities of 40–50% and surface areas ranging from 1 to 100 m^2/g and contain 1–5% organic humic material as coating on the inorganic surfaces. Compost has a 50–80% porosity, similar surface area, and contains 50–80% (by mass) partially humified organic matter.

These porosities and surface areas are similar to those of activated carbon and other synthetic sorbents. The major distinction between natural and synthetic sorbents is that soil and compost have in addition a microbial population of more than 1 billion microbes (bacteria and fungi) per gram, which oxidize organic compounds to CO_2. Another distinction

Table 3. Biofiltration compares favorably with other air-pollution control technologies from a cost standpoint.

	Total cost per 10^6 ft³ of air*
Incineration	$130
Chlorine	$60
Ozone	$60
Activated Carbon (with regeneration)	$20
Biofiltration	$8

*Costs obtained from Reference (8) and converted/updated to 1991 U.S. dollars

	Operating costs	
	Fuel/chemical consumption	Power
Incinerator[†]	$15 per cfm	Negligible
Wet Chemical Scrubbing	Up to $8 per cfm	1 W per cfm
Soil Beds	0	0.6 W per cfm

[†]Costs obtained from Reference (9)

is that moisture in the waste gas stream is beneficial for the microbial oxidation on which the removal efficiency of biofilters depends; moisture in synthetic sorbents, on the other hand, reduces their VOC sorption capacity and removal efficiency. A third distinction is that the costs for natural sorbents are on the order of dollars per ton, as opposed to dollars per pound for synthetic sorbents.

Gases in air flowing through soil pores sorb onto or, as in gas chromatography, partition out on the pore surfaces so that VOCs remain in the soil much longer than does the carrier air. Soil/gas partition coefficients indicate the relative strength of retention. The coefficients increase rapidly with VOC molecular weight and even more with the number of oxygen, nitrogen, and sulfur functional groups in the VOC molecules. Coefficients have been measured for hydrocarbons in dry soils and vary from 1 for methane to 100,000 for octane (4).

However, biofilters operate under moist rather than dry conditions. And even though water competes with most VOCs for sorption, it increases the sorption of water-soluble gases. The soil/gas partition coefficient for octane is then probably on the order of a thousand, and the coefficient for acetaldehyde is around several thousand. Both are relatively independent of the moisture content of an operating biofilter.

In appropriately designed biofilters, the VOCs remain long enough for microbes to oxidize them. If the amounts of VOCs are excessive or the reaction time too short, the VOCs are carried through the bed and released to the atmosphere rather than accumulating in the beds. The gases do not condense in the beds, nor do VOCs dissolve in the pore water (because the pore water is continuously air-stripped by the upward air flow).

Soil and compost have relatively low sorption capacities for gases. Biofiltration relies instead on continuous oxidation of the sorbed VOCs by microorganisms to CO_2 and water. The oxidation regenerates the

When an organic energy source becomes available, the microbes increase their activity and growth rates in response to the input.

sorption capacity, resulting in a steady-state removal of the incoming air pollutants.

Biofilter removal efficiency

Removal efficiency, that is, the fraction of input pollutant removed, is usually of more interest than the removal rate (the organic oxidation rate per unit mass or volume of soil or compost). For example, regulatory agencies may require 90% VOC removal efficiency, and odor complaints may necessitate 99% removal efficiency. The biofiltration removal rate increases proportionally with the input rate and is a less useful parameter.

The steady-state removal efficiency of VOCs by biofilters is approximately first-order with respect to VOC concentration and is rather constant over a wide range of concentrations. The reaction also depends on the oxygen concentration, microbial population, and microbial nutrient supply, but these are usually in excess of what is required to oxidize the VOC input (5).

Over a wide range of VOC concentrations (0–500 ppm), the removal efficiency of the biofilter, RE, is approximately:

$$RE = (1 - C_2/C_1) = 1 - e^{-kt}$$

where C_1 is the input VOC concentration, C_2 is the output concentration, k is an empirical reaction rate constant, and t is the residence time of the carrier air in the bed. Although k varies with the temperature and conditions in the bed, it is mostly a function of the VOC's biodegradability.

To achieve a given removal efficiency, the primary design parameter

is the air residence time in the bed. The residence time of the carrier air for 90% removal of alcohols and aldehydes may range from 30 seconds to a few minutes. For 90% removal of trichloroethylene, the air residence time may be as much as 150 minutes.

At start-up, biofilter beds require an adaptation time for the microbes to adapt to a new VOC input and to reach steady state. For rapidly biodegradable VOCs, the adaptation time is no more than several hours. As the biodegradability decreases, the adaptation time usually increases. For slowly biodegradable gases, the adaptation may take weeks. After start-up, the beds are quite resistent to shock load effects.

The removal efficiency of rapidly biodegradable gases is about 99% if some care is given to moisture control (this is discussed further in a later section). For example, an Arizona soil bed treating extremely odorous air from a rendering plant at 100% relative humidity removes 99% of the odorous VOCs. A Texas soil bed also treating extremely odorous air and relying only on natural rainfall to supply moisture removes 95% of the odors when the bed is too wet or too dry and up to 99% at intermediate moisture contents.

The degradative microorganisms

The organisms responsible for the oxidation of VOCs to CO_2 are the same heterotrophic bacteria and fungi that degrade organic wastes in nature and in wastewater treatment plants and landfills. Heterotrophic organisms (including humans) utilize organic compounds for energy and convert them to CO_2.

The microbial population in soils is rather inactive under natural conditions. When an organic energy source becomes available, the microbes increase their activity and growth rates in response to the input. The potential degradation rate is, therefore, orders of magnitudes greater than the natural, relatively inactive rate.

Bacteria degrade small organic

molecules that are easily ingested into their cells. *Pseudomonas* and *Nocardia* are common bacterial species active in such organic breakdown. Some species such as *Flavobacterium* can adapt to oxidize such compounds as pentachlorophenol. Soils and compost contain roughly one billion bacteria per gram.

Fungi tend to degrade more complicated molecules, and they excrete extracellular enzymes that break down polymers. The fungi population is about 100,000 per gram of soil or compost. The species distribution fluctuates with the type of organic input.

Microbial cultures have been developed in the laboratory from soil cultures that are better adapted to degrading specific compounds, most notably halogenated hydrocarbons. Infusions of these cultures back to field conditions can increase degradation rates, but these increases are generally short-lived. Introduced microbes must compete with, and are usually at a disadvantage when competing with, the native microorganisms already adapted to the field conditions. In addition, native organisms adapt to synthetic organics in the field just as their cousins adapt in the laboratory. Slow degradation rates of organic solids and liquids are usually due to high, inhibitory organic loading rates and oxygen deficiency rather than to the incapability of the native organisms.

Biofiltration may be unsuitable for the more highly halogenated compounds—such as trichloroethylene (TCE), trichloroethane (TCA), and carbon tetrachloride—because they degrade very slowly aerobically. That means long residence times in the beds and large bed volumes, so biofiltration may only be practical for low flow rates of air containing these contaminants. Aerobic breakdown, however, yields complete breakdown to CO_2, H_2O, and Cl^- *(6,7)*, rather than the toxic interme-

diates such as vinyl chloride that are produced by anaerobic degradation. The aerobic degradation of mono- and di-chloromethane is rapid enough to consider biofiltration for moderate flow rates.

To load a biofilter bed so heavily that the sorbate reaches toxic concentrations is virtually impossible. The degradative microbes function at up to 1,000 mg of chlorinated hydrocarbons per kg of soil and up to 100 g of hydrocarbons per kg of

The soil bed in the foreground treats 25,000 cfm of odorous air coming from the building in the background.

soil. Those loading rates are possible when liquids are mixed with soils, but not when the VOCs are added as gases.

Soil vs. compost

The differences between soil and compost are that soil pores are smaller and much less permeable—permeability increases with the fourth power of the pore radius. Soil beds, therefore, require a larger area to process gases at a given flow rate and back pressure. Multilayer soil beds with greatly reduced area requirements are being developed.

Biofiltration may be unsuitable for the more highly halogenated compounds because they degrade very slowly aerobically.

Layering, however, significantly increases the problems associated with controlling the moisture content in the beds.

In soil beds, there is a trade-off between the higher air permeability of coarse soils and the higher sorption capacity of fine-textured soils. For rapidly biodegradable gases, the soil's rate of sorption-degradation often exceeds its gas permeability, so sandy, relatively permeable soils are preferable. Slowly biodegradable gases require long residence times and larger beds per unit of air flow, so low permeability can be advantageous.

Soil bases and minerals neutralize the acidity from the oxidation of acidic VICs such as NO_x, SO_2, NH_3, and H_2S. Preloading of soil beds with lime increases the neutralization capacity. Compost cannot be prelimed to the same extent as soil because this compacts the bed.

Compost is best suited for large air flows containing easily biodegradable VOCs. For slowly biodegradable gases requiring long reaction times in the bed, compost's high permeability may be no advantage. Compost beds must be stirred periodically to prevent caking. Also, some compost is unpleasantly odorous initially.

The moisture content of the beds fluctuates with the relative humidity of the input air. Excessively dry beds stop microbial activity and allow VOCs to flow through untreated. Soil is hydrophilic, so dry soil beds can easily be rewetted. Compost beds, on the other hand, are hydrophobic when dry, and rewetting is tedious and labor-intensive.

Despite some disadvantages of compost biofilters, they have been the overwhelming choice in Europe, where some 500 are in operation. The reasons for their popularity include the smaller land area required, the mild and humid climate (which reduces the incidence of drying), and familiarity.

Operating characteristics

Temperature. The optimum temperature of biofilter beds is about

100°F (37°C). The temperature response curve is rather flat over a range of about 50°F to 140°F (10–60°C), so careful temperature control seems to be unnecessary. The bed temperature is primarily determined by the input gas temperature, since the heat liberated by VOC oxidation is negligible.

The effect of low ambient temperature on biofilters depends largely on the input gas temperature. Low bed temperatures slow the microbial activity, but that is partially offset by the increased gas sorption by solids at low temperatures. In Minnesota, portions of a shallow soil biofilter bed with room-temperature input air freeze when daily high temperatures are below 10°F (–12°C). The VOC removal, as evidenced by odor removal, however, is still adequate.

The upper temperature limit is 140°F (55°C). Microbial activity slows dramatically above 150°F (65°C). Hotter gases can be cooled by water sprays, which offer the additional advantage of raising the relative humidity of the gas.

Shock loads and shutdowns. Biofilters are quite resistant to shock loads. The excesses of oxygen, nutrients, and microbial population absorb sudden VOC increases and VOC changes very well. Shutdowns of up to two weeks have created no problems upon re-start-up. Longer shutdowns may require another microbial adaptation period.

Maintenance

The primary maintenance problem with biofilters is control of the bed's moisture content. The optimal moisture content for soil beds is equivalent to 0.1–0.3 bar (10–30 kPa) water tension, or about 10–25% water-

LITERATURE CITED

1. **Carlson, D. A., and C. P. Leiser,** "Soil Beds for the Control of Sewage Odors," *J. Water Pollut. Control Fed.,* **38**(5), pp. 829–833 (May 1966).
2. **Bohn, H. L.,** "Odor Removal by Biofiltration," in "Recent Developments and Current Practices in Odor Regulations, Controls and Technology," Derenzo, D. R., and A. Gnyp, eds., *Trans. Air Waste Mgmt. Assn.,* pp. 135–147 (1991).
3. **Leson, G., and A. M. Winer,** "Biofiltration: An innovative Air Pollution Control Technology for VOC Emissions," *J. Air Waste Mgmt. Assn.,* **41**(8), pp. 1045–1054 (Aug. 1991).
4. **Bohn, H. L., et al.,** "Hydrocarbon Adsorption by Soils as the Stationary Phase of Gas-Solid Chromatography," *J. Environ. Quality,* **9**(4), pp. 563–565 (Oct. 1980).
5. **Bradford, M. L., and R. Krishnamoorthy,** "Consider Bioremediation for Waste Site Cleanup," *Chem. Eng. Prog.,* **87**(2), pp. 80–85 (Feb. 1991).
6. **Stinson, M. K., et al.,** "EPA SITE Demonstration of BioTrol Aqueous Treatment System," *J. Air Waste Mgmt. Assn.,* **41**(2), pp. 228–233 (Feb. 1991).
7. **Kampbell, D. H., et al.,** "Removal of Volatile Aliphatic Hydrocarbons in a Soil Bioreactor," *J. Air Pollut. Control Assn.,* **37**(10), pp. 1236–1239 (Oct. 1987).
8. **Jaeger, B. and J. Jager,** "Geruchsbekaempfung in Kompostwerken am Beispiel Heidelberg," *Muell und Abfall,* pp. 48–52 (Feb. 1978).
9. **Vaart, D. R., van der, M. W. Vatavuk, and A. H. Wehe,** "The Cost Estimation of Thermal and Catalytic Incinerators for the Control of VOCs," *J. Air Waste Mgmt. Assn.,* **41**(4), pp. 497–501 (April 1991).

FURTHER READING

Pomeroy, R. D., "Biological Treatment of Odorous Air," *J. Water Pollut. Control Fed.,* **54,** pp. 1541–1546 (1982).
Ottengraf, S. P. P., et al., "Biological Elimination of Volatile Xenobiotic Compounds in Biofilters," *Bioproc. Eng.,* **1**(1), pp. 61–69 (Jan. 1986).

filled soil pores, when the average water film is about 0.1 μm thick. Compost beds operate at 40–50% water content by weight, or about 0.2 bar tension.

These moisture contents correspond to about 99% relative humidity of the air in the pores. Humidifying the input air is usually necessary to maintain optimum soil or compost bed performance. Humidity can be raised by water sprays, wet pads, and injection of steam into the input air. In addition, the beds can be wetted by surface sprinklers.

Moisture control by prehumidification is especially important in compost beds. Air flow rates are higher in compost beds and the dry-

ing rate can exceed the downward rate of water flow in unsaturated media. Water control is also more critical in compost because, once dried, it is hydrophobic and can only be rewetted with great difficulty. Soil beds treating low flow rates can often rely on natural rainfall.

Lifetime

The sorption and oxidation of VOC leaves no residues, except for a small amount of humified organic matter and microbial biomass, which are also biodegradable. The lifetime of soil beds for VOC removal is essentially limitless. For VIC removal, the lifetime depends on the soil's capacity to neutralize the acids produced (for example, HNO_3 and H_2SO_4). Soil beds installed in 1964 at Washington sewage lift stations *(1)* have operated continuously with no maintenance, and homeowners next to the pump stations are completely unaware of odors.

Compost beds have a limited lifetime because compost is itself biodegradable. The compost must be replaced after one to five years. **CEP**

H. BOHN is a professor in the department of soil and water science at the Univ. of Arizona, Tucson, AZ (602/621-1646; Fax: 602/621-1647). Previously he worked as a research chemist at the Univ. of California–Riverside and the Tennessee Valley Authority. He is a specialist in the reactions of carbon, nitrogen, and sulfur gases with soils and the thermodynamics of aqueous solutions contacting soils. He received BS and MS degrees from the Univ. of California–Berkeley and a PhD from Cornell Univ. in soil and organic chemistry. He is a Fellow of the Soil Science Society of America, the American Association for the Advancement of Science, and the Air and Waste Management Association.

Control Emissions From Marine Vessel Loading

Understand the safety and operational issues related to emission control system design.

**Gary N. Lawrence
and Stanleigh R. Cross,**
Phoenix Engineering, Inc.

Regulations set by the U.S. Coast Guard *(1)* require safety measures during the loading of marine vessels connected to vapor collection systems. These regulations (which were promulgated in July 1990) immediately impacted all companies involved with the loading of benzene, due to previously enacted U.S. Environmental Protection Agency regulations governing benzene transfer. In addition, regulations issued by the states of California, New Jersey, and Louisiana impose additional marine emission control requirements. These regulations effectively work together — the federal or state environmental rule first requires the collection of the vapors generated from vessel loading, and then the Coast Guard regulation governs the safety features that must be applied to the system.

Currently, only a handful of facilities throughout the U.S. have been affected by the Coast Guard regulations. However, with the pending enactment of additional sections of the Clean Air Act Amendments of 1990, emissions generated during marine loading will come under closer scrutiny. Depending on the vapor pressure of the chemical, a 10,000-barrel barge may emit over one ton of chemical to the atmosphere. Such large volumes make marine loading a prime target for the push to further reduce atmospheric pollution, and it is a good bet that many more companies will be asked to look at the recovery of vapors during the loading of marine vessels.

This article will aid the engineer who may be asked to evaluate the various methods of controlling emissions from vessel loading. It provides some guidance on the requirements of the Coast Guard regulations and briefly outlines some of the technologies that have been used to process the collected vapors. [More detail on technologies for controlling emissions of volatile organic compounds (VOCs) can be found in *(2)*.] Some important design considerations unique to marine systems are discussed to help engineers avoid some of the potential pitfalls. Finally, some estimated costs are provided for two common types of marine vapor control systems.

Hazards and safety features

Marine vapor control systems (MVCSs) can be divided into two sections. The first section is the vapor collection system, which includes the required Coast Guard safety devices and auxiliary items such as blowers or compressors used to move the vapors from the dock. The second section is the end device, which can be either a vapor destruction unit, which destroys cargo vapor by means such as incineration, or a vapor recovery system, which recovers cargo by a nondestructive means such as lean oil absorption, carbon bed adsorption, or refrigeration.

The general safety requirements for MVCSs address three main points of concern:

1. preventing over- or under-pressuring of the vessel;

2. preventing overfilling of the vessel; and

3. preventing and/or eliminating deto-

■ *Dual-inlet safety and injection skid can accommodate the loading of two vessels simultaneously (with the same liquid chemical). The two valves at the inlet provide vacuum relief. Pressure relief is provided by a single valve in the upper portion of the photo. One of the two detonation arresters, which is covered with an insulated jacket, can be seen below and to the right of the pressure relief valve.*

■ *Rear view of a safety and injection skid. The transmitters at the far left end of the skid monitor pressure and send output signals to the high and low pressure alarm and shutdown switches.*

nations that can occur in the vapor collection headers.

Each of these hazards has safety features associated with it. However, to complicate matters, the type of end device associated with the collection system will result in various changes to the safety devices.

1. Preventing over- or under-pressuring of the vessel. This hazard is addressed by pressure sensing devices and relief valves. The Coast Guard regulation requires that pressure sensing devices be installed near the vapor hose connection to the dock. Two switches activate alarms at high and low pressure set points, and the other two switches initiate emergency shutdown procedures at a higher pressure and a lower vacuum. The alarm and shutdown sensors must be independent of each other. The exact set points depend on the specific pressure relief and vacuum relief valves.

In addition, systems that include the injection of inerting, diluting, or enriching gas into the vapor header must have a pressure relief valve with the capacity to relieve the maximum amount of injection gas while holding no more than 2 psig in the vapor col-

lection header. If a blower, eductor, compressor, or other means of moving vapors is included in the vapor collection system, a vacuum breaker must be installed at the dock. This vacuum breaker must have enough capacity to relieve the maximum capacity of the vapor mover while maintaining no more than 1 psig vacuum at the dock.

2. Preventing overfilling of the vessel. There are four alternative methods of sensing or preventing damage to vessels due to overfilling. The vessel may have its own means of preventing overfilling, such as a vessel-mounted level-sensing system that generates an alarm in the event of an overfill. Or the vessel may simply be outfitted with rupture disks or spill valves that open before the vessel reaches a condition that would cause the vessel compartments to rupture. Finally, the vessel may use a shore-mounted device that sends an intrinsically safe signal from the shore to the vessel to monitor the condition of level sensors in each vessel compartment; if an overfill situation occurs in a compartment, the level sensor interrupts the intrinsically safe signal,

which triggers an alarm and shutdown at the shoreside control system.

3. Preventing and/or eliminating detonations within the vapor collection headers. The requirements for this particular hazard are associated with the end device selected. MVCSs that utilize combustion systems must include a detonation arrester accompanied by an enriching, diluting, or inerting injection system located on the dock. To assure that the injection system is operating properly, a hydrocarbon or oxygen analyzer backed up by a second device, such as a second analyzer or system to monitor the ratio of injection gas to collected vapors, is required.

MVCSs that utilize vapor recovery systems are required to have an oxygen analyzer located at the dock that alarms or shuts down loading based on the oxygen concentration in the vapors collected. The MVCS is also required to have the capability of inerting the vapor collection hose or arm and the vapor header prior to loading the vessel. If the MVCS collects vapors from noninert vessels, a detonation arrester or an inerting, enriching, or diluting system with analyzers is required at the

■ *Combination pressure/vacuum relief valve protects the system by ensuring that pressures never exceed 2 psig pressure and 1 psig vacuum; normally some facilities have lower set points.*

dock. An MVCS that collects vapors from more than one dock must have a detonation arrester at each dock.

If an MVCS collects the vapors while vapor balancing to a shoreside storage tank, it must include a detonation arrester at the dock and a second detonation arrester located within the storage tank containment area. The system must also include a storage tank high-level alarm that prevents liquid from entering the vapor collection line.

4. Additional safety requirements. Each MVCS must employ special red and yellow color coding to identify the vapor piping on the shore and on vessels. In addition, specially configured lugs on the vessel and shore piping and matching holes on the vapor hose flanges are needed to prevent the accidental connection of a liquid hose to a vapor line.

At the dock, a quick closing valve is located in the vapor header. This valve must close within 30 seconds

■ *Front view of a safety and injection skid at a liquid loading terminal. The two identical panels at the left house the two pairs of oxygen analyzers (one for each inlet). The single panel to the right of the analyzers acts as a junction box and houses the emergency shutdown button.*

during emergency shutdown conditions in order to isolate the shore facilities from the vessel.

To set or determine vessel filling rates, a means of measuring the liquid flow rate to the vessel is required, and an automatic block valve that can close in 30 seconds if an overfill situation occurs is required in the liquid loading line.

As a final means of determining the safety of the overall system, the Coast Guard regulations require that a safety hazards analysis be completed for the entire system. Several methods of analyzing the system such as hazard and operability (HAZOP), failure modes effects and criticality analysis, fault tree analysis, and others may be used. [For further discussion of these analysis techniques, see (3–5).]

Vapor destruction systems

There are three major types of combustion systems that can operate over the wide flow rate and heat content ranges of marine applications: open flame flares, enclosed flame flares, and thermal incinerators. [Flares are discussed in detail in (6).]

Open flame flares are the simplest and least expensive of the combustion devices. These flares mix the collect-

■ *The vapor collection hoses for these vapor inlets are shown in the upper portion of the photograph. Below the hoses are automatic block valves that close within 30 seconds of an emergency shutdown. Below the block valves are detonation arresters.*

ed vapors with atmospheric air to effect efficient combustion. The flares commonly use forced-air blowers or steam to create turbulent mixing at the point where the vapors exit the flare tip. The mixing causes the hydrocarbons in the vapor to burn without smoke. If the flare gas has a minimum heat content of 300 Btu/scf, meets certain limiting tip exit velocities, and includes a continuous pilot for ignition of the vapors, most environmental regulatory agencies will recognize a combustion efficiency of 98% for open flame flares. Specific flare requirements are contained in 40 CFR 60.18 (7).

Enclosed flame flares are designed to allow combustion of the captured vapors to occur entirely within a refractory lined flare stack. Air for controlling maximum temperatures inside the flare stack is added through forced-air fans or natural-draft dampers at the base of the flare stack. Like open flame flares, smokeless combustion is created by using forced-air fans, which mix a percentage of the stoichiometric amount of air with the vapors as they exit the flare tip(s) located near the bottom of the enclosed flare stack. If enclosed flares are designed to meet the requirements of 40 CFR 60.18, they are considered to be 98% efficient.

Thermal incinerators also combust the vapor in a refractory lined chamber. However, unlike flares, they can maintain a constant combustion temperature by adding auxiliary fuel such as natural gas during times when the vapors have lower hydrocarbon concentration. On the other hand, when the hydrocarbon concentrations are high, thermal incinerators can add air to reduce temperatures in the combustion chamber. By maintaining constant temperatures, thermal incinerators can achieve higher combustion efficiencies. Matters of economics have typically held the combustion efficiency to the range of 99% to 99.99%, although higher lev-

els are attainable. Thermal incinerators have been utilized at facilities where high combustion efficiencies were desirable.

Safety components for vapor destruction systems. The Coast Guard treats the three types of combustion units the same. Each must include the following safety features.

■ *The vertical vessel in the foreground of this blower skid is a knockout pot for the blower, which collects any condensation or liquid and prevents it from reaching the blower. The flare stack is in the background.*

At the inlet, two quick-closing automatic valves are required. These valves must close any time the combustion system reaches an emergency shutdown condition. A closure time of 30 seconds or less is required.

A detonation arrester with a temperature switch mounted toward the combustion device should be installed, along with a liquid seal. It has been customary for flare vendors to arrange the detonation arrester nearest the combustion device since it is most efficient at stopping flash-

backs, and the temperature switch is installed to create a shutdown if it senses a flashback. However, other configurations have been allowed in several instances.

A continuously monitored pilot must also be built into the combustion system. It should be controlled so that automatic valves close and an alarm activates if the pilot is extinguished during vessel loading.

Vapor recovery systems

There are a wide variety of recovery systems that can be used with marine vapor collection systems. However, most vapor collection systems that have been installed to date use refrigeration, carbon adsorption/absorption (CAA), or lean oil absorption or water scrubbing.

All vapor recovery systems share at least one strength and one weakness. The strength is that many facilities are more comfortable with a system that does not involve a flame near the docks. The weakness is the inefficiency when the collected vapors are extremely lean, which can be a major portion of the load for vessels that have been inerted or cleaned before loading.

Conventional refrigeration systems use a refrigerant in an indirect contact loop to cool the collected vapors below their dew point. The condensed vapors are collected and returned either to the storage tank or to a processing unit where they are regenerated. Generally, collected vapors will have a significant amount of water vapor, which causes icing problems for refrigeration systems. This is often solved by installing dual exchangers so that one is processing while the other is defrosting.

CAA systems have been used in several marine applications with good success. These systems will provide satisfactory operation for many applications and are especially successful at facilities that handle a large number of typical hydrocarbons, such as gasoline, gasoline additives, benzene, and so on.

The collected vapors pass through carbon beds, which adsorb hydrocarbons and reject inerts such as air or nitrogen. There are typically multiple carbon beds, with one collecting hydrocarbon while the others are being regenerated. The carbon beds are regenerated by imposing a vacuum on the hydrocarbon-rich carbon bed, which causes the hydrocarbons to desorb. The desorbed hydrocarbons are removed through a vacuum pump and passed through an absorption column, where they are then captured in the recirculating absorption fluid and returned to a storage tank. The distance to a source of absorption fluid should be considered when evaluating carbon adsorption systems.

Lean oil absorption systems contact the collected vapors directly with an absorbing fluid. The vapors are absorbed into the lean oil and returned as a rich oil to a processing facility where they can be removed and the lean oil regenerated for reuse. This system is most practical for facilities that have lean oil with low vapor pressure available and also have the facilities to regenerate the lean oil.

Water scrubbers are attractive when recovering vapors that are soluble in water, such as acrylonitrile. Water is passed over a packed bed, where it is brought into contact with the collected vapors. The hydrocarbon is absorbed into the water, and the inerts are emitted from the scrubber outlet. A facility that uses this system must have a water treatment plant or a means of transporting the contaminated water to a treatment plant.

Safety components for vapor recovery systems. The inlet to a vapor recovery unit that collects vapors that have not been inerted, enriched, or diluted must have a detonation arrester or a flame suppression system. Also, the outlet vent from the vapor recovery unit must have either a detonation arrester or, as an option, an Underwriters Laboratory (UL) or Factory Mutual (FM) approved flame arrester installed in conjunction with a method of proving that the emitted vapors are not combustible.

Authorization for the optional flame arrester must be obtained from the Coast Guard on a case-by-case basis.

The option for a flame arrester can be particularly attractive to facilities that are required by environmental regulations to have continuous emission monitors installed. The monitor can be used to prove that the emissions are not combustible, and the flame arrester will be significantly less expensive than a detonation arrester. The designer should take care to install the flame arrester according to the manufacturer's recommendation, since it will not work if installed incorrectly.

■ *Front view of a blower skid. A detonation arrester, located at the far right, isolates this dock system from the other dock and the truck and train loading area.*

Design considerations and complications

In addition to being familiar with the Coast Guard regulations, it is also wise to be familiar with the components of vessels and with the operation of the shoreside facilities.

Vessel vapor header pressure drop. Many facilities strongly prefer to not include blowers or compressors in the vapor system design. However, facilities that load ships may regret not including such a device to assist in moving the vapors from the dock to the vapor collection device.

The Coast Guard regulations dictate the design parameters that must be followed when sizing the vessel's vapor collection header. Several factors come into play, but a most important

requirement is that the vessel compartment pressure should not be allowed to exceed 80% of the compartment's pressure relief valve set point.

However, it is permissible for the vapor collection header design to use all of the available 80% as pressure drop between the compartment and the point where the vessel vapor piping connects to the shoreside facility. For systems without a blower, this means that no pressure is left when operating at the vessel's maximum design loading rate; the operator is therefore forced to load the vessel at reduced rates, which will extend the vessel's loading time. If a blower had been installed, it would have been able to move the vapors from the dock at design rates regardless of the vessel's piping configuration.

Facility throughput. Facilities that load large quantities across their docks can often afford to absorb the generally higher cost of recovering vapors. This is because the higher cost is offset by the value of the recovered product. However, if the facility has a high throughput of a wide variety of products, the designer should be careful in selecting a vapor recovery system.

Often, vapor recovery systems work well when staying within a range of chemicals but falter when subjected to a wide variety of products. High freezing points can create problems for refrigeration systems, reactivity with carbon can cause high temperatures in carbon systems, and lean oils are often specific to a very narrow range of chemicals. When a large variety of chemicals is to be loaded, the designer should strongly consider combustion devices.

Availability of utilities. Combustion devices require a relatively large amount of fuel in the form of natural gas, propane, or plant fuel gas. During the time when the collected vapors have no heat content, each cubic foot of vapor collected will need to be enriched by 42% with natural gas to maintain a

■ *The blower and motor are in the center of this blower skid for a benzene marine system. A pressure controlling valve can be seen at the left. The knockout pot is at the right, and the motor starter, blower control panel, and heat tracing controls are directly in front of it.*

heat content of 300 Btu/scf (34 kcal/m³). This is the heat content required if the combustion device is a flare.

Carbon adsorption and refrigeration generally require large amounts of electrical power, which if not available will need to be provided for as part of the capital expense of the project. In addition, carbon adsorption systems also require an absorbing medium, which should be recirculated from a relatively large storage tank to prevent changing the tank's contents due to the inclusion of the recovered material.

Pollution problems. The designer of a lean oil absorption system should be particularly aware of the amount of time the collected vapors are inert. A lean oil will absorb hydrocarbons from the vapor as long as there is a driving force. However, if an inert is passed through the absorbing media, it can strip hydrocarbon from the absorbent and actually create emissions to the atmosphere. Therefore, the best absorbent will be one which has a low vapor pressure and also one which in itself is not a pollutant. For instance, a lean oil with benzene as one of its components would have to be evaluated carefully before being used in a marine system.

Refrigeration systems provide a

recovered liquid product that is contaminated by the moisture in the collected vapors. If a vessel has not been inerted, it will have air in its compartment. This air will be in equilibrium with the water in the atmosphere, which is generally humid around marine applications, and can result in significant water content. If the liquid has no value when contaminated with water or if a ready place for the recovered liquid to be purified is not available, it may have to be handled as a hazardous waste. The additional cost of disposing of a hazardous waste could add significantly to the expense of a system.

Condensation. Condensation can add a considerable amount of time to the operation of a system if not properly accounted for in the design. A benzene collection system in the U.S. Midwest was installed with only a small collection pot and no automatic means of removing collected condensate. The knockout pot filled frequently, which created a shutdown of the loading system. Since benzene is closely regulated, the emptying of the knockout pot was complicated, resulting in extended periods of downtime for the system until the oversight was corrected.

Products with high freezing points. Benzene and *p*-xylene are examples of chemicals that easily freeze at ambient conditions. Special care should be taken for such chemicals.

One particularly difficult problem to handle is freezing of the vapors in the collection headers of vessels. In one instance, frozen benzene blocked the vapor collection header of a barge, resulting in an over-pressure that ruptured a compartment.

On the shore side, the designer should consider heat tracing and insulating the entire vapor line. As a minimum, the regulation requires well instrumented liquid collection pots at points where condensate will collect.

Vessel saturation conditions. Several articles have been previously published [for example, (8)] concerning marine vapor control systems, and much has been made of the so-called

Table 1. Estimated installed costs for carbon adsorption/absorption and enclosed flare systems.

	20,000 bbl/h Case		5,000 bbl/h Case	
	CAA	Flare	CAA	Flare
Dock Skid	$234,000	$312,000	$174,000	$242,000
Blower Skid	244,000	246,000	227,000	231,000
Heat Exchanger Skid	133,000	—	73,000	—
End Device	2,177,000	597,000	798,000	374,000
Control System	80,000	75,000	88,000	80,000
Installation	1,632,000	420,000	740,000	273,000
Total	$4,500,000	$1,650,000	$2,100,000	$1,200,000

"S" saturation curve. The S-curve merely refers to the shape of a curve where concentration of the product in the outlet vapors is plotted against time. This curve generally shows a very low concentration of product early in the load, followed by a sharp increase over a short period of time roughly near when the vessel is 75% full, followed by flattening as conditions slowly approach saturation.

This curve adequately predicts the behavior of typical gasoline hydrocarbons. However, when pure chemicals are loaded into barges that carry only a single product, the S-curve design philosophy will be deceiving. Such barges can be saturated at the very beginning of the load and can remain saturated throughout. This is particularly important to the end devices, which for dedicated chemical service vessels must be designed considering the worst case of saturation throughout the load.

Estimated costs

Table 1 summarizes some typical estimated costs. Two types of systems, CAA and enclosed flare, have been selected for comparison. While other end device options exist, these are common choices for recovery or destruction devices.

Each system is sized for two separate flow rates, 5,000 bbl/h (596 m³/h) and 20,000 bbl/h (2,385 m³/h). These flow rates are commonly employed by storage terminals and refineries, respectively.

The installation costs are based on an "ideal" installation: an unobstructed area is available for all equipment and modifications to surrounding structures are not necessary; the required utilities are easily accessed; the end device is located no more than 1,000 ft from the dock; and the interconnecting vapor piping, conduit, and wiring can be easily routed.

The equipment at the dock for each option includes all the required safety components as outlined previously. The blower skid utilizes a simple centrifugal blower with safe-

Literature Cited

1. Code of Federal Regulations, Title 33 — Navigation and Navigable Waters, 33 CFR Part 154, "Oil Pollution Prevention Regulations For Marine Oil Transfer Facilities."
2. Ruddy, E. N., and L. A. Carroll, "Select the Best VOC Control Strategy," *Chem. Eng. Progress,* **89**(7), pp. 28–35 (July 1993).
3. "Guidelines for Hazard Evaluation Procedures," 2nd ed., American Institute of Chemical Engineers, Center for Chemical Process Safety, New York, NY (1992).
4. Freeman, R. A., "What Should You Worry About When Doing a Risk Assessment?," *Chem. Eng. Progress,* **85**(11), pp. 29–34 (Nov. 1989).
5. Ozog, H., and L. M. Bendixen, "Hazard Identification and Quantification," *Chem. Eng. Progress,* **83**(4), pp. 55–64 (Apr. 1987).
6. Niemeyer, C. E., and G. N. Livingston, "Choose the Right Flare System Design," *Chem. Eng. Progress,* **89**(12), pp. 39–44 (Dec. 1993).
7. Code of Federal Regulation, Title 40 — Protection of the Environment, 40 CFR Part 60, Section 60.18, "General Control Device Requirements."
8. Hill, J., "Controlling Emissions From Marine Loading Operations," *Chem. Eng.,* **97**(5), pp. 133–143 (May 1990).
9. "API Recommended Practice for Marine Emission Control System Training," API-R1127, American Petroleum Institute, Washington, DC (Nov. 1993).
10. "API Manual on Installation of Refinery Instruments and Control Systems," American Petroleum Institute, Washington, DC.
11. "Recommended Practice — Installation of Intrinsically Safe Systems for Hazardous (Classified) Locations," Instrument Society of America, Research Triangle Park, NC (1987).

Acknowledgment

The authors acknowledge the assistance of Bill Matthes, of John Zink Co., Tulsa, OK, for help in developing the cost estimates for the carbon adsorption/absorption systems.

G. N. LAWRENCE is a vice president at Phoenix Engineering, Inc., Houston, TX (713/944-7208; Fax: 713/944-1618), where he is responsible for the day to day operation of the company. He was responsible for the design of the first certified marine vapor control system in the United States. Phoenix Engineering is one of the original certifying entities for the U.S. Coast Guard. He received a BS in chemical engineering from the Univ. of Oklahoma and is a licensed professional engineer in the state of Texas. He is a member of AIChE, the National Society of Professional Engineers, and the Texas Society of Professional Engineers.

S. R. CROSS is a design coordinator at Phoenix Engineering. He is involved in the design and fabrication of numerous marine emission control systems, and is responsible for instrument and equipment specification and procurement, engineering estimates, and field expenditure estimates. He is pursuing a BS in construction management from the Univ. of Houston.

ty components as specified by the regulation.

The costs for the CAA unit are based on the following assumptions: two separate systems operating in parallel are installed for the 20,000 bbl/h case while the 5,000 bbl/h case uses a single unit; the circulation tank for the absorbing medium is within 1,000 ft of the CAA unit; heat exchangers after the blower skid maintain vapor temperatures within a range that maximizes hydrocarbon adsorption; an air-cooled heat exchanger cools the seal fluid in the vacuum pumps that regenerate the carbon beds; and all valves are pneumatically actuated and designed to fail in a safe position.

For additional help

Compliance with the Coast Guard regulations will result in a system that provides safe collection of vapors from marine vessels. However, the regulations can be complicated, especially for chemicals other than gasoline, crude oil, and benzene, where a great deal of the design is not specifically published. When designing a marine collection system, the engineer should become very familiar with the regulation, as well as the intent of the requirements. References *(1, 9–11)* should be useful for this purpose.

In addition, the Coast Guard can provide assistance to regulatory "newcomers" through its Hazardous Materials Branch in Washington, DC. **CEP**

Build Flexibility Into Your Operating Permit

A well-thought-out application can help a facility get a permit that includes all the applicable requirements yet allows flexibility to adjust operations as needed, while still maintaining compliance.

T. Ted Cromwell,
Chemical Manufacturers
Association

The centerpiece of the Clean Air Act Amendments of 1990 is Title V, which requires major sources of air emissions to obtain and comply with a federal operating permit. These permits will have more comprehensive terms and conditions than any previous Clean Air Act requirements.

Unfortunately, the permit application and revision process spelled out by the U.S. Environmental Protection Agency (EPA) last August was a complex, four-tiered procedure that did not meet the goals envisioned by either the regulators or the regulated community. EPA requested interested parties to provide additional suggestions or alternatives that would simplify the process, incorporate greater flexibility, and allow public input, and many groups, including chemical process industries (CPI) firms, did so. In response to that input, EPA has decided to issue an alternative proposal for comment, which is expected by the beginning of this month.

While the debate over the Title V process will continue well into 1995, many states have received partial or interim permit program approval and many companies have already started developing their permit application. Regardless of the final outcome, the most important challenge a facility will face is developing a permit application that provides built-in flexibility. Simply put, if you have a permit that allows a variety of adjustments to ongoing activities within existing permit terms and conditions, you will be far less likely to trigger permit revisions in whatever form they ultimately are determined to be.

A previous CEP article *(1)* advised readers on early steps they could take to prepare for the upcoming operating permit program even before the actual requirements were promulgated. Now that the deadlines are imminent — the first group of applications is due in November — this article picks up where that one left off and gives more-concrete guidance on the actual permit application process and some basic strategies that companies should find useful.

Basic provisions

Under Title V, every major source will be required to obtain an operating permit. The Clean Air Act defines a source as major if it emits 100 tons/yr of a criteria pollutant, 10 tons/yr of a single hazardous air pollutant (HAP), 25 tons/yr of any combination of HAPs, or other amounts as defined for nonattainment areas.

The operating permit will codify in one place all of the requirements to which a facility is subject under various standards and regulations. While the permit identifies the terms and conditions under which a source will operate, the specific requirements that a source must implement and comply with are actually documented in the underlying standards. The permit will serve as the primary verification and documentation of a facility's compliance with all applicable requirements of the Clean Air Act.

Title V also addresses requirements that must be met for sources to ensure compliance with the permit. Major sources will not be allowed to operate without a permit. Penalties for violating the terms of a permit are stiff, ranging from $5,000 to $25,000 per day for each violation. Disregard for the requirements and gross negligence could even lead to prison terms for owners or operators.

The permit program envisioned by EPA will be delegated to, and implemented by, the state agencies. The state agency is responsible for interpreting EPA's program and developing a process that substantially reflects all of the elements of it. Then, after EPA approves the state program, the state will phase-in the issuance of permits for all major sources over several years.

Permit fees, which are required by the Clean Air Act, will provide financial support the state may need to develop, implement, and maintain an operating permit program. Fees may be charged to cover any reasonable cost associated with the state's Title V program; they are generally at least $25 per ton of actual or potential emissions. Most state agencies already have established the fee structure and have received EPA's approval to collect these fees.

Even though a permit is issued to cover up to a five-year term, new or modified information will be added and deleted from a permit on a fairly frequent basis throughout the permit's life. As new requirements are developed and implemented, the source and the regulatory authority will have to incorporate these new requirements into the permit. In fact, the Title V permit will likely be considered as an "evergreen document" — continually updated and current to reflect ever-changing terms and conditions.

Permit applications will require intensive preparation and reams of data to ensure that all required information

Permits will be updated frequently as new regulatory requirements are developed.

has been adequately addressed. Enforcement, public notice and comment, and a streamlining of documentation that reflects the terms and conditions that a facility will comply with are all key elements of the Title V application. Understanding what information the permitting authority expects to see included in the permit is the first step a facility owner or operator can take in developing an application. Much of this article focuses on this critical step in the process.

Title V and the CPI

Currently, the CPI are focusing much of their attention on two key areas.

First, the specific requirements that allow for flexibility in the permit must be addressed. An understanding of information and activities that can and should be addressed before and during the permit application submittal will go a long way toward establishing an operating permit that will provide a reasonable amount of flexibility, while

ensuring that the terms and conditions of the permit are acceptable to the regulatory agencies. The types of preliminary efforts needed include:

• an accurate and complete emissions inventory;

• detailed knowledge and documentation of current operating conditions;

• strategic planning to address anticipated future operations; and

• development of alternative operating scenarios in the permit application that will allow current and anticipated operations to proceed in an expedited manner.

Second, the CPI are also carefully evaluating the process for implementing the requirements of the hazardous organic NESHAP (national emission standard for hazardous air pollutants), or HON, regulation and the Title V permit. The integration of the HON, as well as other future maximum achievable control technology (MACT) standards, will be significant. Depending on how a facility interprets "applicable requirements" for the purposes of the Title V permit, incorporating the MACT provisions into the permit could involve a streamlined administrative process, or it could create potential noncompliance situations while the source awaits approval of the necessary permit revisions.

In short, three years from promulgation of any MACT standard, affected sources must demonstrate compliance with that standard. However, the timing for complying with the MACT standards is inconsistent with the timing and compliance requirements for Title V compliance. The critical issue is whether the current permit has more than or less than three years remaining when the MACT standard is promulgated. If less than three years remain in the permit life, a streamlined process will acknowledge the new terms and conditions, but will not require a full review until the permit is renewed. On the other hand, if more than three years remain on the permit, the permit would need to be reopened to incorporate the MACT requirements. EPA has attempted to address this situation (1), although it is not yet clear how the final regulations will ultimately resolve this problem.

Let's now turn our discussion to the contents of the permit application and methods for retaining operational flexibility in the permit. Then we will briefly outline the implementation challenges that remain between Title V and the MACT standards.

Emissions inventory

A detailed and accurate emissions inventory is the cornerstone for all subsequent activities under Title V. It establishes the baseline for determining the range of activities permitted at the facility, and virtually every emission limit, parameter, and description included in the application depends on it. The inventory will be invaluable in helping the facility determine which emission points are subject to the permit regulation, as well as the actual and allowable emissions for each emission point. In addition, some sources not previously subject to federal or state requirements will now be regulated under the Clean Air Act, and the facility will need to accurately know its emissions prior to developing a compliance strategy.

Therefore, the first step in preparing an operating permit application is to develop a comprehensive emissions inventory. (Reference 2 provides guidance on how to do this.) For many plants, this effort should be well underway now; ideally, it should be completed before the state's Title V program is approved.

Every source of emissions in a facility needs to be identified and measured. Developing an accurate inventory will involve a variety of tasks, including stack tests, ambient monitoring, and individual source monitoring. The data generated are then compiled and submitted as part of the permit application.

The data contained in the emissions inventory will help establish the range of operating conditions and parameters, which determines the amount of flexibility allowed in the permit. For example, if a unit has actual emissions of 5 tons/yr and an allowable emission rate of 8 tons/yr, the facility will need to permit that source to operate at any level up to 8 tons/yr. Further, if varying feedstocks are introduced throughout the year (such as in batch operations), the facility will also need to establish the range for each feedstock that is either currently in use or anticipated for use. Without the emissions inventory, the facility and the permitting authority will be unable to clearly evaluate the range of operations and the corresponding emissions. This will likely lead to an inflexible permit that doesn't reflect the range and variability of operations at the facility — one that will severely restrict future operations under the permit.

Designed to accommodate

Determining what the source is currently "designed to accommodate" is the next step in preparing an application. Then, using the emissions inventory and a review of previous and current operations, the facility should identify a range of conditions that may be reasonably anticipated to occur in the future.

Unlike the situation prior to 1990, many activities that involve a small change in operating conditions as a routine matter may now trigger some sort of review under either the Title V process or an underlying standard. Furthermore, a change that triggers an underlying standard will most likely also require a review under the Title V process. This underscores the need to develop permit conditions that will provide the range of options identified by the underlying standard as part of the source's design.

By charting the ranges of such parameters as operating hours, raw materials, frequency of downtime, maintenance, product changes, and other conditions, a facility should be able to establish a series of conditions and ac-

tivities that need to be preserved under the Title V program. While this may seem obvious, the compliance and enforcement conditions established in Title V require continuous compliance with the terms and conditions of the permit. If a plant makes any type of change that is not directly addressed or exempted by the permit, it may be subject to penalties for violating the terms and conditions of the permit, unless it can prove that the authority to make the specified change exists somewhere in the permit.

For example, a source subject to a new source performance standard (NSPS) could have several control options available to meet the standard (say a flare or a catalytic incinerator). The rules define a control efficiency of 98% for both options. If a plant develops the permit to state that it will comply with the terms and conditions of the NSPS, changing from a flare to a catalytic incinerator would still be considered as compliance with the standard. In this case, the change could be made with minimal review and notification, most likely with a seven-day advanced notice of the change. On the other hand, if the plant were to identify a specific control in the permit and then decide to change to another control option, the change would likely trigger a permit revision.

Any activity that a source currently is "designed to accommodate" must either be reflected in the permit, or be deemed insignificant and unnecessary for incorporation into the permit. Many activities mentioned above could be embodied as activities that a facility is designed to accommodate, including: routine maintenance, operational changes already allowed under existing programs, production shifts, raw material switches, and changes in operating hours. These activities must be identified in the permit to ensure that they would not trigger permit modifications. Both the Title V program, as well as underlying programs such as the modification rule for air toxics (Clean Air Act Section 112(g)), demonstrate that these routine activities may be considered modifications unless they are specifically addressed or excluded in the permit application process.

Along with looking back at current or past operations, a facility should also be looking forward to ensure that it will be able to establish enough flexibility in the permit to meet shifting market demands, economic conditions, or new products. This will require strategic planning and a thorough understanding of the company's broader goals and long-range plans for production and operations. The ability to anticipate the types of changes that may take place, within reason and with a high degree of accuracy, will likely provide a considerable competitive advantage to a company.

For example, a plant may generate emission credits through the overcontrol of an emission unit. The facility could opt to sell these credits, use the credits for current netting/offsetting transactions (that is, to offset an increase in emissions elsewhere in the plant so that the net emissions do not increase), or bank the credits for future use. If it anticipates an increase in production or a minor change to a unit during the life of the permit, it should consider banking the credits during the permit-application phase. Once the anticipated change occurs, the facility could then use the banked credits to avoid triggering a modification of the permit.

The ability to anticipate future activities must also include an understanding of the multitude of requirements that a source will be subject to during the life of the permit. MACT standards, nonattainment area requirements, NSPS regulations, and Section 112(g) and (j) provisions will necessitate compliance with many different requirements. Through flexibility, the permit must also anticipate changes required by these standards so that a source implementing the requirements of an underlying standard will not find itself out of compliance. Significantly, a facility that does not anticipate the requirements of these underlying programs could be unable to operate while it awaits the review of preconstruction activities or the approval of some part of the process.

Borrowing from an earlier example, a source meeting NSPS control

All routine changes must be addressed in the permit to avoid triggering permit modifications.

requirements may anticipate that the unit (or perhaps another unit) will become subject to a MACT standard. If the MACT standard control could substitute for the NSPS control, the facility may wish to write into the permit that it would employ the MACT control once the MACT standard becomes effective. Depending upon how the permit program is implemented, this change may require no permit action at all, or it may require only notice (rather than review and approval).

Alternative operating scenarios

Next, the facility will need to use the emissions inventory and the current and anticipated range of activities to establish operating flexibility in the permit application. This includes the typical information that would be included in the permit application, such as identification of a unit, the actual and allowable emissions, monitoring, raw material use, and so on. It may also include the use of "alternative operating scenarios." In fact, many of the actions that are included as a part of the permit application (such as those discussed in the NSPS examples) could be covered under these alternative operating scenarios.

Alternative operating scenarios provide an important mechanism for sources to codify various operating conditions that may be utilized throughout the five-year life of the permit. If a change is covered by an alternative operating scenario in the permit, the plant need only provide a seven-day advance notice to the permitting authority stating that it intends to switch to a particular alternative. As the alternatives would have been preapproved during the permit application process, the seven-day notice would merely represent a formality to document the exact conditions that the source is operating under at any given time.

While this does reflect an opportunity for streamlined and flexible operations in the permit, the scope of activities that could be addressed through alternative operating scenarios is not entirely clear. For example, if a facility broadly defines an "applicable requirement" to mean compliance with a MACT standard, it would be able to provide seven-day notice that it would come into compliance with MACT and schedule this scenario so that it reflects the compliance date established by the underlying MACT standard. If this position can be upheld, this would greatly simplify the coordination between standards (as discussed later).

If, however, the alternative operating scenarios define "applicable requirements" as very specific details from within the underlying standard, then much of the flexibility would be lost. Each of the specific requirements from the underlying standard would need to be codified in the permit. Then each time any of those conditions changed, the facility and the source would be required to either submit seven-day advanced notice of the change or trigger a permit modification procedure. This interpretation of "applicable requirements" would needlessly complicate the permit process with details that are unnecessary and were not envisioned to be included in the permit.

While a full review during the permit application phase will likely be time-consuming and resource-intensive, getting these alternatives approved during the initial application phase will pay off in the long run with a streamlined process for making the changes once the permit is issued.

Coordinating Title V and underlying standards

The second implementation issue is the need for a simple and streamlined approach to meeting the requirements of underlying standards while incorporating these requirements into the Title V permit. At issue is the requirement to reopen the permit to incorporate the HON requirements if more than three years remain in the life of the permit. The reopening could trigger significant delays in implementing the HON and complying in a timely manner.

CPI companies have a particular interest in the methodology for incorporating the HON MACT into the permit. However, the same problems that appear in the process for implementing the HON MACT into Title V will just as surely occur in other MACT standards as they are developed.

Particularly significant is the level of difficulty involved with incorporating new requirements into the permit. Administratively, the timing for submitting complete information into the Title V permit is inconsistent with the timing for completing this same information submittal for the HON. However, as previously discussed, if the HON as a standard were simply

A simple and streamlined approach is needed for meeting the underlying standards while incorporating their requirements into the permit.

referenced as the applicable requirement in the permit application, then the specifics of meeting the HON requirements would not be required in the permit until after the source complies with the HON. A facility would simply go about implementing the HON until compliance is determined and then the HON requirements could be reflected in the permit. Upon the next permit renewal, the specific information would be incorporated into the permit. Prior to that time, the source and the regulatory authority would simply need to look to the HON itself for identification of compliance requirements.

The HON, like the Title V permit, requires compliance certifications, monitoring, recordkeeping and reporting, and all other requirements necessary to ensure compliance with the standard. Nothing would be excluded, the only difference is that the information would not be incorporated specifically into the permit until permit renewal. This would avoid the process of triggering a permit reopening simply to incorporate requirements already being met.

CMA has taken the position that to facilitate the significant emission reductions that will be accomplished by implementing the MACT standards, any activities undertaken to comply with federal or state requirements under the Clean Air Act should be added to the permit through an administrative process. Particularly in light of the environmental benefits provided by compliance with these requirements, every effort should be made to ensure that compliance with the underlying standards is in no way com-

promised by the Title V process. EPA is still undecided about how to accommodate this process.

Closing remarks

Implementation of Title V, including the preliminary gathering of information for submittal of permit applications, is already at full steam. States have either received approval of their permit programs or are close to it, and many companies are already well into the process of putting together emissions inventories and evaluating the range of operations that will need to be incorporated into the permit. Ultimately, a well-thought-out permit application, along with close coordination with the state permitting authority, will greatly improve the odds of having a permit that provides flexibility while ensuring that the goals of the Clean Air Act are met. **CEP**

Literature Cited

1. **Van Wormer, M. B., and R. M. Iwanchuk,** "Prepare Now for the Operating Permit Program," *Chem. Eng. Progress,* **88**(4), pp. 41–49 (Apr. 1992).
2. **Li, R.,** *et al.,* "Develop a Plantwide Air Emissions Inventory," *Chem. Eng. Progress,* **91**(3), pp. 96–103 (Mar. 1995).

T. T. CROMWELL is director of air and water programs in the Regulatory Affairs Div. of the Chemical Manufacturers Association, Washington, DC (202/887-1100; Fax: 202/887-1237). His primary role has been regulatory advocacy relating to the 1990 Clean Air Act Amendments, with a significant focus on the Title V operating permit program and the development and implementation of the Hazardous Organic NESHAP, the principal MACT standard for the chemical industry. Before joining CMA in November 1990, he worked for the state of Virginia, developing environmental impact statements for highway projects. He holds a BS in psychology and anthropology from James Madison Univ.

Understand the Regulations Governing Equipment Leaks

EPA will soon issue standards for leaks of hazardous organic air pollutants from process equipment. Here's a rundown of the requirements.

Richard S. Colyer and Jan Meyer,
U.S. Environmental Protection Agency

Federal regulations requiring the control of air emissions from equipment leaks in the synthetic organic chemical manufacturing industry (SOCMI) have been in effect for many years. The equipment affected by these rules includes valves, pumps, and connectors that can "leak" gaseous or liquid process fluid to the atmosphere.

Existing regulations were developed when there was little experience in controlling equipment leaks and were directed mainly at identifying and controlling the large leaks. Since the establishment of these regulations, experience and knowledge of the control of equipment leaks have increased, making it necessary to refine the regulations, especially in light of the maximum achievable control technology (MACT) requirements under the Clean Air Act Amendments (CAAA) of 1990 for hazardous air pollutants.

This article summarizes the major requirements of the forthcoming equipment-leak regulation for hazardous air pollutants (HAPs). It was developed through negotiation among environmental groups, state and local agencies, industry, and EPA (see sidebar on p.28). An informational notice published in the Federal Register on March 6, 1991, beginning on page 9,315, contains the details of the requirements. Additional information will appear in the Federal Register when the regulation is formally proposed, which is expected to take place in November 1991.

About the rule

The term "equipment leaks" is defined as the loss of process fluid through the sealing mechanism separating the process fluid from the atmosphere. These losses are particularly common around the moving

parts of valve stems and pump shafts. Although most equipment leaks individually are small emission points compared to process vent or storage emissions, the large number of equipment components within a plant makes aggregate equipment leaks a significant emission source.

Existing federal and state regulations have been effective at increasing awareness of the significance of equipment leaks and in stimulating control efforts. These rules basically require that pumps and valves be inspected periodically for leaks with a portable hydrocarbon detector. If a volatile organic compound (VOC) concentration greater than or equal to 10,000 parts per million (ppm), as methane or hexane, is discovered, the component is identified as a "leaker" and maintenance is required to repair the leak. This approach is known as "leak detection and repair" (LDAR).

When it established these existing rules, the U.S. Environmental Protection Agency (EPA) estimated that emissions would be reduced by about 60–70% and that after control, leak frequencies (*i.e.,* the percentage of equipment components within a process unit that leak) would be approximately 5%. Data gathered over the past several years indicate that much lower leak frequencies can be achieved.

A new equipment-leak regulation will be proposed later this year as part of the larger hazardous organic national emission standard for hazardous air pollutants (NESHAP), or HON (for hazardous organic NESHAP). The HON will cover emissions from process vents, storage, transfer, and wastewater from SOCMI sources, as well as equipment leaks. The sidebar on fugitive emission regulations (p. 26) compares the areas of coverage of this HON and existing equipment-leak rules.

The major features of the regulation include:

• a phase-in period for valves and pumps to give plants time to develop and implement effective programs and allow them to focus on larger emission sources first;

• linkage of monitoring frequency with performance, and incentives for using low-emission equipment or reducing the amount of equipment in volatile hazardous air pollutant (VHAP) service;

• retention of provisions in the existing equipment standards for pressure relief devices, sample points, open-ended lines, product accumulator vessels, and compressors, which require installation of equipment or control devices and essentially eliminate emissions from equipment leaks;

• a quality improvement program (QIP) to ensure that certain plants (1) replace (during normal replacement) poorer performing valves with superior performing technologies until the specified performance level is achieved and (2) implement a quality assurance and control program to ensure that all elements of MACT are utilized.

The remainder of this article discusses the specific requirements of the regulation. Readers should refer to the informational notice published in the March 6, 1991 edition of the Federal Register, beginning on page 9,315, for details of the requirements. Section references in this article refer to the sections in the March 6 publication. The section numbering in the actual proposal will be different.

Scope and applicability

The new regulation primarily covers SOCMI facilities, and includes all those processes identified by EPA that use as a reactant or produce one of the 149 organic chemicals contained on the CAAA list of 189 HAPs. These processes are listed in Section 8-2 of the rule. The rule also covers a few non-SOCMI processes and specific chemicals (used primarily as reactants or process solvents): styrene-butadiene rubber production (butadiene and styrene), polybutadiene production (butadiene), chlorine production (carbon tetrachloride), pesticide production (carbon tetrachloride, methylene chloride, and ethylene dichloride), chlorinated hydrocarbon use (carbon tetrachloride, methylene chloride, tetrachloroethylene, chloroform, and ethylene dichloride), pharmaceutical production (carbon tetrachloride and methylene chloride), and miscellaneous butadiene use (butadiene).

An affected "process unit" means equipment that uses a VHAP as a reactant or produces a VHAP or its derivatives as intermediate or final product(s). Included is all equipment associated with the unit process operation, storage and transfer of feed material to the process and final or intermediate products from the process, and operations treating wastewater from the process. The lines within a unit process operation containing process fluids are subject to the regulation; lines and equipment not containing process fluids (such as utilities) are not. A plant site may contain one or more process units.

The rule would apply to both existing and new process units. It categorizes the regulated processes into five groups and uses a staggered implementation scheme, requiring some process units to comply six months after promulgation, while others would have to comply as late as 18 months after promulgation.

The regulation would apply to those pieces of equipment currently regulated in the existing equipment-leak rules, including all valves, pumps, compressors, pressure relief devices, open-ended valves or lines, connectors, closed-vent systems and control devices, sampling connection systems, and product accumulator vessels. The term "connector" refers to all flanged, screwed, or other joined fittings used to connect two pipes or a pipe and a piece of equipment.

What is MACT?

In the Clean Air Act Amendments of 1990, Congress rewrote Section 112, which deals with national emission standards for hazardous air pollutants (NESHAP). Prior to the Amendments, EPA was to set standards for hazardous air pollutants (HAPs) that provided an ample margin of safety to protect the public health. There was much disagreement over the interpretation of these words, and EPA was unable to develop more than a few NESHAP over the last 13 years.

Congress's solution was to shift Section 112 from a health basis to a technology basis followed by a health-based evaluation eight years after promulgation of an applicable standard. Congress defined this technology approach as ". . . the maximum degree of reduction in emissions of the hazardous air pollutants . . . that the Administrator, taking into consideration the cost of achieving such emission reduction . . . determines is achievable for new or existing sources . . ." Previous bills described this as "maximum achievable control technology," or MACT, and this term has carried over despite the change in wording in the final Amendments.

The rule also contains provisions for agitators and instrumentation systems. Although not covered by existing regulations, a limited amount of screening data indicate that agitators could be a significant source of emissions. Even though instrumentation systems are composed of equipment already covered, they are treated as a special class because (1) they are not part of the process proper but monitor process conditions, and (2) the components are closely spaced and do not lend themselves readily to individual LDAR.

The standard applies only to equipment containing or contacting process materials that are 5% VHAP or greater.

In certain chemical plants, particularly batch processes that produce a number of different products, some equipment is used in VHAP service only occasionally. In such cases, implementation of the standard could be difficult and would achieve very little emission reduction. For these situations, equipment that is operated 300 h/yr or less in VHAP service is exempt.

Valves

The first phase of the valve standard (Figure 1), which begins six months after promulgation, specifies quarterly LDAR with a 10,000 ppm leak definition. Plants must monitor valves quarterly and repair any that are leaking at a concentration equal to or greater than 10,000 ppm. During this phase, plants previously unfamiliar with LDAR will be able to develop and assess the necessary changes in operations, maintenance, and training. Phase II also requires quarterly LDAR but specifies a leak definition of 500 ppm.

Phase III includes a base performance level of 2% leaking valves in a process unit in addition to the LDAR requirements of Phase II. This base performance level determines the monitoring frequency or the applicability of the QIP. It was added to ensure a more certain emission performance than achieved through the work practice requirement alone and to create an economic incentive to improve performance. This incentive is less frequent monitoring. If a process unit has less than 1% leaking valves, the monitoring frequency extends from quarterly to semiannually, and if a unit's percent leaking valves drops below 0.5%, only annual monitoring is required.

Conversely, if a unit's percent leaking components is 2% or greater, the monitoring frequency increases from quarterly to monthly. Alternatively, the owner or operator could opt to institute a quality improvement program (described later) instead of monthly monitoring.

Repair of leaks soon after detection is a key feature in the effectiveness of LDAR programs and is a requirment in the proposed rule. The first attempt at repair must be made as soon as practicable after detection but no later than five days after discovery, and repair must be completed within 15 days. Most valve repairs, such as tightening the bonnet bolts, can be performed quickly, and five days should provide sufficient time to schedule simple field repairs that do not require isolation of the valve from the process. Attempting to repair a leak within five days will help identify those valves that cannot be fixed with simple field repair, without shutting down the process unit or without removing the valve from the process.

Some valves cannot be repaired without shutting down the process, which is costly and may result in greater emissions than delaying repair until the next scheduled shutdown. Thus, under certain conditions, repair of these "nonrepairables" can be delayed until the next facility shutdown. Nonrepairable valves, up to a total of 1% of the valves in VHAP service, can be excluded from the Phase III calculation of percent leaking

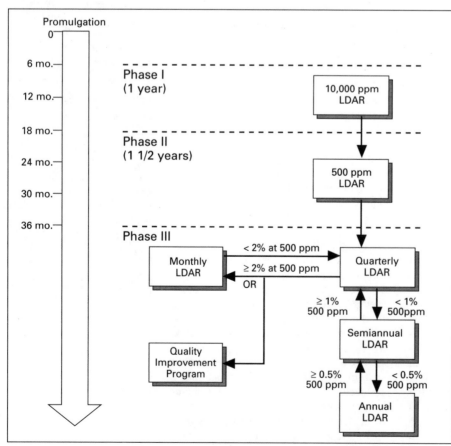

Figure 1. The valve standard will be implemented in phases over a period of three years.

valves for sampling periods after the leaking valve is first identified.

The percentage of leaking valves is calculated using a two-period rolling average. This will ensure that random fluctuations, by themselves, do not force a facility into more frequent monitoring. It also allows a plant to take action, such as increased surveillance on a subset of valves, based on higher-than-normal leak rates in a single period, or to implement supplemental quality assurance programs to ensure that performance remains below base levels. The two-period rolling average also applies to semiannual and more frequent monitoring schedules. When annual monitoring is employed, the average of three out of four years of data is used to calculate the percent of valves that are leaking.

Minimizing the number of components in a process is one of the most effective means for reducing leaks, so the regulation includes an incentive for plants to do this. This incentive is in the form of credits for reductions in the number of valves, which can be applied to the calculation of percent leaking valves and the associated determination of monitoring frequency. The rule allows the owner or operator of the process unit to take a 67% credit for net reductions in the number of valves—*i.e.*, the plant may continue to count two out of every three valves removed from a process unit in the total number of valves when calculating the leak frequency.

The rule exempts valves that are "unsafe to monitor" from routine monitoring requirements, but requires monitoring as frequently as practicable during safe-to-monitor periods. "Unsafe-to-monitor" valves are defined as those that could expose personnel to imminent hazards from temperature, pressure, or explosive process conditions, such as valves at the top of or near high-pressure reactors.

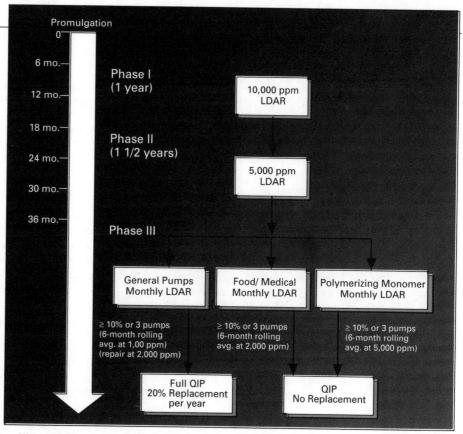

Figure 2. The pump standard will also be implemented in stages.

At existing process units, "difficult-to-monitor" valves are those that can be reached only through extraordinary means. Routine monitoring is not required for valves that require elevating personnel more than 2 m above any permanent support surface. The difficult-to-monitor exemption is available only for existing units, because routine monitoring of valves that have the potential to leak should be considered in the design of new facilities.

Small plants may have limited technical and financial resources, and the emissions from small plants with a quarterly LDAR program may be relatively low. Thus, plants with 250 or fewer valves will not be required to monitor more often than quarterly.

The framework for the pump standard (Figure 2) closely parallels that of the valve standard.

Phase I involves monthly LDAR and a leak definition of 10,000 ppm. Phase II is monthly LDAR and 5,000 ppm. Phase III is monthly LDAR and a leak definition of 1,000 ppm for pumps in general chemical service and a 10% (or 3 pumps) base performance level. However, repair is not required unless screening concentrations are 2,000 ppm or greater, because repair at concentrations lower than 2,000 ppm might result in significant and costly maintenance with little to no emission reduction.

Special considerations exist for pumps in food/medical service and those handling polymerizing monomers. For pumps in food/medical service, leaks are defined as 2,000 ppm. For pumps handling polymerizing monomers, a 5,000 ppm leak definition applies.

Phase III for pumps also contains a QIP provision. If 10% or more of the pumps (or 3 pumps) are found to be leaking, the plant must institute a QIP, with ultimate pump replacement over time. The only exception is for pumps in food/medical or polymerizing monomer service; if these pumps exceed the base performance level, the plant must enter into a QIP, but does not have to adhere to the pump replacement provisions of the QIP.

The 5-day first attempt and the 15-day repair intervals of the valve standard also apply to pumps.

Some pumps in chemical process units are not installed with a spare, and these cannot be repaired without shutting down the process. Thus, the standard has delay-of-repair provisions for pumps in such situations.

Because the variability in pump performance (expressed as percentage of pumps leaking) can be much greater than for valves, pump leak frequencies are to be averaged over a period of several months. A rolling 6-month average is to be used.

Pumps equipped with dual mechanical seal (DMS) systems and sealless pumps (for example, canned or magnetic pumps) can be included in the calculation of percent leaking pumps. For the purpose of this calculation, these pumps are assumed not to leak. The calculation procedure, therefore, gives credit for use of inherently low-leak designs and is intended to provide an incentive to install these designs.

For pumps with DMS systems, the use of light liquid VHAPs as a barrier fluid is prohibited. Heavy liquid VHAPs can be used as barrier fluids because the operation and monitoring requirements for DMS systems are sufficient to minimize any emissions from leaks from the DMS system. Sealless design pumps are subject to a weekly visual inspection and are not subject to an annual performance test requirement to demonstrate no emissions.

QIP for valves and pumps

The quality improvement program provisions require process units not achieving the base performance levels for valves to implement more frequent monitoring or a QIP designed to ultimately achieve the performance levels. Units not achieving the base performance levels for pumps must implement a QIP. If it undertakes these additional measures, a process unit not able to achieve the performance levels is not out of compliance. These provisions allow those plants that do use all elements of MACT but still do not

achieve the base performance levels the flexibility to develop process-unit-specific, cost-effective methods for improving emission performance. In addition, the QIP provides an innovative mechanism for focusing efforts on reducing emissions while avoiding lengthy enforcement actions.

For valves, two alternative QIPs are allowed, although the availability of both of these alternatives is restricted. The QIP may be a program that either demonstrates progress in reducing the percentage of leaking valves, or that implements a technology review and improvement program. The decision of which alternative QIP to use must be made during the first year of Phase III for both new and existing units. Under either alternative QIP, after the process unit has achieved less than 2% leaking valves, the facility may elect to continue the QIP. If the owner or operator discontinues the QIP, however, the QIP may not be used in the future if the unit again exceeds 2%

leaking valves; instead, monthly monitoring must be implemented until the unit has less than 2% leaking valves.

Demonstration of Progress QIP. This alternative allows a plant to design and implement a site-specific program to achieve steady progress in lowering the percentage of leaking valves. Any combination of measures, such as increased maintenance frequency or replacement of components, may be used, provided it achieves the required reductions in the percentage of leaking valves. This alternative QIP will be useful primarily for those process units with leak frequencies only slightly greater than 2%.

Under this QIP, the plant is required to continue quarterly monitoring and to demonstrate an average 10% reduction in the percentage of leaking valves each quarter, based on a rolling average of two quarters of monitoring data. If the facility fails for two consecutive averaging periods to demonstrate at least an overall average reduction of 10% per quarter, the owner or operator must either implement a technology review and improvement QIP or monthly monitoring.

If this QIP is continued after a process unit has less than 2% leaking valves, the plant must continue quarterly monitoring. It does not, however, have to continue to demonstrate the 10% reduction per quarter.

Technology Review and Improvement QIP. This QIP requires data collection and analysis, performance trials, and quality assurance and improvement.

Specific information must be recorded on each valve's design, materials of construction, packing, type, service conditions, and screening values. Engineering evaluations must be conducted to determine the causes of failure of all valves removed from the process unit due to leaks. The plant must then analyze the data to determine if there are specific problem services, operating conditions, valve types, designs, and materials of construction.

A second, and equally important, objective of the analysis is to identify valve designs and operating practices that will operate with less than 2% leaking valves under conditions comparable to those in the process problem areas. Such valve designs or technologies are referred to in this QIP as "superior performing valve technologies." Superior performance is defined as any combination of valve type and process, operating practices, and maintenance practices that achieves the base performance level of the standard.

Performance trials would be required where the data analysis for the process unit does not identify any superior performing valve designs or technologies that can be applied to the unit. Such tests would help determine whether candidate equipment or operation and maintenance procedures can be used in the conditions in the process unit. Small plants—those that have fewer than 400 valves and that are owned by a company with fewer than 100 employees—are exempt from the requirement to conduct performance trials.

The plant must then prepare a quality assurance plan that specifies minimum design standards for each category of valves, includes a written procedure for bench testing of valves, provides an audit procedure for quality control of purchased equipment, and includes procedures to ensure the quality of any rebuilt equipment. The quality assurance plan for replacement valves must be implemented as long as the process unit remains in the QIP.

The regulation does not require mandatory replacement of some proportion of the valves in the process unit each year. On an industry-wide basis, 7–10% of the valves are typically replaced each year for reasons other than leaking. If valves that are replaced, for any reason, are replaced with quality assured valves appropri-

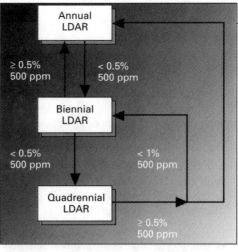

■ *Figure 3. The connector standard involves annual, biennial, or quadrennial leak detection and repair.*

ate to the service conditions, an improvement in emission performance can be achieved at nominal additional cost. This normal replacement of 7–10% of the valves each year was judged to be a reasonable rate of improvement.

The pump QIP. The QIP for pumps is a technology review and improvement QIP. It is similar to the technology review and improvement QIP for valves in that they specify similar requirements and have the same schedules. They differ in several respects due to differences between the two standards and between the equipment.

Unlike the valve provisions, the pump QIP is not an option that the owner or operator can select, and there are no restrictions on when a process unit (or plant) enters a QIP. An owner or operator must comply with the requirements of the QIP whenever the greater of either 10% of the pumps or 3 pumps in a process unit (or at a plant) leak. A process unit must remain in the QIP as long as the percentage of leaking pumps is greater than the base performance level. Theoretically, a process unit could be subject to the QIP provisions more than one time.

As with the technology review and improvement QIP for valves, a comprehensive database must be developed on pump type, design, materials of construction, service characteristics, and screening values. The engineering evaluation requirement differs in that inspections are required for pumps removed from the process due to leaks and for pump

seal designs associated with high failure rates. The data are then analyzed to identify problem areas and services, as well as to identify superior performing pump technologies.

For pumps in general chemical service, pumps or pump seals must be replaced with equipment and/or operations and maintenance practices identified as superior emission performance technology, at a rate of 20% per year. Mandatory replacement does not apply to pumps in food/medical service or to pumps handling polymerizing monomers.

Connectors

Connectors can be significant sources of emissions, even at well-controlled plants, due to the large number of connectors. However, the potential to reduce emissions does exist.

Assuring MACT performance requires less frequent monitoring than is necessary for pumps or valves, because connectors have no moving parts and once repaired they should remain leak-free for extended periods. A number of actions can be taken to reduce or eliminate leaks. In most cases, tightening the flange bolts on flanged connectors will eliminate a leak. In other cases it may be necessary to replace a gasket or to correct faulty alignment of surfaces, but these latter cases are expected to be relatively infrequent.

Therefore, the standard requires annual monitoring of connectors. Process units that demonstrate sustained performance at the level of the standard can monitor less frequently than annually.

For connectors, a leak is defined as a concentration of 500 ppm. This leak definition and a base performance level of 0.5% leaking connectors represents MACT. Process units that have 0.5% or more leaking connectors are required to implement an annual LDAR program for connectors. Process units that have less than 0.5% leaking connectors are allowed to monitor all connectors in a biennial or quadrennial program, as shown in Figure 3.

The 5-day first attempt at repair and the 15-day repair interval apply to connectors as well as to valves and pumps. However, in cases where the connector cannot be isolated from the process and cannot be repaired without shutting down the process, repair can be delayed until the next process unit shutdown.

In some situations, repair of a leaking connector would expose personnel to imminent danger. Therefore, a special provision in the connector standard allows the owner or operator to designate unsafe-to-repair connectors and to delay repair of these connectors until the next process unit shutdown.

It may not be possible to completely prevent nonrepairable connectors, and increased monitoring frequency, if triggered by nonre-pairable components, is likely to be of little benefit. Thus, up to 2% of the connectors can be excluded from the calculation of percentage of leaking if such sources are monitored for leaks within three months after repair and return to HAP service.

Owners or operators have the option of foregoing followup monitoring of disturbed connectors in exchange for not excluding nonrepairable connectors in the calculation of the percentage of leaking connectors. This provides flexibility and encourages owners or operators to develop their own programs to ensure low leak frequencies for connectors. In addition, an owner or operator can switch from one option to the other, provided the new option begins with annual monitoring.

The percentage of leaking connectors is calculated for each monitoring period and is not based on rolling averages of several monitoring periods. This is because properly designed, constructed, and installed connectors should operate for an extended time before failure and exhibit a low frequency of random failures.

A plant can take a partial credit in the calculation of the percentage of leaking connectors for connectors permanently removed from the process unit. If a connector is removed from a process unit by welding the connector or the pipe together, the integrity of the weld must be verified to be eligible for the credit.

As with valves, unsafe-to-monitor connectors are connectors that could expose personnel to imminent hazards from temperature, pressure, or explosive conditions. These connectors must be monitored as frequently as practicable during safe-to-monitor periods.

An alternative approach to setting regulations

Data gathered since the establishment of the existing fugitive emission regulations indicate that very low leak frequencies could be achieved. The data, however, do not reveal specific factors that lead to lower leak frequencies, nor indicate how low leak frequencies could be obtained at all chemical plants. In light of the need to identify MACT for equipment leaks and because of industry's general dislike for strict equipment and work practice standards, EPA turned to the Regulatory Negotiation process to develop a new approach in cooperation with environmental groups, industry, and state and local air pollution agencies.

What is Regulatory Negotiation? Regulatory Negotiation, or Reg Neg, is a formal process introduced at EPA several years ago to improve the regulatory development process, to foster cooperation among the interests affected by the outcome of the regulation, and to develop a better regulation acceptable to the range of interests affected. A negotiating committee is chartered under the Federal Advisory Committee Act and charged with achieving results within a specified time. All committee meetings are announced in the Federal Register beforehand and are open to the public.

Committee members represent the spectrum of competing interests, from the companies and industries affected to environmental groups, as well as EPA and state and local government entities. A neutral facilitator oversees the proceedings and helps direct them to resolution. Other official observers, such as the Office of Management and Budget, are present, as well as members of the public, who are often called upon to provide information and perspective.

The actual negotiating process is dictated by the participants, who develop protocols and agendas for each meeting. Key issues are discussed extensively, and each party is responsible for supporting its views and initiating possible resolutions. Proposals are submitted that not only achieve the goals of the party submitting them, but are also sensitive to the needs of the other parties. These often include "sweeteners" to make the proposal more acceptable. A successfully negotiated regulation is one that all parties can support as a whole, although individual provisions by themselves may be objectionable. The negotiators recognize that to obtain the most desirable provisions, one must trade off some of lesser concern and must negotiate in good faith.

Reg Neg is not a panacea or a preemptive alternative to the more conventional EPA rulemaking process. But given the appropriate ingredients, it can result in a better regulation for all concerned.

The negotiated equipment-leak rule. The committee that negotiated the equipment-leak rule considered the many factors and uncertainties associated with regulating equipment leaks at a wide variety of chemical plants and developed an acceptably balanced approach. The negotiators weighted the need to be flexible, the technical uncertainties, the requirement for MACT standards, and the data limitations. At the final negotiating session, the committee members conceptually resolved all outstanding major issues and over the following several months reached final agreement on the draft regulation and preamble. All committee members have agreed to support the standard providing that EPA proposes and promulgates a regulation and preamble with the same substance and effect as what is contained in the final agreement.

Inaccessible connectors and glass or glass-lined connectors are exempt from the requirement for routine monitoring. Inaccessible connectors are defined as those where monitoring would be extremely difficult, dangerous, or physically impossible. Examples include those that are: buried; insulated in a manner that prevents access by the monitor probe; obstructed by equipment or piping; or not accessible from a 25-ft portable scaffold on the ground and greater than 6 ft above a support surface. For both inaccessible and glass or glass-lined connectors, however, if leaks are detected by visual, audible, or other means, the leaking connector must be repaired no later than 15 days after the leak is detected.

The connector standard allows an owner or operator the option of exempting existing screwed connectors of 2 in. or smaller diameter from routine monitoring, provided these connectors are monitored once after the standard takes effect and after the seal is broken or disturbed.

Other process equipment

Compressors. The standard requires (a) the use of mechanical seals equipped with a barrier fluid system and controlled degassing vents, or (b) enclosure of the compressor seal area and venting of emissions through a closed-vent system to a control device. These systems can provide control efficiencies approaching 100%.

Pressure relief devices. Pressure relief devices in gas/vapor service must be operated with an instrument reading of less than 500 ppmv above background, except during pressure relief. They must operate with an instrument reading of less than 500 ppmv above background within five days after a discharge. As an alternative to rupture disks and other techniques that will achieve less than 500 ppm above background, plants may vent pressure relief devices to a closed vent system connected to a control device.

Sampling connection systems. MACT consists of the closed-purge sampling, closed-loop sampling, and closed-vent vacuum systems in the existing rules for sampling connection systems. Closed-purge sampling systems eliminate emissions due to purging by either returning the purged material directly to the process or by collecting the purge in a system that is not open to the atmosphere for recycle or disposal. Closed-loop sampling systems also eliminate emission due to purging by returning process fluid to the process through an enclosed system that is not directly vented to the atmosphere. Closed-vent vacuum systems capture and transport the purged process fluid to a control device.

Open-ended valves or lines. Emissions from open-ended valves or lines can be eliminated, except when the line is used for draining, venting, or sampling operations, by enclosing the open end by a cap, plug, or a second valve. The control efficiency associated with these techniques is approximately 100%.

Product accumulator vessels. The technique for controlling product accumulator vessels is to connect the vessel to a closed-vent system and control device. Anything in compliance with the requirements for closed-vent systems and control devices is acceptable.

Control devices. Control devices used to comply with the standard must reduce organic emissions by 95% or meet minimum design requirements. Enclosed combustion devices must provide a minimum residence time of 0.5 seconds at a minimum temperature of 760°C. Flares must comply with the requirements of 40 CFR 60.18. Vapor recovery systems must achieve a control efficiency of at least 95%, which is the highest control efficiency that has been demonstrated consistently for such vapor recovery systems as carbon adsorption or condensation units.

Agitators. Although not covered by existing fugitive emissions regulations or any other standard, agitators are included in the standard for several reasons. First, limited screening data indicate that agitators may be a significant source of emissions. Second, agitators are technologically similar to pumps, and, like pumps, emissions can be controlled using seal technology. However, agitators have longer and larger diameter shafts than pumps and produce greater tangential shaft loadings. The performance of pump seal systems, therefore, cannot be used to estimate agitator seal performance.

Thus, only a monthly LDAR program is required. A leak is defined as a concentration of 10,000 ppmv or higher. This program will require replacement of agitator seals with significant leaks and will encourage development of effective bearing and seal systems.

Instrumentation systems. Instrumentation systems typically contain valves 0.5 in. in diameter or less and connectors 0.75 in. or less and are located in confined areas such that monitoring of individual components is generally not feasible. Because these systems provide critical process operating information, they are subject to frequent surveillance and maintenance to assure the reliability of measurements. Thus, leaking equipment in these systems would be readily detected by changes in temperature, pressure, flow rates, or by observation. In addition, it is common practice after maintenance or repair to verify the integrity of these systems by soap bubble testing or pressure checks. Therefore, a routine LDAR program would be redundant and would provide no benefit, and so is not required.

Instead, the standard contains alternative provisions for these systems. Leaking instrumentation systems must be repaired in a timely manner. In addition, monitoring of individual components in instrumentation systems is not necessary if any detected leak is repaired and the repair is verified by soap testing, a pressure check, or any other visible, audible, olfactory, or other means.

Equipment in heavy liquid service. Heavy liquid service is defined as HAP fluids with vapor pressures less than 0.3 kPa at 20°C. Pumps, valves, connectors, and agitators in heavy liquid service, and pressure relief devices in light liquid or heavy liquid service, are excluded from the routine monitoring and inspection requirements. However, if leaks from these sources are detected, the same allowable repair intervals that apply to pumps, valves, connectors, and compressors would apply.

Alternative standards

If EPA establishes work practice, equipment, design, or operational standards, then it must allow use of alternative means of emission reductions if an owner or operator can demonstrate emission reductions equivalent to that achieved by the standards. Generally, alternative means of emission reduction are based on specific circumstances at individual plant sites and must be handled individually. The equipment-leak regulation, however, includes generic alternative standards for two situations—batch processes and enclosed process units.

Batch processes. Equipment used in batch processes may contain process fluids for only a brief period during the batch. Therefore, it would be difficult to schedule monitoring of equipment, and the cost could be much higher than for a continuous process. Moreover, this intermittent operation and low time-in-use of individual components also result in lower annual emissions.

Owners or operators of batch processes have the option of meeting standards similar to those for continuous processes or of periodically pressure testing the batch equipment using either a gas or liquid. If monitoring for leaks is selected, the monitoring frequency is prorated to the time-in-use, and the equipment may be monitored when it is in VHAP service or in use with a surrogate VOC or other detectable compound.

Enclosed units. Processes operated in buildings or enclosures maintained under negative pressure and vented to the atmosphere through a 95% efficient control device are exempt from the LDAR monitoring requirements.

Recordkeeping and reporting requirements

Records of leak detection, repair attempts, and maintenance for leaking

R. S. COLYER is an environmental engineer with the U. S. Environmental Protection Agency, Research Triangle Park, NC. He is responsible for developing national emission standards for hazards air pollutants (NESHAP) and standards of performance for new stationary sources (NSPS). His current projects include standards that are the subject of this article, for gasoline marketing systems, and for ethylene oxide emissions from commercial sterilizers. He has also worked on regulations for hazardous waste treatment, storage and disposal facilities, NSPS for bulk gasoline terminals, and NSPS for residential wood heaters. He holds a BS in biology and environmental science from Centre College of Kentucky, an MS in environmental science and an MEng in environmental engineering from the Univ. of Louisville, and an MBA from Duke Univ.

J. MEYER is an environmental engineer with the U.S. Environmental Protection Agency, where she develops NESHAP and NSPS. She is presently working on the hazardous organic NESHAP of which the equipment leak standards will be a part. Previously, she worked on NESHAP for benzene emission sources. She holds a PhD in environmental engineering from the Univ. of North Carolina at Chapel Hill.

equipment are required by the standard. These records consist of the information needed to document compliance with the standard and for evaluating the effectiveness of repair efforts.

For components subject to equipment standards, records of the dates of installation, equipment repair, and equipment modifications are required. For closed-vent systems and control devices, records of the location and type of equipment, the design specifications, and the monitoring parameters are required.

These records must be maintained for a period of two years in a readily accessible recordkeeping system and made available to the EPA upon request. This system may be maintained by physically locating the records at the plant site, or by accessing the records from a central location by computer.

The standard requires an initial report, semiannual reports of leak detection and repair efforts, and notifications of initiating monthly monitoring or a QIP for valves or pumps.

The initial report notifies EPA that the process unit is (or will be) subject to this regulation and identifies the process unit, number and type of equipment components, and the method of compliance for each piece of equipment. The planned monitoring schedule for each type of equipment in each phase must also be reported.

The semiannual report must include information on: the numbers of leaking and nonleaking components; leak frequencies of pumps, valves, and connectors; attempts to repair; reasons for delay of repair; process unit shutdowns; and changes in the information submitted in the initial report. The reports must also include the results of performance tests conducted within the reporting period for equipment subject to a 500 ppm performance standard and notifications of initiation of monthly monitoring or QIP for valves or pumps.

The rule's significance

The negotiated equipment-leak regulation is designed to meet the objectives of all affected parties, and it is supported as a whole by all the members of the negotiating committee. It is a tight, but flexible, standard that provides disincentives and rewards where appropriate and creates innovative approaches such as QIPs that protect the industry from penalties when base performance levels are not achieved.

This is a precedent-setting regulation under the Clean Air Act Amendments of 1990, and is testimony that EPA, industry, and environmental groups can work together to develop a complete rule for a complicated emission source. **CEP**

Estimate Fugitive Emissions from Process Equipment

Various methods, with varying degrees of accuracy, can be used to quantify emissions from equipment leaks. Here's how to decide which to use.

Joanne R. Schaich,
Eastman Chemical Co.

The U.S. Environmental Protection Agency (EPA) has defined equipment-leak fugitive emissions as emissions from sources such as pumps, valves, flanges, compressors, open-ended lines, pressure-relief devices, and sample connection systems. At some plants, 70–90% of the air emissions have been estimated to be the result of fugitive emissions.

Plants must calculate these emissions for reports required by Title III of the Superfund Amendments and Reauthoriztion Act (SARA Title III), for new source permitting, and, in some cases, for state emission inventories. Furthermore, regulatory agencies will base future regulations on the amount of fugitive emissions calculated and on the projected impact of these emissions on the community.

To estimate fugitive emissions, plants can use a variety of methods, including SOCMI factors [developed by EPA as averages for the synthetic organic chemicals manufacturing industry (SOCMI)], leak/no-leak factors, stratified factors, EPA correlation curves, and process-specific curves. Depending on the complexity of the method, testing may be involved. Testing, which consists of screening or bagging (or both) of equipment, can improve the emissions estimates from a plant and provide insight into how to reduce emissions from equipment. Generally, the more detailed and accurate the estimating method selected, the more it will cost.

A strategy flow chart developed by the Chemical Manufacturers Association (CMA), Figure 1, puts the different methods into perspective and summarizes some of the decision criteria. Basically, the approach involves first using the simplest technique, SOCMI emission factors, to estimate the emissions for the various sources of leaks. If the plant is not satisfied with these values, then screening can be conducted. If the resulting estimates still seem inadequate, bagging can be conducted, and emissions are calculated using generic or process-specific correlations. Once the plant is satisfied with the accuracy of the emission estimates, the data are combined to yield the total fugitive emissions for the plant.

SOCMI emission factors

EPA developed average SOCMI emission factors based on studies of the petroleum refining industry, 24 different types of SOCMI production facilities, and six maintenance facilities. Table 1 summarizes these average emission factors.

The first step in using this method (and for most of the other methods) is to obtain

Table 1. Average SOCMI emission factors for estimating fugitive emissions.

Equipment	Service	Emission Factor (kg/h/source)
Valves	Gas	0.0056
	Light Liquid	0.0071
	Heavy Liquid	0.00023
Pump Seals	Light Liquid	0.0494
	Heavy Liquid	0.0214
Compressor Seals	Gas/Vapor	0.228
Pressure-relief Devices	Gas/Vapor	0.104
Flanges	All	0.00083
Open-ended Lines	All	0.0017
Sampling Connections	All	0.0150

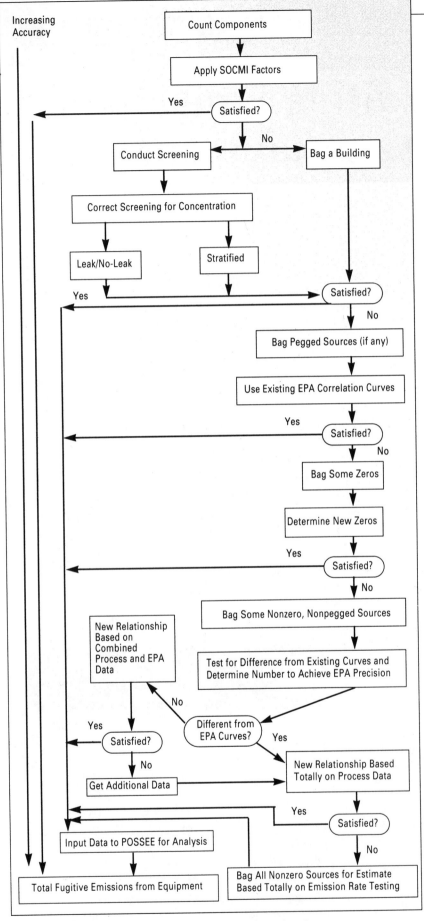

Increasing
Accuracy

Count Components

↓

Apply SOCMI Factors

↓

Yes ← Satisfied? → No

Conduct Screening ← → Bag a Building

↓

Correct Screening for Concentration

↓   ↓

Leak/No-Leak   Stratified

Yes   ↓

→ Satisfied? → No

Bag Pegged Sources (if any)

↓

Use Existing EPA Correlation Curves

↓

Yes ← Satisfied? → No

Bag Some Zeros

↓

Determine New Zeros

↓

Yes ← Satisfied? → No

Bag Some Nonzero, Nonpegged Sources

↓

New Relationship Based on Combined Process and EPA Data

Test for Difference from Existing Curves and Determine Number to Achieve EPA Precision

↓

No ← Different from EPA Curves? → Yes

Yes ↓

Satisfied? ↓ No

Get Additional Data → New Relationship Based Totally on Process Data

↓

Yes ← Satisfied? → No

Input Data to POSSEE for Analysis

↓

Total Fugitive Emissions from Equipment   Bag All Nonzero Sources for Estimate Based Totally on Emission Rate Testing

■ *Figure 1. The engineer has a variety of options for estimating fugitive emissions. Source: CMA.*

a count of the plant's equipment that could release fugitive emissions. This is called a component or equipment count. If detailed engineering drawings are unavailable or if a rough estimate is needed, the CMA guidance manual (see the sidebar) provides a number of ways to develop an estimated equipment count.

For each type of equipment, multiply the emission factor by the number of pieces of that equipment to obtain the emission rate for that source type. Do the same for each of the other types of sources in the process unit. Then add the emission estimates for all types of equipment to obtain an estimate of the fugitive emissions from that unit.

This simplified method is based on the following assumptions:

1. The leak frequency at a particular plant is similar to that of the average process in EPA's studies.

2. All components emit at rates similar to those of like equipment in petroleum refinery processes.

Past CMA studies indicate that the SOCMI factors overestimate emissions by 10 or even 10,000 times the actual emissions found during testing.

Leak/no-leak factors

In the leak/no-leak approach, EPA defines a leak as a source with

J. R. SCHAICH, a senior environmental affairs representative with Eastman Chemical Co., Kingsport, TN, is responsible for coordinating air permits for the firm's plants in Arkansas, South Carolina, and Tennessee. She worked as a process design engineer for four years before moving into the Environmental Affairs Dept. of Eastman Chemical Co. A 1981 graduate of the Univ. of Tennessee with a BS in chemical engineering, she is a member of AIChE and has served as chair of the East Tennessee AIChE Local Section.

a measurable emission concentration of 10,000 ppm or greater as measured at the interface of the leak. Plant personnel use a hand-held analyzer to screen each component to determine whether emissions are above or below 10,000 ppm.

Then, in a manner similar to the SOCMI method, the plant calculates emissions by multiplying the number of components by an emission factor. There are now, however, two emission factors, one for sources screening above 10,000 ppm (called leaking sources) and one for sources screening below 10,000 ppm (nonleaking sources), as listed in Table 2.

Stratified factors

The stratified factors method requires more precise data than the leak/no-leak approach because emission factors are available for three leak ranges: 0–1,000, 1,001–10,000, and over 10,000 ppmv. These factors are given in Table 3.

As with the leak/no-leak approach, the plant screens each component and determines how many sources fall into each of the three ranges. The plant then generates an emission estimate by multiplying the number of components in each screening-value range by the appropriate emission factor.

The assumptions underlying both the leak/no-leak and the stratified factor approaches are as follows:

1. Components within each range emit at rates similar to like equipment in petroleum refinery processes.

Table 2. Leaking and nonleaking emission factors for estimating fugitive emissions.

Equipment	Service	Leaking (≥10,000 ppm)	Nonleaking (<10,000 ppm)
		Emission Factors (kg/h/source)	
Valves	Gas*	0.0451	0.00048
	Light Liquid	0.0852	0.00171
	Heavy Liquid	0.00023†	0.00023
Pump Seals	Light Liquid	0.437	0.0120
	Heavy Liquid	0.3885	0.0135
Compressor Seals‡	Gas	1.608	0.0894
Pressure-relief Devices	Gas	1.691	0.0447
Flanges	All	0.0375	0.00006
Open-ended Lines	All	0.01195	0.00150

*The leaking and nonleaking emission factors for valves in gas/vapor service are based on the emission factors determined for gas valves in ethylene, cumene, and vinyl acetate units during the SOCMI maintenance study.

†Leaking emission factor is assumed to be equal to the nonleaking emission factor because the computed leaking emission factor (0.00005 kg/h/source) was less than the nonleaking emission factor.

‡Emission factor reflects existing control level of 60% found in the industry.

2. The EPA factors assume that less than half of the sources will register zero during screening for most equipment types. Data from CMA member companies, however, indicate much higher percentages of zero-screening sources.

3. Plants need to develop response factors for their type of instrument and calibration gas to match EPA's data. A response factor is the ratio of the actual concentration of the compound to the observed concentration on the detector.

Correlation curves

The next level of refinement in calculating fugitive emissions involves correlation curves. Mathematical correlations estimate emissions by providing a continuous function over the entire range of screening values instead of discrete intervals.

Help from the CMA

The Chemical Manufacturers Association has developed a series of videotapes and a guidance manual entitled "Improving Air Quality: Guidance for Estimating Fugitive Emissions from Equipment." This manual illustrates how to calculate fugitive emissions and how to conduct screening and bagging tests. (For more information, contact CMA, 2501 M Street NW,

Washington, DC 20037-1303; 202/887-1100.)

To help industry better estimate fugitive emissions and effectively reduce them, CMA developed the Plant Organizational Software System for Emissions from Equipment (POSSEE). POSSEE helps plants set up a testing program, provides a format for data entry, and enables facilities to develop

plant-specific emissions estimates.

POSSEE is part of a long-term CMA project to collect and compile data from plants that use the system into an industrywide database. Analysis of the database will lead to the development of better fugitive emission estimating techniques and help plants find ways of reducing fugitive emissions from equipment.

Table 3. Stratified emission factors for estimating fugitive emissions.

Equipment	Service	Emission Factors (kg/h/source) Screening-Value Ranges, ppmv		
		0–1,000	1,001–10,000	Over 10,000
Valves	Gas/Vapor	0.00014	0.00165	0.0451
	Light Liquid	0.00028	0.00963	0.0852
	Heavy Liquid	0.00023	0.00023	0.00023
Pump Seals	Light Liquid	0.00198	0.0335	0.437
	Heavy Liquid	0.00380	0.0926	0.3885
Compressor Seals	Gas/Vapor	0.01132	0.264	1.608
Pressure-relief Devices	Gas/Vapor	0.0114	0.279	1.691
Flanges, Connections	All	0.00002	0.00875	0.0375
Open-ended Lines	All	0.00013	0.00876	0.01195

EPA has published four correlation curves, which are shown in Figure 2, that relate screening values to leak rates. The data for these curves were derived from tests where the components were enclosed and their leak rates were measured. The equations for the curves are given in Table 4. To use them, one needs to have component counts by source and screening concentrations; some leak-rate measurements are recommended.

This approach has the potential for the greatest accuracy. The database supporting the curves, however, had very few points below 100 ppmv, and, as a result, the low ends of the curves are not well defined. Recent data from CMA member companies show that

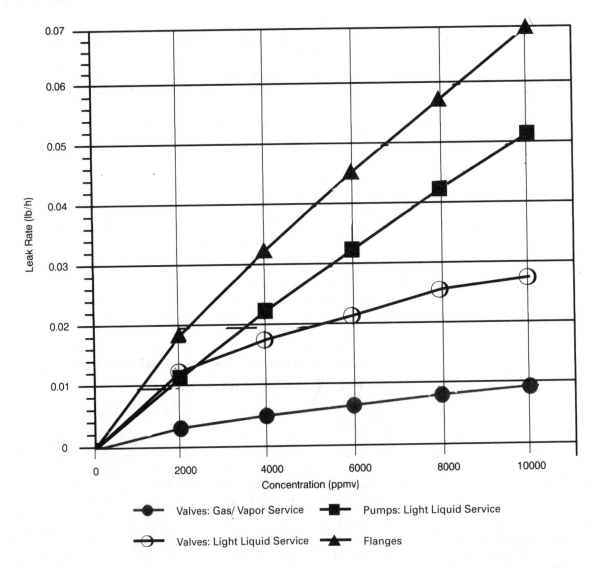

Figure 2. EPA has published correlation curves for estimating fugitive emissions from valves, pump seals, and flanges.

few components screen above the instrument's lower limit of detection. Therefore, making some leak-rate measurements at very low and at zero-screening values when using the curves is important. (Otherwise, EPA has set the "default-zero" screening value, which is applied to equipment components screening between 0 and 8 ppm, at 8 ppm.)

Leak-rate measurements are also necessary for processes that have "pegged" sources (*i.e.*, those sources whose screening value is known to be above the instrument's maximum reading but whose actual screening value is unknown).

The following assumptions are involved with using these published curves:

1.The curves accurately relate the screening concentration to the leak rate over the range of observed screening concentrations.

2. The default zero value and associated emission rate accurately reflect the emissions from those sources whose screening values were between 0 and 8 ppm.

Unit-specific correlations

The unit-specific-correlation method results in the most accurate fugitive emission estimate. However, it requires additional work, including obtaining more screening and bagging data.

Using this method, a plant determines the actual emission rates of equipment in three ranges: those that screen below the screening instrument's lower level of detection, those within its range of detection, and those greater than its upper detection limit (pegged sources). This is done by bagging, which involves isolating the emission source in an enclosure or tent and measuring the leak rate. Generally, fewer than 30 bags will be adequate to develop new equations.

From these bagging data, the plant calculates a leak rate vs. screening value correlation curve that is specific to the unit tested.

More information on all of these estimation methods is available in CMA's fugitive emissions guidance manual (see the sidebar). **CEP**

Table 4. Equations for transforming screening values to leak rates.*

Source Type	Equation†
Valves	
Gas/Vapor Service	$lb/h = 3.766\ (10^{-5.35})(ppmv)^{0.693}$
Light Liquid Service	$lb/h = 8.218\ (10^{-4.342})(ppmv)^{0.47}$
Pump Seals	$lb/h = 2.932\ (10^{-5.34})(ppmv)^{0.898}$
Flanges	$lb/h = 2.02\ (10^{-4.733})(ppmv)^{0.818}$

*These equations are based on the maintenance study after removing pegged sources. EPA recommends using the curve for pumps in light liquid service for all other source/service categories not listed in this table.

†lb/h = nonmethane leak rate; ppmv = maximum screening value using an organic vapor analyzer.

FURTHER READING

Code of Federal Regulations, Title 40, Part 60, Appendix A. Reference Method 21, "Determination of Volatile Organic Compound Leaks."

Chemical Manufacturers Association, "Improving Air Quality: Guidance for Estimating Fugitive Emissions from Equipment," 2nd ed., Washington, DC (1989).

U.S. Environmental Protection Agency, "Protocol for Generating Unit-Specific Emission Estimates for Equipment Leaks of VOC and VHAP," Research Triangle Park, NC, Publication No. EPA-450/3-88-010 (1988).

Kittleman, T. A., *et al.,* "Fugitive Emissions Field Testing of Three Processes: Test Results and Conclusions," presented at the Air Pollution Control Association Annual Meeting (June, 1987).

Stelling, J. H. E. (Radian Corp.), "Emission Factors for Equipment Leaks of VOC and HAP," prepared for the U.S. Environmental Protection Agency, Research Triangle Park, NC, Publication No. EPA-450/3-86-002 (1986).

U.S. Environmental Protection Agency, "Fugitive Emission Sources of Organic Compounds—Additional Information on Emissions, Emission Reductions, and Costs," Research Triangle Park, NC, Publication No. EPA-450/3-82-010 (1982).

Blacksmith, J. R., *et al.* **(Radian Corp.),** "Problem-Oriented Report: Frequency of Leak Occurrence for Fittings in Synthetic Organic Chemical Plant Process Units," prepared for the U.S. Environmental Protection Agency, Research Triangle Park, Publication No. EPA-600/S2-81-003 (1981).

Langley, G. J., and R. G. Wetherhold (Radian Corp.), "Evaluation of Maintenance for Fugitive VOC Emissions Control," prepared for the U.S. Environmental Protection Agency, Research Triangle Park, NC, Publication No. EPA-600/S2-81-080 (1981).

Wetherhold, R. G., and L. P. Provost (Radian Corp.), "Emission Factors and Frequency of Leak Occurrence for Fittings in Refinery Process Units," prepared for the U.S. Environmental Protection Agency, Research Triangle Park, NC, Publication No. EPA-600/2-79-004 (1979).

ACKNOWLEDGMENTS

The author is grateful for the assistance of Debbie Stine and Karen Fidler for their contributions to this article. She would also like to recognize all of the participants responsible for the development of these estimations methods during the last decade, including EPA and members of the CMA Fugitive Emissions Work Group. In addition, she thanks T. A. Kittleman for his technical review of the article.

Environment: Air, Water, and Soil **143**

Control Fugitive Emissions from Mechanical Seals

Proper selection and operation of mechanical seals will become increasingly critical as new air-pollution regulations are implemented. Advanced mechanical seal technology can help meet the challenges posed by fugitive emissions control requirements.

William V. Adams,
Durametallic Corp.

Because of their widespread use in the chemical process industries (CPI), mechanical seals for rotating equipment deserve careful evaluation as a means to control fugitive emissions. As a result of the 1990 Clean Air Act Amendments, it is anticipated that specific regulations governing fugitive emissions of volatile organic compounds (VOCs) from general service process pumps and other rotating equipment will be promulgated in 1993. The regulations are expected to be implemented in three phases over a period of four years, with increasingly stringent requirements.

How seals rate today

As shown in Table 1 *(1)*, there is an enormous spread in emission rate indices for various methods of pump emissions control. Depending on the type of design and the nature of the application, mechanical seals can exhibit emission levels that range from very low to effectively zero.

The Seals Technical Committee (STC) of the Society of Tribologists and Lubrication Engineers (STLE) established an Emissions Working Group (WG-3) to monitor existing and pending legislation and to evaluate current seal technologies that can be able to comply with legislative regulations. The WG-3 group recently published a set of guidelines *(2)* that relate recommendations for selecting sealing systems to the specific gravity of the pumped fluid and the target level of allowable emissions, Figure 1.

A key inference to be drawn from this chart is that current mechanical seal technology is more than able to control fugitive emissions at or below the levels expected to be called for in the upcoming regulations. Even a single seal, the most economical type available, can control emissions to less than 1,000 ppm if it embodies today's advanced design technology.

A survey of 107 pump seals in light-hydrocarbon service at three refinery facilities in the Los Angeles basin was conducted last year in accordance with U.S. Environmental Protection Agency (EPA) Method 21 *(3–4)*. Although 99% of the seals tested were single seals, 91% of the total sample had emissions of less than 10,000 ppm. Moreover, 83% of the seals

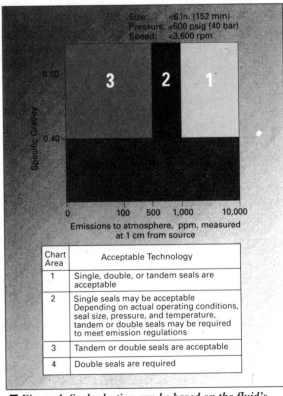

Chart Area	Acceptable Technology
1	Single, double, or tandem seals are acceptable
2	Single seals may be acceptable Depending on actual operating conditions, seal size, pressure, and temperature, tandem or double seals may be required to meet emission regulations
3	Tandem or double seals are acceptable
4	Double seals are required

■ *Figure 1. Seal selection can be based on the fluid's specific gravity and the maximum allowable VOC emission levels.*

leaked less than 1,000 ppm. From a practical standpoint, perhaps the most significant finding was the fact that only 9% of the seals accounted for 91% of the total emissions encountered in the survey.

Statistical analysis of the survey data and follow-up investigations of leaking seals led to three key conclusions of immediate relevance to plant and process designers.

1. Excessive seal leakage is a direct symptom of the misapplication of a seal and/or improper operation of the seal or its associated rotating equipment.

2. Few seals leak abnormally, and these can be readily identified and corrected.

3. A strong correlation exists between the level of seal leakage and the mean time between failure (MTBF) of its associated equipment.

The correlation between low emissions and pump MTBF cannot be overemphasized. The follow-up investigations revealed that proper seal selection was just one of several factors determining the overall performance of rotating equipment. Other significant issues included the selection and installation of environmental control systems, the mechanical condition of rotating equipment, proper start-up and operating procedures, and effective troubleshooting techniques. All are related to both reduced emissions and extended pump MTBF.

How to reduce fugitives

Obviously, the first step toward bringing rotating equipment into compliance with clean air requirements is the identification of "problem" pumps and/or seals—*i.e.,* those leaking in excess of effective regulations. In addition to identifying the problem units, the cause(s) of subpar performance should be determined as well. Simple replacement or upgrading of a mechanical seal to bring a unit into compliance might well ignore the underlying source of the problem. For example, it may be necessary to upgrade the mechanical condition of the associated rotating equipment.

If changeout of a seal is called for, upgrading the existing seal chamber with an enlarged-bore retrofit seal chamber as spelled out in the revised ANSI B-73 stan-

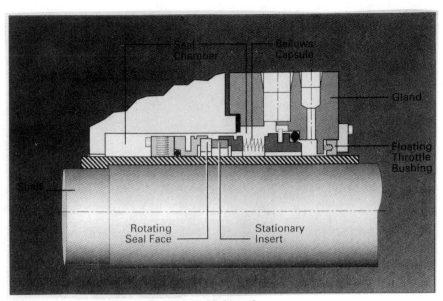

■ *Figure 2. Single seals, such as this welded metal-bellows seal with a floating throttle bushing, have fewer parts than other types of mechanical seals.*

dard and API Standard 610, seventh edition, should be considered. Extensive field and laboratory evaluations have shown that, on average, seal life is doubled when a properly designed and applied seal is operated in an enlarged-bore seal chamber *(5).*

Basic sealing guidelines

As indicated in Figure 1, a fair degree of freedom exists in the selection and application of various mechanical designs, including single seals. What will be required is a clear grasp of the capabilities and limitations of each type of seal and the manner in which it is operated.

Single seals. Single seals, Figure 2, which have a minimum number of components and usually require a minimum of support equipment, are highly economical and reliable systems. Laboratory and field test data show that emissions from single seals can readily be controlled to levels below 1,000 ppm. In many cases,

Table 1. The amount of leakage from a centrifugal pump varies considerably depending on the type of seal installed.

Seal Type	Leakage Index
Packing with no sealant	100
Packing with sealant	10
Single mechanical seal, flushed	1.2
Tandem seal	0.15
Double seal	0.004

emission levels less than 100 ppm can be achieved.

In single seals, the pumped product is generally used to provide the lubricating film between the rubbing seal faces. With this type of arrangement some trace amounts of product will escape to the atmosphere unless the atmospheric side of the seal is vented to a vapor recovery system.

Tandem seals. Tandem seals, Figure 3, are becoming increasingly popular as a method for containing leakage in the event of a failure of the primary seal. The primary inner seal is cooled and lubricated by the product, usually through a bypass line from the pump discharge. During normal operation, the secondary outer seal operates at atmospheric pressure in a clean, externally supplied sealing fluid. Because of its function and pressure relationship to the pumped product, the sealing fluid in this instance is usually referred to as a buffer fluid. Tandem seals can provide zero emissions of product to the environment, especially if the specific gravity of the product is lower than that of the buffer fluid, the product is immiscible with the barrier fluid, and the buffer fluid is not a hazardous air pollutant.

The buffer fluid isolates the product from the atmosphere but cannot contaminate the product because of its lower pressure. Therefore, the secondary seal is seldom the first to fail. If the primary seal fails, the leakage to the secondary seal system will contaminate the buffer fluid and increase the pressure in the cavity between the two seals. The pressure rise can be registered on a gage or can activate an alarm, and the outer seal will assume the primary sealing role until repairs can be made. A supply tank or reservoir support system is used to maintain the buffer fluid.

If the buffer fluid in the supply tank is a hazardous air pollutant, emissions from the outer seal must meet regulation limits. To qualify for an exemption from monthly monitor-

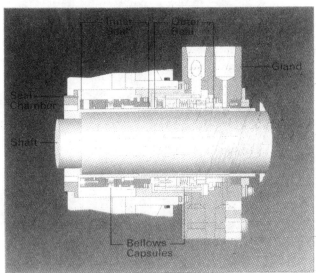

■ *Figure 3. In a tandem seal, the secondary outer seal serves as a backup in case the primary inner seal fails.*

ing requirements, the tandem seal system must be connected to a suitable vapor recovery system.

Recently, seal manufacturers have introduced tandem seals that are designed primarily for emissions control applications. One such device was developed specifically for use with enlarged-bore seal chambers and features metal-bellows compression units and heavy-cross-section, flexibly mounted stationary seal faces to improve resistance to distortion caused by pressure and temperature extremes.

Double seals. A double mechanical seal, Figure 4, is basically two conventional seals installed either back to back or face to face in a seal chamber. The two seal assemblies operate in a buffer fluid (sometimes also called a barrier fluid) maintained at a pressure higher than that of the pumped product. Double seals create an artificial environment in the seal chamber in order to extend seal life in conditions where the pumped fluid is toxic, abrasive, volatile, or corrosive. A clean barrier fluid, such as water, is circulated between the two seals at a pressure higher than that of the product. The inner seal prevents the barrier fluid from entering the product, while the outer seal prevents the barrier fluid from leaking to the atmosphere. In addition to acting as an effective barrier between the product and the atmosphere, the barrier fluid serves

as a lubricant for both seal faces.

Double seals are recommended for products with specific gravities less than 0.4 because of the inadequate lubrication qualities of these fluids for single or tandem seals. They may be connected to their own self-contained auxiliary barrier fluid systems or pressurized from an external source (as outlined in ANSI Piping Plan 7332/API Plan 32). Using a supply tank limits product dilution to a few gallons if the inner seal fails.

Double seals are also the proper choice when suitable vapor recovery systems do not exist.

Pressure reversals caused by an upset in the process system or a failure of the outer seal can unseat the inner seal faces and leak product into the barrier fluid unless the inner seal is balanced in both directions and the inner stationary insert is held in place by a retainer.

Seal design factors

In addition to basic mechanical seal configurations, several key design factors will affect emissions and MTBF. Parameters such as face width, face loading, balance ratio, and face deflection should be reviewed in detail with the mechanical seal manufacturer, especially in critical applications. Special hydraulic balancing may be required, for example, for the effective sealing of light hydrocarbons that must be handled near their vapor pressure.

The application of finite element analysis (FEA) is extremely useful for minimizing emissions and maximizing MTBF. Where high temperatures and/or pressures could lead to excessive seal face deflection in a conventional seal design, FEA may be the only feasible way to develop a suitable geometry that controls deflection and provides the correct balance between seal face lubrication and leakage. In such situations, it is also advisable to work closely with the seal manufacturer and to specify maximum pump operating conditions and any cyclic conditions that may occur. Major seal manufacturers have completed this kind of design analysis for standard products and can respond quickly with solutions for off-design conditions.

Proper material selection constitutes another key to low emissions and longer MTBF. From the standpoint of emissions control, some materials such as Stellite (an alloy of chrome and cobalt, with small amounts of nickel, iron, and tungsten), Ni-Resist (a high-nickel, austenitic cast iron), and sprayed-on tungsten carbide coatings should be considered obsolete as face materials. These materials are extremely prone to heat checking and galling when subjected to upset or cyclic conditions. The resultant irregularities of the seal mating surfaces measurably affect seal emissions.

Problems have also been linked to the improper application of polytetrafluoroethylene (PTFE) and flexible graphite secondary seals. Both of these materials have relatively high gasket-sealing factors (6) compared with elastomers and, therefore, require higher mechanical loads in order to achieve low emissions. Instead, using the appropriate O-rings is recommended in services requiring low emissions.

Preferred seal face materials have a high modulus of elasticity, good heat-transfer properties, and low coefficient-of-friction characteristics. Opposing seal faces of carbon vs. silicon carbide is the best combination of materials for most critical applications. The carbon should be a premium resin-filled grade. If the seal faces are running in a light hydrocarbon, the silicon carbide should be a fine-grained, reaction-bonded grade. Silicon carbide exhibits a wide range of chemical resistance and great forgiveness to upset conditions that may occur with a process cycle. Also, momentary periods of pump cavitation or dry running are more readily tolerated.

Tungsten carbide (with nick-el binder) is also widely used as a mating face with carbon. Again, it is advisable to consult with a seal manufacturer to obtain the optimum material selection for specific applications.

Environmental controls for sealing systems

Once the appropriate design and materials of construction have been established for a mechanical seal application, environmental controls become the next priority factor determining the proper operation and service life of the seal. Seal flushing involves introducing a fluid through the seal chamber at a flow rate sufficient to maintain the correct seal environment. Circulation of a fluid through the seal chamber carries away both the frictional heat generated at the seal faces and the heat generated by turbulence around the seal.

For fugitive emissions control, maintaining an adequate boiling point margin (BPM) in the seal chamber is essential. BPM is the difference between the seal chamber temperature and the initial boiling point (IBP) temperature at the seal chamber pressure. Inadequate BPM can create numerous adverse effects (7). Seal-generated heat can initiate flashing of light hydrocarbons between the seal faces, which in turn can cause dry-running, unstable conditions leading to material failures. Thus, it is imperative that a BPM of at least 9°C (15°F) or at least a 170-kPa (25-psi) pressure margin be maintained above the IBP of the fluid in the seal chamber. With low-specific-gravity fluids that exhibit rapid pressure increases with small temperature increases, a BPM higher than 9°C will result in longer seal life and reduced emissions. BPM can be increased by cooling the seal chamber or increasing the seal chamber pressure.

The proper amount of cooling is determined by performing a heat balance around the seal. This is done in basically the same way for single-seal and dual-seal arrangements; however, the amount of heat generated from each seal must be considered. Every seal generates heat at the seal faces. In some cases, heat soak (i.e., heat absorbed) from the pumped product must be compensated for as well. To assure the greatest accuracy, the actual developed heat load should be determined with the assistance of the manufacturer of the specific seal in question.

Bypass flush. If an analysi shows that cooling is required for a single seal or the inner seal of a tan-

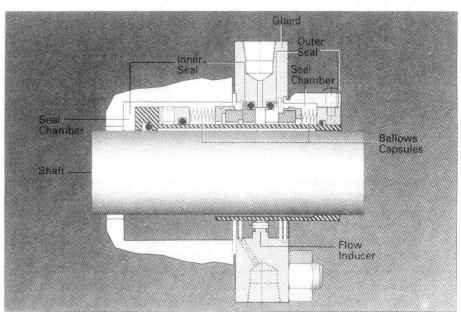

■ *Figure 4. A double seal is basically two conventional seals installed back to back or, as here, face to face.*

dem seal, it is usually done by circulating a fluid through the seal chamber by a bypass flush from the pump discharge. The most commonly used flushing method is the product, or bypass, system (ANSI Plans 7311 and 7313, API Plans 11 and 13), where flushing is achieved using pipelines to recirculate the product to the seal faces. The pumped fluid used to flush the seal must be clean and at a temperature well below the product's vapor pressure at the seal chamber pressure.

An orifice may be required in the flush line to limit flow to that necessary to cool the seal without erosive damage to the seal components. In some cases, where the product is near its IBP at chamber pressure, the seal chamber pressure may be increased by not using an orifice in the bypass flush line, but instead installing a floating bushing in the throat of the chamber. Such a bushing will increase the seal chamber pressure and reduce damaging flashing and popping of the product at the seal faces.

External flush. In some applications, a compatible fluid from a clean, cool, external source can be used as the flush for single seals to maintain a proper environment (ANSI Plan 7332, API Plan 32). The flush should be free of debris that could migrate between the faces, plug the springs of pusher-type seals, and hang up O-rings. The flush must be at a pressure at least 100 kPa above the product pressure and at a flow sufficient to remove seal-generated heat and to prevent the pumped fluid from enter-

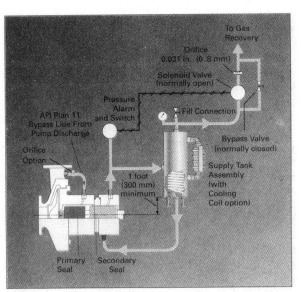

■ *Figure 5. Induced circulation system for tandem seals.*

ing the seal chamber. A velocity of 4.6 m/s (15 ft/s) at the throat is recommended. A close-clearance floating bushing installed in the bottom of the seal chamber can reduce the flush rate by increasing the velocity at the throat for a given flow.

Induced circulation systems. Tandem- and double-seal systems require special consideration. All liquids absorb gases up to the liquid's saturation point. Because the saturation point is inversely related to the temperature of the liquid, the buffer or barrier fluid's temperature rise should be limited or gas molecules may come out of the liquid. The bubbles can cause foaming and lead to a loss of the liquid's heat-transfer capability and a loss of lubrication at the seal faces. The maximum temperature rise should be limited to 20°F for VOCs and 40°F for other compounds. In addition, the temperature should be maintained at least 50° below both the fluid's flash point and the limits of the

seal's materials of construction.

Environmental controls for all dual seals (tandem and double) can involve the induced circulation of a buffer fluid through the seal chamber and an external supply tank or reservoir. Cooling is usually accomplished by adding coils to the supply tank assembly. In addition to cooling coils, supply tanks can also be furnished with high- and low-level switches to actuate suitable alarm devices.

Induced circulation systems are recommended with all double and tandem sealing systems equipped with a supply tank. A device may be built into the seal assembly or a small external pump may be used.

Figure 5 shows an induced circulating system for tandem seals (ANSI Plan 7352, API Plan 52). A circulating feature induces flow of the buffer fluid through the seal chamber from a supply tank while the pump is rotating. The supply tank is maintained at atmospheric pressure and often contains cooling coils to dissipate the sealing system heat load.

In some systems, an orifice in the piping from the supply tank to a gas-recovery unit allows normal primary seal weepage to vent and not accumulate in the buffer fluid. In the event of a primary seal failure, the orifice will not handle the excess flow and a supply tank high-pressure switch can activate an alarm that will notify the operator to schedule seal maintenance.

Figure 6 shows a pressurized supply tank with induced circulation for double seals (ANSI Plan 7353, API Plan 53). A built-in circulating shroud pumps the barrier fluid through a supply tank, maintained typically at a pressure 170 kPa (25 psi) above the product pressure. The circulating shroud must be sized for adequate head and flow to provide proper lubrication and cooling. Optional level and pressure switches may be used in this system as well.

Buffer or barrier fluids. Ideally, a buffer or barrier fluid exhibits the following characteristics:

W. V. ADAMS is vice president for engineering, research, and quality assurance for the Durametallic Corp., Kalamazoo, MI (616/381-2650; Fax: 616/381-8368). He has served on the ASME B-73 Chemical Pump Standards Committee for the last five years and is past chairman of the Seals Technical Committee of the Society of Tribologists and Lubrication Engineers (STLE). Currently, he is chairman of the WG-3 Emissions Task Force of STLE and is Technical Committee coordinator for all of STLE. He holds a BSME from Western Michigan Univ. and is the author of a number of articles on the design, application, and failure analysis of mechanical seals.

- presents no health, fire-safety, or environmental hazards;
- is compatible with seal materials of construction;
- does not foam, coke, or form sludge;
- has a flash point above the maximum operating temperature of the system;
- has a boiling point above the maximum operating temperature;
- exhibits good heat-transfer characteristics;
- is compatible with the process fluid (in the case of double seals).

There are five basic types of buffer or barrier fluids: 1) water, 2) ethylene glycol (EG) and EG and water mixtures, 3) alcohols, 4) diesel fuel or automatic transmission fluid (ATF), and 5) process fluids such as raffinate. Selection of a barrier fluid is based on the fluid's physical properties at the seal's operating pressure and temperature and on the environmental compatibility of its leakage. In the case of double seals, the barrier fluid must also be compatible with the process. (Commercial products such as EG and ATF contain additives for rust control and other functions that may not be compatible with the process.) If a hazardous fluid is used as the barrier fluid, the outer seal must meet emission limits.

Instrumentation

Proper instrumentation will be essential for seal installations involving critical emission control standards. The equipment installed will vary with local plant procedures and regulations.

In many cases, on-site temperature, flow, level, pressure, and leak indicators may be adequate if the data are recorded by the operator on regularly scheduled tours and a system is in place to act promptly on the operator's observations. In critical remote applications, alarms may be necessary. Instrumentation can warn the operator of immediate or impending problems so that corrective action can be taken and repairs

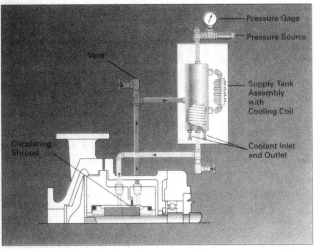

■ *Figure 6. Pressurized supply tank with induced circulation for double seals.*

scheduled before emissions levels and durations exceed what is allowed by the regulations.

Economic considerations

Notwithstanding the mandate of environmental regulations to reduce fugitive emissions, leaking seals can impose a substantial economic penal-ty. Product loss from a seal leaking more than 10,000 ppm amounts to 437 g/h or some 3,800 kg/yr. At an average refinery product cost of 32¢/kg [based on data from the EPA (11), updated to 1991 dollars using a 4%/yr inflation rate], that loss amounts to roughly $1,215/yr per seal.

Upgrading seal performance to 1,000–10,000 ppm will result in a product saving of 3,535 kg/yr or $1,131/seal per year. Thus, economic considerations alone would argue for the upgrading of severely leaking seals.

An analysis of the effect of reducing leakage from the 1,000–10,000 ppm range to absolute zero, however, shows a saving of only $90/year. This further reduction is not justified on the basis of product savings or for any expected EPA regulations. **CEP**

LITERATURE CITED

1. **Lipton, S.,** "Fugitive Emissions," *Chem. Eng. Prog.,* pp. 42-47 (June 1989).

2. **Adams, W. V., et. al.,** "Guidelines for Meeting Emissions Regulations for Rotating Machinery with Mechanical Seals," STLE Publication SP-30 (Oct. 1990).

3. U.S. Code of Federal Regulations, Title 40, Part 60, Appendix A, Reference Method 21, "Determination of Volatile Organic Compound Leaks."

4. **U.S. Environmental Protection Agency,** "Protocols for Generating Unit-Specific Emissions Estimates for Equipment Leadk of VOC and VHAP," EPA-450/3-88-010 (Oct. 1988).

5. **Battilana, R. E.,** "Better Seals Will Boost Pump Performance," *Chem. Eng.,* pp. 106-114 (July 1989).

6. **Swick, R. H.,** "Designing a Leak-Proof Gasket," *Machine Design,* pp. 100–103, (Jan. 22, 1976).

7. **Will, T. P., Jr.,** "A Powerful Application and Troubleshooting Method for Mechanical Seals," *Proc. 2nd Intl. Pump User's Symp.* (1984).

8. API Standard 610, 7th edition, American Petroleum Institute, Washington, DC (1989).

9. **Bloch, H. P.,** "Pump Reliability Improvement Methods for Economy," seminar by H. P. Bloch Process Machinery Consultant (1990).

10. **Flitney, R. K., and B. S. Nau,** "Reliability of Mechanical Seals in Centrifugal Process Pumps," BHRA International Fluid Sealing Conf., paper A2, Cannes, France (1982).

11. **U.S. Environmental Protection Agency,** "VOC Fugitive Emissions in Petroleum Refining Industry—Background Information for Proposed Standards," NTIS No. PB83-157743 (Apr. 1981).

Minimize Fugitive Emissions With a New Approach To Valve Packing

Most packing systems will meet EPA emission requirements— the question is for how long. This approach can keep emissions low far longer than conventional stem-sealing systems.

Ronald Brestel, *et al.*, Valve Packing Development Team
Fisher Controls International, Inc.

The issue of how long a valve packing system will be able to meet the U.S. Environmental Protection Agency (EPA) emission requirements is crucial because short packing life means frequent replacement and more frequent monitoring. In addition, packing systems are challenged by conditions common in chemical process industries plants—such as when the operating conditions are severe, when the process fluid is prone to leakage, and when the valve strokes frequently. And, EPA leak limitations must be met even while the valve is stroking.

Newly developed valve packing technology, incorporating antiextrusion rings and live loading in an integrated valve packing system, can meet all of these challenges. This technology can be applied to both polytetrafluoroethylene (PTFE) and graphite packing systems for sliding stem and rotary valves. These applications of the technology have been performance tested to demonstrate that the packing will continue to meet EPA requirements for stem leakage—with little or no maintenance—for 10–50 times as long as conventional stem-sealing systems.

This packing technology was developed specifically for use in control valves, although the concepts are equally applicable to manually operated valves. Because of the frequent stem movement and the need to keep stem friction low for smooth actuator operation, control-valve packing presents a more difficult challenge than manual-valve packing. The packing systems described in this article are designed to maintain their seal even while the valve stem is moving.

We should note that there is some disagreement within the packing community over some of the principles on which this approach to packing is based. However, the application of these principles does result in valves that have lower leakage rates and longer service lives than conventional packing. The information presented here should provide the engineer a stronger basis for evaluating valve packing systems.

Packing principles

As the result of a four-year research and development program, four major packing-design principles were developed. A valve-packing system that follows these principles (plus other supporting design criteria) will have a long, low-maintenance service life and will meet EPA leak requirements throughout that life. These principles are:

1. prevent the packing from extruding out of the packing area by installing less pliable antiextrusion rings on either side of the packing;

2. keep the valve stem aligned with stem bushings installed near the packing;

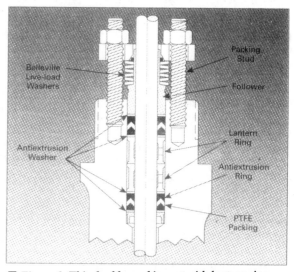

■ *Figure 1. This double-packing set with lantern ring incorporates the features required for low leakage and long life.*

3. minimize the adverse effects of thermal cycling by using only the minimal amount of packing required to effect a seal;

4. apply a constant and proper packing stress with live-load springs; the proper stress depends on the type of valve and the type of packing system.

In addition, the stem must be protected from physical damage and corrosion in order to provide a good sealing surface. Other metallic components of the packing system must also be corrosion resistant to assure that they remain functional. Packing elements themselves must be selected to minimize voids (air spaces) within and between them to minimize consolidation.

Some of these principles may be inconsistent with many of the types of valve packing now widely used in the CPI (for example, braided filament graphite systems). The validity of these principles, however, has been demonstrated by comparative performance testing of scores of different packing designs.

Any valve-packing system must be tested to evaluate its sealability and service life. These tests should be conducted at the process temperature and pressure and include thermal cycling. It is also important to recognize that accelerated performance testing of packing in a laboratory cannot duplicate plant conditions of vibration, chemical attack, and the aging of packing materials.

Principle 1

Prevent packing from extruding out of the packing area by installing less pliable antiextrusion rings on either side of the packing.

Packing must deform in order to seal the space between the stem and the packing bore. This deformation is effected by pushing against the packing with a follower ring. This axial load stresses the packing axially and causes it to bulge radially.

Unless special measures are taken to contain it, packing material will gradually be lost from the packing system. This loss can occur from erosion and extrusion.

Erosion loss occurs because the stem rubs against the packing and slowly scrapes off tiny pieces of packing. It can be minimized by keeping the stem smooth and free of corrosion.

Extrusion loss occurs because the stress imposed on the packing, although necessary to effect a seal, can also force the packing material past the follower and out of the packing area. Extrusion loss can occur with any material that deforms under load. Because of its tendency to cold flow, PTFE packing is more susceptible to extrusion loss than most commonly used packing materials.

As packing is lost, stress on the packing lessens, which weakens the seal between the stem and packing. The seal can be reestablished by retightening the packing (moving the follower) frequently, but only as long as sufficient packing material remains in the packing area.

A live load (*i.e.,* a spring) can alleviate, but not eliminate, the stress reductions caused by material loss. Live load is discussed in more detail later.

Antiextrusion ring. Packing loss by extrusion can be prevented by containing the packing with another element. Unlike the packing itself, this antiextrusion ring need not contain the process fluid; it must fit on the stem only tightly enough to retain the packing material.

The antiextrusion ring must meet two key criteria:

1. it must fit closely to the stem but without damaging the stem;

2. it must be less pliable than the packing itself so that it transfers the load from the follower into the packing, yet pliable enough to form a seal to contain the packing material.

The specific design required for an antiextrusion ring depends on the type of valve and packing and is discussed later.

Principle 2

Keep the valve stem aligned with stem bushings installed near the packing.

Sealing a valve stem that does not remain concentric to the valve bore requires that the packing continually change its shape. This places greater demands on a packing system.

A packing system of "graded pliability" components can help keep the stem aligned. The most pliable material, the packing, is retained by a less pliable antiextrusion ring, which is in turn retained by a very hard bushing. The bushing serves to keep the stem aligned and prevents extrusion of the inner components of the packing system. A PTFE-lined packing follower can also serve as

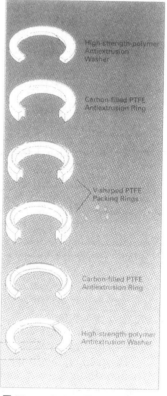

■ *Figure 2. An effective PTFE packing arrangement for rotary valves consists of V-shaped packing rings held in place by high-strength-polymer and carbon-filled PTFE antiextrusion rings.*

a bushing to help keep the stem aligned.

Principle 3

Minimize the adverse effects of thermal cycling by using only the minimal amount of packing required to effect a seal.

If loss of packing material is a major cause of valve leaks, it would appear that excess packing should be used. In fact, performance tests show that using excessive packing reduces packing life and increases stem leakage.

Stress measurements in packing demonstrate that the maximum radial packing deformation, and thus the position of the seal between the packing and stem, occurs over a very limited length of the packing arrangement. The length of this seal does not increase with increased packing height, so using additional pliable material adds nothing to the sealability of the packing. Additional packing can, however, detract from the sealability and service life of the packing in several ways.

First, additional packing material adds to the stem area being squeezed by the packing and thus increases friction. In the case of high-friction graphite packing, this additional friction can increase the required actuator size and add to erosion loss of packing material.

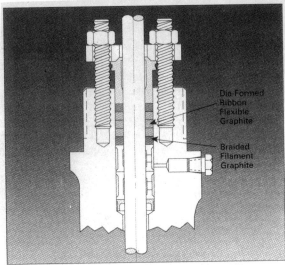

■ *Figure 3. In a traditional graphite packing set, the flexible graphite packing is retained by braided filament graphite.*

Second, using more packing material merely increases the potential for consolidation. Consolidation, the loss of packing volume due to the elimination of voids in and between packing rings, has the same effect on packing sealability as the loss of packing material itself.

Third, the adverse effect of thermal cycles on extrusion loss will increase if there is more packing material trying to expand out of the packing area. The effect of thermal cycles on the performance of PTFE packing was demonstrated in two tests.

In the first test, a PTFE packing set containing five V-shaped packing rings was installed with a torque of 100 in.-lb on the packing flange studs. Table 1 shows the effect of tempera-

ture cycles on the packing stress as measured by follower torque. The follower torque was not readjusted during the tests.

As the table shows, packing load increased as the PTFE tried to expand when heated. Packing material, however, was lost by extrusion. When the temperature dropped to ambient, the packing load dropped to near zero because of the loss of material, and the packing leaked. Each subsequent time the packing temperature was increased, the load increased, but not to as great a level as in the previous temperature cycle, again demonstrating the loss of packing material. After a few thermal cycles, enough packing material was lost so that even the thermal expansion would not increase the load enough to prevent leaking. The packing material that extruded out of the packing area was clearly visible when the system was taken apart for inspection.

A similar test used only two V-shaped PTFE packing rings, but retained them with antiextrusion rings. The results of this test are also shown in Table 1.

To date, this test has continued for three thermal cycles and 750,000 mechanical cycles with the same results. There has been virtually no loss of packing material and no leakage in excess of EPA limits, in spite of the fact that springs have continually kept the packing under stress during the entire test.

Principle 4

Apply a constant and proper packing stress with live-load springs. The proper stress depends on the type of valve and the type of packing system.

Some controversy has surrounded the use of a live load, or springs, to keep a relatively constant stress on packing. Some studies have shown that live loading reduces packing life and does not contribute to improved sealing.

Indeed, there are circumstances

Table 1. The effects of thermal cycling on PTFE packing.

	Temperature Cycle Number					
	0	1	1	2	2	3
Five V-shaped packing rings						
Temperature, °F	70	300	70	300	70	300
Torque, in.-lb	100	110	10*	60*	10*	50*
Two V-shaped packing rings retained with antiextrusion rings						
Temperature, °F	70	300	70	300	70	300
Torque, in.-lb	100	110	80†	110	80†	110

*Leak rate 5–10 times the rate allowed by EPA regulations.

†Reduced torque caused by consolidation of the packing material.

where live loading can be counter-productive to packing service life and sealing efficiency. For example, in the absence of antiextrusion rings, packing that is constantly stressed will continuously lose material by extrusion. As the packing volume decreases, the spring expands and reduces both the stress on the packing and the integrity of the seal.

Performance tests on graphite packing containing elements of braided filament graphite also show that the filament often breaks down when subjected to the high live load required for many severe-service applications. In addition, live loading can cause stem-alignment problems unless the packing set contains stem-alignment bushings near the packing.

Lightly stressed packing can provide a seal, and, if the valve application is not severe (i.e., there are no thermal cycles, the operating temperature is low, and the system contains high-viscosity lubricating fluids) it may even meet EPA standards. But when leak rates must be low in spite of severe process conditions, live loading is essential to maintaining the correct stress in the packing. The live load will not contribute to rapid extrusion loss of packing material and short packing service life if 1) it is used in conjunction with the correct antiextrusion system, and 2) the load applied is correct for the packing system (type of packing, type of valve, and type of antiextrusion ring).

For example, if the load is too high for the packing set, it can extrude the packing past the antiextrusion rings and/or increase stem friction. If the load is too low, the packing will not deform sufficiently and the valve will leak.

Live loading can also somewhat alleviate the adverse effects of thermal cycling. If there is room for the live-loading spring to compress, it can relieve the packing of some (but not all) of the excess stress of thermal expansion and reduce extrusion loss. With a conventional packing gland, however, the thermal

expansion and resultant stress increase must all go to increasing friction and extrusion loss.

Belleville washers. Belleville washers are commonly used as the springs for live loading. Basically a formed metal washer with its inside diameter pushed higher than its outside diameter, a Belleville washer is a compact disk spring that can provide the high loads needed for live-loaded packing. Usually incorporated into the packing follower, Belleville washers must be sized and arranged specifically for each packing system. It is essential that the right number of the right-size washers be used, and used in the correct arrangement in order to provide the correct load.

To make sure that they are assembled correctly when packing systems are installed, replaced, or adjusted, the Belleville washers can be pre-assembled as a set for each valve. There should also be some indication within the live-loading system to tell a mechanic how far the washers should be compressed.

For valves that do not have enough space on top of the stem for the correct size and number of Belleville washers, multiple sets of Belleville washers can be provided on the packing studs.

The Belleville washers must also be protected from corrosion. Because corroded washers can lose their

strength and corrosion products (rust) can constrict their movement, they should be made of a high-alloy, corrosion-resistant material.

Applying the principles

Packing designs that do not conform with these principles can still meet EPA standards. They may even have a long service life if the operating conditions are not too severe.

Performance tests, however, show that when a packing system implements all of the principles, it will meet EPA leak standards under severe process conditions where other packings would fail. A packing system that applies all of these principles can also meet those EPA leak standards—even while the valve is stroking—for an extended period of time. The service life of such packings usually extends beyond the time the valve needs to be repaired for some reason other than to replace the packing.

PTFE packing for sliding-stem valves

PTFE is one of the most versatile and widely used packing materials. It has a very low coefficient of friction, it deforms easily, and it can typically handle temperatures up to 450°F.

PTFE, however, has two problems that must be overcome in order for the packing to have a long service life and to continue to effect a satisfactory seal throughout that life. First, PTFE tends to cold flow and extrude through small openings or along the stem even under low loads. At the high loads required to seal around a valve stem, it extrudes out of the packing area very quickly. Second, PTFE has a coefficient of thermal expansion 10 times greater than steel, which significantly increases packing stress and extrusion loss during thermal cycles.

After performance tests using dozens of antiextrusion ring systems, it was determined that a two-component system would be highly effective at

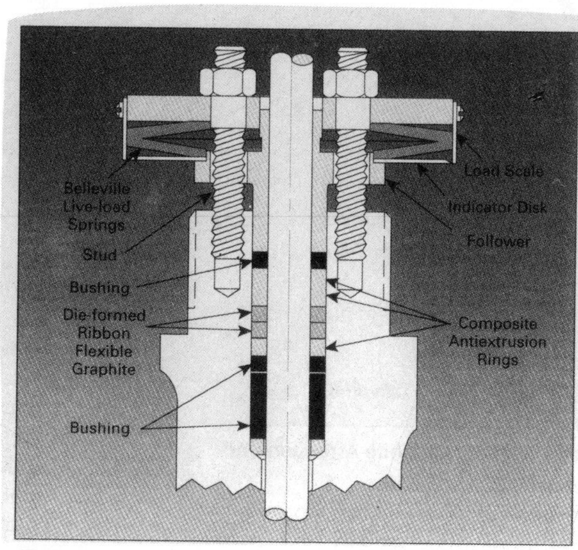

Belleville Live-load Springs
Stud
Bushing
Die-formed Ribbon Flexible Graphite
Bushing
Load Scale
Indicator Disk
Follower
Composite Antiextrusion Rings

■ *Figure 4. A flexible-graphite/amorphous-carbon composite ring is effective at retaining flexible graphite packing.*

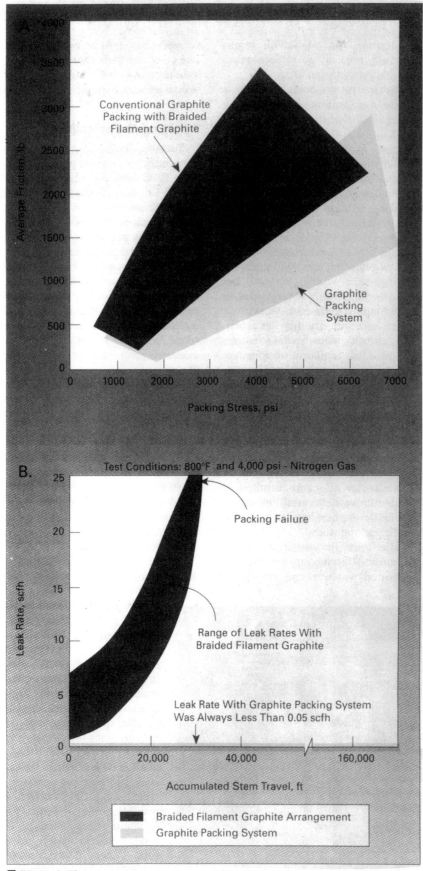

confining the PTFE packing. Antiextrusion rings made of carbon-filled PTFE installed on both sides of the PTFE packing will confine it and wipe the stem of any material picked up from the environment or from the packing. Reduced packing erosion, as well as the effectiveness of antiextrusion rings, can be further improved if the stem is polished to an RMS 4 finish or better.

Because the antiextrusion ring is PTFE based, it won't scratch the stem. Its PTFE content, however, can extrude slightly, albeit much less than the pure PTFE packing rings. To further contain the antiextrusion ring, close-fitting, antiextrusion washers should be added to the outside of the packing set. The packing set would then look like the one shown in Figure 1.

PTFE packing for rotary valves

Conventional wisdom teaches that a rotary valve is easier to seal than a sliding-stem valve because 1) rotary valves have less packing-to-shaft travel distance for each stroke of the valve, thus decreasing loss of packing material due to friction, and 2) dust and other abrasives are not "dragged in" from the environment as the shaft slides in and out of the packing, so the sealing surface between stem and packing stays cleaner.

The radial shaft deflection inherent in rotary valves can, however, make executing the four packing principles more difficult. As the shaft rotates, it deflects within the bearing system and pushes radially against the packing. As the shaft moves radially, the packing has to deform radially to continue to effect a seal. This deformation can only be achieved if the packing is highly stressed. Generally, with PTFE packing, a rotary valve needs higher stress levels than sliding-stem valves to deform the packing sufficiently to meet EPA leak standards in severe-service applications and while the valve is rotating. The exact load required must be determined experimentally.

Because of this high live load, plus the additional high stress generated by

■ *Figure 5. The composite ring system produces lower and more consistent stem friction than a conventional graphite packing system (a) . . . and it has a much longer service life(b).*

shaft deflection, preventing extrusion loss is more difficult for rotary valves than for sliding-stem valves. In addition, rotary valves can be limited in the space available for packing. Performance tests using a variety of antiextrusion ring designs demonstrate that long service life and low leakage rates can be achieved with a packing set consisting of two V-shaped PTFE packing rings contained by two carbon-filled PTFE antiextrusion rings and then further contained with two scarf-cut antiextrusion washers made of a high-strength polymer.

In order to retain the carbon-filled PTFE rings at the high stress required for a rotary valve, the antiextrusion washers must fit tightly on the stem. A scarf-cut washer can provide essentially zero clearance. The scarf cut opens so that the washer slips over the shaft without gripping the shaft. The antiextrusion washer must also be hard enough so that it will not extrude. Figure 2 illustrates how these components can be arranged in a rotary valve. The components are similar to those used in the sliding-stem PTFE system except for the design of the outer retaining washers.

Graphite packing

Sliding-stem valves. Although PTFE is the preferred packing for most applications, graphite is generally used when the application exceeds the limits of PTFE or when fire-safe packing is required (for example, a packing that will not melt or decompose when exposed to the high temperatures of a fire).

In a conventional graphite packing set, the flexible graphite packing is retained on each side with braided filament graphite, as shown in Figure 3. Under moderate conditions of temperature and pressure, this packing system is often satisfactory. But the braided filament graphite can fail under the severe conditions of temperature and pressure often encountered by a graphite-packed valve.

Flexible graphite packing rings require a very high minimum stress—about 2,000–3,000 psi—to deform the graphite enough to fill voids around

the stem and packing bore and effect a seal. If braided filament graphite surrounds the packing, however, the filaments can add sufficient stem friction to prevent all of the load from reaching the flexible graphite packing rings. Thus, even more load—up to 4,000 psi or more—can be required to effect a seal. This high load further increases stem friction.

The purpose of surrounding the flexible graphite rings with the braided graphite filament is to prevent the flexible graphite from extruding under a load, but the braided filament graphite itself can break down under high load. The filament is also porous, so it provides a poor seal for retaining process fluid. In addition, the binders that are necessary to fabricate braided filament graphite can be incompatible with high process temperatures and speed the degradation of the braid.

Performance tests using dozens of antiextrusion ring systems demonstrated that a ring made of a composite of flexible graphite and amorphous carbon would be highly effective at retaining the flexible graphite packing. In addition, the composite ring, although less pliable than the packing, should itself be retained by an even harder carbon bushing. The composite ring confines the flexible graphite, while the harder carbon bushing confines the composite ring and serves as a stem guide, as illustrated in Figure 4.

This article was written by the Valve Packing Development Team at Fisher Controls International, Inc., Marshalltown, IA. The coauthors are Ronald Brestel, Charles Wood, P.E., and Wilber Hutchens of the design engineering group; Ronnald Funk, Frank Jensen, and Ali Zarghami of test engineering; Will Wears and Larry Pothast, design drafters; and Clarence Scheffert, William C. Scheffert, and Virgil Pepper, test technicians. Together, this group has more than 200 years of valve design experience at Fisher Controls.

In addition to preventing the extrusion of the flexible graphite, the composite rings wipe the stem. This keeps particles of flexible graphite contained in the packing area and prevents exterior contamination from entering the packing area. As with PTFE sliding-stem packing, reduced packing erosion (and the effectiveness of antiextrusion rings) can be further enhanced if the stem is polished to an RMS 4 finish or better.

The minimum packing stress required to deform the flexible graphite ring and produce a seal between the stem and packing bore is about 2,000–3,000 psi. To maintain the seal against process pressure, however, the axial stress on the packing must exceed the process pressure. The exact load required for each packing system must be determined experimentally.

The friction of any graphite packing system can be quite high and require large actuators. But the valve packing described here produces substantially lower and more consistent stem friction than a conventional graphite packing system with its multiple graphite rings and braided graphite filament, Figure 5a. It also has an exceptionally long life—over 160,000 ft of accumulated stem travel, compared with about 25,000 ft for a braided filament graphite arrangement, Figure 5b.

This packing system is designed to have a long service life and a low leak rate in high-pressure, superheated-steam service and other severe-duty applications. For applications that need the high-temperature performance or fire safety of graphite packing but must also meet the EPA requirements for stem leakage, wafers of PTFE can be inserted between the packing elements. The PTFE wafers add sufficient lubrication and sealing to reduce the leak rate to EPA standards.

Rotary valves. The packing principles also apply to graphite packing systems for rotary valves. With the high friction of graphite packing, however, rotary valve packing systems using graphite must be designed, and the packing material specified, to minimize packing damage due to erosion. **CEP**

Reduce Fugitive Emissions Through Improved Process Equipment

These pointers will help CPI plants meet present and future emission limits.

Sydney Lipton,
Consultant

The Clean Air Act Amendments of 1990 mandate reductions in emissions from chemical process industries (CPI) plants. [The requirements of the Act have been described in other numerous articles, such as *(1),* and seminars, so details will not be reviewed in this article.] Here we focus on those provisions relevant to equipment control, specifically regarding fugitive emissions, or equipment leaks. Fugitive emission sources are non-point sources such as valves, flanges, and pumps. An earlier group of articles *(2–5)* discusses fugitive emissions at length. This article provides additional information about the selection of new equipment and the upgrading of existing equipment.

Clean Air Act overview

The Clean Air Act requires significant fugitive emissions reductions. The control of hazardous air pollutants is normally cited as the primary concern, but other areas must also be considered when assessing the requirements for improved fugitive emission control.

Title I of the Act is concerned with nonattainment and the reduction of volatile organic compound (VOC) emissions. Since many CPI plants are in nonattainment areas, the restrictive nonattainment regulations also impact VOC fugitive emission control systems. The combination of VOC requirements and hazardous air pollutant control in Title III of the Act suggests that the selec-

■ *This fire test is capable of evaluating gasketing and packing materials to API and company specifications. (Photo courtesy of Garlock, Inc.)*

tion of equipment controls should be based on severe emissions limitations.

Title III also impacts upon emission controls through considerations of safety risk management and accidental release control *(6–8).* With increasing attention on plant worker safety and exposure of workers and the public to pollutant releases, risk assessments are mandated under the Clean Air Act *(7).*

Consequently, fugitive emission controls should also be selected for fire resistance and mechanical strength to meet risk assessment objectives. All accidents are investigated and risk assessments are emphasized, so fugitive emission control systems contributing to an accident would be included in a revised risk analysis as potential problem elements. Since satisfactory performance of emission control equipment under adverse conditions is necessary, control systems capable of meeting stringent service conditions are required to eliminate consideration as risk elements in an overall risk assessment analysis.

The Act's Title V covers permitting and permission to operate, which is required under Titles I and III. A permit also includes a compliance program to meet stringent regulatory targets, including fugitive emission limits. [Van Wormer and Iwanchuk (9) provide guidance regarding the permitting program.]

Fire safety

Few emission control systems can be tested adequately to ensure safe operations in a process unit. However, certain valve packing fire tests and a new gasket fire test are satisfactory for assessing performance under fire conditions.

Valve fire safety tests. Valve fire tests were originally designed to evaluate leakage of the process fluid across or through a typical gate, ball, plug, or butterfly element when the valve is blocked. Although the tests were not developed to observe leakage to the atmosphere, they are satisfactory for evaluating packing performance under severe conditions. In addition to evaluating packing performance and line-side leakage, a critical aspect of the test is the ability of the valve to close or block a process line under fire conditions. Poor valve closure performance in this situation may contribute to the development and expansion of a fire. Thus, a valve must have both fire-resistant packings and internals that can maintain a seal and mechanical integrity under adverse conditions.

Three valve fire safety tests are widely used and typify the various tests conducted in North America and Europe. These are the American

Table 1. General valve applications in the CPI.

Valve Type	Industry Segments
Gate	Petroleum Refining
Globe	Chemical Manufacturing and Petroleum Refining
Plug	Chemical Manufacturing
Ball	Chemical Manufacturing and Petroleum Refining (small)
Butterfly	Chemical Manufacturing and Petroleum Refining
Control Valves (Automatic Operation)	
Reciprocating (Sliding Stem)	Chemical Manufacturing and Petroleum Refining
Quarter-Turn	Chemical Manufacturing and Petroleum Refining
Valve Materials	
Metal	Petroleum Refining
Metal and Plastic	Chemical Manufacturing and Petroleum Refining*

* Special safety requirements

Petroleum Institute's API-607, the Exxon EXES 3-14-1-2A, and the British Standard OCMA FSVI tests. The Exxon and British tests are similar in requirements and test materials; the API test is less severe and is conducted with water, which is a less desirable medium.

Existing valves must be modified where necessary, and new valves must be specified.

The Exxon fire test, Exxon EXES 3-14-1-2A, was developed to simulate field conditions, and it provides a realistic assessment of the valve, internals, and packings under severe conditions. This test is recommended to thoroughly assess the fire resistance of valves and valve packings. Test descriptions are provided in the Exxon Standard and elsewhere (10). Briefly, the Exxon test requires heating the valve body to 1,200°F for a minimum of 15 minutes and then quenching the valve with water to simulate fire fighting effects. The valve is closed from the open position during the test and leakage through the valve seats is checked. In addition, leakage from the valve stem during the fire test should not result in flames greater than 4 in. high. Under fire con-

ditions, the stem leakage within this tolerance is very small.

The fire test is a pass-or-fail test and valve or packing acceptance is based upon test results. Consequently, test performance requires the presence of a purchaser's representative to ensure that the test is conducted properly and that all materials are thoroughly evaluated. While a few vendors have fire testing capability with kerosene, the severity of the test and the importance of the test results necessitate verification by the consumer. Since fire tests for qualification are conducted infrequently, consumer representation should not be a problem from a manpower perspective.

Gasket fire safety tests. A system for testing flange gaskets similar to that described for valves has not been developed. Placing flame sources around a flange introduces mechanical changes that do not permit a satisfactory assessment of the gasket. Flames impinging on flange bolts result in bolt stress changes that affect flange and gasket stresses. Leakage under these conditions does not accurately reflect gasket capabilities. Furthermore, standardizing a flange gasket test is difficult because other mechanical variables, such as bolt age, bolt stress, flange face alignment, and face roughness, will affect gasket performance.

The Materials Technology Institute of the Chemical Process Industries (MTI) has developed a test, termed Fire Simulation Screening Test (FIRS), that is used to evaluate gaskets under simulated fire conditions. The FIRS test involves placing a load and elevated temperatures on a test gasket using heating elements in a test rig resembling a bolted flanged joint *(11)*. The test involves imposing a temperature of 1,200°F on the gasket and flange and using an inert gas in the system to determine leakage rates under the simulated fire conditions. The FIRS procedure is a reliable method of screening or assessing gasket performance under simulated fire conditions. Gaskets should meet the FIRS criteria to minimize potential risks in actual fire situations. Failure in this simulated fire test should automatically disqualify a gasket for flammable and potential safety-risk services.

Emission data and emission factors

While some emission data and emission factors have been compiled, a comparison of emissions from various valves, packings, gaskets, and the like is not available. The Chemical Manufacturers Association (CMA) has collected emission data on various equipment sources, and the data should be analyzed during the next year, indicating those equipment items with the lowest emission rates.

Pending standards for hazardous pollutant controls will limit emissions from valves and flanges to 500 ppm and emissions from pump seals to 1,000 ppm. The regulation will require the facility owner or operator to periodically monitor equipment components and repair any found to be leaking at or above those concentrations. [For more information on the regulations, see Colyer and Meyer *(2)*.]

While the 500 ppm monitoring concentrations for valves and flanges may not appear difficult to meet, practical considerations require operating in the 1–25 ppm range. Low emissions levels are also recommended for pumps, where small

Table 2. Improved emission control considerations in existing block valves.

Valve Condition Before Packing
Straight Stem
Unmarked Stem
Unmarked Packing Box Wall
Unused Bleed-off Holes Plugged

General Packing
Graphite in Flammable and Potential Safety-Risk Services
Polymeric in Nonflammable and Non-Safety-Risk Services

Quarter-Turn Valve Alternatives
Packing
Extended Packing Box with Graphite or Polymer Packings
External Bellows
Nonflammable and Non-Safety-Risk Services
Flammable Services if Acceptable in Fire Test

changes in a seal can rapidly change the emission rate.

Valve emission control

Valves have been a major source of fugitive emissions in the CPI. Extensive changes in valves and packings have occurred over the last decade to reduce atmospheric emissions.

Table 1 summarizes industry valve usage. In general, gate valves predominate in petroleum refineries and quarter-turn valves predominate in chemical manufacturing plants. Globe valves are used in both types of facilities, but are not used as standard manual control valves in refineries or petrochemical plants. Sliding stem and quarter-turn control valves are used in both industry segments.

Valves are frequently lined with or constructed of polymeric materials.

For safety reasons, metal valves have long been a standard in the petroleum refining and petrochemical manufacturing industries, with full polymer valves limited to nonflammable services. Where ball or butterfly valves are installed in petroleum refineries and petrochemical plants, the valves must be fire-safe. This requires special metal inserts that ensure full line blockage under fire conditions. For conformance with the Clean Air Act, polymer-lined valves, plastic-seated valves, and valves containing polymeric packings should be fire tested in potential fire-risk situations.

Theoretically, valves containing nonflammable fluids should not present a problem. However, the potential for an external fire from another source affecting a non-fire-safe valve should be evaluated in accordance with risk assessment considerations.

Existing valve emission control

When attempting to reduce valve emissions, existing valves must be assessed and modified where necessary, and new valves must be specified. Valves must be in a satisfactory mechanical condition to ensure optimum packing performance and minimal emissions. Considerations for existing block valves are summarized in Table 2 and for existing control valves in Table 3.

Packing removal. Packing

Table 3. Improved emission control considerations in existing control valves.

Valve Condition Before Packing
Same as for Block Valves

General Packing
Graphite in Flammable and Potential Safety-Risk Services
Polymer in Nonflammable and Non-Safety-Risk Services Below 400°F
Graphite in Nonflammable and Non-Safety-Risk Services Above 400°F

Recommended Installation
Single Graphite Set
Dual Packing Set with Bleed-off Only Where Activation Source is Satisfactory and
    Bleed-off Requirements are Acceptable

removal (for inspection and repacking) with a pick is laborious and frequently results in marring of the stem and packing box wall. Where possible, high-pressure water jets are preferable for removing packing, because these jets are specifically designed for this service and do not mar the stem or wall surfaces. Moreover, packing removal is accomplished in significantly less time compared with manual packing removal.

Valve condition before packing. An important consideration in achieving low emissions is the valve's conditions prior to packing (or repacking). The valve stem must be straight, unmarked, and with an appropriate roughness based upon the selected packing. The packing box wall should also be unmarked and any unused bleed-off ports plugged. Wall roughness is determined by the type of packing, but the wall is normally rougher than the stem. Inspection and repairs are undertaken to meet these objectives.

Packings. Graphite valve packings are recommended in all flammable fluid services and those services with a potential safety risk. When there is a high probability of fire from an external source, nonflammable service valves should also have graphite

packings. In nonflammable and non-safety-risk services polymeric packings are acceptable.

Graphite packings are widely available in a variety of configurations. Where preformed graphite rings are installed, braided end rings are necessary. Thus, the standard graphite packing set consists of two braided end rings and three preformed rings, for a total of five rings.

In quarter-turn valves, deeper or extended packing boxes are recommended to accommodate the installation of a complete packing set.

Additional rings do not improve emission control. The graphite packing density should be in the range of 70–76 lb/ft^3 to permit compression and adjustment over time (12).

Pressure spring glands are available to maintain compression pressure on a packing set. This pressure was necessary with volume shrinkage in asbestos packing. Insufficient field emission data are available to support the assumption that spring loading the standard five-ring graphite packing sets is superior to non-spring-loaded graphite packings (12, 13).

In control valves with various polymeric packing ring assemblies, spring loading appears to be advantageous. Where a full graphite set is installed in a control valve, the need for spring loading has not been established.

In addition to the standard five-ring graphite packing sets, modified graphite packings have been developed to reduce emissions below the leakage rates in standard five-ring sets. One original modified packing set has 11 rings and is complicated to install. A newer version that has seven rings and is considerably easier to install, shown in Figure 1A, is now available.

Emission control with this simplified packing set appears to be superior to the standard five-ring preformed graphite sets and equivalent to the original 11-ring set. Emission reports from various installations seem to support the manufacturer's emission control claims.

Quarter-turn valves. Many quarter-turn valves have shallow packing boxes, which present a more difficult sealing problem than deep boxes. Where a shallow box has a depth of three standard rings, a graphite packing set consisting of two special "thin" braided end rings with two standard preformed graphite rings can be installed.

Another alternative is a special version of the modified graphite packing set, termed a short stack, which is illustrated in Figure 1B. The short stack arrangement is considered superior in emission control to the four-ring modified standard set described in the previous paragraph.

Rather than modifying the packings for shallow boxes, in some valves the packing box depth can be extended. For new valves, a number of manufacturers will supply deeper boxes upon request. With existing valves, some of the packing boxes can be extended, although this requires changes in the valves. For quarter-turn valves, an external bellows seal can be installed to replace the packing. The bellows are internally pressurized with process fluid and a plug on the outside (called a telltale) indi-

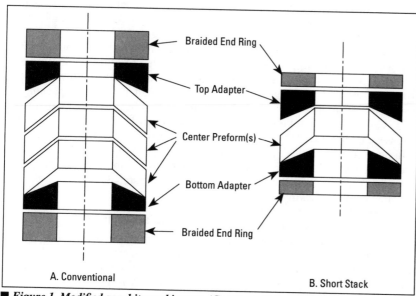

Braided End Ring

Top Adapter

Center Preform(s)

Bottom Adapter

Braided End Ring

A. Conventional

B. Short Stack

■ *Figure 1. Modified graphite packing set. (Courtesy of Garlock, Inc.)*

cates bellows integrity. There is no leakage from this system. This device should be fire-tested to ensure emission control under adverse conditions.

Control valves. Emission control in existing control valves requires a carefully installed packing system. General mechanical requirements are similar to those for standard gate valves. However, the long stem must be absolutely straight for a satisfactory installation. In addition, the stem should not bend during operation.

Packings should be selected to meet emission and safety risk criteria. Many control valves have packings and O-rings with maximum practical operating temperatures of about 400–425°F, but a risk analysis may suggest installation of a fire-safe, graphite packing. For satisfactory emission control, one graphite packing set at the outboard end is normally sufficient, with the intervening packing box volume filled with spacer rings. One packing set in an existing valve minimizes the need for additional activation power. Maximum emission control with one graphite set is generally achieved with the modified seven-ring graphite set described earlier.

Where one packing set is not completely effective (for example, due to a slightly curved stem or some other fault), dual packing sets with a bleed-off between the sets will control emissions. This requires additional activation power and significantly increases costs. Moveover, the disposal system for the bleed-off of vapor or liquid must be carefully designed and frequently presents problems, since the receiving pressure should be maintained close to atmospheric levels. If a vacuum system is installed, air may be drawn into the bleed-off circuit. Bleed-off circuits can be expensive depending upon design and disposal method.

Emission control in new valves

Emission control considerations for new valves are shown in Table 4. Where graphite packing is indicated in the table, either a standard five-ring preformed graphite set or a modified

seven-ring set can be installed. The latter apparently has a lower emissions factor.

Normally, bellows valves are installed in highly toxic services to minimize exposure to toxic materials. With the large emission reductions that will be required in the near future, increasing attention is being given to bellows valves to significantly reduce and control emissions at very low levels.

Table 4. Emission control considerations in new valves.
Gate and Globe Valves
Graphite
Bellows with Graphite Backup
Quarter-Turn Valves
Extended Packing Box with Graphite Packings
Bellows Available for Certain Valves
Control Valves (Automatic)
Heavier Nonflexing Stem
Graphite Packing
Bellows with Graphite Backup

However, an important consideration is the need for backup packing with a bellows. Leaking bellows do occur *(14)*, so packing is needed to prevent large releases to the atmosphere when a bellows fails. Many valve manufacturers supply bellows valves with a packing backup as a standard product. Installation of a satisfactory metal bellows requires careful analysis and specification of the bellows to ensure a long operating life. Moreover, bellows valves are expensive, and in valves larger than 3 in. the cost difference between bellows valves and packed valves increases rapidly *(14)*.

In quarter-turn valves, deeper or extended packing boxes are recommended to accommodate the installation of a complete packing set. These valves and packings should be fire-safe, as previously described. Most valve manufacturers have fire-safe models that meet the Exxon and British OCMA standards, which simplifies purchasing requirements. In addition, some valve manufacturers

have lengthened the packing box to permit the installation of dual packings with a bleed-off connection. Where the dual packing system with bleed-off can be installed, emissions can be controlled to very low levels. The dual packings should consist of graphite packing sets.

An external bellows system can be obtained as original equipment on various quarter-turn valves. In addition, sealed valves of differing design are becoming available. However, all of these valves should be fire-tested in accordance with previous comments.

Control valves. An important consideration in control valves is ensuring that the valve stem is straight and does not flex. Valve stem irregularities and movement affect packing efficiency, which results in increased emission rates. Improved designs will lengthen packing life and reduce emissions losses. Packing systems can be single or dual sets based upon the packing box design. Control valve suppliers have numerous models with long packing boxes that are satisfactory for dual packing systems. Where possible, bleed-off connections between packing sets for discharge to a receiver are desirable to minimize leakage. The graphite packing sets are either a standard five-ring preformed graphite set or a modified seven-ring set.

Frequently, bellows control valves are installed to prevent emissions to the atmosphere. The bellows control valves should have backup packing in the event a bellows leaks or fails. Most control valve manufacturers supply bellows control valves with backup packing sets. These sets should have graphite packings specified by the purchaser.

Flanges and gaskets

Another major source of fugitive emissions is flanges. With the elimination of asbestos gaskets, a wide variety of other gasket materials is now available commercially. However,

Table 5. Typical flange gaskets available commercially.

1. Polymeric
2. Carbon or Graphite Fiber
3. Spiral-Wound/Graphite-Filled
4. Double Metal Envelope with Graphite or Other Fillings
5. Graphite Chemically Bonded to Flat Metal Sheet
6. Graphite Mechanically Bonded to Flat Metal Sheet (Tang Type)
7. Graphite Bonded To Corrugated Metal

Table 6. Major mechanical variables that affect joint sealing of flanges.

Parallel Flange Faces
Flange Mechanical Design
Flange Face Surface Conditions
Bolt Specifications
Proper Bolt Tightening Procedures
Equal Bolt Stress in all Bolts
Meeting Bolt Design Stress Requirements

subjected to fire or heat from an external source, a fire-safe gasket should be selected.

Some of the gaskets currently available are shown in Table 5. Many of these gaskets have been tested through procedures supported by MTI or the Pressure Vessel Research Council (PVRC). The test procedures, test fixtures, and various test results have been described in numerous articles and reports (11, 15–17). The test methodology includes a fire test and also evaluates leakage. Fire testing indicates that polymeric and some fibrous materials cannot withstand high temperatures (the typical fire simulation test temperature is 1,200°F) or control emissions. Most of the other gaskets appear satisfactory, particularly Numbers 3, 4, 6, and 7. Gaskets not

gaskets are subject to the same safety considerations as packings and identical safety and risk assessments. Consequently, gaskets should be tested for fire-safety. Where a gasket is in nonflammable service but may be

included in this list should be tested and evaluated before use.

While actual gasket performance can be compared in standardized test fixtures, specific gasket characteristics are not the only variables that affect flange emission rates. The other variables are mechanical and may have a significant impact on gasket performance. A brief summary of certain mechanical variables is presented in Table 6.

The need for parallel flange faces is imperative and flange face alignment is an industry standard The correct mechanical flange design and specifications are necessary to ensure adequate strength under design bolt load conditions. These are determined by American National Standards Institute (ANSI) Standards B16.5 and B31.3. Also, flange face surface conditions specified in the ANSI standards are very important — flanges must be free of rust and imperfections.

Bolt specifications covering the size, material, and number of bolts must be carefully designed in accordance with ANSI standards to achieve design gasket compression and maintain bolt integrity. In addition, bolts must be tightened in a particular sequence and with procedures specified in the ANSI standards or the flanges, and in turn the gaskets, will be affected. The final bolt stress should be equivalent on each bolt to ensure a constant compression stress across the entire gasket face. Various techniques are available to indicate the actual bolt stress, and these should be employed, as the manual tightening of large numbers of bolts frequently results in bolt stress variations. A hydraulic bolt tightening system is available. In addition, the

LITERATURE CITED

1. **Davenport, G. B.,** "Understand the Air-Pollution Laws that Affect CPI Plants," *Chem. Eng. Prog.,* **88**(4), pp. 30–33 (April 1992).
2. **Colyer, R. S., and J. Meyer,** "Understand the Regulations Governing Equipment Leaks," *Chem. Eng. Prog.,* **87**(8), pp. 22–30 (Aug. 1991).
3. **Schaich, J. R.,** "Estimate Fugitive Emissions from Process Equipment," *Chem. Eng. Prog.,* **87**(8), pp. 31–35 (Aug. 1991).
4. **Adams, W. V.,** "Control Fugitive Emissions from Mechanical Seals," *Chem. Eng. Prog.,* **87**(8), pp. 36–41 (Aug. 1991).
5. **Brestel, R.,** *et al.,* "Minimize Fugitive Emissions With a New Approach to Valve Packing," *Chem. Eng. Prog.,* **87**(8), pp. 42–47 (Aug. 1991).
6. **Freeman, R. A.,** "Documentation of Hazard and Operability Studies," *Plant/Operations Prog.,* **10**(3), pp. 155–158 (July 1991).
7. "Clean Air Act Amendments of l990," HR Report 101-952, Title III, Section 112 Item(r), Section 303, Section 304 (Oct. 26, 1990).
8. **Kelly, W. J.,** "Oversights and Mythology in a HAZOP Program," *Hydrocarbon Proc.,* **70**(10), pp. 114–116 (Oct. 1991).
9. **Van Wormer, M. B., and R. M. Iwanchuk,** "Prepare Now for the Operating Permit Program," *Chem. Eng. Prog.,* **88**(4), pp. 41–49 (April 1992).
10. **Lipton, S., and J. Lynch,** "Health Hazard Control in the Chemical Process Industry," John Wiley, New York (1987).
11. **Bazerqui, A.,** *et al.,* "A Gasket Qualification Test Scheme for Petrochemical Plants, Part I — Test Methods and Application Results, Part II — Quality Criteria and Evaluation Schemes," ASME/NSME Pressure Vessels and Piping Conference, Honolulu (July 23–27,1989).
12. **Lipton, S.,** "Fugitive Emission Control in Packed Valves," *Chem. Eng. Prog.,* **86**(8), pp. 70–76 (Aug. 1990).
13. **Harrelson, A.,** "Live-Loaded Packing," Fugitive Emissions Seminar, Valve Manufacturers Association of America, Houston (Sept. 25–26, 1991).
14. **Snyder, P.,** "Chevron's Experience with Fugitive Emissions Regulations and the Cost of Compliance," Fugitive Emissions Seminar, Valve Manufacturers Association of America, Houston (Sept. 25–26, 1991).
15. **Bazerqui, A., and G. Louis,** "Tests with Various Gases in Gasketed Joints," Experimental Techniques, Society for Experimental Mechanics (Oct. 1988).
16. **Bazerqui, A., and G. Louis,** "Predicting Leakage for Various Gases in Gasketed Joints," Experimental Techniques, Society for Experimental Mechanics (Oct. 1988).
17. **Bazerqui, A., and J. R. Payne,** "Evaluation of Test Methods for Asbestos Replacement Gasket Materials," MTI Project No. 47 Final Report, Materials Techology Institute of the Chemical Process Industries (Oct. 1988).
18. **Adams, W.V.,** "Sealing of Fugitive Emissions," *Chem. Eng.,* **98**(5), pp. 189–198 (May 1991).
19. **Parker, K.,** "Mechanical Seals Meet Fugitive Emissions Challenge," *Chem. Proc.,* **54**(12), pp. 28–35 (Dec. 1991).
20. "Mechanical Seals," *Industrial Lubrication and Tribology,* pp. 3–6 (May/June 1990).
21. "Pump Endures for Five Years After Three Hours of Dry Running," *Chem. Proc.,* **55**(1), pp. 59–60 (Jan. 1992).
22. **Fegan, S. D.,** "Select the Right Zero-Emission Pump," *Chem. Eng. Prog.,* **86**(9), pp. 46–49 (Sept. 1990).

use of special torque nuts permits rapid bolt tightening and consistently accurate bolt stress development, and these nuts significantly reduce installation time.

A new gasket is available commercially that is made of graphite bonded to corrugated metal. This gasket is fire-safe and apparently has very low leakage rates in standardized tests. Since it is new, comparative data on this gasket may not be available in the CMA emissions study. However, various companies are installing this gasket based on satisfactory plant performance. Other gasket alternatives are spiral-wound/graphite-filled and mechanically bonded graphite sheet (tang type). The tang gasket is a metal sheet with controlled holes punched in the sheet that move the metal cutouts aside as though they were hinged. The sheet after punching has a large number of small vertical metal cutouts adjacent to each hole that mechanically bond the graphite to the metal sheet. When one of these potentially low-emission gaskets is installed properly and the flange is bolted in accordance with design conditions, emissions are extremely small or nil. In addition to these various considerations for a satisfactory flange joint, piping vibrations should be minimized and inordinate stresses should not be placed on flanges through poorly supported piping systems.

Table 7. Ten-year life cycle cost comparison of dual-seal and sealless pumps.

| | Dual-Seal | Sealless | |
		Canned	Mag Drive
Pump Installation*	Base	Lower	Slightly Lower
Utility Consumption	Base	Slightly Higher	Higher
Recordkeeping (EPA)[†]	Base	Lower	Lower
Maintenance[‡]	Very High	Base	Slightly Higher
Seal Fluid	Base	Nil	Nil
Plot Area	Base	Much Smaller	Slightly Lower

* Includes pump cost, plus costs for installation of pump, auxiliaries, alarms, and footings.
[†] Includes inspection.
[‡] Includes costs for four sets of dual seals plus installation; assumes a two-year seal life (shorter life further increases costs).

Centrifugal pumps and seals

Much has been published about centrifugal pump seals, seal improvements, and installation recommendations to improve seal performance [for example, (4, 18)]. Dual seals, both double and tandem, have extremely low emissions concentrations and are frequently installed to minimize emissions (19). A new dual system is also available that consists of a standard outboard seal and an inboard pumping seal (20). The pumping-seal fluid pressure is essentially atmospheric, which is similar to that in low-pressure tandem seals. Although a small amount of seal fluid is pumped into the process continually, the daily flow rate is quite small. However, the seal and process fluids must be compatible. When dealing with high-purity streams, contamination at any level is a concern, so careful analysis of each installation is required.

In contrast to sealed pumps, sealless pumps completely eliminate emissions on a continual basis, meeting all requirements of the Clean Air Act, including maximum achievable control technology (MACT). These pumps, termed canned or magnetic pumps, have improved significantly through the years as size and capacity have increased substantially. However, poor experience with these pumps, which has been reported frequently in various articles, has been a drawback. Generally, adverse experiences have been the result of fluid vaporization in the pump, thereby ruining the pump bearings. In many cases, the pump characteristics and fluid considerations were not fully understood, yielding poor results. Properly designed sealless pump systems perform well, and numerous examples of sealless pumps operating in severe service for many years are available (21).

There are two types of sealless pumps — the canned pump, shown in Figure 2, and the magnetic drive

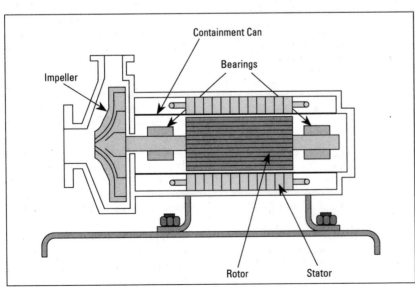

■ *Figure 2. Simplified drawing of a canned sealless pump.*

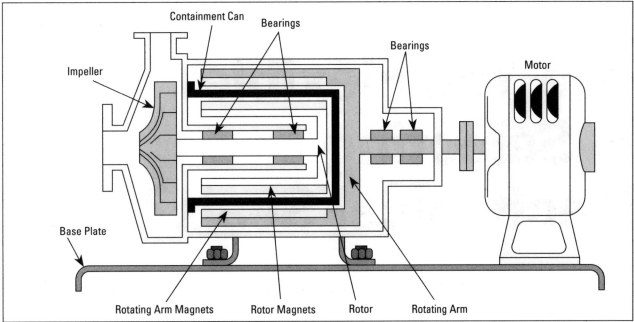

■ *Figure 3. Simplified drawing of a magnetic drive pump.*

pump, shown in Figure 3. The canned pump is essentially an induction motor with a wound stator and a non-wound metal rotor. In the magnetic drive pump, permanent magnets are mounted in the periphery of the rotor and a rotating arm with permanent magnets connected to an external motor drives the rotor through a magnetic coupling. The outer arm is separated from the main pump rotor by a nonmagnetic casing. The extended shaft from the motor to the rotating arm is supported by bearings in the pump housing. The canned pump, in contrast, is self-contained.

S. LIPTON is now a consultant living in Port Townsend, WA (206/385-9571). He is retired from Exxon Research and Engineering Co., where he worked in petroleum and chemical process design, process development, and operations during a 32-year career with the company. During the last 18 of those years he was involved with environmental technology, concentrating on environmental health and exposure controls. His consulting efforts have concentrated on emission and exposure controls. He earned his BS in mechanical engineering and MS in science from Stevens Institute of Technology. He is a member of AIChE.

Installation of sealless pumps should be considered in both existing and new units. Newer sealless pumps are designed with ANSI dimensions and can readily be installed as replacement pumps. In addition, a variable-frequency drive with these pumps can eliminate a control valve, further reducing emissions. [Fegan *(22)* discusses magnetically coupled and canned-motor pumps further.]

While sealless pumps obviously eliminate emission sources, general plant economics may dictate pump and seal selection. Normal economic comparisons are generally based on initial installation cost with brief evaluations of efficiencies. This has been relatively satisfactory for the comparison of essentially identical pumps and seals produced by various vendors. However, these comparisons can be misleading, particularly regarding emission characteristics. It is preferable to compare the alternatives on a life-cycle-cost basis.

An assumed life of 10 years for properly installed sealless pumps is not an unreasonable basis for comparison. A brief qualitative comparison of sealless and dual-seal pumps on this basis is shown in Table 7. Canned pump costs over a 10-year period are considerably lower than costs for a dual-sealed pump. An important factor in this cost comparison is mechan-

ical seal life, which is assumed to be two years. In some plants, this may be considered rather short, and in other plants it may be rather long. Consequently, the economic comparison in Table 7 depends on mechanical seal operating life. (For a two-year seal life, at least four sets of seals must be purchased and installed over the 10-year comparison period.)

A quantitative pump comparison may show a fivefold difference over the life period. Moreover, an expanded and detailed comparison may reveal other factors that can further increase the difference. For example, detailed installation labor charges may be considerably less for canned pumps if they are carefully estimated, since canned pumps do not require shaft alignment, base-plate drainage, etc. and are more easily handled during installation. Also, monitoring and recordkeeping can be significantly more costly for sealed pumps than for sealless pumps.

Extending the method

This cost-comparison technique can be applied to other emission control systems as well. It is a more effective analytical tool when comparing alternative systems than restricting economic evaluations to a comparison of initial installation cost. **CEP**

Are You Flexible in Selecting Mechanical Seals?

Hard-faced double and tandem shaft seals can effectively control emissions, but so too can flexible face seals.

Charles Wells,
Rotoflex, Inc.

The time is ripe for chemical engineers to compare costs and benefits associated with controlling emissions through traditional hard-faced vs. flexible face seals. Armed with a basic understanding of flexible face sealing technology (the new), we shall compare them with hard-face seals (the traditional) in the area of fugitive emissions, the cost associated with flush and barrier fluids, and finally, installation and maintenance costs.

If not now, when?

1993 marks the year when the first phase of the Environmental Protection Agency (EPA) Clean Air Act Amendments of 1990 will begin to take effect. Chemical process plants across the United States will be subject to unannounced EPA inspections, resulting in fines and mandatory, 24-hour equipment changes if unacceptable amounts of volatile organic emissions are detected.

This applies to organic manufacturing plants whose process stream contains more than 5 ppm of organic material and has flow rate of greater than 0.02 liters per min. The EPA estimates in its current proposed regulations (Congressional Register 63) that this would include more than 400 organic manufacturing industries, 75% of which are located in New Jersey, Louisiana, and Texas.

The Clean Air Act will be enforced in three phases: 1993, 1995, and 2000, so organic chemical processors may not need to change equipment for another year or two. In many cases, however, it makes good economic sense to meet the most stringent requirements now instead of gradually upgrading over the next five years. In order to meet the new most achievable control technology (MACT) standards, organic chemical processors must evaluate all process sealing mechanisms, including pump seals, flange gaskets, sight glasses, connectors of all types, pressure relief valves, storage tanks, and so forth.

Pump shaft seals in particular are the "achilles heel" of emissions. Seals are a major cause of emissions, and because they are dynamic, they are more difficult to control. Mechanical seals are designed with a constant, controlled leak which can result in an emissions greater than those accepted by current legislation.

Essentials of today's alternatives

Tandem and double seals. Tandem and double seals are attempts to rectify this situation by providing means to dilute the leaking fluid into a barrier fluid. The chemical process industries (CPI) have generally accepted double seals as the *de facto* standard for meeting MACT standard. (In fact, double mechanical seals are identified in CFR 63 as a key method of applying MACT.) These seals effectively control emission, because material that is leaked from the process in double or tandem seals is diluted substantially by the barrier fluid (1,2). However, the barrier fluid is then contaminated with product and must often be treated as hazardous material.

As previously discussed, hard-faced seals are designed with a controlled leak-

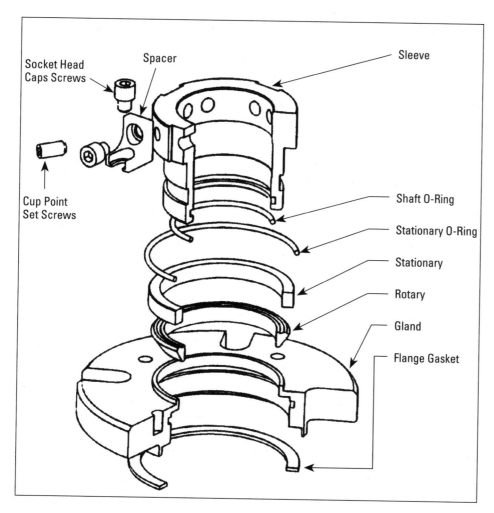

Socket Head
Caps Screws

Spacer

Sleeve

Cup Point
Set Screws

Shaft O-Ring

Stationary O-Ring

Stationary

Rotary

Gland

Flange Gasket

■ *Figure 1.*
Mechanical seal
components.
With regard to the
"rotary" component
shown, the marketplace
now affords the user
the serious option of
choosing a flexible
material or a hard one.

age which cools and lubricates the seal face. Traditional hard-face double seals control emissions through dilution. Essentially, these seals still feature a controlled leak. However, the process product leaks through the primary seal into a barrier fluid instead of into the atmosphere. The barrier fluid, which now contains a limited amount of product, leaks from the secondary seal into the atmosphere.

This solution works well to control emissions. However, the barrier fluid is contaminated with product and must be carefully disposed of or reclaimed. For many organic chemical producers, this is outweighed by the fact that the barrier fluid guarantees seal lubrication and adequate heat removal.

Another benefit is that double seals (tandem mode) provide superior protection against failure of the primary seal, because leaked product is captured in the barrier fluid. In traditional double seal designs, failure of the primary seal will result in barrier fluid flowing into the product. For sensitive applications, such as food or pharmaceutical processing, contamination of the product would be costly. Tandem seals were designed with a pressure load in the barrier chamber less than that of the stuffing box, so that if the primary seal should fail, product would flow into the barrier fluid instead of vice versa.

Most highly volatile pumping applications require a vapor recovery system to contain spillage in the event of a leak. The vapor containment system itself is subject to EPA inspections and must be monitored daily according to 1990 Clean Air Act amendments. Most double seals have an alarm system on the barrier circuit to provide an indication of seal failure.

Flexible face seals. Tandem and double seals ought to be weighed against an alternative: flexible face seals, as shown

C. WELLS is vice president, marketing and engineering, for Rotoflex, Inc., Pleasanton, CA (510/463-8800; Fax: 510/463-8835). He holds a BS in chemical engineering from Vanderbilt Univ, Nashville, and an MS in chemical engineering and PhD in electrical engineering from Washington Univ., St. Louis. Dr. Wells holds several patents in process control and graphics, and has published more than 50 technical papers in leading professional journals.

in Figure 1. Flexible face seals provide a mechanism in which the the original base seal has less leakage than hard-face seal technology and therefore bypasses the need for barrier fluids. These seals are designed without the controlled leak conventional hard-faced seals. They do not require an external flush and therefore also eliminate the need for the ancillary equipment typically required with hard-faced sealing systems.

A single flexible faced seal guarantees less than 10 ppm leakage from a properly installed seal. Conventional double seals, however, do leak barrier fluid at about the same rate as single flexible faced seal. Such leakage is from the outside seal. It also leaks barrier fluid into the pump stuffing box, thereby presenting a tradeoff for those concerned about contamination. Barrier fluid must be compatible with the product being pumped. The double seal always has a higher pressure in the barrier chamber than the stuffing box.

A single flexible face seal is not suitable for hazardous or toxic applications; however, a double seal version of the flexible face seal is better suited for such conditions, which has been tested in both the "double mode" and "tandem mode." The tandem mode is when the pressure in the barrier chamber is less than the stuffing box pressure.

Materials. Due to restrictions on sliding speeds and limited ability to absorb frictional heat, conventional flexible seals can not be used in most dynamic applications. Additionally, rubber rotaries would effectively seal only within limited temperature ranges and are susceptible to sudden pressure fluctuations due to sudden stoppage of heavy loads during routine operations.

Recently, a new seal pair geometry and mechanical arrangement allow a flexible elastomeric rotary to be used effectively in rotating mechanical seals. The key to effective flexible face seals lies in providing contact surfaces with the correct geometry and pressures to assure a static and dynamic seal.

While several materials can be used for the seals hard face, the exact characteristics of the elastomer will determine flexible face seal success.

The most common include:
• perfluoroelastomer;
• vinylidene fluoride hexafluoropropylene tetrafluoroethylene;
• ethylene polypropylene diene monomer (EPDM); and
• Hydrogenated nitrile butadiene rubber (HNBR).

HNBR, a fully saturated nitrile rubber, is the newest elastomer, and has only been commercially available since 1984. It is appropriate to use when you need to resist oils, light hydrocarbons, sulfur bearing fluids, and black liquors.

Why fugitive emissions are expensive

Organic fugitive emissions are costly for two basic reasons. First, and foremost are the looming fines and inconveniences if a plant does not conform to the Clean Air Act standards. Secondly, the plant must contend with the cost of losing organic material.

Let us first consider the issue of government regulation and potential fines. According to the EPA Protocols for Generating Unit-Specific Emission Estimates *(3)*, working mechanical seals typically leak more than 0.437 kg/h of volatile organic material, which translates to an excess of 10,000 ppm of organic material around the seal face. In California, the Bay Area Quality Management District has specified that by 1993 "average" seals, that is, those leaking in excess of 10,000 ppm—must be replaced with seals that leak less than 1,000 ppm. In most states, government officials may arrive at random and unannounced to test for organic emissions. Regardless of the threat of random government testing, however, organic chemical processors should reduce emissions to increase profitability.

Tradeoffs to consider

Consider a typical chemical process plant which transports light hydrocarbons through 200 pumps. If 100% of these pumps are leaking at 10,000 ppm, the plant would be emitting 87.4 kg/h of organics (200 pumps × 0.437 kg/h). This translates to 2.3 tons of organic material lost through emissions per day. If the plant pays $0.25 per pound for these organic materials, it loses $415,300 through emissions.

Flexible face seals are designed to eliminate "controlled measurable leakage", because they do not require a measurably thin film between the rubber rotary and the hard-faced stationary. This design virtually eliminates emissions and makes the seals more cost-effective.

In our 200-pump plant scenario, if the pumps were converted to non-leaking, flexible seals, the discharge to the environment would be 87.0 kg/h less and the plant would emit only 0.4 kg/h or organics. While converting to new seals would cost $200,000, because of the two-ton, $400,000/yr reduction or organic emissions, the seal investment would be paid back in less than six months directly from the savings in lost product. If one considers the cost of conventional double seals and the maintenance of the barrier fluid system, the payback time is even smaller.

Factor hidden costs in your decisions

While the financial savings available through reducing emissions is clear, the cost of flush or barrier fluid required by mechanical seals and ancillary equipment (such as tanks and heat exchangers) must be factored into the equation.

If the process fluid is clean, it can serve as a flush and barrier fluid source. In most cases, however, between 1.5 and 3.0 gal/min of water or another fluid must be used. If our hypothetical 200-pump plant pays only $0.001 per gal for water, it

Table 1. The cost of maintaining tank and pumping systems to control seal leakge.

COOLANT COSTS (200 pumps);

• Cooling water costs: 0.75 gal/min at $0.001 gal	$77,760/yr
• Tank system maintenance: $600 per pump	$12,000/yr
TOTAL COOLANT COSTS	**$89,760/yr**

would cost an annual rate of $780 per pump, or $156,000 annually for all 200 pumps. Of course, if water causes crystallization or coagulation of the process fluid and another flush must be used, the plant could incur up to 10 times the cost.

The plant must also pay for pumping and temperature control equipment. The typical hard-face seal requires the flush to remain close to 125°F. Since most pumps operate at 200°F, the flush must be water-cooled with clean, cool water. The cost of maintaining tank and pumping systems must also be factored into the equation. For the 200-pump plant, the costs are significant, as Table 1 shows; these figures do not include the initial capital expenditure for purchasing equipment, which could exceed $3,000 per pump.

Chemical manufacturers can significantly reduce emissions through the use of double seals. These seals leak substantially less than single mechanical seals and can generally meet the current and proposed EPA Clean Air Act Standards. However, these seals require a continuous monitoring of the performance of the barrier fluid circuit and treatment of the barrier fluid itself as a hazardous material.

The advantage of hard-face seals is that the condition of the lubrication and heat removal are accurately controlled with the external barrier fluid circuit. However, the cost of this ancillary equipment in many cases far exceeds the original cost of mechanical seals. Flexible face sealing technology provides one solution for this imbalance. Flexible face seals are designed without controlled leakage; therefore, they bypass the need for elaborate methods of capturing emissions. In doing so, flexible face seals eliminate the costs of purchasing, processing, and disposing of flush water and barrier fluids.

Installation and maintenance

Hard-faced seals—single, tandem, or double—can be readily installed by plant maintenance personnel with several years experience. A number of seal manufacturers are making strides towards developing seals which are easier to install, require fewer trips back to the manufacturer for rebuilding, and therefore result in less pump downtime. Some manufacturers can even turn the job of tearing down a pump in order to maintain hard-faced seals from a job taking a crew of three a full day to one that a single person can do in 20 min.

By comparison, flexible face seals do not require as many parts and therefore, they have fewer points of failure and fewer parts to replace. As a result, the pump needs to be torn down less frequently in order to maintain the seal.

This low maintenance is attributed to the fact that flexible face seals do not require springs or bellows for maintaining contact between the seal stationary face and rotary. Additionally, while hard-faced seals require the faces to be periodically relapped in order to retain a smooth finish, an elastomeric material running against a silicon carbide face does not have to be polished in order to create a good seal-ing surface. And while flexible faced seals require replacement of the rotary, this can generally be accomplished at the plant during a regularly scheduled plant shutdown with a common tool kit.

You decide

While hard-faced seals are standard in the CPI, flexible face sealing technology is new and emerging as a viable alternative to pump shaft sealing. Mechanical seals with elastomeric rotating faces provide one approach to solving shaft sealing problems. The materials, when properly designed for rotating applications, can operate at conditions found in 90% of all American National Standards Institute (ANSI) pump applications.

Keys to your decision-making are in calculating water, power, and steam savings vs. the cost of seal retrofitting, and how far you can bring the payback period down vs. potentially lost product. **CEP**

Literature Cited

1. **Schaich, J. R.,** "Estimate Fugitive Emissions from Process Equipment," *Chem. Eng. Progress,* **87,** 8, p. 31, (Aug. 1991).

2. **Adams, W. V.,** "Control Fugitive Emissions from Mechanical Seals," *Chem. Eng. Progress,* **87,** 8, p. 36, (Aug. 1991).

3. **EPA-450/3-88-010** (Oct. 1988).

Further Reading

Bashta, T. M., and A. N. Shvestkov, "End Face Seals on the Rotating Shafts of Hydraulic Machines with an Elastic Sealing Unit," 10, pp. 51–55, 56, *J. Russian Eng.* (1976).

Develop an Effective Wastewater Treatment Strategy

Follow these steps to a successful wastewater management program for a new or existing CPI plant.

Lisa A. McLaughlin and Hugh S. McLaughlin,
Waste Min Inc.
Kimberly A. Groff,
List Inc.

Water quality regulations developed under the U.S. Environmental Protection Agency's (EPA's) National Pollutant Discharge Elimination System (NPDES) permitting program have put an increased emphasis on the control of toxic wastewater discharges from industrial facilities. And, the regulations are likely to result in more stringent discharge permits and monitoring requirements for chemical process industries (CPI) plants.

New NPDES requirements include chemical-specific limits for the protection of aquatic biota and human health, as well as whole-effluent toxicity testing. Site-specific biological criteria for individual water bodies may also be imposed on industries discharging into surface waters in the near future. These new regulations are implemented by increased restrictions and monitoring requirements in new and renewed NPDES permits.

As a result, many CPI plants must meet very stringent effluent limitations, regardless of the technologies available for treatment. Wastewater discharge permits are based on requirements that protect the receiving waters — technical feasibility and cost effectiveness do not directly enter into the procedure for developing wastewater discharge standards. Thus, many facilities may find themselves facing "mission impossible" to meet the necessary discharge requirements.

Many plants need, or may need in the future, to take another look at their wastewater management strategy in order to comply with wastewater discharge limitations. This article is intended to help chemical engineers determine when their wastewater management strategy may be faltering, and it outlines a procedure for developing a successful new strategy.

Evaluating compliance

The ultimate goal of any wastewater treatment system is to comply with the regulations in a cost-effective manner. Compliance will depend on the ultimate outlet for the wastewater. Although for most manufacturing operations the outlet is typically a municipal sewer that leads to a publicly owned treatment works (POTW), there are other options — some are the plant's choice, others are dictated by law. The most common outlets for wastewater discharges include:

• *Discharge to surface water.* Effluent from wastewater treatment operations is piped directly to a surface water body and is subject to NPDES regulations. Effluent limitations depend on ambient water quality criteria, the conditions of the receiving stream, and the amount of mixing available. Discharge to surface water is usually a viable outlet for effluents that contain benign contaminants or that have been treated to a level guaranteeing that the receiving stream will not be impacted.

• *Discharge to the sewer.* Effluent from wastewater treatment operations is sent to the sewer, which is connected to a POTW. Wastewater is subject to the pretreatment regulations of the municipality. This is typically a good outlet for effluents containing constituents that the POTW will effectively degrade, principally biodegradable organics of moderate strength. The capacity of the POTW to accept the waste needs to be considered.

• *Off-site disposal.* Effluents and other residuals (sludge) from wastewater treatment operations are transported to an off-site treatment facility. The level of pretreatment required for off-site disposal is determined by the handler. This is the most appropriate disposal method for low-volume, high-toxicity effluents and residuals. Effluents and residuals in this category are usually prohibited from discharge via any other outlet (NPDES outfalls or municipal sewers).

Compliance evaluations can take one of two forms. The first type of evaluation is a straightforward assessment of whether or not a plant's wastewater treatment operations are currently meeting effluent discharge limitations. A thorough review of the permit requirements and all compliance monitoring data will indicate whether or not there may be immediate compliance problems and, if so, the degree of response required.

For example, if a facility is consistently out of compliance on a critical parameter on an NPDES permit, such as the average daily concentration of a priority pollutant, this is clearly more urgent than an occasional minor excursion on one of the many composite parameters such as biochemical oxygen demand (BOD). The former situation will require an immediate revision of a facility's wastewater management strategy, which may involve significant modifications to or even the complete replacement of existing treatment units. The latter situation might be easily addressed by a simple audit of flows to the wastewater treatment units and some minor revisions to standard operating procedures.

The second type of evaluation is conducted at facilities that may be in compliance with current limitations and regulations but are uncertain as to whether or not the facility will meet newer, more-restrictive discharge limitations that may be imposed when NPDES permits are renewed. These facilities should examine their current permits and allow adequate time before permit renewal to determine whether or not they will be able to meet the anticipated discharge limitations. Engineers need to be aware of regulatory

trends that may impact effluent limitations in the future. A good example of this is the requirement for whole-effluent toxicity testing, which is currently being imposed on many plants that have permitted NPDES discharges to surface waters.

There are several ways that engineers can keep current with new and proposed regulatory requirements. First, they should develop an open channel of communication with their plant's environmental compliance manager. Too often, compliance managers will not "trouble" engineering staff with proposed regulatory requirements until they are promulgated and

implementation deadlines are imposed. This can leave too little time for replacement of, or modification to, facilities to meet the new requirements. If engineers indicate their need to know of any *proposed* regulations that may impact their operations, environmental compliance managers are usually more than happy to accommodate them.

If a facility does not have an environmental compliance manager *per se,* engineers have to take steps to keep current with regulatory trends. This can be done by attending environmental conferences (such as those held by the Hazardous Materials Control Research Institute and the Air and Waste Management Association) and meetings of their professional associations (*e.g.,* AIChE, ASME, ASCE, etc.). As more and more capital is being

■ *These sequencing batch reactors are part of the biological treatment component added to Schenectady Chemical Co.'s existing wastewater treatment facility. Photo courtesy of Malcolm Pirnie, Inc.*

allocated by the CPI to environmental control technologies, professional associations have been quick to respond with both seminars and workshops on environmental topics. Subscriptions to the *Federal Register* and the Bureau of National Affairs' publications also can help a facility's engineering staff stay on top of changes in wastewater discharge requirements.

As effluent limitations regulate the discharge of the wastewater treatment facility, the "track record" of current wastewater management practices with current or potential future limitations determines whether or not existing operations require modification or replacement. If a facility is currently out of compliance, or is in danger of such at permit renewal time, a new effluent management strategy needs to be developed.

It is as important to understand the specifics of how the wastewater is produced as it is to know what contaminants are present.

This article provides a step-by-step guide for practicing engineers to follow in developing an appropriate wastewater management strategy for a specific industrial facility. A guide to the decisions involved in developing this strategy is shown in Figure 1. Although a compliance evaluation is indicated as the starting point, the remainder of this logic diagram is applicable whenever modifications to or replacement of wastewater treatment unit operations are contemplated.

Characterizing the wastewater

The most important step in developing a wastewater management strategy is to completely characterize the wastewater. Wastewater is typically characterized in terms of permit requirements and, as a result, is usually only monitored at the final discharge point. In addition, the number of parameters being monitored routinely is often not comprehensive, due to both the expense of monitoring and the quick turnaround times required for submission of monitoring reports to the regulating agencies.

Although compliance monitoring does indicate both the current compliance status and the degree of process control being achieved, it is not an adequate starting point for the cost-effective design of a wastewater treatment system. Instead, both the sources to and effluent from current wastewater treatment operations need to be adequately characterized.

Even though effluent data alone <u>can</u> be used to determine whether or not unit operations can be added to the end of a treatment train to bring it into compliance, only by completely understanding the overall manufacturing process can the engineer identify the most appropriate and cost-effective method of upgrading a waste treatment facility. Depending upon which parameter is out of compliance, the solution may be as simple as optimizing the current treatment process or adding a small packaged treatment unit to an individual process, or as drastic as total treatment train replacement or loss of manufacturing capability for certain products.

Sources of wastewater in a CPI plant should be characterized according to the manufacturing processes from which the various waste constituents originate. It is as important to understand the specifics of how the wastewater is produced as it is to know what contaminants are present. It is common for contaminants that are not detected in a screening of waste streams to appear, as if by magic, in the composite wastewater. This is due to unsteady-state manufacturing processes, where certain

Whole-Effluent Toxicity Testing

NPDES permits control toxic pollutants through chemical-specific limits or "free from" statements, such as "the waters of the state shall be free from substances in concentrations that are toxic to aquatic life." Chemical-specific limits are designed to prevent the discharge of chemicals above their toxic threshold. Since chemical-specific testing cannot predict the biological effects of effluent constituents, whole-effluent toxicity testing was developed.

The whole-effluent toxicity test involves exposing a selected test organism to a known concentration of sample (or effluent dilution percentage in the case of industrial outfalls) for a specified period of time. The acute toxicity of the sample is expressed as a concentration of a specific compound or as a dilution percentage of effluent lethal to 50% of the organisms, which is denoted by LC_{50}. The value of LC_{50} is determined through a statistical analysis of mortality as a function of concentration. The lower the value of the LC_{50} the more toxic the wastewater. The

chronic test measures the ability of the organism to reproduce and grow while in contact with the effluent.

Many new or renewed NPDES permits now include testing provisions and limitations for whole-effluent toxicity. In the event of several failed bioassays, the EPA may require dischargers to conduct a toxicity reduction evaluation (TRE). A TRE is an analysis that determines the cause of and the remedy for effluent toxicity that is in violation of state or federal water quality requirements. (The article following this one discusses TREs in more detail.)

It should be noted that interpretation of whole-effluent toxicity testing results can be very difficult. This is because the compounds contributing to toxicity may vary daily or even hourly in complex effluents. In addition, toxicity testing can be very expensive. Therefore, facilities should have qualified personnel conduct testing and interpret results. This will ultimately reduce compliance costs.

waste streams are discharged only at discrete times.

It is much more expensive, time-consuming, and error-prone to find waste streams by chasing "back up the pipe" than it is to understand the manufacturing processes and predict the generation of wastes. In addition, a review of the manufacturing process will provide the knowledge base needed to evaluate the best place to reduce, recover, or treat individual waste streams. For example, a small stream of concentrated organics combined with a small stream of toxic metals combined with a large stream of utility blowdown water is a formidable — and costly — treatment challenge. However, straightforward treatment technologies are available to treat each stream individually.

Therefore, the list of constituents that should be assayed and quantified in both the influent and effluent will depend on the characteristics of the manufacturing process and should be determined on a case-by-case basis. In general, the constituents to be analyzed should be selected to assess compliance with current and potential future regulatory requirements, the options for treatability of the individual wastewater sources, and the potential for modifying the manufacturing process to reduce, eliminate, or modify the contaminants of concern. Examples of what to analyze for are presented in Table 1.

To assess whether current treatment systems require modification or replacement, engineers need to be aware of not only the target compounds for treatment but also additional constraints that individual treatment processes may have. For example, in the case of an activated-sludge wastewater treatment system, which can effectively treat highly biodegradable organics, the presence of inhibitory compounds such as metals should be assessed. Another example is the need to analyze for surfactants in metal waste streams, because these compounds can chelate with metals and deteriorate conventional metal-removal technologies. A third example is evaluating the presence of salts when con-

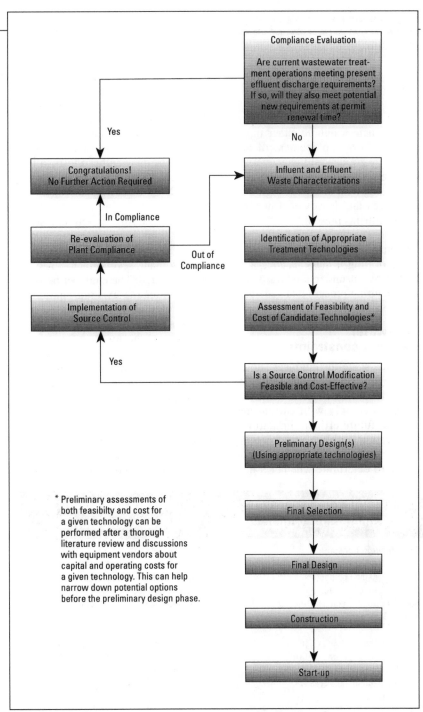

■ *Figure 1. Developing a wastewater treatment strategy involves these steps.*

sidering recycling, since equipment restrictions (such as the need to prevent scaling) may limit the level of salts that can build up during recycle. Thus, the list of constituents to analyze for should be developed with both current and potential treatment technologies in mind.

There is a close relationship between the options for treatment and the idiosyncrasies of manufacturing processes. For example, in selecting a treatment technology, the engineer must consider such factors

as whether the discharge of a pollutant from a given plant operation is continuous or intermittent and whether there are any secondary constituents present that may inhibit a potential treatment technology. Thus, it is critical that chemical engineers participate in the development of wastewater treatment approach — not because they enjoy the challenge, but because they understand the manufacturing process.

Finally, wastewater should also be characterized in terms of flow rate. A comprehensive understanding of flow rates and patterns of flow to wastewater treatment operations is critical when designing either modifications to or replacement of any system. A good waste characterization should account for all components of the final discharge including all sources and losses of water and the constituents present.

The best way to develop a complete characterization of the waste is by performing a mass balance augmented by an understanding of the manufacturing process that generated the individual wastewater streams.

Wastewater treatment constraints

At this stage the engineer should have a good idea of the ability of the current wastewater treatment system to comply with current and potential future effluent limitations. If there is a problem with meeting either, further evaluation will be needed to determine whether a new treatment system will be necessary or if compliance can be achieved through modifications to the existing plant. The approach to evaluating treatment options will be highly dependent on whether the resulting treatment facility will be built from scratch or added to an existing facility.

A new facility essentially starts with a blank piece of paper, which simplifies the design process. In an existing plant, several factors need to be considered. The engineer needs to ask whether the existing equipment can be used and if so, what the economics are. The design engineer should also be aware of potential biases toward existing equipment by wastewater treatment personnel and previous corporate commitments — for example, if a facility has recently made a significant capital investment in a particular treatment system. In that case, one must determine whether modifications to the current treatment train could bring the system into compliance, even if another treatment technology would have been more effective and cost-efficient when starting from scratch. In addition, existing plants may require continuous wastewater treatment. If so, construction and startup of a new facility becomes more difficult.

Engineers can choose from three basic strategies for construction at the outset of design. These include conversion while maintaining continuous wastewater treatment to the facility, switching over with a short-term disruption (typically a shutdown), or building a parallel treatment plant. The best strategy for a given plant will depend on the manufacturing demands, regulatory time constraints, and the practicality and economics of using existing equipment.

Selecting treatment technologies

The next stage in the design process is to conduct a preliminary screening of technologies to achieve compliance. Technologies should be categorized into those that work, those that have the potential to work, and those that have no place for the particular application.

Viable technologies should be identified for each of the individual wastewater streams. Then the streams that use the same technologies are combined, on paper, to create composite waste treatment trains. The resulting wastewater treatment trains are then compared to the current manufacturing and waste treatment practices to identify possible candidates for waste segregation and independent treatment.

The problem associated with combining two waste streams that require different technologies is that the cost of treating the combined stream is virtually always more expensive than the individual treatment of the separate streams. This is because the capital cost of most waste treatment operations is proportional to the total flow of wastewater (recall the sixth-tenths rule for capital costs that applies to the scaleup of tankage-intensive treatment opera-

Table 1. Examples of common pollutants and other parameters for which effluents should be characterized.

Constituent/Parameter	Description
Volatile Organic Compounds Acid-Extractable Organics Base/Neutral-Extractable Organics Metals, Total and Metals, Soluble	Priority pollutants. Concentrations of these compounds are typically regulated on both sewer and NPDES permits.
Biological Oxygen Demand (BOD) Chemical Oxygen Demand (COD) Total Organic Carbon (TOC) Total Suspended Solids (TSS) Temperature pH	Conventional pollutants. Permissible levels/values are also typically regulated on both sewer and NPDES permits.
Whole-Effluent Toxicity (LC_{50})	A relatively new parameter, it is usually only evaluated for NPDES permits.
Surfactants	Potential interfering agents.
Ammonia Nitrate, Nitrite Phosphorus	Nutrients. Determination is needed to adequately evaluate the potential for biological treatment.
Sulfate Chloride Sodium	Inorganic salts. Potential interfering agents.
Flow rates	Necessary in order to perform a mass balance on the facility.

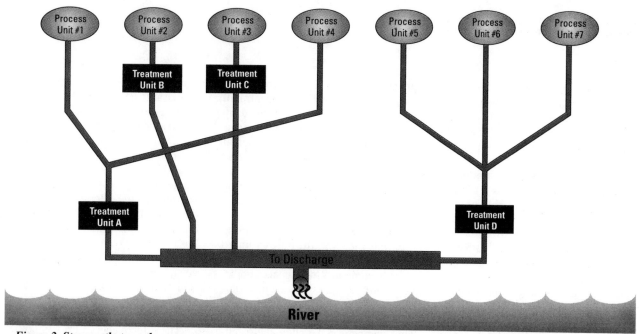

Figure 2. Streams that use the same treatment technology can be combined and other streams can be treated at the source.

tions, such as biological treatment or sedimentation), and the operating cost for treatment increases with decreasing concentration for a given mass of contaminant.

Thus, if two waste streams use the same treatment, then combining them results in improved economies of scale for the capital investment and similar operating costs. In contrast, if two different treatment operations are required, combining the two streams results in higher capital costs for both treatment operations. In addition, if the streams are combined prior to treatment, both treatment operations see lower contaminant concentrations for the same net contaminant mass, which results in higher operating cost per pound of contaminant removed.

Wastewater treatment technologies have historically been drawn from three areas of engineering practice:
• the expertise associated with the treatment of domestic waste;
• the precipitation technology for the removal of soluble metals and the clarification of potable water; and
• the separation unit operations of the CPI, such as steam stripping.

In addition, various speciality technologies have been commercialized to address the needs of certain industries. Most of these technologies are variations on fundamental treatment processes that have been adapted to the specific size and operating requirements of a particular

application. Examples of these include UV-promoted oxidation by hydrogen peroxide or ozone (UV/chemical oxidation), wet air oxidation, and membrane separation.

Both UV/chemical oxidation and wet air oxidation are variations on common water and wastewater treatment operations. UV/chemical oxidation uses ultraviolet radiation to promote and accelerate the oxidation of

Combining streams that require the same type of treatment results in improved economies of scale.

soluble refractory organics by either ozone or hydrogen peroxide. Such oxidation processes have been used for generations in both drinking water and municipal wastewater treatment for disinfection, the removal of soluble refractory organics, and/or odor control. Wet air oxidation uses high water temperatures and pressures to oxidize organic wastes in the absence of chemical oxidants or air.

Membrane separation can also be considered an evolutionary product of standard water and wastewater treatment practices. Filtration has been employed for decades in the form of granular media filters (such

as sand, anthracite coal, and diatomaceous earth) and cartridge filters. As the need for ultrapure water grew and requirements for total suspended solids (TSS) became more stringent, filtration devices with smaller and smaller pore sizes were developed. Now, membranes not only can segregate contaminants by size on a molecular level, but they also can be treated so that they will only permit compounds with certain chemical characteristics (such as polarity) to pass through them.

It is imperative that all plant resources participate in the development of the facility's wastewater treatment strategy. Once a conceptual process is correctly selected, then the subsequent efforts can be delegated to traditional engineering resources, such as engineering and construction firms and suppliers of pre-engineered equipment packages.

Wastewater treatment technologies can be divided into separation processes and transformation processes. Separation processes concentrate the contamination into one stream, leaving a lower residual concentration in the original wastewater. However, there are disposal considerations associated with the concentrated stream that must be considered. Examples of separation processes are filtration (all the way from coarse screens to reverse osmosis), activated carbon, ion exchange, and any other unit operations that use thermal and concentration gradients as driving forces to separate chemicals (such as air and steam stripping).

Transformation processes alter the physical and/or chemical form of the contamination to make it more acceptable for disposal. The two most common examples are thermal oxidation to carbon dioxide and water and biological oxidation to carbon dioxide, water, and new biomass. However, transformation processes do not destroy the contamination, they just convert it into another form. For example, although contaminants are transformed into carbon dioxide, water, and biomass during biological processes, the disposal of biomass must be addressed. Each application should be scrutinized to confirm that the disposal of the new forms or byproducts are correctly accommodated in the overall process design.

In general, separation processes offer the advantage of preserving the option of recycling to the manufacturing process, since the potentially reusable contaminant remains intact. Transformation processes are favored whenever the value of the recovered contaminant is negligible or the cost of recovery and purification is prohibitive. Many wastewater treatment processes combine both separation and transformation

processes, such as on-site steam stripping to recover a concentrated mixed organic stream, followed by off-site incineration as the final disposal method. Similarly, biological digestion is a transformation process that converts soluble organics into biomass, which is then separated by gravity filtration.

After narrowing down the field of potential candidates, the remaining technologies should be evaluated based on their effectiveness (ability to reliably attain treatment goals), implementability (availability of materials and services), and costs (capital and operation and maintenance). A thorough comparison, based on the specific application at hand, should indicate which primary technologies are the best candidates for a given waste stream. [References (1) and (2) discuss how to make such comparisons.]

Once the appropriate primary technologies have been identified for the contaminants of concern, the design engineer can begin to develop a complete treatment train. In doing so, he or she should keep in mind the primary and secondary effects of treatment. For example, if a wastewater has a high content of biodegradable organic constituents and trace quantities of priority pollutant metals that do not inhibit biological treatment but are above discharge standards, biological treatment may remove these metals to the degree

necessary through adsorption onto biomass. If not, additional treatment, such as ion exchange, could be added to an individual high-strength stream prior to combining it with other streams.

Hidden costs associated with various technologies should also be considered. In the above example, the disposal of sludge containing heavy metals and the regeneration or disposal of ion exchange resins need to be addressed. In addition, auxiliary equipment (dewatering operations) will be required to handle the sludge. Handling and disposal of sludge can double wastewater treatment cost.

The effectiveness of a technology will need to be accounted for as well. For example, although biological oxidation may remove most biodegradable organics, carbon polishing may still be necessary to meet discharge limitations if drinking-water-quality effluent is required.

The following sections describe some common wastewater treatment technologies and typical applications.

Carbon adsorption

Carbon adsorption is a separation process whereby predominantly organic contaminants are physically attracted to the surface of activated carbon particles. Its effectiveness for the removal of a specific compound depends upon the following characteristics of the compound: carbon chain length, aromaticity, polarity, and aqueous solubility. Typically, in wastewater applications the less soluble a compound, the higher its affinity to the carbon surface is and the more effective activated carbon treatment becomes. Carbon adsorption is also favored by increasing aromaticity and chain length.

Activated carbon is supplied in either granular or powdered form. Powdered activated carbon is typically added directly to mixed basins in biological systems for spot treatment of organic surges. The carbon adsorbs organics that may be toxic or inhibitory to the biological degradation process.

L. A. McLAUGHLIN is the president of Waste Min Inc., Groton, MA (508/448-6066; Fax: 508/448-6414). She has nine years of experience in environmental chemistry, chemical and environmental engineering, and groundwater remediation. Her previous employers include Polaroid, Goldberg-Zoino Associates, and Environmenal Science and Engineering, Inc. She recently started her own company, Waste Min Inc., to focus on waste minimization technologies for industrial applications. She received an AB in chemistry and environmental studies from Dartmouth College and an ME in chemical engineering from Rensselaer Polytechnic Institute.

H. S. McLAUGHLIN is the vice president of Waste Min Inc. Until recently, he was a senior process engineer with ENSR Consulting and Engineering. He has 16 years' experience as a process engineer, working for a variety of industries on industrial waste treatment, process development applications, and waste minimization. He has a BS in chemistry from Harvey Mudd College, an MS in chemical engineering from the Univ. of Southern California, and a PhD in chemical engineering from Rensselaer Polytechnic Institute. He is a registered professional engineer in Massachusetts, New York, and Puerto Rico.

K. A. GROFF is director of environmental technology for List Inc., Acton MA (508/635-9521). Her previous employer was ENSR Consulting and Engineering. She has 10 years' experience in industrial and wastewater engineering and groundwater remediation. She holds a BA in biology from the Univ. of Delaware, an MS in environmental science from Drexel Univ., and a PhD in environmental engineering from Georgia Institute of Technology.

Granular activated carbon (GAC) is more commonly used as a separate unit operation to remove specific organic compounds in a configuration that resembles filtration. (In fact, GAC is also an excellent filter.) However, the accumulation of solids on the carbon will rapidly deteriorate the performance for the adsorption of organics. Thus, GAC typically requires filtration to remove any solids prior to the carbon adsorbers.

GAC treatment may be carried out in a batch, column, or fluidized-bed operation. The usual contacting systems are fixed-bed, in either the up-flow or down-flow mode. Designs typically specify two or more carbon adsorbers in series, with individual contactors usually sized to permit a minimum residence time of 15 minutes.

The amount of carbon needed to treat simple waste streams can be estimated from adsorption isotherm data. An adsorption isotherm is a batch test designed to determine the equilibrium relationship between organic compounds in the aqueous phase, measured as concentrations, and the organic compounds adsorbed onto the carbon, expressed as the weight of contaminant adsorbed per unit weight of carbon. However, dynamic column tests are required for estimating the amount of carbon needed to treat more-complex streams. Once the carbon usage rate has been determined, the carbon contactor should be sized based on the frequency and practicality of carbon replacement.

Activated carbon does not destroy contaminants in wastewater, it simply transfers and concentrates them from the liquid phase to the solid phase. Consequently, further treatment and/or handling of spent activated carbon is necessary and should be factored into estimates of operating costs. Operating costs for carbon systems depend on options for carbon regeneration and/or disposal and replacement. Generally, regeneration and reuse is preferable to disposing of spent carbon as a hazardous waste (typically via incineration).

However, the feasibility of on-site or off-site regeneration of spent carbon needs to be determined on a case-by-case basis. Feasibility depends on such factors as:

- the contaminants — the more easily a compound can be desorbed, the more attractive regeneration becomes;
- the quantity of spent carbon — regeneration is usually only economically feasible for large wastewater operations with high carbon usage rates;

A concern with the use of air strippers is the fate of the contaminants once they are transferred to the vapor phase.

- the type of regeneration required — steam vs. thermal desorption vs. solvents; and
- the waste code classification of the spent carbon — for example, if the spent carbon is classified as a Resource Conservation and Recovery Act (RCRA) "F" waste, regeneration may not be viable.

All of these factors have economic implications that must be examined before one can determine whether regeneration will be feasible.

Air and steam stripping

Air stripping is a separation technology that transfers volatile substances from solution in liquid (*e.g.*, water) to solution in gas (*e.g.*, air). This transfer takes place because the contaminant in solution exerts an equilibrium vapor pressure proportional to its concentration in the aqueous phase. A proportionality constant known as the Henry's Law constant is used to predict the vapor phase concentration based on the contaminant's aqueous phase concentration. The Henry's Law constant is an indicator of a contaminant's combined volatility and solubility, and it determines whether or not the compound can be air-stripped out of solution. Contaminants with high Henry's Law constants are more easily stripped out of an aqueous solution than those with

low constants. [For more on Henry's Law, see the articles by Carroll *(3, 4)*.]

There are a number of basic equipment configurations for air stripping, including diffused aeration, countercurrent packed columns, and cross-flow towers. The major cost of air stripping is the capital cost of the equipment, which is a function of the height and diameter of the air-stripping tower. These dimensions are selected based on results of pilot-scale treatability studies. The main operating cost for an air stripper is for the power to run the blower. Other factors that may influence maintenance costs include cleaning and replacement of packing due to scaling and/or bacterial fouling and maintenance of vapor-phase control systems.

A major concern with the use of air strippers is the fate of the contaminants once they are transferred to the vapor phase. Air permitting requirements must be reviewed, as air strippers may represent a significant source of air pollution. In addition, regulatory agencies tend to frown on multimedia transfer with subsequent discharge into the environment.

To avoid this, a plant can use an air stripper as the source of combustion air for a boiler. In this manner, the organics are destroyed during the combustion process. However, if the contaminants are chlorinated organics, hydrochloric acid will form in the flue gas, which will cause concerns for the refractories and require air emission permits and controls.

An alternative to air stripping is steam stripping, where the vapor phase is steam instead of air. Due to the higher temperatures, the Henry's Law constants are typically an order of magnitude higher (Henry's Law constants are not actually very constant — they depend on many factors, especially temperature), making the removal process easier.

Following stream stripping, the steam can be condensed and the contaminants returned to the liquid phase. Depending on the contaminant, a concentrated organic-rich phase that is a good candidate for recycle or off-site recovery may be formed. At a minimum, the contaminant has been concentrated by the steam stripping process, which should improve the economics of subsequent treatment or disposal.

Biological treatment

Biological treatment involves using the organic contaminants as a food source for microorganisms, which convert the contaminants into additional biomass, carbon dioxide, water, and nonbiodegradable organic byproducts. The process may be carried out under aerobic or anaerobic conditions. Because many compounds in industrial wastewaters can be toxic to microorganisms, effluent pretreatment may be necessary. It is important to know the physical and chemical properties of the constituents to be treated because other removal mechanisms, including adsorption and stripping, may also take place in biological systems.

Biological treatment processes are typically divided into two categories: suspended-growth systems and fixed-film systems.

Suspended-growth systems are more commonly referred to as activated-sludge processes, of which there are several variations and modifications. The basic system consists of a large basin into which contaminated water and air or oxygen are introduced. Microorganisms are present in the aeration basin as suspended material, and they metabolize soluble organics. The biological flocs are later separated by gravity settling. Sequencing batch reactors enable the activated-sludge process to be carried out in a single reactor. Anaerobic fermenters are used to treat high-strength wastes. During anaerobic biological treatment, energy may be recovered from the methane that is generated.

Fixed-film processes differ from suspended-growth systems in that the microorganisms attach themselves to an inert support medium. Biological towers (or trickling filters) and rotating biological contactors are the most common aerobic forms of fixed-film processes. Fixed-film systems can have lower costs because oxygen transfer is generally more efficient

LITERATURE CITED

1. Corbitt, R. A., "Standard Handbook of Environmental Engineering," McGraw-Hill, New York (1990).
2. Metcalf & Eddy, Inc., "Wastewater Engineering: Treatment, Disposal, Reuse," 2nd edition, revised by G. Tchobanoglous, McGraw-Hill, New York (1979).
3. Carroll, J. J., "What is Henry's Law?," *Chem. Eng. Prog.,* **87**(9), pp. 48–52 (Sept. 1991).
4. Carroll, J. J., "Use Henry's Law for Multicomponent Mixtures," *Chem. Eng. Prog.,* **88**(8), pp. 53–58 (Aug. 1992).

FURTHER READING

Eckenfelder, Jr., W. W., "Industrial Water Pollution Control," 2nd edition, McGraw-Hill, New York (1989).
Eckenfelder, Jr., W. W., and D. L. Ford, "Water Pollution Control: Experimental Procedures for Process Design," The Pemberton Press, Jenkins Publishing Co., Austin and New York (1970).
Okoniewski, B. A., "Remove VOCs from Wastewater by Air Stripping," *Chem. Eng. Prog.,* **88**(2), pp. 89–93 (Feb. 1992).

and does not require much aeration equipment. In addition, fixed-film systems tend to be easier to operate and are less susceptible to impairment from large toxic slugs. Fixed-film degradation may also be carried out under anaerobic conditions using support media such as activated carbon.

Capital costs for biological treatment depend on the size of the reactor and the type of system. Operating costs depend on the power requirements for air supply, mixing, and nutrient feed. As mentioned previously, additional costs associated for the treatment of wasted biomass, or sludge, will also be incurred.

The most appropriate type of biological treatment system for a particular application (aerobic or anaerobic, fixed-film or suspended-growth) needs to be determined on a case-by-case basis.

Assessing the cost and optimizing the design

Once an appropriate treatment system has been developed, the next step is to assess its cost. This is done by determining what factors are driving the total cost of treatment.

Depending on the situation, end-of-pipe treatment may be inappropriate or too costly for the waste stream in its current form, and source control prior to wastewater treatment may be more cost-effective. Source control measures may include the segregation of dilute wastewater from concentrated streams, or (if a particular chemical is driving the cost of treatment) chemical or raw material substitution. The prioritization of the factors driving costs will allow the engineer to target source control efforts to those areas of the plant where the greatest cost savings can be attained. A detailed analysis of source control measures is beyond the scope of this article.

If source control measures are identified and implemented, the waste should be recharacterized to evaluate compliance and redefine treatment goals. [For more details, see Reference *(2)*.]

Design, construction and startup

Once the conceptual wastewater treatment upgrade is developed, subsequent efforts resemble other capital projects, with detailed design, construction, and startup phases. The plant engineer should recognize that changes to the basic treatment strategy are increasingly more difficult to accommodate as the project gets closer to completion.

The requirement to not "change horses in the middle of the stream" is why the initial effort to identify all the processing flexibility and alternative treatment approaches is critical to the overall success of a wastewater treatment strategy. Once the strategy is chosen, the fate of the entire manufacturing facility is decided.

Process engineers should take pride that they have the resources to contribute to such an important effort. Additionally, the manufacturing organization should realize the gravity of the situation and provide the necessary cooperation to assure the timely and cost-effective development of the optimal wastewater treatment strategy. **CEP**

Zero Discharge: What, Why, and How

"Zero discharge" means more than "nothing at all out of the plant."

Mike E. Goldblatt,
Karen S. Eble, and
Jennifer E. Feathers
Betz Industrial

Because of the growing public concern for the quality of the environment and in response to increasingly stringent environmental regulations, many chemical process industries (CPI) companies are actively seeking ways to minimize waste generation. In many cases they are working toward a goal of "zero discharge."

This article outlines the various definitions of zero discharge, discusses why it is important for companies to consider zero discharge as a goal, and highlights some generic ways of reducing wastewater discharges.

Ideal vs. real processes

In discussing zero discharge, it is useful to conceptualize a theoretically "perfect" or "ideal" industrial process. Figure 1 illustrates this concept, where all raw materials are converted into usable products, and all processing aids (including catalysts, solvents, water, and the like) are fully regenerated to their original quality, recycled, and reused. The ideal process, while creating no waste, would also operate profitably.

However, since the completely perfect or ideal process does not exist, we need to look at a real world process, such as that depicted in Figure 2. Let us assume that the process is an integrated petroleum refinery. The raw material (crude oil) is acted upon with the help of processing aids (steam, water, solvents, catalysts, additives) to produce products (fuels,

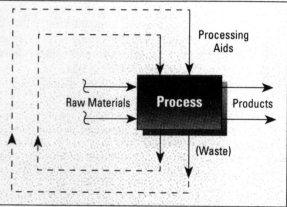

■ *Figure 1. The ideal process would generate no unusable wastes.*

lubricants, petrochemicals). A small fraction of the raw material is lost as waste; consumed or degraded processing aids, including water, also become part of the waste stream.

Today's challenges for engineers are to strive for the best process designs that emphasize waste minimization, choose raw materials or reactants that reduce the volume of unusable or problem byproducts or processing aids, improve processing steps to minimize waste generation, and optimize operations to minimize waste.

Defining "zero discharge"

"Zero discharge" is a term used to promote conservation of the quality of the environment. However, it has different specific meanings in different contexts.

True zero discharge implies that all reactant materials are converted into products, and all processing aids are reused and no waste is produced. This refers to the ideal process.

The following are several real-world working definitions of zero discharge:

1. Zero discharge is the elimination of certain priority pollutants or toxic substances in the wastewater effluent from a facility. These priority pollutants would include banned or regulated substances that tend to concentrate in the food chain (such as DDT) or heavy metals or other pollutants that would be toxic to aquatic life in the receiving waterways.

2. Zero discharge means that no water effluent stream will be discharged from the processing site. All wastewater, after secondary or tertiary treatment, is converted to a solid waste by evaporation processes, such as brine concentration followed by crystallization or drying. The solid waste may then be shipped to a landfill. Note that this implies a tradeoff where wastes that might have been dispersed in a receiving waterway are concentrated in solid form. There may also be an effect on air emissions when all liquid waste is eliminated.

3. A somewhat looser definition incorporates parts of the first two definitions. Here, zero discharge means that while wastewater volume discharge may not be small, it is relatively safe. Some concentration of dissolved solids or salts may have occurred, but this amount is not harmful to the receiving waterway. Tertiary treatment, including disinfection and dechlorination, has rendered the effluent suitable for discharge to a pond or stream used for recreational purposes.

These three working definitions cover most aspects of zero discharge discussions. Definition 1 might apply to a chemical manufacturer pursuing a different reaction pathway to a product so as not to generate a particular objectionable intermediate byproduct waste. Definition 2 might apply to the new construction of a power plant in an arid climate where a highly integrated water system with maximum recycle is in place; wastewater is minimal and may be applied to ash handling or evaporated in a small pond. Definition 3 may cover instances where an industrial or municipal wastewater is reused as feedwater or for recreational purposes. These three definitions are not all-inclusive, but they

do address the issues most often associated with zero discharge or wastewater reduction efforts.

It is important to be clear in one's definition of zero discharge and, more importantly, to understand the driving force behind the pursuit of zero discharge.

Why pursue zero discharge?

With each passing year, the successful operation of chemical processing and other industrial facilities becomes increasingly dependent upon efficient water use and management. This becomes more challenging as virgin water supplies are depleted and environmental preservation (pollution prevention) is given higher priority. These con-

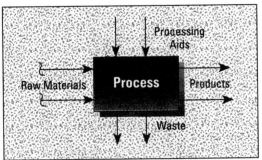

■ *Figure 2. In real processes, some of the raw materials and processing aids end up as wastes.*

siderations lead to legislation, which is drawn up to allow prudent water use while promoting preservation of natural water resources.

CPI plants may be driven to pursue zero discharge for a variety of reasons:

1. Reduced availability of fresh water. In areas where there is large population growth, such as Florida and California, limited water supplies serve a larger user base and are at risk of being depleted. In some areas, water rights may be in dispute. For example, southern California is dependent on water from the Colorado River, but Arizona's increasing water needs have resulted in a decrease in California's share of this resource. Natural droughts increase the shortages. And where groundwater (well water) is the primary source of fresh water, increased pumping from the wells may deplete the aquifer faster than natural replenishment can take place. Zero liquid discharge minimizes consumption of fresh water, so that zero discharge efforts should help relieve fresh water availability limitations.

2. Discharge permit compliance. The Clean Water Act establishes a national policy to restore and maintain the chemical, physical, and biological integrity of the nation's waters. In support of this, the U.S. Environmental Protection Agency (EPA) administers the National Pollutant Discharge Elimination System (NPDES) for permitting point-source discharges to waterways *(1)*. Table 1 is an example of NPDES discharge permit limitations for a typical petroleum refinery. Permits vary regionally and are influenced by the impacts of wastewater discharge on the receiving waterway, so a permit for discharge into a small stream is normally more stringent than a permit for discharge into a waterway that provides more dilution.

Efforts to reduce wastewater generation in some cases may help compliance with NPDES permits by allowing for more complete on-site wastewater processing due to hydraulic load reduction. Longer residence time of wastewater in on-site aeration basins, for instance, allows for more complete digestion of organic pollutants and results in less volume of pollution discharged. Elimination of wastewater by implementing zero liquid discharge would, of course, obviate the need to comply with increasingly stringent NPDES permits.

Planning zero discharge into the construction of a new facility opens up more possibilities with regard to site selection. If a receiving waterway is not needed for wastewater effluent, site location is less limited.

3. Legislation banning priority pollutants. Such substances may be banned or phased out based on scientific data that show that the substance

Table 1. Typical petroleum refinery NPDES permit limits.

Effluent Characteristics	Discharge Limitations	
	mg/L	lb/d
Biological Oxygen Demand (BOD$_5$)	15	21
Total Suspended Solids (TSS)	24	34
Chemical Oxygen Demand (COD)	150	213
Oil and Grease	10	14
Phenols	0.2	0.3
Ammonia as N	9	13
Sulfides	0.16	0.2
Total Chromium	0.16	0.2
Hexavalent Chromium	0.02	0.03
Free Cyanide	Report	
Maximum Temperature	115°F	
pH	6–9	

may cause harm even at low concentrations. Regional considerations may also affect the banning of a substance. For example, the same substance may be considered harmful at low levels in relatively stationary waters such as the Great Lakes, yet deemed not harmful in flowing waters where dilution and natural

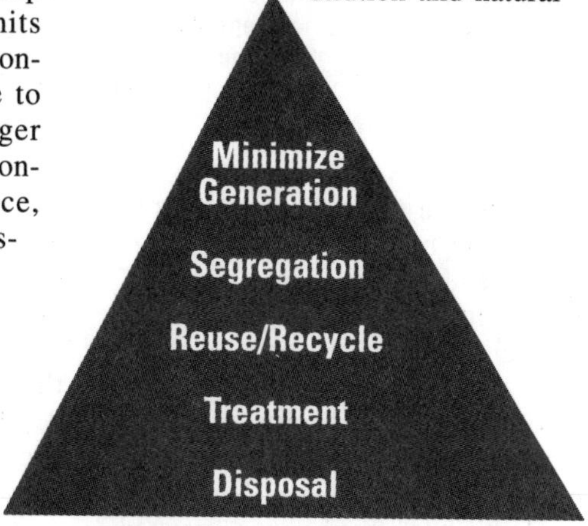

■ *Figure 3. Waste and wastewater management hierarchy.*

processes minimize impact and where bioaccumulation (the accumulation of the substance in the food chain) cannot occur.

4. Economics. Purchased water costs and wastewater treatment and disposal costs for a CPI plant can be significant. Thus, the savings associ-

ated with minimized site makeup water and wastewater flows can justify capital expenditures to minimize, if not completely eliminate, wastewater flows. Furthermore, zero discharge can save money on real estate costs in the case of new facility construction, since location near a suitable receiving waterway would not be necessary.

5. "Good neighbor" policy. Noncompliance with discharge standards costs a great deal not just economically but also in community trust and support. On the other hand, demonstration of a good faith effort to reduce wastewater or pollutant discharges shows sensitivity to not only the natural environment, but to the community as well.

How to approach zero discharge

Figure 3 illustrates how one can approach the task of designing a zero liquid discharge facility. Starting at the top of the triangle, the designer takes steps to incorporate minimization of waste and wastewater generation into the process design. Next, segregation of waste streams allows for reuse of some wastewater streams with little or no reprocessing required. Processes are designed and physically located (where possible) to facilitate reuse. Wastewater treatment processes and final disposal by evaporative processes complete the picture.

The same concepts apply when retrofitting an existing facility for zero discharge or wastewater minimization.

Reducing wastewater generation. Reduction of waste and wastewater generation may be achieved by careful planning in the selection of process methods, raw materials, and operating conditions:

• *Alternative process method.* Use of an alternative method of producing the desired product may result in a decrease of byproduct or waste to be disposed of.

• *Alternative raw materials.* Frequently different raw materials that can be used in a given process are available. Each raw material has different properties, and these differences may result in different operating conditions, products, byproducts, and discharges. In addition, the use of water in the process may vary due to the contaminants in the raw material.

• *Purer feeds.* Not surprisingly, raw materials containing fewer contaminants will require the disposal of fewer contaminants or smaller volumes of contaminated material. For example, if a plant switches to a cleaner source of water as makeup to the cooling system, contaminant levels will be reduced; increasing recirculating cooling water cycles of concentration is then feasible, thus decreasing blowdown volume while maintaining the same acceptable contaminant concentration.

• *Alternative operating conditions.* Optimizing operating conditions may result in lower levels of byproduct or less waste produced.

Segregation and reuse of wastewater streams. Water streams that are used to clean up process streams, or water wash, may be reused to clean up progressively more contaminated streams.

For example, double desalting in oil refineries involves the progres-

Table 2. Zero liquid discharge technologies.

Unit Operation	Description	Comments
Reverse Osmosis (RO)	Concentrates ionic species via semipermeable membrane producing a purified permeate tream and a concentrate stream.	Organics and solids foul membranes. Two-stage RO can produce boiler-quality water.
Electrodialysis (ED) Electrodialysis Reversal (EDR)	Concentrates water with 1,000–10,000 ppm dissolved solids (such as cooling tower blowdown or RO concentrate) by use of applied direct current in conjunction with semipermeable anionic and cationic selective membranes.	Typical concentration of 80% on charged ions. Silica and organics are not removed. EDR is applicable as an intermediate purification/concentration step.
Vapor Recompression Evaporation	Evaporator that economically uses energy to produce a pure distillate while highly concentrating the wastewater.	A high-solids (100,000 ppm) waste stream is produced.
Air Fin Cooling	Process coolers in which fans drive air across fins.	Good for high process temperature cooling. Reduces cooling tower heat load, which reduces evaporation of cooling water.
Sidestream Softening	Lime/soda ash softener softens slipstream from recirculating cooling water, reducing scaling potential.	Allows increased reuse (less blowdown) of cooling tower water.

sive removal of contaminants from the crude oil stream. The water stream and the crude oil essentially run countercurrent to each other through two desalter, or wash, vessels. The "clean" water stream is initially contacted with crude oil exiting the first desalter vessel as it enters the second desalter. This crude oil has already been desalted

The costs for implementing zero discharge in an existing plant are much greater than for a new facilty.

to some extent. After the water picks up some salts, it is put in contact with the incoming crude oil in the first desalter vessel. The once-used water is too dirty to clean up the crude oil exiting the second desalter vessel, but it is unsaturated enough to pick up some of the salts from the raw crude.

Treatment and reuse of final effluent wastewater streams. Increasingly stringent discharge requirements can necessitate wastewater cleanup to the point where it may be reusable as is, as a portion of cooling tower makeup, for instance. Advanced treatment and processing that removes dissolved as well as suspended solids may produce boiler quality water from the wastewater. These advanced

treatments may include precipitation softening, multimedia
filtration, carbon adsorption,
deionization, reverse osmosis,
or distillation.

Equipment used for zero liquid discharge

Once all steps to minimize
and reuse wastewater streams
are taken, the remaining
wastewater is normally treated and then disposed. In the
case of zero liquid discharge,
all wastewater effluent is
reprocessed and reused.
Treatment and processing for
reuse require consideration of
the following factors:

- the method best suited for
  the specific contaminants
  to be removed;
- capital costs of equipment
  and installation, piping,
  and pumps required for
  water conveyance;
- operation and maintenance
  costs;
- costs of influent water and
  effluent water discharge; and
- the variability of quality and
  flow rate expected of the water
  to be treated.

In addition to standard methods of
secondary biological wastewater
treatment and tertiary solids and
organic removal and disinfection,
several newer water conservation and
reuse technologies are seeing wider
application where zero liquid discharge is the goal. Table 2 summarizes the applicability of some of the
technologies.

The article by Rosain immediately following this one covers water
reuse in more detail and provides
some "how-to" guidance on selecting
technologies and equipment for
water reuse.

Economics

Evaluation of the economics of
water conservation and reuse or zero

■ *Figure 4. Water processing costs as a function of wastewater effluent flow rates.*

discharge should take into account
the following factors:

- availability and cost of supply
  water;
- restrictions on and costs of
  discharge water;
- recycle stream characteristics
  and effects on production or
  product quality;
- purchase and operating costs of
  water purification equipment;
  and
- whether the effort involves a
  new or existing facility.

In the case of a new facility, the
engineer plans for the optimal integrated design by considering utility
water as a commodity chemical of
significant value. This elevation of
the importance of utility water (vs.
its traditional lowest priority status)
in the design phase allows for integration of water systems. This
results in the best disposition of
lower-quality, medium-quality, or
refined-quality water, and places
importance on recycle and on waste
(water) minimization.

Independent power producers

(such as cogeneration plants) have
used integrated water system design
for zero discharge in many new projects in the U.S. The water-related
equipment in these facilities is
designed, sized, and physically located for maximum reuse and recycle of
wastewater. This minimizes use of
raw water and minimizes the flows
to the more costly processes used to
evaporate wastewater in zero liquid
discharge facilities.

In the case of an existing plant,
costs are more likely to be prohibitive. The existing facility may have
been designed with utility water use
and disposal given little importance.
While some degree of reuse, recycle,
or reprocessing may be relatively
easy to implement, complete implementation of true zero liquid discharge could require extensive repip-
ing or costly unit operations.

Example: cost of wastewater reduction

Figure 4 qualitatively illustrates
water and wastewater treatment costs

as a function of liquid effluent discharge flow rate at a hypothetical CPI plant built before the passage of the Clean Water Act. Although not shown, one should also note that raw water and sewer fees increase linearly as a function of effluent flow rate.

Large quantities of water are used and disposed of, but in recent years raw water quality and availability have deteriorated. Discharge permitting has become more stringent and plant expansions have increased water use, on-site wastewater processing flows, and discharge flows. Future plant expansions will cause hydraulic overload of existing wastewater processing equipment. The objectives are to reduce wastewater flows as well as maintain or improve wastewater quality (reduce total volume of pollutant discharge), and to minimize the costs of doing so or perhaps save money. Some or all of these characteristics are ingredients of many actual waste or wastewater minimization projects.

In general, zero discharge is costly with regard to zero discharge unit operations, but minimizes incremental water purchase and sewer charges. The term "zero discharge unit operations" typically refers to anything that contributes to the zero discharge goal, for example the unit operations shown in Table 2. Conversely, maximum wastewater flows obviate the need for costly zero discharge unit operations, but increase incremental water purchase, wastewater treatment, and sewer costs.

Let us now consider this facility, starting at Point 5 on Figure 4. Point 5 represents current operations. Wastewater effluent flow is high.

If the effluent flow gets much higher, there may be a step change in costs for processing, represented by Point 6. This is because it would be necessary to expand the wastewater treatment facilities.

Starting again at Point 5 and moving left toward Point 4 represents cost reductions due to optimization efforts that require only operational changes or minimal expense. These may include decreasing cooling tower blowdown by finding and reducing uncontrolled discharges, thus reducing wastewater flow and influent water demand. (Recall that influent and effluent charges decrease proportionately to flows.)

Moving further left, the step increase in costs for treating the water at Point 3 may be attributable to capital projects that allow further reduction of water usage and wastewater production. Examples include the installation of facilities to allow segregation, possible reprocessing, and reuse of intermediate quality waste streams, or the installation of a sidestream softener to allow for higher recycle of cooling tower water (and lower blowdown).

Proceeding left toward Point 2, a large step change would be attributable to installation of zero discharge equipment such as electrodialysis, brine concentration, evaporation-crystallization, or drying.

In many zero discharge cases, absolute zero liquid discharge need not be attained. The last small portion of highly concentrated wastewater may find use in ash conditioning or other on-site disposition. This could eliminate the need for the final crystallization or drying unit operation represented by Point 1. ◆

Literature Cited

1. Davenport, G. B., "Understand the Water-Pollution Laws Governing CPI Plants," *Chem. Eng. Progress,* **88**(9), pp. 30–33 (Sept. 1992).

Further Reading

Blake, D., *et al.,* "Zero Discharge: A Goal Whose Time Has Come?" *Water Environment and Technology,* pp. 58–61 (Oct. 92).

Hammer, K.P., and F.J. Campo, "Zero Wastewater Discharge From a 30-MW Cogeneration Plant," presented at the American Power Conference, Los Angeles, (Apr. 1990).

Makansi, J., "IPP Reports Top Performance, Zero Discharge Challenges," *Power,* pp. 77–78 (Dec. 91).

Peterson, D., and T. Bradham, "Case Study: Zero Liquid Discharge Water Treatment System at the Stratton Energy Project Biomass Power Plant," presented at the 51st Annual Meeting International Water Conference, Pittsburgh, PA (Oct. 21–24, 1990).

Rossiter, A. P., *et al.,* "Apply Process Integration to Waste Minimization," *Chem. Eng. Progress,* **89**(1), pp., 30–36 (Jan. 1993)

M. E. GOLDBLATT is a project engineer with Betz Industrial, the water management div. of Betz Laboratories, Inc., Trevose, PA (215/355-3300; Fax: 215/953-2473). He has six years of experience in water treatment, most recently concentrating on wastewater minimization, water reuse, and water use optimization. Previously, he worked as a process engineer at Exxon Chemical's Baton Rouge facility. He holds a BS in chemical engineering from the Univ. of Massachusetts and an MS in chemical engineering from Clemson Univ. He is a member of Tau Beta Pi, AIChE, and the Water Environment Federation. He is a registered Professional Engineer in Pennsylvania.

K. S. EBLE is a project engineer with Betz Industrial. She has six years of water treatment experience at Betz, specializing in cooling systems. Before joining Betz, she worked at Arco Petroleum Products Co. for six years and at Arco Chemical Co. for two years. She has a BS in chemical engineering from Drexel Univ.; she has also completed the AIChE course on fouling of heat-transfer equipment and a course in corrosion engineering at the H. H. Uhlig Corrosion Laboratory at Massachusetts Institute of Technology. She is a member of AIChE.

J. E. FEATHERS is a project engineer with Betz Industrial. She has five years of water treatment experience, four of which have been with Betz, where she specializes in water treatment and reuse. She has a BS in mechanical engineering from Vanderbilt Univ. She is a member of the American Society of Mechanical Engineers, the National Petroleum Refiners Assocation, and the Technical Association of the Pulp and Paper Industry.

Reusing Water in CPI Plants

This systematic approach will enable companies to develop and implement an integrated water reuse program for a new or existing facility.

Robert M. Rosain,
CH2M Hill

Increasing regulatory pressure for expanded wastewater treatment, waste minimization, and direct water conservation have forced many chemical process industries (CPI) plants to look seriously at water reuse. Many companies are finding that water reuse, in some form, can be cost-effective for existing plants, can open up previously closed opportunities for new plant siting and, in some cases, can even increase product quality and plant reliability.

These may appear to be lofty claims, but they are being realized as water reuse is studied further and applied to more industries. In the past, the availability of an abundant water resource and modest waste discharge restrictions created a "use it and discharge it" mentality. But now this is shifting to "use it and reuse it." Treatment technologies are available to allow the use and reuse of water of virtually any quality.

This article presents a systematic approach to evaluating water reuse for a CPI plant. It briefly reviews the major water reuse issues, and then discusses establishing a plant information database, evaluating water reuse strategies and technical options, developing and selecting water reuse alternatives, implementating a reuse plan, and measuring success.

The issues

Some of these water reuse issues were discussed in the preceding article by Goldblatt *et al.* A few additional considerations are also important, and several points are worth emphasizing again. The key issues involve:

Regulatory considerations. Waste-water discharge restrictions in many areas of the country are forcing a serious evaluation of CPI plant water reuse. It is often easier and more economical to treat and reuse a wastewater than to treat it to meet increasingly stringent discharge limitations. Furthermore, recent regulatory initiatives in many parts of the country, particularly in the arid Southwest, are also requiring water reuse in some fashion, often by mandatory reductions on raw water consumption.

Costs. The advantages of water reuse come with a price, sometimes a substantial one (historically the single biggest deterrent to water reuse). However, it still may be far more cost-effective to treat and reuse a previously discharged effluent than to treat it to meet the stringent water quality and toxic standards being imposed in new National Pollutant Discharge Elimination System (NPDES) permits. For new plants, dealing with new water use and development regulations also often requires a significant amount of project schedule time, which usually translates into added project costs.

Product quality. The impact of water quality on product quality must be carefully evaluated. For example, is product quality dependent on plant water quality? Could a lesser quality water source (*e.g.,* a reuse source) be substituted and still enable product quality goals to be met? Conversely, would a higher quality water source improve product quality? If treatment for reuse is required, such treatment, or perhaps only a slightly enhanced treatment, might yield a water quality better than the current raw water quality, thereby enhancing product quality.

Operations. The impact of water reuse

on plant operations and production can also be important. For example, are plant operations currently affected by seasonal fluctuations in raw water quality from the raw water source or water purveyor? Treatment and reuse of plant-generated wastewaters may reduce these fluctuations and dependency on this source. Water reuse will, however, likely increase the complexity of plant operations due to the need for additional water treatment unit operations.

Technology. The technology to treat and reuse water is available now and numerous examples exist where treatment and reuse has been the norm for many years. It may, however, be necessary to pilot test some aspects of the reuse treatment system to optimize the process and verify sizing and economics.

Residual wastes. The management of residual wastes is an operations and regulatory issue that must usually be addressed with water reuse systems. Examples are the disposal of sludge associated with a precipitation treatment process and the air quality impacts associated with a cooling tower using a higher TDS makeup water source.

Establishing a database

Before starting an evaluation of water reuse, it is essential to define the current base line of information on plant water use and wastewater generation. Traditionally, this area has not received a high priority in most operating plants, and an informational database is often lacking or nonexistent. For new plants, the design of a water management data acquisition and a data management system should be incorporated into the overall plant design from the start. This is far more cost-effective than retrofitting it later.

A plant water/wastewater database should include the following information:

• raw water use and process water quality requirements;
• current water treatment capability and costs;

Table 1. Typical plant water uses.

Product Formulation
• as a reactant or intermediate (liquid or steam)
• for washing, rinsing, or extraction

Cooling
• direct contact cooling
• via heat exchangers, condensers, and cooling towers

High-Purity Water Makeup Systems
• for boiler feedwater
• for laboratory operations
• for general process use

General Plant Service Water
• hose bibbs for general maintenance and housekeeping
• pump seal water
• sample condensers

Waste Conveyance/Transfer
• boiler ash sluicing
• air scrubber system

Potable/Sanitary Service

Fire Protection

• process wastewater generation (flow and composition);
• current wastewater treatment capability and costs; and
• flow and mass balances for the water/wastewater management system.

Water use and quality requirements

CPI plants are significant consumers of water. Consequently, most plants are located near an abundant and inexpensive source of fresh water. As such, there has often been no real incentive to conserve or reuse water to any significant degree, and there has been little interest in fully defining water quality requirements for specif-

ic plant processes. However, when considering water reuse for a specific process area or for the whole plant, this information becomes essential.

Table 1 lists some of the major water-consuming functions in a CPI plant.

Developing a database for an existing plant will require looking at unit piping and instrumentation diagrams (P&IDs), field verifying piping connections, and looking for nondocumented water uses. For new plants, the plant design documents should include a water use or water balance diagram, and up-to-date P&IDs and plant piping drawings.

It is also necessary to document the water quality requirements for each plant water use. For example, boiler feedwater requirements are dictated by the boiler manufacturer, or by accepted engineering practice considering the boiler pressure, temperature, and chemical treatment regime. In turn, the boiler feedwater treatment system is designed based on these water quality requirements and the quality of the raw water source. Where water is used in product formulation or manufacture, water quality requirements may be based on product quality specifications.

A critical evaluation of the water quality requirements for each specified use is paramount to a successful water reuse program. As water is reused, its dissolved and/or suspended solids content will increase through the pickup of dissolved contaminants or through the evaporation and subsequent concentration of solids present in the source. Although water can be treated to remove these contaminants, if a particular process can use such water with little or no treatment, the overall economics of the water reuse system will improve.

Consequently, it's critical to seriously question the defined water quality specifications for each process. It is not unusual to find little basis for the specifications other than "...that's what we've always specified..." or "...we need the best water we can get..." Clearly, product manufacture and plant operations will require particular water quality specifications for several areas of the plant, but they should be specific and technically based. For example: feedwater to a 600-psig boiler does not need complete demineralization whereas feedwater to a 1,500-psig boiler does; cooling towers can usually accept a wide variety of makeup water qualities and do not have to use only "fresh" water; and waste conveyance and plant washdown water can use a relatively low quality water.

In evaluating water quality requirements, don't forget plant metallurgy. For example, stainless steel piping is susceptible to stress corrosion cracking from high chloride streams and high ammonia streams are incompatible with copper alloy heat exchangers.

Current water treatment capabilities and costs

The plant water/wastewater database should also include the plant's current raw water treatment capabilities and the associated water treatment costs. For example, is there a central water treatment plant or smaller water treatment plants at each unit? What are the flow and contaminant removal capabilities of these plants? What are the costs of operating these plants and how are they allocated to the units? If the plant receives raw water from a municipality or water district, what are the unit costs?

In assigning costs, remember to include the pumping, storage, and distribution system as well as treatment.

Wastewater generation

CPI plants are large wastewater generators. In addition to water use data, the plant database should inventory all wastewater discharges on a unit- and process-specific basis. Some information is routinely collected due to discharge permit monitoring requirements. However, the monitoring requirements necessary to develop and implement a water reuse system normally go well beyond those needed for permit compliance.

Wastewater is generated from three sources within a plant:
• unit processes engaged in product manufacture;
• utility support operations; and
• stormwater collected in and around the plant unit areas (that is, contaminated stormwater).

Table 2 lists some of the more prevalent wastewater sources found in a typical CPI plant. A full inventory of plant wastewater sources, flows, and compositions can take a considerable effort.

Often, wastewater drains are connected and routed to common sumps, making it difficult to monitor and sample the individual streams. If the

Table 2. Typical plant wastewater sources.

From Product and Manufacturing Operations
• reaction product water
• kettle and column washdowns and cleanouts
• product washing and rinsing

From Utility and Support Operations
• boiler and heat-recovery steam generator blowdown
• cooling tower blowdown
• boiler ash sluicing
• air scrubber blowdown
• general plant housekeeping
• pump seal water (including vacuum pumps)
• vacuum eductors
• water treatment plant wastes (softeners, demineralizers, filters)

Stormwater
• contaminated through contact with plant processes
• uncontaminated

individual streams are small and of a similar origin, it may be acceptable to monitor and eventually deal with the common stream. If not, stream segregation should be considered. The ultimate goal of a flow and sampling study is to develop a flow and mass balance around the plant and plant business units. The specifics of this effort are discussed in more detail later.

Wastewater treatment capabilities and costs

The plant's wastewater treatment capabilities should be inventoried in terms of individual unit operations for the plant as a whole (and for each business unit, if applicable). Essential information should include the hydraulic capacity and contaminant removal capability, any constraints on the wastewater treatment process, whether it can adequately handle current flows, the maximum plant capacity with respect to flow and load, and the treated effluent quality and the factors upon which quality is contingent.

The costs associated with this treatment should be documented in terms of $/1,000 gal treated and $/contaminant removed (e.g., $/lb chemical oxygen demand [COD] removed). Include all operating costs, the cost of unamortized capital improvements, and permit and legal fees. These data can be used later to define the cost of treatment on a business-unit basis.

Flow and mass balances

Flow and mass balances are essential for the design and operation of a water reuse or water management system. Historically, plant water management has not been tightly controlled, and hence, for existing plants the data necessary to develop such a balance are often not available. As indicated earlier, obtaining water use, water quality, and wastewater flow and composition data can be a significant undertaking, but unfortunately there is no alternative if one wants to develop an efficient and cost-effective water reuse system. For new plants, these monitoring and data acquisition systems can be, and should be, incorporated into the plant design.

Measuring flow rates. Water and wastewater flow rates can be measured directly or indirectly.

Table 3 indicates some of the more common methods of direct flow measurement. Regardless of the type of device chosen, the ability to measure or calculate average, minimum, and maximum flow rates, and to define the duration of surges and/or spikes in the flow, are essential.

Indirect flow measurement can be accomplished through difference (that is, subtracting one flow measurement from another) or through a mass balance measurement. The latter method is often chosen if a flow meter cannot be installed on the stream in question, or if routine chemical monitoring for another reason will yield sufficient data to calculate flow.

Usually the chemical species chosen to determine flow by the mass balance technique is one that is conserved through the plant system. For example, potassium is a good ion to track because it has a high solubility, it generally does not leave the system, and it does not volatilize. In addition, a species such as potassium is ideally suited to automated monitoring with an ion-specific electrode. Calcium, on the other hand, would not be a good ion to track due to its potential insolubility.

Measuring compositions. In order to complete the mass balance, the water and wastewater compositions will have to be determined at critical junctions in the system or be calculated by difference. First, the necessary physical and chemical parameters to be monitored must be defined. Generally these include species that will impact the system operation in some way if allowed to build up. Table 4 lists typical species

Table 3. Typical flow meters and their applications.	
Type	**Applications and Characteristics**
Magnetic	All fluids, especially for slurries and streams with high total suspended solids (TSS) concentrations.
Ultrasonic Transmissive Type	Requires TSS < 3%; for larger pipe sizes; electrodes contact fluid.
Doppler Type	Requires TSS > 2%; for smaller pipe sizes; exterior mounting.
Vortex Shedding	Clean fluids of low viscosity; highly accurate.
Turbine	Clean fluids; wide flow range; highly accurate.
Propeller	Clean fluids or slurries; wide flow range.
Orifice Plates	Clean fluids; restricted rangeability; moderate accuracy.
Venturi Tubes	Clean fluids or slurries; restricted rangeability.
Nozzle Tubes	Clean fluids or slurries; restricted rangeability.
Pitot Tube	Clean fluids; moderate accuracy.
Wedge	For slurries or sludges.
Rotameter	Clean fluids; low flow or metering.
Weirs, Flumes	Gravity, open channel flow.

monitored in a water management system.

All of the flows in the system need not be monitored for all species. Rather, monitoring can be done selectively based on the need for the information at each particular point in the system.

Data management. For most plants, the amount of data obtained through a comprehensive monitoring program can be significant. Using a

Table 4. Typical species monitored in a water management system.		
Cations	**Anions**	**Other**
Ca	HCO_3	SiO_2
Mg	CO_3	pH
Fe	SO_4	Total Suspended Solids (TSS)
Na	Cl	Total Dissolved Solids (TDS)
	NO_3	Chemical Oxygen Demand (COD)
		Biological Oxygen Demand (BOD)
		NH_3

Note: Other parameters may be added if required (for example, specific heavy metal or organic species).

standard database program for data handling, manipulation, and reporting is generally preferred and easily applied.

For the system flow and mass balances, it is also advisable to set up a computer program to perform the computational tasks. This can be as simple as a spreadsheet program or as complicated as a custom, interactive model. The choice depends on how the flow and mass balance model will be used and the degree of automation desired. [For more information on data management, see (1).]

Process control. Some plants that rely on timely and accurate data to control their water management system utilize an automated data acquisition system linked to an on-line flow and mass balance model. The model is contained within the plant's computer control system. The computer continuously receives data and periodically updates the status of the water management system in real time.

These systems can represent a substantial investment, but they are no different than those routinely specified for unit operation and process control. Water management system operators need accurate and timely data to optimize their water reuse system in the same manner that production control operators need accurate and timely data to optimize production and process control.

Goals and strategies

Before developing specific water reuse alternatives for a plant, it's appropriate to review water reuse goals, water reuse strategies, and applicable water treatment technologies.

A plant's ultimate water reuse goals will guide the development of water reuse alternatives. Each plant may have a unique set of water reuse goals based on its specific situation. In general, the goal of any water reuse program should be to match water reuse opportunities with plant-specific water quality requirements to achieve cost-effective reuse con-

sistent with regulatory requirements. Some typical water reuse goals include:

• *Minimize raw water consumption.* Real or enforced water use restrictions may drive a plant to reuse water. This can be accomplished by reusing plant effluents, using secondary-treated effluent from a nearby municipal wastewater treatment plant (a growing option in many areas of the country), or even reusing the effluent from another nearby industrial process plant.

• *Minimize effluent discharge.* Water reuse will, almost by definition, minimize a plant's effluent discharge, and developing and adopting waste minimization strategies may accomplish the same result without implementing a specific water reuse effort. It should be noted that waste minimization often results in a more concentrated effluent, often requiring more extensive treatment than previously used to meet discharge permit conditions.

• *Zero liquid discharge.* Water quality based discharge standards may, in fact, require a plant to treat its effluent to achieve contaminant levels lower than the raw influent water drawn from the same water source. When faced with this situation, it is probably more cost-effective to treat and reuse the total plant effluent than to treat it for discharge. This is known as zero liquid discharge. Although treatment will likely be extensive and relatively costly, zero liquid discharge will eliminate the need for an NPDES permit for process effluents and the associated costs. [The preceding article by Goldblatt *et al.* discusses zero discharge more extensively.]

Once the plant database has been established and goals set, applicable water reuse strategies should be reviewed. General strategies applicable to all water reuse systems include:

• *Cascade reuse* — Cascade reuse seeks to reuse, directly or with minimal treatment, the effluent from one

unit operation as the raw water source for another unit operation. Examples include routing floor drains (after oil/water separation) to cooling tower makeup and saving the fast rinse portion of a demineralizer regeneration sequence to use as dilution water for subsequent regenerations.

• *Waste minimization* — Waste minimization opportunities abound based on past water use practices. For example, switching to mechanical seal pumps can significantly reduce pump seal water flow that normally goes to a process sump and the wastewater treatment plant. Use mops and brooms instead of water and hoses to maintain plant areas.

• *Source reduction* — Source reduction includes ways to reduce or eliminate the use of water (or any source material) in the process itself. Examples include: countercurrent rinsing; product washing to minimize fresh water use; the use of a baghouse to capture fugitive dust instead of using a water scrubber; and using secondary municipal effluent in place of raw water where appropriate (*e.g.,* cooling tower makeup). The latter option has been successfully employed for many years by numerous plants across the country.

Segregated vs. end-of-pipe treatment. End-of-pipe treatment has traditionally been the strategy of choice in the CPI. However, stream segregation and treatment is sometimes more effective. In this way contaminants can be effectively removed before being combined and diluted with other effluents. This strategy may apply to individual unit operations, groups of unit operations, or whole business units. Segregated treatment may be particularly effective if the removal of only one or two contaminants will allow the waste stream to be reused directly, or will reduce the size or complexity of the end-of-pipe treatment system. [McLaughlin *et al.* (2) discuss segregated treatment further.]

In reality, an integrated water reuse system will likely employ a combina-

Table 5. Water reuse technologies grouped according to their constituent removal capabilities.

Suspended Solids, Oil, Tar

Gravity Separation
Coagulation, Flocculation, Clarification
Flotation (Dissolved-Air or Induced-Air)
Coalescing Filters
Granular Media Filtration
Membrane Filtration

Phenols

Solvent Extraction
Wet Air Oxidation
Biological Oxidation, Aerobic
Activated Carbon
Chemical Oxidation

Cyanide/Thiocyanate

Steam Stripping
Biological Oxidation
Alkaline Chlorination
Ion Exchange
Chemical Precipitation

Desalination (TDS Removal)

Mechanical Evaporation
Evaporation Ponds
Reverse Osmosis
Electrodialysis
Crystallization
Spray Drying

Ammonia

Steam Stripping
Biological Nitrification
Ion Exchange Using Clinoptilolite
Air Stripping
Breakpoint Chlorination

Dissolved Organics

Biological Oxidation, Aerobic
Biological Oxidation, Anaerobic
Chemical Oxidation
Activated Carbon
Wet Air Oxidation
Incineration

Heavy Metals

Chemical Precipitation
Ion Exchange
Activated Alumina
Reverse Osmosis
Electrodialysis

Sludge Handling/Disposal

Thickening
Vacuum Drum Filters
Filter Press
Belt Filter Press
Centrifuge
Incineration
Thermal Drying
Solidification/Stabilization
Landfill Disposal

tion of segregated and end-of-pipe treatment systems to achieve cost-effective water reuse.

Water reuse technologies

Fortunately, water reuse is not technology-limited. There is an abundance of successful, proven technologies to pick from, as well as many emerging technologies that show great promise for the future. Table 5 lists some commonly applied water treatment technologies.

As previously indicated, effective water reuse will lead to a concentration of dissolved and suspended contaminants that must ultimately be removed before discharge or recycle.

In plants that can reuse water extensively (or in zero liquid discharge plants), a final treatment process will include desalination to remove dissolved inorganic ions. Although inherently expensive, this process can yield a water source of higher quality than the plant's fresh source water and may even enhance product quality in some way.

Developing and selecting water reuse alternatives

Plantwide water reuse alternatives should be developed around the appropriate strategies and technologies that meet process (reuse) water quality requirements. Alternatives

will likely involve a combination of unit operation, business unit, and end-of-pipe treatments.

Figure 1, a generic plant water reuse flow diagram, illustrates the integration of water treatment and reuse that is possible.

Every plant will have a unique set of alternatives. Once conceived, they should be evaluated and ranked based on a set of established criteria that will permit the selection of the best alternative. Selection criteria might include costs, proven performance in a similar application, reliability, complexity, impact on product quality, impact on operations, secondary waste generation, and regulatory compliance. A sensitivity and risk analysis may also be appropriate.

Implementation

Implementation issues include phased development, pilot testing, design and construction, and operations.

If the chosen water reuse alternative is complex and/or highly integrated with plant operations, it may be wise to consider a phased implementation. This can minimize the impact on plant operations and costs, as well as allow the plant to monitor and adjust the reuse system, if necessary, before proceeding with the next phase.

Although water reuse technologies are generally proven, pilot testing may be necessary to obtain specific design data for the application planned. Bench-scale tests may be sufficient if scale-up factors at this level are well established. If not, field pilot tests may be warranted. Such testing should be given the same attention and priority that pilot testing of a new production process would receive.

Design and construction can follow traditional paths. If the planned installation is small, consider a turnkey performance specification approach. If the installation is large and highly integrated with plant operations, it is probably advisable to

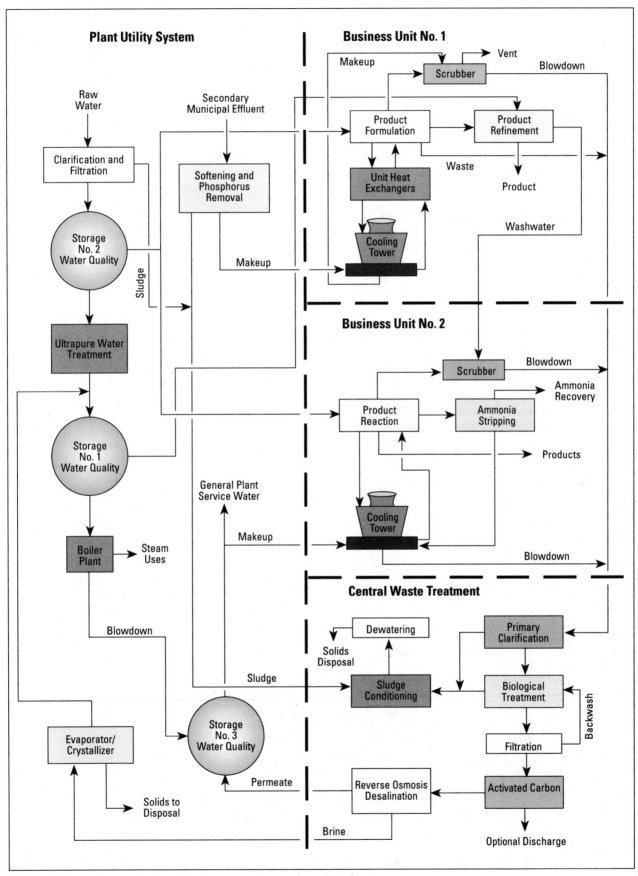

■ *Figure 1. Example of an integrated water reuse diagram for a CPI plant.*

retain a design architect/engineer and keep the plant staff intimately involved in the project design; construction can then proceed by one or more traditional paths.

Implementation of a water reuse system from an operations standpoint deserves some serious consideration. Assuming a highly integrated, plantwide system, an operations and management infrastructure must be present to staff and operate the facilities. This may mean a significant upgrade from past practices.

Above all, a visible management commitment to the water reuse plan must exist for it to be successful. Unfortunately, water and wastewater treatment systems have often not been given the status, budgets, or management attention that other operating units have achieved. However, when water reuse is inextricably linked to the production processes, this must change. This could mean, among other things, establishing a new labor union classification for water reuse system operators — something that would provide a defined career path for these

operators so that they don't attempt to "bid out" of the job at their earliest opportunity to higher "status" or higher paying jobs in the plant. A comprehensive training, and possibly certification, program for water reuse system operators is also essential to the continued success of the facility.

Measuring success

The success of a water reuse program can be measured in many ways, some tangibly, others more subjectively. However, the basic question is: "Does the reuse system meet regulatory requirements in a cost-effective manner without compromising product quality and other plant systems?" Specific considerations are:

Costs. What are the life cycle costs of the water reuse system? What are the unit cost impacts on the products produced? Don't forget to take a cost credit for raw or treated water use reduction where appropriate.

Regulatory compliance. Does the water reuse system meet all current and foreseeable regulatory standards? Consider raw water consumption standards, wastewater discharge standards, air-quality standards, and solid waste residual standards.

Product quality. Has the unit or plant product quality been affected through water reuse? Has it been improved through an improved quality makeup water source?

Operations. Has the plant's production capability been affected in any way? Has it improved through a decreased dependence on the plant's previous water source?

Image. Has the plant's community image been enhanced through water reuse, and are the plant employees supportive of management efforts in this area? Can the water reuse system be pointed to as a positive step in the plant's commitment to environmental improvement? **CEP**

R. M. ROSAIN is senior project manager and assistant director of Industrial Water and Wastewater Engineering for CH2M Hill, Bellevue, WA (206/453-5000; Fax: 206/462-5957). With 20 years' experience in industrial water and wastewater engineering, he has designed and managed projects for a wide variety of industries and clients including the CPI. His projects have specifically included integrated wastewater treatment and reuse systems, physical-chemical treatment systems for organic and inorganic contaminants, cooling water treatment systems, desalination systems, ultrapure water treatment systems, corrosion investigations, and hazardous waste investigations and treatment systems. Mr. Rosain has also been involved in the analysis of regulatory policy and standards as they apply to industrial discharges. He holds a BS in chemistry from Western Illinois Univ., an MS in chemistry from the Univ. of Idaho, and an MS in environmental science and applied water chemistry from Washington State Univ. He is a member of AIChE, the Air and Waste Management Association, and the American Society of Mechanical Engineers.

Literature Cited

1. **McMorris, R. L., and R. Gravely,** "Managing Data from Large-Scale Continuous Monitoring Projects," *Chem. Eng. Progress,* **89** (3), pp. 111–115 (March 1993).
2. **McLaughlin, L. A.,** *et al.,* "Develop an Effective Wastewater Treatment Strategy," *Chem. Eng. Progress,* **88**(9), pp. 34–42 (Sept. 1992).

Further Reading

Adams, C., "Source Control for Wastewater Discharges for the Organic Chemicals Industry," *The National Environmental Journal,* **2**(5), pp. 28–34 (Sept./Oct. 1992).

Benforado, D. M., *et al.,* "Pollution Prevention: One Firm's Experience," *Chem. Eng.,* **98**(9), pp. 130–133 (Sept. 1991).

Berglund, R. L., and C. T. Lawson, "Preventing Pollution in the CPI," *Chem. Eng.,* **98**(9), pp. 120–127 (Sept. 1991).

Cable, J. K., *et al.,* "Integration of RCRA Corrective Action with Clean Water Act Compliance," *Env. Progress,* **11**(2), pp. 85–90 (May 1992).

Freeman, H., *et al.,* "Industrial Pollution Prevention: A Critical Review," *J. Air Waste Management Association,* **42**(5), pp. 618–656 (May 1992).

Geishecker, E. P., "Reuse of Contaminated Water," presented at the First Biennial Conference of the National Water Supply Improvement Association, Washington, D.C. (June 1986).

Meier, D. A., and K.E. Fulks, "Water Treatment Options and Considerations for Water Re-Use," presented at the National Association of Corrosion Engineers Corrosion 1990 Meeting, Las Vegas, NV (Apr.1990).

Pojasek, R. B., "For Pollution Prevention: Be Descriptive, Not Prescriptive," *Chem. Eng.,* **98**(9), pp. 136–139 (Sept. 1991).

Strauss, S. D., "Water Management for Reuse/Recycle," *Power,* **135**(5), pp. 13–28 (May 1991).

Sundberg, S. R., *et al.,* "Industrial Cooling with Reclaimed Water," Proceedings of the Industrial Wastes Symposia. 63rd Annual Water Pollution Control Federation Conference, pp. 96–105 (Oct. 7–11, 1990).

Wett, T., "Successful Source Reduction Means a New State of Mind About Waste," *Chem. Proc.,* **53**(3), pp. 46–52 (Mar. 1990).

Choose Appropriate Wastewater Treatment Technologies

Understand the various physical, chemical, thermal, and biological processes and how they can be used to treat different types of industrial wastewaters.

Dannelle H. Belhateche,
Environeering, Inc.

Even though complex chemical manufacturing processes can be operated effectively, consistent successful wastewater treatment continues to elude even some of the most sophisticated chemical process industries facilities. Industrial wastewater treatment has been slow to develop, and in some respects has not kept up with advances in manufacturing technology.

One reason for this may be an over-reliance by industrial facilities on biological treatment to solve all of their wastewater problems. Properly operated biological treatment is certainly an inexpensive, relatively simple way to deal with many different types of industrial wastes. However, it cannot alone solve all waste problems.

An earlier *CEP* article *(1)* outlined a procedure for developing an effective wastewater treatment strategy. It involves first characterizing all the plant's wastewater streams and evaluating the treatment needs of each stream with respect to the applicable regulatory standards. Then streams requiring the same type of treatment are combined and treated as composite streams. This approach improves the cost-effectiveness of the overall treatment scheme.

This article discusses the various wastewater treatment technologies in more detail and includes tables that compare their applications, advantages, and disadvantages. It also provides guidance on when to apply what type of treatment to which waste streams. This information can help bridge the gap between where the plant needs to be, in terms of effluent quality, and where it is, in terms of wastewater characteristics.

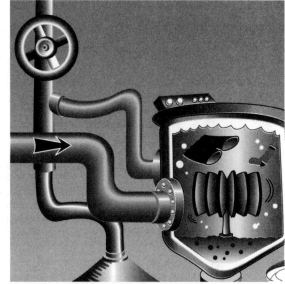

Wastewater characteristics

Table 1 lists common characteristics of industrial wastewaters and the inherent hazards that wastewater treatment is intended to address. In general, wastewater can be characterized based on its bulk organic parameters, physical characteristics, and specific contaminants.

The bulk organic wastewater quality parameters measure the amount of organic matter present in a waste stream. Typical parameters are total organic carbon (TOC), chemical oxygen demand (COD), biochemical oxygen demand (BOD), and oil and grease (O&G) or total petroleum hydrocarbons (TPH). These parameters do not measure a specific single identifiable compound, but rather a group of constituents.

TOC, COD, and BOD measurements indicate the amount of organic matter present in a waste stream that requires stabilization, or oxidation. BOD measures organic compounds that are amenable to biological treatment, while TOC and COD measure, respectively, the amount of or-

Table 1. Typical industrial wastewater contaminants.

Parameter	Concern
Bulk Organic Parameters	
TOC	Can be toxic; depletes oxygen
COD	Can be toxic; depletes oxygen
BOD	Depletes oxygen in receiving waters
Oil and Grease/TPH	Damages vegetation and wildlife
Physical Parameters	
TSS	Turbidity; toxic to aquatic life
pH	Acidity or alkalinity is toxic to aquatic life
Temperature	Toxic to aquatic life
Color	Aesthetic; destroys algae
Odor	Can be toxic to aquatic life and humans; aesthetic
Redox Potential	Can be toxic to aquatic life
Contaminant-Specific Parameters	
NH_3/NO_3	Toxic to aquatic life; eutrophication
Phosphates	Eutrophication
Heavy Metals	Toxic to aquatic life and humans
Surfactants	Toxic to aquatic life and humans; aesthetic
Sulfides	Toxic to aquatic life and humans; aesthetic
Phenol	Toxic to aquatic life and humans; aesthetic
Toxic Organics	Toxic to aquatic life and humans
Cyanide	Toxic to aquatic life and humans

Key:
TOC = Total Organic Carbon
COD = Chemical Oxygen Demand
BOD = Biochemical Oxygen Demand
TPH = Total Petroleum Hydrocarbons
TSS = Total Suspended Solids

Effluent quality

In addition to knowing the characteristics of a specific wastewater, one must also understand the treatment requirements. Table 2 summarizes the most common effluent quality requirements that are often addressed in wastewater permits. The parameters that must be monitored and their permissible discharge levels are determined based on federal and state water quality programs. The federal wastewater permitting program, the National Pollutant Discharge Elimination System (NPDES), is found in (2); state water quality standards are found in individual state regulations.

Specific effluent requirements depend on the wastewater's ultimate discharge point. For example, a discharge to an environmentally sensitive ecosystem (such as an endangered habitat) or to a pristine or drinking water source will have to meet much more stringent limitations than a discharge to a setting that has already been negatively impacted. In the latter case, the discharge would simply have to not contribute further to the decline of water quality.

Discharges to a publicly owned treatment works (POTW) receive additional treatment before being discharged to the ultimate receiving point. In addition, contaminants in individual waste streams become diluted at the front of the treatment plant. These factors can result in less stringent effluent

ganic carbon, and the amount of carbon that could theoretically be oxidized to carbon dioxide plus any oxidizable inorganic material. For example, a waste stream with a high COD or TOC and a low BOD indicates an organic waste that is not amenable to biodegradation; on the other hand, a high COD and a low TOC indicates that an inorganic oxidizable species is present (inorganic COD is usually not amenable to biological treatment).

The O&G and TPH parameters indicate the presence of oil or petroleum hydrocarbons, which can be either dissolved or in a free-phase state. These parameters are useful for measuring the organic quality of a wastewater only when such compounds are expected to be present in large quantities, such as in refinery wastewaters.

Physical characteristics of wastewater include total suspended solids (TSS), pH, temperature, color, odor, and sometimes oxidation/reduction (redox) potential. Some of the physical characteristics reflect aesthetic qualities of a wastewater (such as a fishy odor or dark color), whereas other characteristics, such as pH and temperature, can have a negative impact on receiving water bodies.

Specific contaminants of interest in wastewater can be organic or inorganic in nature. Table 1 lists a few common examples. The particular constituents of concern vary from one facility to another, and the quantity of each permissible in an effluent is determined by the regulatory agencies.

Table 2. Typical industrial wastewater effluent limitations.

Parameter	Concentration, mg/L
COD	300–2,000
BOD	100–300
Oil and Grease/TPH	15–55
TSS	15–45
pH	6.0–9.0
Temperature	Less than 40°C
Color	2 color units
NH_3/NO_3	1.0–10
Phosphates	0.2
Heavy Metals	0.1–5.0
Surfactants (Total)	0.5–1.0 total
Sulfides	0.01–0.1
Phenol	0.1–1.0
Toxic Organics (Total)	1.0 total
Cyanide	0.1

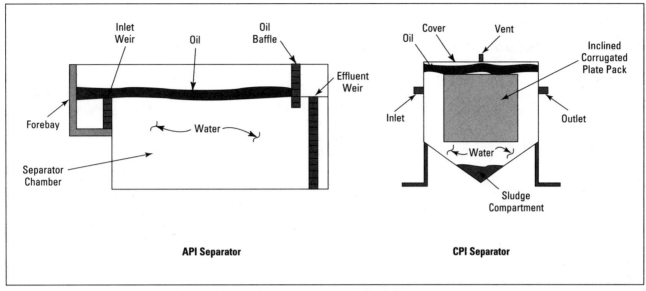

■ *Figure 1. The most common devices used for gravity separation are API and CPI separators.*

limitations if the POTW is relatively sophisticated and receives other industrial discharges or has a large treatment capacity. However, the recent trend is for individual dischargers to provide complete end-of-pipe treatment for their waste streams, rather than just pretreatment prior to treatment in a POTW. POTWs have to meet their own effluent quality limitations, and many don't want to take responsibility for industrial dischargers.

Types of treatment

Once you've characterized your wastewater and you know your treatment requirements, the next step is to reconcile the two. Even if you've hired a consultant, you still need to evaluate the alternatives given to you. To do so, you need a basic understanding of the different types of treatment processes available and what they can do, so that you can select those that are right for your facility.

Wastewater treatment is generally classified into four levels — primary, second, tertiary, and quaternary treatment. Each treatment level is aimed at removing a more specific class of contaminants. Primary treatment involves simple physical processes that remove suspended solids and entrained oil from a wastewater stream. Secondary

treatment is designed to remove soluble material from waste streams that cannot be removed by simple physical means. There are numerous secondary treatment technologies available, depending on the constituents to be removed. Tertiary and quaternary treatment are used to polish an effluent to remove a specific contaminant that is not removed in the first and second treatment steps.

In addition to these four levels of treatment, there are four classes of treatment technologies — physical, chemical, thermal, and biological. The specific processes selected to remove contaminants from the bulk aqueous phase can be used either singularly or together in various combinations.

PHYSICAL TREATMENT

Physical treatment processes bring about a physical change in the properties of the contaminant(s) while the chemical nature of the compounds remains unaffected. The physical properties of the contaminants are manipulated to facilitate the removal of pollutants from the bulk wastewater stream.

The physical treatment processes typically employed in industrial wastewater treatment (Table 3) are gravity separation, air flotation, oil coalescing, evaporation, filtration, activat-

ed carbon adsorption, air or stream stripping, and liquid/liquid extraction.

Gravity separation is used as primary treatment to remove free oil and solids entrained in the bulk waste stream. Air flotation is used as a secondary oil removal process to remove finely dispersed oil droplets and mechanically emulsified oil. Oil coalescing can be used either as a primary or secondary oil removal process, removing free-phase oil, fine oil droplets, and some emulsified oil. Evaporation and filtration are employed when the bulk waste stream contains a high concentration of solids or other phase-separable material. Evaporation reduces the amount of bulk liquid to be ultimately disposed of. Activated carbon adsorption, air or steam stripping, and liquid/liquid extraction are secondary, tertiary, or quarternary treatment processes aimed at removing specific organic contaminants from a waste stream, based on the physical and chemical nature of the contaminants.

Gravity separation

Gravity separation *(3, 4)* is used to treat wastewater streams where contaminants can be separated from the bulk waste stream as a result of their specific gravities being more or less than that of water, which is 1.0. Oils,

Table 3. Summary of physical wastewater treatment processes.

Process	Applications	Advantages	Disadvantages
Gravity Separation	Free Oil, Suspended Solids	Inexpensive. Low maintenance. No mechanical elements. Relatively safe. Easy to operate. Low energy costs.	Volatile emissions. Does not remove dissolved constituents. Waste oil and waste sludge disposal.
Air Flotation	Emulsified Oil, Fats, Grease, Finely Suspended Solids	Breaks mechanical emulsions. Inexpensive. Relatively safe.	Volatile emissions. Does not remove dissolved constituents. Does not break chemical emulsions. Requires chemical additives. Froth disposal. Difficult maintenance. High energy cost.
Oil Coalescing	Free and Emulsified Oil	Low maintenance. No mechanical elements. Breaks mechanical emulsions. Enhanced oil removal. No chemical additives. Relatively safe. Easy to operate. Low energy costs.	Does not remove dissolved constituents. Waste oil disposal. Does not break chemical emulsions.
Evaporation	Volatile Organics, Reduction of Water Volume	Reduces volume of waste. Removes dissolved contaminants. Relatively safe. Easy to operate.	Volatile emissions. High energy costs. Susceptible to fouling. Difficult to maintain.
Filtration	Free and Emulsified Oil, Suspended Solids	Removes some dissolved contaminants. Relatively safe. Easy to operate. Low energy costs.	Susceptible to fouling. Backwashing required. Odors, bacterial growth. High maintenance.
Adsorption	Organic Compounds, Some Inorganic Compounds	Removes dissolved contaminants. Relatively safe. Easy to operate. Low capital costs.	Susceptible to fouling. Odors, bacterial growth. High maintenance. Regeneration or disposal required.
Stripping	Volatile Organics, Some Semivolatile Organics	Removes dissolved contaminants. Somewhat safe. Low capital costs.	High energy costs. Difficult to maintain. Susceptible to fouling. Volatile emissions. Does not remove inorganics.
Extraction	Organic Compounds, Some Inorganic Compounds	Removes dissolved contaminants. Somewhat safe. Easy to operate. Low capital costs.	Volatile emissions. High energy costs. Difficult to maintain. Additional waste streams to be treated.

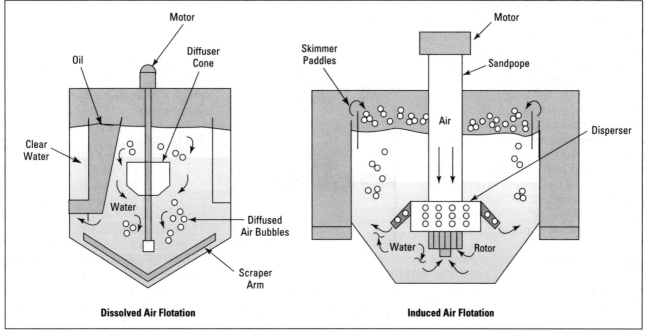

■ *Figure 2. Air flotation is an enhanced gravity separation method.*

with specific gravities ranging from about 0.8 to 0.95, are separated into an upper floating layer, while solids, with specific gravities typically ranging from 1.05 to 2.6, settle into a bottom sludge layer. The heavier or lighter the constituent, the faster it will settle to the bottom or float to the surface. Typically, gravity chambers known as API separators or corrugated plate interceptor (CPI-type) oil/water separators (Figure 1) are used.

API separators are usually rectangular concrete basins built into the ground. For effective separation to occur, the wastewater must reach quiescent conditions. How long this takes depends on the length, width, and depth of the basin. The design constraint is the basin's length — it must be sufficient to give the slowest moving particle in the waste stream time to completely settle or rise. If the wastewater reaches the effluent chamber before all of the particles or oils have dropped to the bottom or risen to the surface, the contaminants will be carried out into the effluent.

Because of these space requirements for inground API basins, CPI-type separators are becoming more popular.

CPI separators utilize media inserted into the gravity chamber to increase the surface area over which the wastewater flows. The additional surface area increases the effective length that the wastewater travels without changing the external dimensions. (In some cases, the corrugated plate packs can be retrofitted into existing API basins to improve separation.)

Recent regulations governing emissions from industrial wastewater treatment plants now require separators to be covered to reduce air emissions and odors. Retrofitting covers on existing API basins can be extremely costly because the covers have to be explosionproof and they must be vented to prevent the buildup of explosive gases under the cover. Most CPI separators come equipped with vented covers (an added benefit of which is that the covers can easily be removed for maintenance).

API basins present another problem just by being open — they are susceptible to dumping, which can lead to mixing of incompatible wastes. The basins are usually linked to secondary treatment processes, which can be adversely affected by dumping, and at

great cost. In addition, the inground concrete basins are susceptible to leakage and cracks, and can be the cause of soil and groundwater contamination.

Air flotation

Dissolved-air flotation (DAF) and induced-air flotation (IAF) are accelerated gravitational separation methods (Figure 2). Air flotation *(4–8)* is useful for removing fats, grease, and oily materials that would naturally float given ample time but that present a problem for conventional gravity separators.

Oily material entrained in wastewater can become mechanically emulsified by turbulent mixing, pressurization, or centrifugal pumping. Emulsified material will not separate out rapidly under quiescent conditions because the material has been dispersed so finely into the bulk waste stream that a stable suspension occurs.

This dispersed oil can be removed using dissolved gases, in the form of micron-size bubbles, to form agglomerates between finely dispersed particles. The introduction of the air bubbles reduces the overall specific gravity of the agglomerates; agglomerated material then floats to the surface, where it forms a scum layer that can be removed by skimming.

In the most common form of DAF, air under pressure is introduced at the bottom of an open basin, and as the air bubbles (now at a lower pressure) rise to the top of the basin, agglomerated material floats to the surface. In other DAF configurations, the wastewater itself is pressurized and supersaturated with air, then the wastewater pressure is allowed to reduce to atmospheric conditions, causing the excess dissolved gases in the wastewater to float to the surface. In IAF, air is introduced through a rotor and dispersed using air diffusers.

Air flotation is most effective when the air bubble size is small — 2 mm is typical. Often coagulation aides are necessary to increase the size of agglomerates and facilitate flotation. The process is not effective for chemically emulsified wastes, such as desalting and amine treating wastes, and wastes containing surfactants; in these cases, the chemical emulsion must first be broken to free the entrained material.

Oil coalescing

Pilot-scale fibrous-bed coalescers (4, 8) have been used in wastewater treatment for the removal of residual secondary oil/water emulsions that can not be separated in gravity separators. The fibrous-bed coalescer consists of a fixed filter element made of fibrous or other oil-attracting material. Oil droplets have a great affinity for these fibers, which act to coalesce oil droplets and break emulsions. Some fibers are less than 0.5 microns in diameter. When oil/water emulsions are forced through the element, the micron-size oil droplets grow and migrate through the element to rise to the surface of the water. The floated oil is then removed from the water surface by skimming.

The efficiency of the unit depends on operating parameters such as influent oil concentration, fiber size, flow rate, oil wettability, oil droplet size, and solids concentration. The process usually requires extensive pretreatment and is susceptible to solids and biological fouling.

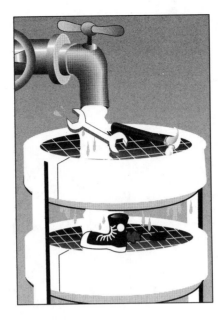

Evaporation

Evaporation involves the vaporization of a liquid from a solution or a slurry. The objective is to concentrate a solution composed of a volatile solvent and a less-volatile solute. Therefore, this technology is useful for waste streams that contain a high solids concentration, or when minimization of the volume of the bulk liquid phase is necessary to reduce disposal costs.

Concentration is accomplished by driving off the solvent as vapor. This requires the transfer of sufficient heat (the latent heat of vaporization) from a heating medium to the process fluid to vaporize the volatile solvent; boiling heat transfer is most commonly used.

The function of evaporation equipment is to provide the means to transfer heat to the liquid and allow the vaporization process to occur. The three most common types are the rising-film, falling-film, and forced-circulation evaporators; heat exchangers, flash tanks, pumps, and ejectors are common auxiliary equipment used in evaporation systems. Important considerations in the design of an evaporator are heat-transfer characteristics, energy requirements, and the potential for corrosion, foaming, entraining, salting, and scaling.

Evaporation is an expensive technology both in terms of capital and operating costs. A bottoms stream and a condensate are produced, both of which require further processing. Solar evaporation energy costs are less, but the cost of land for evaporation requirements may be high. Evaporation lagoons often require bottom liners and must include provisions for emptying, so that they are considered a treatment practice, rather than a disposal practice as in the past.

Filtration

Filtration systems (4–8) can separate free and emulsified oil from refinery wastewaters. They can be used independently or in conjunction with gravity-type separation or air flotation systems. The removal mechanism in filtration is by direct filtration and induced coalescence of oil globules on the filter medium. The process can also break mechanical emulsions.

Materials such as glass, porous ceramic, metals, plastics, sand, anthracite, and graphite have been used as filter media. Specific filter materials must be tested on each waste stream to determine their effectiveness. The filter medium is selected for a particular waste stream based on its affinity for the specific entrained oil.

Filtration can also be used to remove suspended solids from wastewater prior to further treatment for specific contaminants. The filter media most commonly employed for this application are sand and other inert solid materials.

Filtration systems are susceptible to fouling and plugging of the filter medium, especially in waste streams where solids are present. The filters must be backwashed frequently, and the backwashed wastewaters reprocessed through the wastewater treatment system.

In open filters, bacterial or algal growth on the filter medium can result in reduced filter efficiency and odor and pest problems. The influent must therefore be pretreated with biocides.

Adsorption

Adsorption (8–10) involves the use of high-surface-area activated carbon for the surface adsorption of organic contaminants from wastewaters. Activated carbon adsorption can be used to remove a wide variety of organic, and sometimes inorganic, contaminants. The process is relatively nonspecific and is used as a broad-spectrum secondary treatment operation. It should generally be considered for organic contaminants that are nonpolar, have low solubility, or have high molecular weights.

Activated carbon adsorption is usually performed using powdered activated carbon (PAC) in complete-mix reactors or granular activated carbon (GAC) in column or fluidized-bed reactors (Figure 3). The concentration of the contaminant in the effluent will usually be close to zero until the point of "breakthrough," or saturation of the carbon adsorption sites, when the influent and effluent concentrations will become equal. Adsorber columns are typically operated in pairs so that when breakthrough occurs in one column, the other column is brought on-line and adsorbs contaminants while the first column is regenerated. Because changes in influent wastewater charac-

teristics can result in premature exhaustion of the carbon, the effluent must be continually monitored to indicate adsorber efficiency.

The most common operational problem with carbon adsorbers is the development of excessive head loss as a result of solids accumulation or biological growth on the carbon. Biological growth can lead to the formation of hydrogen sulfide by sulfate-reducing bacteria under anaerobic conditions. Therefore, pretreatment of the waste to remove solids and add biocides is typically required.

Another problem is premature exhaustion of the carbon. The internal pores in the carbon can become blocked by high-molecular-weight contaminants, making the carbon unavailable to the contaminants of interest. Changes in influent pH can also result in a chromatographic effect, whereby previously adsorbed species are replaced by compounds with higher adsorption energies, resulting in concentration spikes in the effluent.

Air/steam stripping

Stripping (9, 11, 12) is a physical treatment process in which dissolved contaminants are transferred from the liquid phase into a vapor stream (Figure 4). The driving force for mass transfer is the concentration gradient between the liquid and gas phases.

In air stripping, the governing equilibrium relationship is Henry's Law (13, 14). Therefore, the process is only applicable to contaminants that are volatile under normal atmospheric conditions. Air is introduced into the wastewater to remove relatively volatile dissolved organic contaminants. The residual contaminated off-gas stream must be further treated to remove the contaminants before release to the atmosphere.

Steam stripping utilizes live steam as the gas phase. In this case, the vapor/liquid equilibrium between water and the organic compound is the key equilibrium relationship. Steam stripping can remove some semivolatile or more soluble compounds that are not easily removed

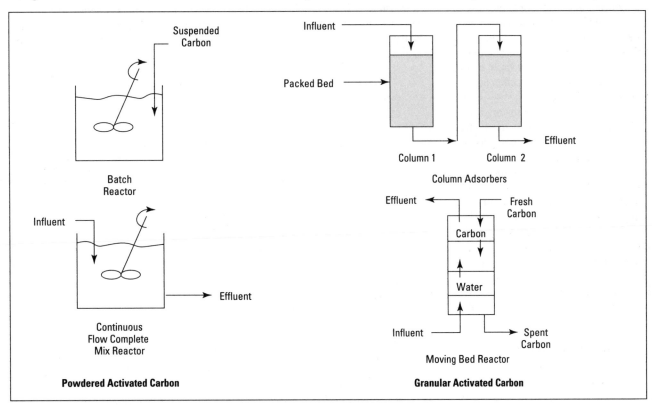

■ *Figure 3. Activated carbon can be used to remove contaminants by adsorption.*

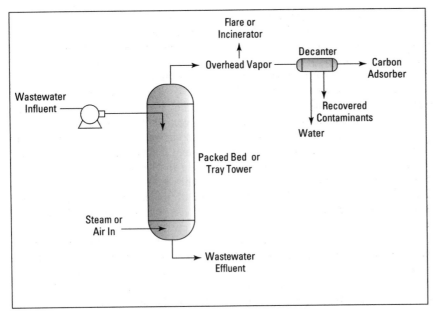

■ *Figure 4. In air and steam stripping, contaminants are transferred from the liquid phase to the vapor phase.*

by air stripping. The overhead vapor from steam stripping is then condensed for recovery of stripped compounds.

Stripper efficiency will depend on the concentration of contaminants in the incoming waste stream and off-gases, the surface-area-to-volume ratio, the air-to-water ratio, and the amount of solids. Various types of stripper media are available to provide different surface-area-to-volume ratios — plastic packings are the most commonly used.

Operational problems most frequently encountered with strippers are media plugging or fouling and loss of sufficient air pressure. Stripping produces an off-gas stream that usually contains high concentrations of organic compounds. This gas stream must then be further treated before it can be discharged to the atmosphere. Off-gas treatment is typically accomplished by flaring, incineration, or activated carbon adsorption, depending on the nature of the contaminants and applicable environmental regulations.

Liquid/liquid extraction

Liquid/liquid extraction, or solvent extraction *(15),* is the separation of constituents from a liquid solution by contact with another liquid in which the constituents are more soluble. The second liquid, or extracting solution, is usually immiscible in water at the extraction conditions. The constituents are transferred from one liquid to the other but are unchanged chemically. The extraction solution is selected such that it can be reused and easily separated from the bulk wastewater stream.

The solvent extraction process (Figure 5) consists of four basic components:

• contact between wastewater and solvent;

• separation of extracted wastewater and solvent;

• treatment of solvent to remove extracted contaminants; and

• treatment of wastewater to remove residual solvent.

Extraction systems range from simple single-stage mixer-settlers to multi-stage extraction columns. Separation of solvent and contaminant can be accomplished by air or steam stripping, distillation, evaporation, or a second solvent extraction step. Treatment of the extracted wastewater can be by air or steam stripping, carbon adsorption, or biological degradation.

Process efficiency will depend on the equilibrium distribution of the contaminant between the wastewater and the extracting solvent and the kinetics of mass transfer between the two liquid phases. To increase efficiency, the solvent-to-waste ratio can be in-

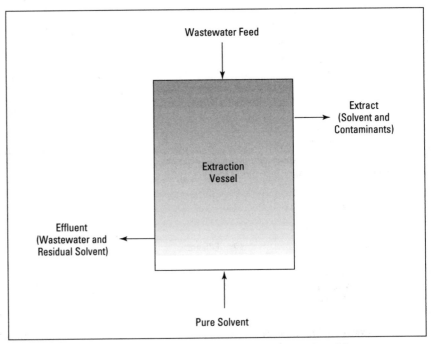

■ *Figure 5. In liquid/liquid extraction, contaminants are transferred from one liquid to another.*

creased, or multiple extraction stages can be employed.

The primary safety concerns with the extraction process are the flammability and toxicity of the solvent. And because residual solvent will be present in the treated wastewater, care must be taken when selecting a solvent so that this residual does not present additional effluent quality problems.

Extraction is applicable for constituents that are completely miscible in both the solvent and water. It is used in applications where recovery of the contaminants is desired for reprocessing or waste minimization purposes. At present, the process is used for removal of phenols and acetic acid from industrial wastewaters. Typical extraction solvents include crude oil, light oil,

benzene, toluene, isopropyl ether, butyl acetate, methyl isobutyl ketone, and methyl chloride.

CHEMICAL TREATMENT

Chemical treatment processes manipulate the chemical properties of the contaminants to facilitate removal of the pollutant from the bulk wastewater stream or to decompose the compound within the waste stream. The end products of the chemical destruction process can then either be easily separated from the waste stream, or are innocuous and pose no effluent problems if they are left in solution (such as carbon dioxide and water).

The chemical treatment processes most often employed in industrial wastewater treatment (Table 4) are

chemical precipitation and coagulation, electrolytic recovery, ion exchange, reverse osmosis, and chemical oxidation and reduction.

Precipitation and coagulation are primary or secondary treatment processes that remove metals or other settleable materials from the waste stream. Electrolytic recovery, ion exchange, and reverse osmosis are secondary, tertiary, or quarternary treatment processes aimed at removing metallic compounds from the bulk waste stream, for recovery or disposal. Oxidation and reduction are secondary or tertiary treatment processes that degrade or otherwise alter the contaminant (usually an organic or heavy metal compound) based on its specific chemistry, to reduce its toxicity.

Table 4. Summary of chemical wastewater treatment processes.

Process	Applications	Advantages	Disadvantages
Chemical Precipitation	Inorganics, Metals	Removal of dissolved constituents. Low maintenance. Metals recovery. Somewhat safe. Low energy costs. Easy to operate.	Volatile emissions. Proper handling and storage of reactants is required. Waste sludge disposal. Selective removal. Requires chemical additives.
Electrolytic Recovery	High-Concentration Organics, Inorganics, Metals	Removal of dissolved constituents. Metals recovery. Relatively safe. Easy to operate. No waste sludge.	High capital and operating costs. Selective removal. Difficult maintenance. High energy cost. Susceptible to fouling.
Ion Exchange	Low-Concentration Organics, Inorganics, Metals	Removal of dissolved constituents. Metals recovery. Relatively safe. Easy to operate. Water can be reused.	High capital and operating costs. Selective removal. Difficult maintenance. High energy cost. Susceptible to fouling.
Reverse Osmosis	Low-Concentration Organics, Inorganics, Metals	Removal of dissolved constituents. Metals recovery. Relatively safe. Easy to operate. Water can be reused.	High capital and operating costs. Selective removal. Difficult maintenance. High energy cost. Susceptible to fouling.
Chemical Oxidation/Reduction	High-Concentration Organics, Some Inorganics	Removes dissolved contaminants. High degree of treatment. No waste streams.	High capital and operating costs. Selective removal. Difficult maintenance. Difficult to operate. High energy costs.

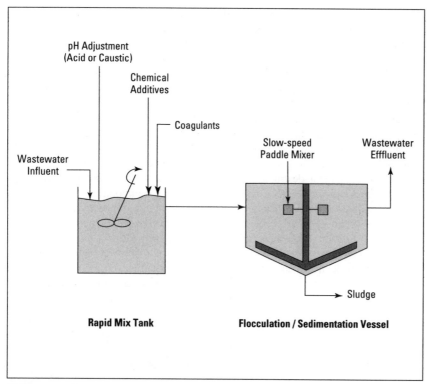

■ Figure 6. Chemical precipitation is useful for removing toxic metals from wastewater.

Chemical precipitation

In chemical precipitation (9, 15), soluble contaminants are converted to insoluble forms by chemical reactions, or by changes in the composition of the solvent that diminish the contaminants' solubility. Precipitated solids can then be removed by coagulation, flocculation, and settling or filtration.

The process has wide applicability for the removal of toxic metals — including arsenic, barium, cadmium, chromium, copper, lead, mercury, nickel, selenium, silver, thallium, and zinc — from aqueous waste.

The chemical precipitation process is implemented by adding a chemical precipitant to the metal-containing waste in a continuously stirred reactor (Figure 6). The dissolved metals are converted to an insoluble form by a chemical reaction between the soluble metal and the precipitant. Flocculation or slow-speed mixing, with or without coagulant aides, is then used to agglomerate the finely dispersed solids. The flocculated wastewater is routed to a primary clarifier, where the solids are allowed to settle out of solution.

Cementation is an electrochemical precipitation process in which the metal of interest is displaced from solution by a metal higher in the electromotive series (15). This process is used to remove and recover reducible metallic ions, such as in the precipitation of silver from photographic processing solutions and copper from printed-circuit solutions. Cementation offers significant economic advantages over other treatment processes because valuable metals (such as gold, silver, or copper) can be recovered.

Because of the nature of the chemistry involved, chemical precipitation is most effective when the concentrated waste streams are segregated from the bulk wastewater stream. Many metal-containing waste streams are either highly acidic or highly alkaline and must be neutralized prior to the precipitation process. The heat given off by this exothermic reaction can result in a substantial temperature rise in the reactor and may cause localized splashing and emission of volatile materials. In addition, corrosion of treatment equipment must be considered. The addition of a lime slurry to the reaction tank prior to the addition of an acidic waste can coat the tank and allow the dissipation of the heat of reaction throughout the wastewater in the tank. Cooling of the tank, either externally or by aeration, is often required, and adequate ventilation should be provided.

Electrolytic recovery

The electrolytic recovery process (9, 15) is used primarily to recover metals from process waste streams or rinse waters, either for their economic value or to meet effluent discharge criteria. It has been used to recover copper, zinc, silver, cadmium, gold, and other heavy metals.

Electrolytic recovery involves oxidation and reduction reactions that take place at the surface of conductive electrodes (cathode and anode). The electrodes are immersed in a chemical medium, and an electric potential is applied. The metal ion is reduced to its elemental form at the cathode, while gaseous products, such as oxygen, hydrogen, nitrogen, or chlorine, evolve at the anode. The major process equipment consists of the electrochemical reactor containing the electrodes, a gas venting system, recirculation pumps, and a power supply.

The process is most effective on highly concentrated metal-containing waste streams (> 1,000 ppm). Electrolytic recovery in more dilute wastewaters can result in cathodic polarization, which resists the migration of the metal toward the electrode. The process is not effective for recovery of nickel because of nickel's low standard reduction potential and the high stability constants of its complexes.

Ion exchange

Ion exchange (8, 9) is not often used for wastewater treatment. When it is employed, it is used for water reuse purposes. It is used primarily for the removal of "hardness" ions, such as calcium and magnesium, in water treatment, and for the removal of iron and manganese from groundwater.

The ion-exchange process involves the exchange of an ion held on a solid surface by electrostatic forces with an ion of similar charge that is present in the bulk waste stream. Synthetic solid resins with soluble ionic functional groups attached are typically used. Cation exchangers are those resins containing exchangeable cations, whereas anion exchangers contain exchangeable anions.

The purified water can be recycled back into the industrial process. The contaminants can be recovered by regeneration of the ion-exchange resin using caustic or acid solutions. The process is usually carried out cyclically, alternating between in-service and regeneration modes.

Ion exchange is well-suited to the detoxification of large flows of wastewater containing relatively low levels of heavy metal contaminants. High concentrations of organic contaminants (100 ppm or greater) can foul the resin, requiring frequent regeneration.

Reverse osmosis

Osmosis involves the migration of water across a semipermeable membrane separating solutions of different concentrations. The membrane is permeable to water, but it retains dissolved contaminants such as salts or high-molecular-weight compounds. In osmosis, water normally migrates across the membrane from the dilute side toward the more-concentrated side, until both sides of the membrane have equal concentrations. In reverse osmosis (8, 9), pressure is applied to the more-concentrated side to drive water into the less-concentrated side, in reverse of the normal osmotic flow — hence the term reverse osmosis (Figure 7).

The membranes employed generally have very high water permeability and very low salt permeability. Cellulose acetate and aromatic polyamides are two of the most common membrane materials. The membrane material is typically fabricated into tubular or spiral-wound modules, which are then packed inside a pressure vessel. Feed is introduced under pressure and flows axially or radially through the modules.

A key advantage of reverse osmosis is that it produces a purified high-quality effluent that can be reused, while the waste stream that requires disposal is concentrated and reduced in volume.

The degree of treatment achieved with a reverse osmosis system depends on the membrane type, the concentration of dissolved solids, feed temperature, and applied pressure. The most important factor in reverse osmosis performance is membrane fouling caused by suspended materials in the wastewater or materials that precipitate out during treatment (usually due to decreased solubility in the concentrated solution). This problem can be avoided by pretreating waste streams or by recycling the treated effluent to dilute the influent.

Chemical oxidation/reduction

The oxidation and reduction treatment process (9) consists of a pair of reactions in which the molecules of one reactant lose electrons (oxidation) while the molecules of the other reactant gain electrons (reduction). Oxidation and reduction reactions are important in the treatment of metal-containing wastewaters and for inorganic toxins. They are also important in the treatment of many organic wastes, such as phenols, pesticides, amines, mercaptans, and chlorophenols.

Oxidation and reduction reactions utilize simple equipment to mix the wastewater and the treatment chemical. Some reactions are so rapid that they can be conducted in a pipeline reactor, whereas others may require several hours and are conducted in batch reactors or continuous stirred reactors in series. Typical oxidation reactants are sodium hypochlorite, hydrogen peroxide, calcium hypochlorite, potassium permangate, and ozone. Reducing agents include sulfur dioxide and sodium borohydride.

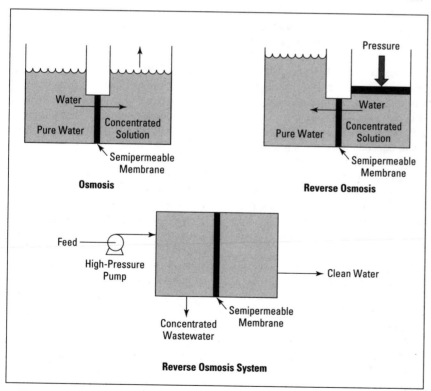

■ *Figure 7. Reverse osmosis employs a semipermeable membrane to concentrate the waste and recover purified water.*

Table 5. Summary of thermal wastewater treatment processes.

Process	Applications	Advantages	Disadvantages
Wet Air Oxidation	High-Concentration Organics, Toxic Compounds	Removal of dissolved constituents. Destruction process. No secondary waste treatment.	High energy costs. High capital and operating costs. Difficult maintenance. Somewhat unsafe. Difficult to operate.
Supercritical Oxidation	High-Concentration Organics, Toxic Compounds	Removal of dissolved constituents. Destruction process. No secondary waste treatment. Can be self-sustaining.	High energy costs. High capital and operating costs. Difficult maintenance. Somewhat unsafe. Difficult to operate.
Incineration	High-Concentration Organics, Toxic Compounds	Removal of dissolved constituents. Destruction process.	High energy costs. High capital and operating costs. Difficult maintenance. Somewhat unsafe. Difficult to operate. Off-gas treatment.

Oxidation, with or without the use of catalysts, is used to remove organic compounds from wastewater. Hydrogen peroxide and ozone are used frequently to oxidize toxic organic compounds that are resistant to biological treatment. Cyanide oxidation is accomplished using either sodium hypochlorite or calcium hypochlorite.

Reduction is used most often to reduce chromium from Cr^{+6} to Cr^{+3}, which is less toxic and can be precipitated out of solution readily.

Oxidation and reduction reactions conducted in the aqueous state are generally safe. However, violent reactions and even explosions can occur with strong oxidants, such as hydrogen peroxide, under certain conditions. The storage and handling of strong oxidizing and reducing agents requires safety precautions because of the instability of the concentrated solutions. The solutions are also extremely corrosive and require specific materials of construction, such as special alloys or coated steels.

THERMAL TREATMENT

Thermal treatment processes utilize elevated temperatures to bring about the decomposition of contaminants. Metallic species are decomposed to their elemental form either as ash or

pure gases, while organic compounds are decomposed to carbon dioxide, water, and halides or halogen gases. Because the conditions of destruction of the contaminant are generally at or near the critical point of water, the entire bulk waste stream undergoes decomposition rather than separation of contaminants of concern. Purified water results from the condensation of produced steam, or by returning the waste stream to normal ambient pressures and temperatures.

Because the energy cost of bringing a bulk aqueous waste stream to critical temperatures and pressures is high, thermal processes are generally reserved for the treatment of solid waste streams. However, in some cases, thermal treatment is selected as the only viable treatment technology for an aqueous waste.

The thermal treatment processes employed in industrial wastewater treatment (Table 5) are wet air oxidation, supercritical oxidation, and liquid injection incineration.

Wet air oxidation and supercritical oxidation are both used primarily for the removal of specific organic contaminants. Supercritical oxidation is used when the oxidation reaction produces enough energy to maintain the reaction.

Wet air oxidation is often used as a primary or secondary treatment process to degrade recalcitrant organic compounds into a form that is more readily biodegradable using conventional biological treatment processes. Incineration of liquids is used only for special wastes that cannot be treated using other technology, either as a result of environmental restrictions or because no other form of detoxification is feasible (such as with wastes containing high concentrations of heavy metals that cannot be precipitated effectively).

Wet air and supercritical oxidation

Wet air oxidation (15) occurs when an organic or oxidizable compound in an aqueous bulk phase is mixed with air at temperatures of 150–325°C and pressures of 2,000–20,000 kPa. Most organic compounds are oxidized stoichiometrically to carbon dioxide and water.

The process has been used to treat organic-containing waste streams that are too dilute for incineration but too toxic for biological degradation. It is also used to regenerate spent activated carbon and to pretreat refractory wastes to a more readily biodegradable form. Wastes containing organic cyanides and sulfides, halogenated aromatic compounds, and heterocyclic compounds containing oxygenated nitrogen are easily oxidized. The process is not suitable for wastes that contain primarily inorganic contaminants.

Supercritical oxidation exploits the properties of water at its critical temperature and pressure (374°C and 25.3 MPa). Under these conditions, organic substances become completely miscible and salts are insoluble.

Both wet air and supercritical oxidation can be self-sustaining in terms of energy requirements because of the exothermic nature of the oxidation reactions. There is, therefore, a maximum heating value of the waste (Btu per pound or gallon) to be treated. More dilute wastes can be treated by supplementing with fuel.

Reactor and heat exchanger design involves consideration of materials that can withstand the maximum temperature and pressure of the reaction, such as specialized steels. Reaction efficiencies are on the order of 99.99% to 99.9999%. The effluent usually contains some low-molecular-weight organic compounds that are readily biodegradable. The effluent will also contain suspended solids if the influent had a high metal content.

Problems associated with these treatment processes are the formation of byproducts due to incomplete oxidation if the reaction conditions are not right and the treatment of off-gases containing low-molecular-weight organic compounds. These volatile organic compounds can be removed by activated carbon adsorption or fume incineration.

In addition, the formation of solids in the reactor may lead to plugging.

Liquid injection incineration

Liquid incineration (15) is a thermal destruction process whereby the waste liquid is directly burned or injected into a flame or combustion chamber in an incinerator. The heating value of the waste and the energy needed to bring the waste to combustion temperatures are the critical factors that must be considered. Although the process results in the complete destruction of the waste, the energy required and the operating and maintenance requirements are high. However, for some waste streams, such as those containing certain heavy metals, polychlorinated biphenyls (PCBs), or dioxin, incineration is the only viable technology.

Table 6. Summary of biological wastewater treatment processes.

Process	Applications	Advantages	Disadvantages
Activated Sludge	Low-Concentration Organics, Some Inorganics	Removal of dissolved constituents. Low maintenance. Destruction process. Relatively safe. Low capital costs. Relatively easy to operate.	Volatile emissions. Waste sludge disposal. Somewhat high energy costs. Susceptible to shock loadings and toxins. Susceptible to climatic changes.
Aerated Lagoons, Stabilization Ponds	Low-Concentration Organics, Some Inorganics	Removal of dissolved constituents. Low maintenance. Destruction process. Relatively safe. Low capital costs. Low energy costs. Easy to operate. Infrequent waste sludge.	Volatile emissions. Susceptible to shock loadings and toxins. Susceptible to climatic changes. High land requirement. No operational control.
Trickling Filters, Fixed-Film Reactors	Low-Concentration Organics, Some Inorganics	Removal of dissolved constituents. Low maintenance. Destruction process. Relatively safe. Relatively little waste sludge.	Volatile emissions. Susceptible to shock loadings and toxins. Susceptible to climatic changes. Relatively high capital and operating costs. Susceptible to fouling.
Anaerobic Degradation	Low-Concentration Organics, Chlorinated Organics, Inorganics	Removal of dissolved constituents. Destruction process. Treats chlorinated wastes. Produces methane. Reduced sludge generation.	Susceptible to shock loadings and toxins. Susceptible to climatic changes. Relatively high capital and operating costs. High energy costs if no methane recovery.

Organic and aqueous liquids go through three phases before actual oxidation occurs — they are heated, vaporized, and then superheated to ignition temperature. At the same time, the liquid must be in intimate contact with oxygen. Oxygen reacts with the oxidizable species present to release energy. As this occurs, there is a sudden rise in temperature, completing the oxidation reaction.

Liquid injection incinerators are usually refractory-lined chambers equipped with primary, and frequently secondary, combustors. Operating temperatures are in the range of 1,000–1,700°C. Residence times vary from milliseconds to seconds. The liquids are injected through atomizers, which break the liquid into fine droplets and control the rate of flow of liquid. Most liquid incinerators do not produce solids or ash, and air-pollution problems are minimal, although scrubbers must be provided to remove the acid gases.

The physical, chemical, and thermodynamic properties of the waste, such as corrosivity, ignitability, polymerization, heat of combustion, viscosity, and reactivity, must be considered in the design of incinerator systems. Typical systems consist of storage tanks, mixers, pumps, atomizers, combustors, refractory, heat-recovery and quench systems, and air-pollution control equipment. Other systems consist of a liquid-injection burner and utilize waste liquids as a source of auxiliary fuel.

The degree of treatment achieved in incinerators is on the order of 99.99% and higher for most organic constituents of interest. However, the production of incomplete combustion byproducts has been a concern in the incineration of certain waste streams.

BIOLOGICAL TREATMENT

Biological treatment processes utilize biological and biochemical mechanisms to bring about a chemical change in the properties of the contaminants of interest. The chemical properties are altered under the action of a wide variety of microorganisms to cause the decomposition of the com-

pound within the bulk waste stream. The decomposition, or metabolism, of organic compounds produces more biological mass as well as energy needed by the microorganisms to sustain life. In other words, the contaminants serve as "food" for the microorganisms. Inorganic compounds, such as ammonia and sulfate, are also utilized in metabolic processes as alternative oxygen sources (or oxidants).

The major objective of biological treatment is to stabilize the organic matter in a waste stream so that biological degradation does not occur in the wastewater distribution system or in the receiving water body. This is done to protect the receiving water from the depletion of oxygen or other critical nutrients, and to prevent the buildup of waste products.

The ideal end products of microbial degradation are more cell mass, carbon dioxide, water, halides, elemental nitrogen and sulfur, heat, and excess energy. Often, decomposition of organic compounds is not complete, and low-molecular-weight compounds, such as alcohols, ketones, and organic acids, are formed. However, these compounds are usually of low toxicity to microbial or aquatic life and are easily further biodegraded under proper conditions.

The biological treatment processes typically encountered in industrial wastewater treatment (Table 6) include activated sludge processes, aerated lagoons or stabilization ponds, trickling filters or fixed-film reactors, and anaerobic processes.

Aerated lagoons and stabilization ponds are used when there is a large land area available. Activated sludge processes are used when less land is available and it is desirable to reduce the amount of time required for degradation of wastes. Trickling filter or fixed-film processes can be accomplished in either small vessels or large open areas, depending on the desired flow rate. They are often preferred to the activated sludge process because of lower demand for mechanical aeration equipment. Anaerobic treatment processes are used when it is desirable to reduce the amount of sludge that must be disposed of, and in the degradation of chlorinated or halogenated organic wastes.

Activated sludge

In the activated sludge process (6, 7), a mixed liquor consisting of suspended microorganisms, dissolved oxygen, organic compounds, and nutrients is continuously mixed or aerated (Figure 8). (The term "activated" refers to the living microorganisms, as opposed to inert material.) The suspended biological matter comes in contact with the oxygen, nutrients, and organic matter to degrade organic compounds to carbon dioxide and water (ideally), which remain in the liquor. The aeration tank, where degradation occurs, is followed by a clarifier or settling step, where the suspended microbial matter is removed from the bulk waste stream. The treated water is then discharged, or further treated in tertiary treatment processes such as activated carbon adsorption to remove residual organic and inorganic contaminants.

A portion of the settled biomass, or sludge, is recycled back to the aeration tank, while the rest is disposed of, or "wasted." Often the sludge that is wasted is further treated to stabilize the biological matter (called sludge digestion) or to reduce the volume of the sludge (called sludge thickening). In some cases, wet air oxidation or in-

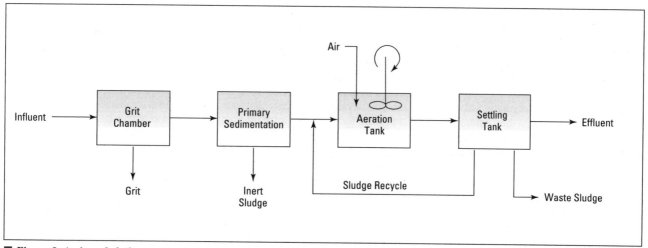

■ *Figure 8. Activated sludge processes degrade wastes faster than aerated lagoons.*

cineration can be used to dispose of the sludge.

In sludge digestion, which can be aerobic or anaerobic, the sludge is allowed to self-degrade in a reactor that is mixed (aerobic) or unmixed (anaerobic). The microorganisms consume their own biomass when a lack of food sources is available, thus reducing the amount of material to be disposed of and stabilizing the waste as growth and activity of the starving organisms decreases.

Sludge thickening involves removing the entrained water from the sludge to concentrate the biomass. This is accomplished using belt or filter presses.

The practice of recovering and recycling a portion of the active biomass improves the biodegradation kinetics in the aeration tank. As organisms are recycled through the system, the overall sludge age, or sludge retention time, increases. This allows a more mature, acclimated biomass to develop. Microorganisms become naturally adapted to a waste stream the longer they are exposed to a particular food source. In addition, a large variety of microorganisms can proliferate in an activated sludge. Certain microorganisms have slower growth rates and would be "washed out" of the system if shorter sludge retention times were employed. After the development of a stable, varied biomass population, the activated

sludge process is self-sustaining as long as a food source, oxygen, and certain nutrients, such as nitrogen and phosphorous, are provided.

Treatment of concentrated waste streams can generate a great deal of excess biomass that must be wasted. Treatment of dilute streams, on the other hand, may never require wasting, or may need it only infrequently.

Because microorganisms are living organisms, they are naturally susceptible to climatic changes, shock loadings, the presence of toxins, and other process upsets. When an industrial facility employs a biological treatment system, the impact of every operational change or process upset must be considered. The efficiency of the biological process will depend heavily on the characteristics of the incoming waste stream. To prevent shock loadings, equalization or other forms of pretreatment, such as pH adjustment, are performed. Microbial activity is optimum only in a narrow pH range of 6.0 to 8.0.

The biomass suspended in the mixed liquor can also have adsorption capabilities as well. Organic and inorganic compounds can adhere to the surface of agglomerations of biomass and thereby be removed from the waste stream. If an adsorbed compound is biodegradable, it is later decomposed. If it is not, it will be removed from the system along with the wasted sludge.

The presence of excessive free oil in a waste stream will coat the surface of the biomass, preventing the transfer of necessary oxygen to the microorganisms and thereby reducing the efficiency of the system. For this reason, many activated sludge plants are preceded by gravity-type oil/water separators and dissolved-air flotation units.

The presence of high concentrations of toxic metals may also present a problem, as they may precipitate and concentrate in the sludge and inhibit microbial activity. Their presence in the waste sludge can also present disposal problems. Therefore, pretreatment to remove these compounds is recommended.

The design and effective operation of biological treatment systems must take into consideration the complex nature and interrelationships between living organisms. The activated sludge process is relatively safe, although the presence of strippable organic compounds in the waste stream may result in excessive emissions from an aeration tank. Many systems use blowers, which must be monitored and usually require silencers. If floating surface mixers are used, maintenance of the mixers can be quite difficult and dangerous if a proper retrieval system does not exist. The same is true for air diffusers located at the bottom of tanks (inground or aboveground). A safe method of retrieving the diffusers for maintenance and repair must be installed — the cost and time consideration of emptying an aeration tank usually requires that maintenance be performed on active systems.

Recently, a new configuration of the activated sludge system has emerged, called the sequencing batch reactor (SBR). In this process, a single tank is used for aerobic degradation of contaminants, sedimentation of the biomass, clarification of the effluent, and sludge digestion. The process involves a tank, mixers, aerators, and a control system. The tank is allowed to fill with wastewater and the contents are mixed for a period of time to suspend the biomass and stabilize the waste. Air is supplied by diffused aerators, or by mixing. The air or mixers are then shut off, and the biomass is allowed to settle to the bottom of the tank by gravity. The clarified effluent is discharged from the top of the tank. The settled sludge self-digests anaerobically for a period of time before the next fill cycle begins.

Aerated lagoons and stabilization ponds

In their simplest form, aerobic stabilization ponds (6, 7) are large, shallow earthen basins that use natural processes involving both algae and bacteria to stabilize organic material present in a waste stream. The ponds are shallow so that both light and oxygen can penetrate at all depths. Algae and bacteria are suspended and proliferate in the pond. The proliferation of algae is encouraged because they release oxygen in the photosynthesis process that can then be used by the bacteria for the stabilization of organic matter.

Operational problems associated with stabilization ponds include the production of offensive odors and pests (flies and mosquitoes) and reduced treatment efficiency during colder weather. In addition, if insufficient oxygen is present, the pond will become anaerobic, increasing odors from putrefying wastes and hydrogen sulfide production. There is also no way to control activity in the ponds and large land areas are required.

For many years, stabilization ponds were an appealing biological treatment process because of the lack of operational controls required. However, the ponds were susceptible to overflow from severe rainstorms and washout of biomass. In addition, seepage of toxic compounds into underlying soils has reduced the use of earthen basins.

Aerated lagoons (Figure 9) are similar to stabilization ponds, except that a constant supply of oxygen, either from aeration or mixing, is provided to maintain aerobic conditions in the basin. Sludge that is produced settles in certain areas of the lagoon and forms a blanket on the bottom. The sludge blanket becomes anaerobic and the sludge self-digests. The introduction of oxygen eliminates the production of offensive odors, and allows smaller land surface areas to be used because the lagoons can be deeper than stabilization ponds.

Lagoons have their own limitations and disadvantages. They need to have sludge removed from the bottom from time to time to restore hydraulic capacity. This presents operational, safety, and disposal problems because of the type of equipment required to remove sludges from an active pond (it is usually not feasible to drain the pond prior to cleanout). When certain conditions prevail, the solids in the lagoon do not settle and can be carried over into the effluent. The proliferation of algae in aerated lagoons is not desired because the algae do not settle but remain suspended in the bulk liquid. And, because the sludge is not wasted, toxic concentrations of metals and recalcitrant organics will accumulate, which can inhibit biological activity and poses a disposal problem during lagoon cleanout.

The mechanism of degradation in stabilization ponds and aerated lagoons is the same as for activated sludge, however, the sludge retention time is extremely long (on the order of months

■ *Figure 9. Aerated lagoons require large land areas.*

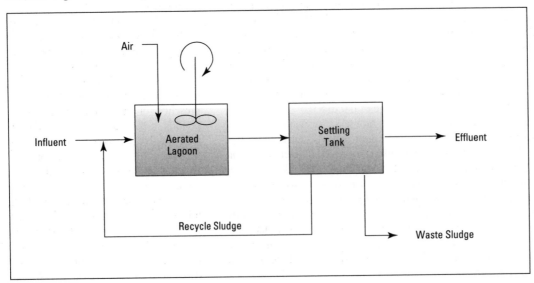

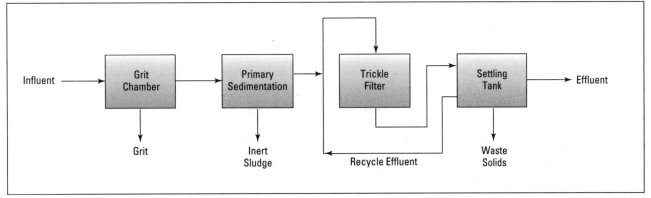

■ *Figure 10. Trickling filter processes can be done in small vessels or large open areas.*

to years). For this reason, they are often used for degradation of recalcitrant compounds, which would otherwise require extremely large aeration reactors.

Fixed-film reactors

An alternative to aerobic suspended growth biological processes is the fixed-film process *(6, 7)*. In fixed-film biological processes, the microorganisms are not suspended, but rather remain attached to a fixed surface. The fixed microbial population forms a slime layer that absorbs nutrients from the moving waste stream as it passes over the film. The controlling process is the rate of diffusion of food sources and oxygen into the slime layer. Certain compounds can also be adsorbed onto the fixed film and be removed that way.

The most common fixed-film processes involve trickling filters or packed-bed reactors called biotowers. The medium used for the attachment of microorganisms can be rocks, wood, sand, or plastic. It must either be permeable to water or allow water to drain through channels formed in between the individual medium elements.

In trickling filters, the wastewater is distributed over the top of the medium and is allowed to trickle to the bottom of the filter (Figure 10). The term "trickling" filter describes the slow movement of water through the medium, which is required to provide sufficient time for diffusion of contaminants into the fixed film to occur. The

medium must be randomly packed to provide complex pathways for the water to follow. Alternatively, in some packed-bed reactors, the medium is submerged in water and the reactor is designed to provide sufficient hydraulic retention time for diffusion of the contaminants to occur.

In nonsubmerged beds, oxygen is provided by natural air convection, while aeration must be provided to submerged beds. The difference in temperature between the top and the bottom of the column draws air through the medium. Problems with this system can arise in summer when the temperature difference between the ambient air and the bulk waste stream is minimal.

As the microorganisms attach to the medium surface and absorb food sources from the wastewater, a slime layer builds up. This layer reaches an optimum width, beyond which further growth cannot be supported because the rate of diffusion drops off sharply with the thickness of the slime layer. The microorganisms closest to the medium surface may die off from a lack of food and oxygen, causing sloughing of the slime layer. In addition, turbulent water and air flows can strip the slime layer from the medium surface.

A solids settling unit may therefore be required downstream of a fixed-film reactor to collect the sloughed solids. The solids are not usually recycled in a fixed-film process, except during the early stages of startup, when the pro-

duction of a slime layer is occurring. Sloughed material cannot be reattached to the medium — the slime layer building process must be started all over again. For this reason, occasional supplements of live organisms are employed to stimulate the replacement of sloughed films.

The effluent from fixed-film processes is usually recycled. This is done to improve treatment efficiency and to dilute the strength of incoming wastewater, because the surface of the slime layer is very susceptible to any toxins in the waste stream.

Because the rate of diffusion of compounds through the film surface is slow, fixed-film reactors must be very large or accept only low hydraulic loading rates — typically from 1 to 10 m^3/m^2 of filter area per day. Therefore, they are often used as roughing filters upstream of a conventional aerobic treatment process, such as activated sludge. The filter removes a certain portion of the contaminants prior to the second biological step, where the remainder of the stabilization process occurs. They are also used upstream of biological nitrification processes (which convert ammonia to nitrite) to remove a certain portion of the organic material, which allows the nitrification microorganisms to preferentially proliferate.

For this use, smaller filters with high flow rates can be employed. The size requirements for the activated sludge process are also reduced and a solids settling step is not required, as sloughed solids will be carried over into the activated sludge process. The major advantages of the fixed-film process are the lower energy requirements for mixing and aeration and the reduction in the amount of sludge that must be disposed of.

As with all aerobic biological processes, the fixed-film processes are relatively safe and easy to operate. However, when used in an industrial facility, care must be taken to protect the system from shock loadings or upsets in process conditions. The development of an effective slime layer may take months. Therefore, the loss of an established microbial slime in the filter can cause effluent parameter upsets over a long period of time until a stable system has been reestablished.

Anaerobic degradation

In general, anaerobic treatment processes (6, 7) are similar to aerobic processes, except that the biological degradation of contaminants occurs in the absence of oxygen. The end products of anaerobic degradation are methane, carbon dioxide, and a variety of low-molecular-weight compounds.

Anaerobic processes do not produce as much biomass as aerobic processes because much less energy is obtained from anaerobic metabolism. This is often considered an advantage because of the reduced costs for the handling of waste biological solids. However, because the energy obtained from anaerobic metabolism is less, the process kinetics are much slower than aerobic processes, requiring degradation times on the order of months for some constituents, compared to days for aerobic processes.

The production of methane from anaerobic processes is considered an advantage because it can be captured and used for fuel. Often, heating of anaerobic processes is required, especially during the winter, so the production and recovery of methane can help reduce operational costs. At the same time, the production of methane poses a safety problem, as the buildup of methane in biological reactors can result in disastrous explosions.

Literature Cited

1. **McLaughlin, L. A.**, *et al.*, "Develop an Effective Wastewater Treatment Strategy," *Chem. Eng. Progress*, **88**(9), pp. 34–42 (Sept. 1992).
2. Code of Federal Regulations, Title 40 — Protection of the Environment, Part 122 (40 CFR 122).
3. "Design and Operation of Oil-Water Separators," API Publication 421, 1st ed., part of the "Monographs on Refinery Environmental Control series," American Petroleum Institute, Washington, DC (Feb. 1990).
4. **U.S. Environmental Protection Agency**, "Environmental Assessment Data Base for Petroleum Refining Wastewaters and Residuals," U.S. EPA, Washington, DC, EPA/600/2-83-010 (Feb. 1983).
5. **Beychok, M. R.**, "Aqueous Wastes from Petroleum and Petrochemical Plants," John Wiley & Sons, London (1967).
6. **Metcalf & Eddy, Inc.**, "Wastewater Engineering — Collection, Treatment, Disposal," McGraw-Hill, New York (1972).
7. **Metcalf & Eddy, Inc.**, "Wastewater Engineering —Treatment, Disposal, Reuse," McGraw-Hill, New York (1979).
8. **Azad, H. S.**, "Industrial Wastewater Management Handbook," McGraw-Hill, New York (1976).
9. **Weber, W. J.**, "Physicochemical Processes for Water Quality Control," John Wiley & Sons, New York (1972).
10. **Stenzel, M. H.**, "Remove Organics by Activated Carbon Adsorption," *Chem. Eng. Progress*, **89**(4), pp. 36–43 (Apr. 1993).
11. **Okoniewski, B. A.**, "Remove VOCs from Wastewater by Air Stripping," *Chem. Eng. Progress*, **88**(2), pp. 89–93 (Feb. 1992).
12. **Bravo, J. L.**, "Design Steam Strippers for Water Treatment," *Chem. Eng. Progress*, **90**(12), pp. 56–63 (Dec. 1994).
13. **Carroll, J. J.**, "What is Henry's Law?," *Chem. Eng. Progress*, **87**(9), pp. 48–52 (Sept. 1991).
14. **Carroll, J. J.**, "Use Henry's Law for Multicomponent Mixtures," *Chem. Eng. Progress*, **88**(8), pp. 53–58 (Aug. 1992).
15. **Freeman, H. M.**, "Standard Handbook of Hazardous Waste Treatment and Disposal," McGraw-Hill, New York (1989).

Further Reading

Perry, R. H., and D. W. Green, eds., "Perry's Chemical Engineers' Handbook," 6th ed., McGraw-Hill, New York (1984). The following sections are most relevant to wastewater treatment: Sec. 26, Waste Management; pp. 11-31 to 11-43, Evaporators; Sec. 14, Mass Transfer and Gas Absorption; Sec. 15, Liquid-Liquid Extraction; Sec. 16, Adsorption and Ion Exchange; pp. 17-22 to 17-27, Reverse Osmosis.

The microorganisms that perform anaerobic metabolism utilize other compounds, such as sulfate, carbon dioxide, nitrate, and certain organics, as the oxidizer instead of oxygen. These microorganisms have developed from a completely different genera than aerobic microorganisms. (They also differ from facultative microorganisms, which can use different oxidant sources depending on whether oxygen is present or not.)

Not only do anaerobic microorganisms not require oxygen, but they cannot survive in environments where oxygen is present, even in low concentrations. As a result, anaerobic reactors must be specially designed to prevent the infiltration of oxygen. The oxygen that is normally present in wastewaters must be depleted by aerobic degradation before anaerobic processes can begin.

Anaerobic degradation is gaining popularity because of the recalcitrant nature of certain synthetic organic contaminants present in industrial waste streams. In many cases, chlorinated compounds cannot be degraded by aerobic microorganisms until the chlorine has been released from the molecule,

Table 7. Wastewater characterization checklist.

Question	Analyses Required
1a. Does the manufacturing process involve inorganic raw materials, byproducts, or end products?	Total metals Alkalinity COD Total dissolved solids Other specific contaminants
1b. Does the manufacturing process involve organic raw materials, byproducts, or end products?	TOC BOD (COD optional) Oil and grease or TPH Other specific contaminants
2. Does the process generate waste streams that are acidic or caustic?	pH Buffering capacity
3. Does the process generate high-temperature waste streams?	Temperature
4. Does the waste stream contain entrained solids?	Total solids Total suspended solids Total dissolved solids Turbidity
5. Does the waste stream contain nitrogen compounds?	NH_3 NO_3 Total Kjeldahl nitrogen
6. Does the waste stream contain cyanide compounds?	Total cyanide Reactive cyanide
7. Does the waste stream contain sulfur compounds?	Sulfides Sulfates Sulfites
8. Does the waste stream contain phosphorous compounds?	Phosphates
9. Does the waste stream contain surfactants or have excessive foaming?	Surfactants
10. Does the waste stream contain any toxic compounds?	Total toxic organics Toxic metals

because these organisms lack the proper enzymes to break the chlorine bond. Anaerobic treatment to remove chlorine followed by aerobic treatment to completely degrade the compound is required.

Anaerobic treatment processes can be carried out in reactor configurations similar to aerobic processes, provided that the reactors are hermetically sealed and methane recovery is provided. Recently, dome- or egg-shaped reactors have been constructed. These reactors are efficient at maintaining the right thermal conditions for optimized growth.

The temperature of the influent wastewater is critical to anaerobic microorganisms and must be maintained in the range of 30–40°C for mesophilic organisms and 50–75°C for thermophilic organisms. (Bacteria are clas-

sified based on the temperature range in which they function best.) Changes of more than 2°C per day will upset the microbial population.

The pH of the wastewater and the mixed liquor are also critical. The optimum pH for the methane-producing bacteria (methanogens) is 7.0. The production of organic acids during metabolism can result in pH drops that will stop biological activity. Therefore, the alkalinity, or buffering capacity, of the wastewater is also an important consideration.

Anaerobic microorganisms are also very susceptible to toxins (even in low concentrations), changes in the composition of the incoming wastewater, and buildup of degradation byproducts in the liquor. In addition, the quality of the methane gas produced by anaerobic degradation will depend

on the type of substrate being degraded. Reaction conditions must be monitored closely to maintain efficient treatment.

SELECTING APPROPRIATE TREATMENT PROCESSES

Once you have a basic technical understanding of the various processes employed in industrial wastewater treatment, you can select the process (or processes) that best suits your needs.

The first step in selecting the proper treatment process is to characterize the wastewater using the indicator parameters discussed earlier (Table 2). The choice of parameters can be made by answering the questions in Table 7.

After the wastewater has been characterized, the decision flow diagram found in Figure 11 can be used to de-

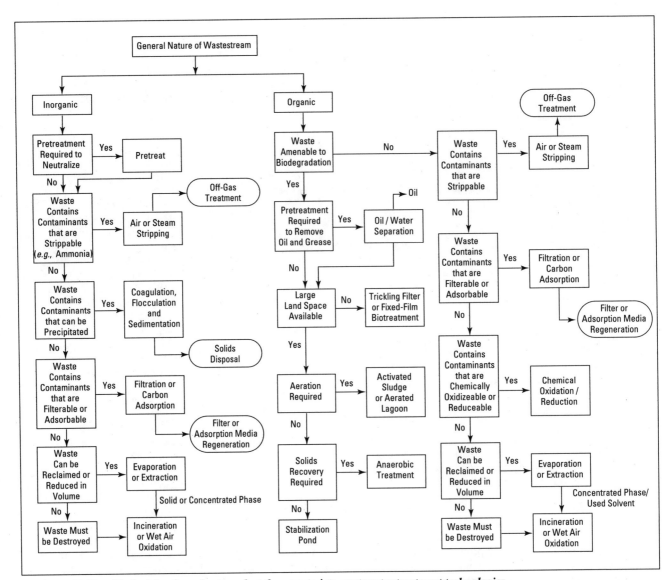

■ *Figure 11. Use this decision flow chart to select the appropriate wastewater treatment technologies.*

termine the first round of processes that should be considered.

Beyond this point, a detailed screening should be performed that takes into consideration technical, economic, regulatory, safety, reliability, and operability factors. This screening will identify a smaller group of processes with a high likelihood of success for your particular waste stream.

After narrowing the list down to these technologies, bench-scale, and sometimes pilot-scale, treatability and feasibility studies should be performed. Such studies:

• confirm the appropriateness of that particular process for a given waste stream;

• develop and collect data necessary to determine the treatment efficiencies that can be expected; and

• provide engineering and cost information necessary for full-scale application of the technology.

In addition, information generated during the screening and testing process is useful in the education of operators, technicians, environmental personnel, and managers. This can help ensure the long-term continued success of the wastewater treatment system. CEP

D. H. BELHATECHE is a project engineer with Environeering, Inc., (713/578-5800; Fax: 713/578-5875), an environmental consulting and engineering firm located in Houston, TX, and serving primarily the Gulf Coast area. She specializes in the areas of industrial wastewater treatment, hazardous and solid waste treatment, and air quality. In the wastewater field, she has conducted numerous projects at industrial facilities that involved troubleshooting wastewater compliance problems and providing engineering and technical services to implement corrective measures, and she has conducted wastewater plant operator training and prepared written operating manuals. She holds a BS in biology with a minor in environmental engineering from the Illinois Institute of Technology and an MS in environmental engineering from the Univ. of Houston. She is a member of the Texas Hazardous Waste Management Society and the Air and Waste Management Association.

Use Biomonitoring Data to Reduce Effluent Toxicity

Bioassay data may allow a plant to eliminate toxicity while avoiding an expensive toxicity reduction evaluation (TRE).

Alison M. Martin,
James M. Montgomery,
Consulting Engineers, Inc.

Previously, pollutant limits in National Pollutant Discharge Elimination System (NPDES) permits were based on treatment technology standards. In 1984, the U.S. Environmental Protection Agency (EPA) issued a new policy under which pollutant limits are now based on the quality of the receiving water. To assess the toxicity of an effluent to receiving waters, bioassays are conducted. The requirement to perform biomonitoring has already been written into many NPDES permits and is being incorporated into virtually all new permits.

When biomonitoring indicates toxicity or the presence of specific toxic pollutants in an effluent, EPA and many state agencies require the discharger to perform a toxicity reduction evaluation (TRE). A TRE is an analysis to determine the cause for effluent toxicity that is in violation of state or federal water quality standards and identify a course of action to achieve compliance by reducing toxicity or chemical concentrations to acceptable levels.

TREs can be extremely expensive and time-consuming. Performing all of the initial toxicant characterization tests on complex wastestreams can run upwards of $30,000 *for one sample.*

Obviously, it is best to avoid a mandated TRE if at all possible. Taking appropriate action before biomonitoring requirements are incorporated into an NPDES permit can save a company hundreds of thousands of dollars.

This article provides some guidance on how to reduce toxicity and avoid a TRE. First, some brief background on biomonitoring is presented. Next, the basic procedures for conducting a TRE are outlined. Even if a TRE is not required, the techniques involved may be used on a voluntary and less-structured basis. Finally, strategies for avoiding a full-scale, formal TRE are discussed.

Biomonitoring basics

Biomonitoring is done using species that occur in the receiving waters or closely related species. Fish, invertebrates, and plants may all be considered for biomonitoring. Typical freshwater species include fathead minnows (*Pimephales promelas*) and daphnids (*Ceriodaphnia dubia, Daphnia pulex,* and *Daphnia magna*). Occasionally, algae (*Selenastrum capricornutum*) may be used *(1).* Popular marine (or salt water) species include sheepshead minnows (*Cyprinodon variegatus*) and mysid shrimp (*Mysidopsis bahia*).

The required toxicity measurements may be acute, chronic, or both. Acute toxicity is a measure of the organisms' survival rate. Chronic toxicity is a measure of survival as well as effects on organism growth and reproduction. In general, acute tests are quicker to run and less expensive than chronic tests.

The acute toxicity test generally determines the concentration of effluent mixed with dilution water that results in the death of 50% of the organisms at the end of a specified time period, denoted LC_{50}. (The lower the LC_{50}, the more toxic the effluent is.) This test can be run for different time periods, usually 24, 48, or 96 hours.

EPA's chronic toxicity tests typically measure growth and reproduction rates over a seven-day period. The parameter resulting from this test is typically the "no observable effect concentration," or NOEC. The NOEC is the greatest concentration of

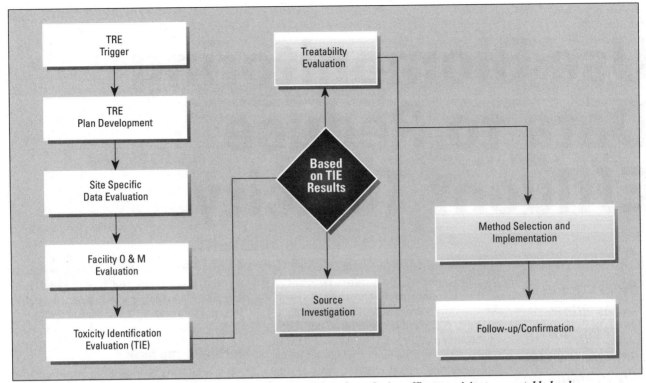

■ *Figure 1. A toxicity reduction evaluation is a systematic procedure for reducing effluent toxicity to acceptable levels.*

effluent at which organism reproduction and growth is not affected by any of the compounds present.

The type of toxicity test, species used, and frequency of biomonitoring vary widely. Frequently a discharger is required to perform biomonitoring once per quarter for a year, and if the results of these tests are satisfactory to the regulating agency then no additional biomonitoring will be required for the life of the permit.

TREs are triggered when an effluent fails its toxicity criteria. Generally, regulators do not initiate a TRE the first time an effluent does not meet its toxicity requirement because the toxicity could have been due to a one-time circumstance. Short-term intensive biomonitoring is typically implemented as a first step — usually, six to eight biomonitoring tests are conducted over a two-month period. If toxicity persists, then a TRE is launched.

Conducting a TRE

A toxicity reduction evaluation

is a procedure whereby through investigation, analysis, and implementation, effluent toxicity is reduced to acceptable levels. This article provides an overview of the procedure; details can be found elsewhere *(2–5)*. A flow sheet summarizing the steps

generally required in performing a TRE is shown in Figure 1.

TRE plan development. Industrial facilities show an extremely wide variation in effluent characteristics and variability. Therefore, a specific TRE must be developed for

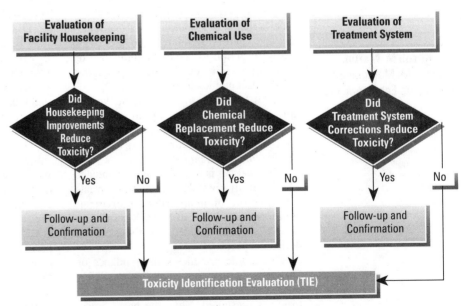

■ *Figure 2. An evaluation of a plant's operations and maintenance practices is an important part of a TRE.*

each facility. The TRE plan will include the objectives of the TRE (reducing the stream toxicity to a designated level), the type of toxicity indicator tests to be conducted (usually LC_{50} or NOEC), and the general site information acquisition and sampling plan. It is important to recognize that the biomonitoring tests performed as part of the NPDES permit may not be the same as the toxicity indicator tests performed during the TRE. The discharger must be aware that compliance with the facility's NPDES permit is the ultimate goal of a TRE.

Site-specific data evaluation. The next step is to obtain all relevant site-specific data that may be useful for conducting the TRE. These data usually include:
- regulatory objectives;
- facility monitoring data (NPDES, state, and in-house);
- products manufactured;
- chemicals used (quantities, as well as toxicity and biodegradability information on the chemicals and their breakdown products);
- process flow diagrams;
- facility site plan, and water and process piping lines;
- wastewater treatment plant schematic and operating data; and
- historical NPDES or monitoring reports.

Operations and maintenance (O&M) evaluation. A good evaluation of the O&M practices at the facility is essential to the success of the TRE. Frequently, O&M changes can reduce whole-effluent toxicity to the point that no further reduction is necessary. The O&M evaluation usually includes evaluation of plant housekeeping practices, evaluation of chemical use, and an evaluation of the waste treatment system, as shown in Figure 2.

Toxicity identification evaluation. The toxicity identification evaluation, or TIE, is the heart of the TRE. This is also the most elaborate and expensive phase of the TRE. It is during this phase that potential toxicants are characterized and then either further toxicant identification is performed or treatability testing (that is, testing to determine what

treatment can eliminate or reduce the toxicity) is carried out. The TIE is discussed in more detail later.

Method selection and implementation. Once the TIE has been completed, the next step is the selection and implementation of a method that will result in acceptable wastestream effluent quality. If a treatability approach was chosen during the TIE phase, then a treatment process unit will be designed, constructed, and brought on-line. If during the TRE toxicants were identified, then, in general, either source control or improvements in

waste treatment operations, or the introduction of a new waste treatment unit process, will be used to reduce effluent toxicity. This portion of the TRE procedure usually involves evaluations of alternatives based on treatment performance, process reliability, ease of implementation, and costs.

Follow-up and confirmation. It is important that after a treatment method has been implemented, follow-up and confirmation be performed. An appropriate biomonitoring plan should be adopted to

confirm that effluent toxicity levels have been reduced to acceptable levels. This step is particularly important if the organisms used during the TIE are not the same as those specified in the facility's NPDES permit. The biomonitoring used to confirm reduced toxicity would probably be performed on an accelerated timescale with the permit-specified organisms prior to returning to the NPDES biomonitoring requirements. Additionally, effluent tests for the presence and concentration of the toxicant(s) responsible for effluent toxicity may be desirable.

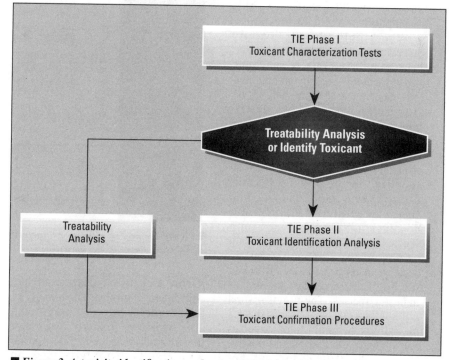

■ *Figure 3. A toxicity identification evaluation (TIE) is the heart of a TRE.*

Toxicity identification

Figure 3 outlines the steps involved in a TIE. The first step consists of toxicant characterization tests (Phase I). This is followed by a decision to either undergo a treatability approach or identify the toxicants (Phase II). The last step in the TIE consists of toxicant confirmation (Phase III).

TIE Phase I: toxicant characterization

The Phase I tests are designed to identify the physical and chemical class(es) of compounds responsible for effluent toxicity. Samples are collected over a period of time to determine the qualitative and quantitative variability of toxicity. Normally, acute LC_{50} tests are preferred for this phase of the testing due to the relative simplicity of the test. If the observed toxicity is acute, using the acute tests during the TIE is entirely appropriate. To use acute methods for measuring chronic toxicity, whole-effluent acute toxicity must be measurable and it must be assumed that the acute and chronic toxicities are caused by the same chemical. The Phase III confirmation procedures of the TIE will verify this assumption.

Where only chronic toxicity effects are observed, chronic tests must be run. The EPA recently published methods whereby chronic toxicity can be used for the TIE (2, 6). However, for purposes of illustration, we will assume that toxicity can be measured by the acute test.

If possible, the test organisms chosen will be the same as those specified in the NPDES permit; *Ceriodaphnia dubia* and *Daphnia magna* are often used. Occasionally, however, toxicity is based on a species (such as trout) that will not be used as the TIE test species. In such an instance, it is useful to use species of the same class to serve as indicator species. The use of several different species as TIE test species may also be helpful. Multiple species can indicate the causes and magnitudes of the effluent toxicity. For example: minnows are more sensitive to un-ionized ammonia than daphnids; sunfish are more resistant to metals than goldfish, minnows, and daphnids; and fish are more sensitive to chlorinated hydrocarbon insecticides than daphnids (4).

The toxicant characterization tests are designed to alter or render biologically unavailable a group of toxicants, such as oxidants or cationic metals. By performing a matrix of these tests (usually over a two-day period), certain classes of toxicants in the effluent may be identified.

Samples are treated to

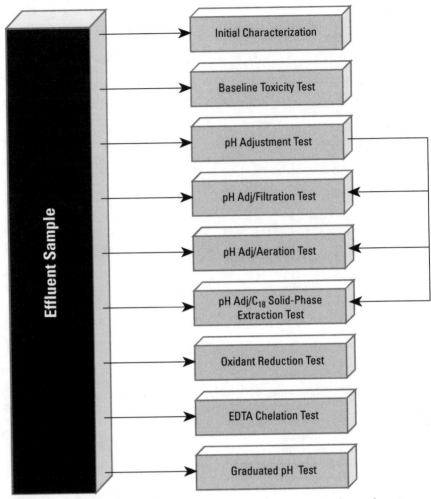

■ *Figure 4. The TIE effluent characterization tests involve a series of procedures to identify the source of the toxicity.*

remove toxicity associated with specific chemical groups. The toxicity of the sample is recorded before and after treatment. Toxicity results for the treated samples are compared with results of toxicity tests on untreated effluent — any differences indicate that the constituent or class of constituents removed is likely the cause of toxicity. Figure 4 shows the tests typically conducted during Phase I of the TIE.

Initial characterization test. The initial effluent characterization is carried out on Day 1 and usually involves the following parameters:

By performing a matrix of tests, classes of toxicants may be identified.

- temperature;
- pH;
- chlorine;
- hardness;
- alkalinity;
- conductivity;
- ammonia;
- total organic carbon (TOC);
- dissolved oxygen (DO); and
- the initial toxicity test.

The initial toxicity test provides an estimate of the 24-hour LC_{50} in order to determine exposure concentrations for the remaining Phase I tests. The remaining tests should result in LC_{50} values approximately equal to or greater than the initial toxicity value. Therefore, knowledge of this value establishes an expected range for LC_{50} and can eliminate running unnecessary dilutions.

Baseline toxicity test. This test, performed on the second day of testing, serves two purposes. First, it provides a baseline toxicity measurement against which the remaining Phase I tests are compared. Secondly, the results of this test, when compared to the results of the initial toxicity test, may reveal useful information about the effluent toxicants. If the toxicities of the two samples are substantially different, then the two-day testing schedule may require revision.

pH adjustment test. The toxicity of many compounds is affected greatly by pH. The pH of a solution can influence polarity, solubility, and volatility of a compound. Inorganic and organic acids and bases are affected most. During the pH adjustment test, aliquots of the effluent are adjusted to pHs of 3.0 and 11.0. Portions of the pH-adjusted samples are used for other tests (pH-adjustment/filtration, pH-adjustment/aeration, and pH-adjustment/C_{18}-extraction). The remaining portions of the two pH-adjusted samples are held until the aforementioned tests have been completed. Then, the aliquots are returned to the initial pH of the effluent sample and toxicity tests are conducted.

pH-adjustment/filtration tests. These tests provide information on toxicants associated with filterable material. They are performed on unaltered effluent and on the two pH-adjusted effluent samples. The samples are filtered through 1.0-μm glass-fiber filters, and then the pH-adjusted sample filtrates are returned to the initial effluent pH. Toxicity tests are then performed on the sample filtrates.

pH-adjustment/aeration test. This test determines the presence of toxicants that are volatile or oxidizable. The test is conducted on both unaltered effluent and on the two pH-adjusted samples. Initially, these tests are performed by sparging with air. If toxicity decreases, then subsequent tests may be performed using nitrogen sparging to determine if toxicity reduction was due to sparging or oxidation.

pH-adjustment/C_{18} solid-phase extraction test. This series of tests determines toxicity associated with

the presence of nonpolar organics or metal chelates. The sample is passed through an adsorption column that contains a solid sorbent. The solid phase octadecyl sorbent is nonpolar and will extract nonpolar compounds from the effluent. This test is performed on filtered effluent and on the pH-adjusted filtered effluent.

Oxidant reduction test. This test gives an indication of toxicity associated with oxidants. Some common species detectable by this method include chlorine, ozone, chlorine dioxide, monochloramines and dichloramines, bromine, and iodine. During this test, varying amounts of sodium thiosulfate

---

It may be possible to treat the effluent for certain classes of compounds without identifying the causative agent.

($Na_2S_2O_3$) are added to aliquots of the effluent. The sodium thiosulfate reduces the oxidants present, reducing effluent toxicity if the toxicity was due to the oxidants.

EDTA chelation test. The EDTA (ethylenediaminetetraacetic acid) chelation test determines the presence of many cationic metals, including aluminum, barium, cadmium, cobalt, copper, iron, lead, manganese (+2), nickel, strontium, and zinc. The addition of EDTA to an effluent containing metal cations will result in the formation of largely nontoxic complexes, thus reducing measurable toxicity. The test is performed by adding varying amounts of EDTA to effluent aliquots, and then performing the toxicity tests.

Graduated pH test. The graduated pH test is performed to determine if ammonia is causing any or all of the effluent toxicity. The un-ionized form of ammonia (NH_3), which is present in solutions at pHs greater than 7.0, is toxic to many forms of aquatic life. Toxicity tests are carried out on effluent aliquots that have been adjusted to pHs of 6.0, 7.0, and 8.0. If toxicity increas-

es with increasing pH, ammonia may be present.

TIE Phase II: toxicant identification

Based on the results of the Phase I testing, either treatability testing or toxicant identification can be carried out. If one or two of the tests significantly reduced toxicity, then it may be possible to treat the effluent for those particular classes of compounds without actually identifying the causative agent(s). For example, if the oxidant reduction test resulted is an effluent that had no toxicity, and the other tests did not affect toxicity, then treatability testing using reducing compounds could be carried out without identifying the specific toxic chemical.

Generally, it is less expensive to proceed directly into treatability testing than to identify the source of toxicity. However, the Phase I test results must be examined carefully to ensure that nothing has been overlooked.

If treatability after Phase I is not recommended, then Phase II of the TIE is implemented, wherein the toxicants are identified. Phase II testing builds on the results obtained in Phase I. Specifically, identification can be carried out when effluent toxicity is related to nonpolar organics, ammonia, or cationic metals. For nonpolar organics, tests consist of extractions and concentrations of the toxic fractions, high-performance liquid chromatography (HPLC) fractionation, and identification through a gas chromatography/mass spectroscopy (GC/MS) scan. Two tests are used to confirm ammonia as a toxicant: an equitoxic solution test and a zeolite test. The presence of cationic metals can be verified with atomic adsorption spectroscopy (AAS) and/or inductively coupled plasma atomic emission spectroscopy (ICP).

Certainly, chemical species other than those listed above may be responsible for the effluent toxicity. Thus, there are no hard-and-fast rules concerning the methodology for identifying toxicants. Ingenuity and flexibility are required to determine effluent toxicants during Phase II of the TIE.

TIE Phase III: confirmation

After toxicants have been preliminarily identified, it is necessary to confirm that they are the true cause of effluent toxicity. This is essential because effluent manipulations performed during Phases I and II may alter the effluent, resulting in erroneous conclusions about its toxicants. Additionally, the Phase III procedures will reveal the variability of the causative agents.

Many different approaches can be used during Phase III. A few include:

• correlation approach — determining a linear relationship between toxicant concentration and effluent toxicity;

• symptom approach — examining similar toxicity symptoms between two different effluents containing the same toxicant, and comparing organism exposure to the suspected toxicant and effluent;

• species sensitivity approach — obtaining LC_{50} values for effluents of different toxicity for a species and comparing them as a ratio; LC_{50} values for the same effluents on different species should result in the same ratio;

• spiking approach — increasing toxicant concentration to determine if toxicity increases;

• mass balance approach — performing a mass balance on the toxicant after it has been removed from the wastestream to determine if all effluent toxicity is due to the toxicant in question (this can only be done when the toxicant can be removed from the effluent); and

• deletion approach — testing effluent toxicity after the toxicant has been removed from the effluent.

Other approaches can be used to ensure that the toxicant(s) identified is(are) indeed the cause of effluent toxicity. The approach chosen will

depend in part on the nature of the suspected toxicants.

Avoiding a TRE

The best way to avoid a TRE is to collect your own data, and if toxicity is present, conduct your own investigation prior to actually having biomonitoring as a compliance requirement. Typically, in-house studies may take a bit longer than full-scale formal TREs but they are usually much less expensive. Some tips that may be useful in conducting a toxici-

---

You may be able to avoid a TRE by gathering data and conducting an in-house toxicity investigation.

---

ty investigation are discussed below.

1. Start early. Even if your permit is not due for renewal for several years, it's never too early to begin collecting data. Use organisms that are typically written into permits for discharges that use the same receiving waters as your facility, or use those that are written into the permits of nearby plants. Monitor the whole plant effluent at least quarterly for signs of toxicity. If toxicity appears and persists, you will need to investigate the cause.

2. Examine possible permit negotiation. Occasionally, certain organisms proposed in a draft permit may be extremely sensitive to a substance that is present in the effluent due to circumstances beyond the plant's control. In such cases, regulators may agree to remove or substitute those organisms with others.

For example, a leachate treatment system for a hazardous waste landfill was designed and constructed. The effluent from the treatment system was to be discharged to a nearby lake. The draft permit for the treatment system effluent included acute testing on fathead minnows and

Ceriodaphnia dubia and chronic testing on minnows, Ceriodaphnia, and green algae, and it required 100% survival. Toxic effects were seen only in the algae. However, it was determined that the algae was sensitive to low background levels of copper that were indigenous to the area. When the landfill operators presented this information to the regulators, the draft permit was modified.

3. Examine equipment cleaning and maintenance practices. In addition to an examination of general plant housekeeping and chemical usages, it's a good idea to track equipment cleaning and maintenance. This is particularly true if the toxicity is not always apparent, but rather intermittent. Chemicals used for vessel cleaning, and the cleaning method itself, can contribute to effluent toxicity. Switching to different surfactants, detergents, solvents, or other cleaning chemicals and/or using smaller amounts may reduce toxicity. For example, replacement of a detergent that only partially degrades in a wastewater treatment plant with a highly degradable soap may reduce effluent organic loading as well as toxicity. Pretreatment of equipment cleaning streams may be required to reduce toxicity.

For example, a large specialty organic chemical manufacturer had a low-flow stream with a very high organic load from a tote bin wash area to the biological wastewater treatment plant. The manufacturer received returned tote bins that its product had been shipped in and cleaned the bins for reuse. The bins sometimes still contained significant amounts of the original products, which were poured directly into the waste treatment drain system. The bins were cleaned using high-strength detergents and pressurized water. The resulting wastewater was extremely variable in nature, as the facility manufactures over 400 products — three wastewater samples collected on different occasions were quite different from one another in appearance, odor, organic levels, and pH. Due to the detergents used to clean the bins, the wastewater from this area was usually a very stable emulsion with no free phase. The toxicity of this stream was high.

A waste minimization study identified a ferric chloride/cationic polymer mixture to coagulate and flocculate organics in the wastewater and a pretreatment unit employing this mixture was installed. Additionally, plant personnel switched to a different type of detergent that was more biodegradable than the previous type, and the washing procedure was altered to use less detergent and generate less wastewater. These changes all resulted in a significant decrease of the organic load and toxicity of this waste stream.

4. Look hard at the wastewater treatment plant operation. The whole operation should be examined, including equipment, operations, and chemicals used. In fact, on occasion it has been found that addition of seemingly innocent chemicals, such as water treatment polymers, has been the source of effluent toxicity. Sometimes, adjustment of the wastewater treatment plant operation can reduce effluent toxicity. This is particularly true in the case of biological treatment plants. For example, if ammonia is suspected as a toxicant, check to see if it is added upstream for pH adjustment or nutrient addition. Ammonia toxicity can also be problematic in plants that operate their biological wastewater treatment plants at a low sludge retention time (SRT), which is a measure of the microorganism age. Increasing SRT usually stimulates the growth of nitrifying microorganisms that can

convert ammonia to nitrate and thus reduce toxicity.

It has recently been found at some plants that effluent toxicity has resulted from the biological treatment system itself. Substances that are toxic to the toxicity-indicator organisms are secreted or excreted by the biomass organisms. It may be possible to reduce toxicity in this case by altering the biomass population through SRT adjustment or addition of chemicals that will inhibit the growth of certain organisms.

For example, bench-scale testing of a biological treatment process for a specialty chemical manufacturer was performed. Toxicity testing on the resulting effluent revealed that the effluent was extremely toxic to *Ceriodaphnia dubia*. Subsequent testing showed the presence of cyanide in the effluent. Previous testing performed on the full-scale plant effluent had never shown toxicity or any measurable cyanide. It was determined that organic nitrogen-containing compounds produced by the facility had been partially metabolized to cyanide under the three-day hydraulic residence time (HRT) conditions in the bench-scale unit. The actual full-scale facility operated with a 12-day HRT; most likely, the immediate cyanide metabolite produced is further oxidized to ammonia, then to nitrate under these long HRT conditions. Therefore, when the new treatment facility was designed, a large HRT was required to ensure that no residual cyanide remained in the effluent.

5. Explore all plant water sources. Water infiltration from unlikely sources, such as non-contact cooling water or stormwater, is always a possibility. This is especially true at older facilities. Cooling water may be treated with biocides to prevent biological growth in cooling towers, and these chemicals would certainly cause some amount of toxicity if present in the plant effluent. Certain types of algae, especially the blue-green variety (*Cyanobacteria*), secrete extremely toxic substances. If a facility's cooling water is drawn from a lake, pond, or river, the

LITERATURE CITED

1. **Reed, D.,** "Biomonitoring Wastewater", *Pollution Engineering*, **24**(7), p. 45 (1992).
2. **Norberg-King, T.J.,** *et al.,* "Methods for Aquatic Toxicity Identification Evaluations: Phase I Toxicity Characterization Procedures," EPA/600/6-91/003, U.S. Environmental Protection Agency, Duluth, MN (1991).
3. **Mount, D., and L. Anderson-Carnahan,** "Methods for Aquatic Toxicity Identification Evaluations: Phase II Toxicity Identification Procedures," EPA/600/3-88/035, U.S. Environmental Protection Agency, Duluth, MN (1989).
4. **Mount, D., and L. Anderson-Carnahan,** "Methods for Aquatic Toxicity Identification Evaluations: Phase III Toxicity Confirmation Procedures," EPA/600/3-88/036, U.S. Environmental Protection Agency, Duluth, MN (1989).
5. **U.S. Environmental Protection Agency,** "Generalized Methodology for Conducting Industrial Toxicity Reduction Evaluations (TREs)," EPA/600/2-88/070, U.S. EPA Risk Reduction Engineering Laboratory, Cincinnati, OH (1989).
6. **Norberg-King, T.J.,** "Toxicity Identification Evaluation: Characterization of Chronically Toxic Effluents, Phase I," EPA/600/6-91/005, U.S. Environmental Protection Agency, Duluth, MN (1991).

A. M. MARTIN is a supervising engineer with James M. Montgomery, Consulting Engineers, Inc., Metairie, LA (504/835-4252; Fax: 504/835-8059). For the last four years, she has specialized in waste minimization, emissions control, and industrial wastewater treatment. She received her BS from the Univ. of California at Berkeley, and her ME and PhD from Tulane Univ., all in chemical engineering. She is a registered Professional Engineer (chemical) in Louisiana, and is a member of Tau Beta Pi, Sigma Xi, Omega Chi Epsilon, and AIChE.

indigenous organisms present should be examined. If an effluent's intermittent toxicity can be linked to rain events, stormwater contaminants could be the culprit. Stormwater can pick up any toxic components present at the facility, including pesticides or herbicides used nearby.

6. Be a detective. Follow up any leads, no matter how remote. Don't make preliminary judgements about the cause of toxicity — you may be following the wrong path. It may be impossible to determine a single compound responsible for toxicity, as toxicity may vary with a gross parameter, such as chemical oxygen demand (COD) or TOC. Finding the cause of whole-effluent toxicity can be a frustrating puzzle, but thoroughness, good judgment, and persistence will result in future permit compliance. **CEP**

Stormwater Management in Industrial Facilities: An Integrated Approach

Edgar J. Oubre, Robert M. Howe and J. David Keating, Jr.

AWARE Engineering Inc., 8552 Katy Freeway, Suite 300, Houston, TX 77024

Numerous existing and proposed regulations have forced owners/operators of industrial facilities to reevaluate current methods for management of stormwater which falls within the confines of their facilities. Most have found it necessary to integrate their current stormwater and wastewater management practices into a comprehensive environmental compliance program. The development of an integrated stormwater management program should focus on the following objectives:

- *compliance with state and federal regulations*
- *the minimization of the volume of stormwater requiring treatment.*
- *the minimization of the impacts on process/production unit operations*
- *the minimization of the impacts on wastewater treatment unit operations.*

A well designed and properly implemented stormwater management program can reduce capital and operating costs and enhance effluent quality.

INTRODUCTION

The development of a stormwater management program that can be readily and effectively integrated into the overall environmental management plan for a facility can be expensive and complex. The implementation of such a program should address the specific requirements associated with stormwater management, as well as the mutual requirements of plant operations and associated environmental programs. The following project implementation profile has been developed to address the concerns mentioned above:

- Design Philosophy Development
- Management System Options Development
- Detailed Management System(s) Evaluation
- Project Implementation

The most crucial elements of this project approach are the first three phases, as they form the basis for, and can drastically affect the implementation of, the subsequent phases of the project. This paper will focus on these developmental phases of the project and will only briefly discuss project implementation.

DESIGN PHILOSOPHY DEVELOPMENT

The first stage in the development of a stormwater management program, or any engineering project, is the development of a design philosophy. This phase of program development should focus on defining the problems to be addressed, gaining an understanding of the existing systems, and determining the overall objectives, both regulatory and design, to be employed in the subsequent phases of the project. The major elements associated with the development of a design philosophy for stormwater management programs should include:

- Regulatory Requirement Assessment
- Facility Audit
- Area Definition Map

The regulatory assessment results in the establishment of the regulatory compliance objectives to be addressed in the project design philosophy. The facility audit addresses the problem definition and existing system evaluation objectives of this project phase. The culmination of this phase of the project is the development of an overall plan drawing of the facility, the area definition map, which depicts the design philosophy to be adopted by the subsequent phases of the project.

Regulatory Requirement Assessment

Because state and federal regulations are generally the catalysts that compel owners/operators of industrial facilities to reevaluate and modify their existing stormwater management practices, it is essential to assess the applicability of those regulations when developing the design strategy. Regulations which affect stormwater management in industrial facilities include, but are not limited to, the following:

- EPA National Pollution Discharge Elimination System Permit Regulations (40 CFR 122)
- EPA General Provisions for Effluent Guidelines and Standards (40 CFR 401)
- EPA Resource Conservation and Recovery Act (40 CFR 260 to 266)
- Federal and State Water Quality Standards

When assessing regulatory requirements related to stormwater management, it is important to first determine which regulations apply to the specific industry with which a facility is associated (i.e., petroleum refining or chemical manufacturing). Once the comprehensive list of pertinent regulations has been compiled, the sections/subsections of these regulations that are applicable to the specific facility must be identified.

These specific requirements are then applied to the individual components of the existing stormwater and wastewater management systems at the facility. In most cases, individual companies which operate industrial facilities have adopted their own interpretations and regulatory compliance requirements. These facility interpretations/requirements are combined with the results of the regulatory review described above, and the regulatory compliance objectives for the project are developed.

In general, the regulatory compliance objectives for a stormwater management program include definitions which classify the waters which are discharged from a facility. The following key definitions have been employed on past projects:

- Process Wastewater—". . . any water which, during manufacturing or processing, comes into direct contact with or results from the production or use of any raw material, intermediate product, finished product, by-product, or water product." [1, 2]
- Stormwater—". . . storm water runoff, snow melt runoff, and surface runoff and drainage." [1]
- Stormwater Discharge Associated with Industrial Activity—". . . the discharge from any conveyance which is used for collecting and conveying storm water and which is directly related to manufacturing, processing or raw materials storage areas at an industrial plant. . ." [1]
- Best Management Practices (BMPs)—". . . BMPs also include treatment requirements, operating procedures, and practices to control plant site runoff, . . ." [1]
- Contaminated Runoff—"The term 'contaminated runoff' shall mean runoff which comes into contact with any raw material, intermediate product, finished product, by-product or waste product located on petroleum refinery property." [2]
- Non-contaminated Runoff—Stormwater discharges from parking lots and administrative buildings along with other discharges from industrial lands that do not meet the regulatory definition of "associated with industrial activ-

ity" can be treated as "clean stormwater" or non-contaminated runoff.

Facility Audit

In order to determine the implementation strategies required to achieve the regulatory compliance objectives, it is often necessary to perform a detailed evaluation (audit) of the existing facility. The facility audit is an essential element in project definition and design philosophy development. The facility audit consists of an onsite investigation that can include any or all of the following tasks, depending on the size and scope of the project:

- Evaluate existing stormwater management system(s)
- Evaluate current operating, maintenance, and spill prevention and control procedures
- Review facility general environmental guidelines and design standards
- Collect/review existing stormwater system data
- Review existing discharge permit limits
- Identify potential sources of stormwater contamination
- Identify non-stormwater discharges
- Identify/review ongoing projects which may impact stormwater management

The results of this evaluation are utilized to develop a facility watershed map and a flow schematic that together depict the existing stormwater and wastewater management systems.

Area Definition Map

The final step in the design philosophy development phase of a stormwater management program is the development of what is referred to as an area definition map. The area definition map incorporates the regulatory definitions related to the stormwater and wastewater management systems and the findings of the facility audit into one comprehensive drawing. This map delineates the facility's watersheds into area classifications as set forth by the regulatory definitions. In previous projects, these area definitions included the following broad categories:

- Process areas—areas within the facility in which the manufacturing or processing of any raw material, intermediate product, finished product, by-product or waste product takes place.
- Contaminated areas—areas within the facility such as tank farms, product storage areas, material handling areas, lay down yards and other areas that are associated with an industrial activity but not directly associated with manufacturing or processing.
- Clean areas—areas within the facility which are not associated with industrial activity; where there is little or no potential for contamination of stormwater under normal operating conditions.

The area definition map can be further subdivided into more specific subcategories as required at each facility.

MANAGEMENT SYSTEM OPTIONS DEVELOPMENT

At this point in the project development, potential stormwater management options and a conceptual design basis can be developed. The two major concerns of any stormwater management program are the transportation (conveyance) and the storage (impoundment) of the stormwater which falls within a facility. There are several methods for analyzing stormwater systems to determine the peak runoff rates and runoff volumes required to design stormwater conveyance and impoundment systems.

Storm Event Selection

A key element in estimating runoff rates and volumes is the selection of storm events to be utilized in the analyses. Statistical rainfall data, such as is published in the National Weather Service's Technical Paper No. 40, are typically used in runoff analysis.

Existing regulations provide little guidance in the selection of storm events for the design of stormwater management facilities. Storm events with durations equivalent to the time of concentration of a watershed are utilized to determine the peak runoff rates used to size stormwater conveyance systems. The duration of these storms is usually short, 10 minutes to 1 hour, because the watersheds being analyzed are small. Longer duration storm events, typically 24 hours, are used to determine the volume of stormwater requiring impoundment. During the options development phase of a project, storm events with return frequencies between 1 and 10 years and between 10 and 100 years are generally utilized to size the conveyance and impoundment systems, respectively.

Stormwater Management System Alternatives

The alternatives for management of stormwater vary depending on the regulatory compliance objectives and the layout and configuration of the existing stormwater and wastewater management facilities. The development of stormwater management system alternatives should consider the following general concepts:

- Source control and waste minimization to eliminate or reduce the amount of pollutants entering the stormwater system
- Segregation of process, contaminated or potentially contaminated areas from clean stormwater systems, wherever practical
- The minimization of contaminated/potentially contaminated stormwater volumes through the use of stormwater controls, such as:
  —installation of curbs or containment levees
  —modification of existing drainage barriers, (i.e., raising perimeter roads or levees)
  —modification/improvement of existing drainage patterns
- Addition of lift station and impoundment facilities to control the discharge of stormwater to the wastewater treatment facility, and
- Local impoundment of contaminated/potentially contaminated stormwater in product storage and tank farm areas during storm events.

In most cases, several options can be developed employing one or more of the concepts presented above. A preliminary evaluation of the options is performed to narrow down the field of potential candidates. Rudimentary hydrologic analyses, usually employing the Rational Method, are conducted to develop initial conveyance and impoundment system designs for the various options. This analysis is usually performed using several storm events to develop a matrix of sizing criteria for each option (i.e., varying storm duration and return frequency).

The options are then evaluated with regard to the following criteria:

- Constructability
- Operability
- Environmental and Safety Concerns
- Operating and Capital Costs
- Implementation Schedule

In most cases, this evaluation becomes an iterative one. The conceptual design basis and area definitions may require some fine tuning once the first run through the options evaluation process has been completed. The list of candidate options is then narrowed, and one or more feasible options is recommended for further detailed evaluation.

DETAILED MANAGEMENT SYSTEM(S) EVALUATION

The selected option(s) is then subjected to a more detailed analysis for refinement of the various elements of the management system. This analysis usually consists of modelling of the stormwater management system to predict peak runoff rates and runoff volumes.

Computer Modelling

For smaller watersheds where modifications to the drainage area and conveyance systems are not complex, straight forward analyses using the Rational Method may suffice. When major modifications to the drainage area (i.e., curbing and paving) and modifications to the conveyance system (i.e., installation of new interceptor sewers and lift stations) are proposed, a more detailed model should be considered.

The selected model should be capable of the following:

- simulating the runoff from a watershed for any prescribed rainfall pattern
- routing the runoff through the conveyance system
- simulating pressure flow or surcharge, backwater conditions, flooding, transfer of flow by weirs, orifices or pumping facilities and on- or off-line storage facilities
- simulating large sewer systems with large numbers of subcatchments (watersheds), channels/pipes and junctions

The stormwater model which has been utilized in past projects is the United States Environmental Protection Agency Storm Water Management Model (SWMM).

At this point in the analysis, the storm events to be simulated are selected. For modelling the conveyance system, a statistical storm event with a return frequency of 10 years is typically selected. As stated previously, the duration of the storm event will vary depending on the time of concentration for the watershed(s) being modelled so that peak runoff rates can be determined. Less intense storm events of longer duration, typically a 25-year, 24-hour event, are simulated to determine stormwater impoundment volume requirements.

The outputs of the SWMM model include inflow hydrographs at selected inlets, outflow hydrographs at specified junctions, and conduit and junction output summary tables which include the location and duration of flooding in the system. The outflow hydrograph from the peak runoff analysis is then utilized to size the stormwater transfer system (lift station and pumps). The outflow hydrograph from the simulation of the 24-hour storm event is used to size the stormwater impoundment system.

The output from the peak runoff simulation is also evaluated to determine the location and duration of localized in-plant flooding. This is an important evaluation as the occurrence of flooding may adversely affect plant operations. If the flooding in some areas proves to be unacceptable, the conveyance system sizing (channels/pipes and/or transfer pumps) should be modified and the simulation rerun.

At the conclusion of the storm event simulation, storm sewer modifications, equipment and impoundment system sizes are defined and capital and operating costs are developed.

Risk Assessment

The system option(s) should at this point be subjected to a risk assessment. This risk analysis should address such items as events which exceed design capacity, equipment failure and operation errors. The results of this analysis may lead to modification of the design basis and the area definition map.

Selection of Management Program and Design Parameters

At the conclusion of the management system option(s) analysis, one stormwater management system, and its associated design basis is selected for design and construction.

PROGRAM IMPLEMENTATION

Having finalized the developmental phases of the stormwater management program, the next step is the refinement and implementation of the program through the following project phases:

- System Design Package
- Detailed Design Engineering
- Construction

The details of the project implementation phases of a project are more site specific than the development phases, so only a brief discussion of these phases will be presented.

System Design Package

The development of a system design package for the stormwater management system would include: design basis, P&IDs, equipment data sheets, tank data sheets, instrument data sheets, operating philosophy and instrumentation/control philosophy. This provides a refinement of the system design sufficient to:

- Develop a cost estimate of sufficient detail to support an appropriations funding request.
- Develop an implementation sequence for determining cash flow projections.
- Develop a construction/implementation schedule, recognizing continued operation of plant production facilities.

Detailed Design Engineering

Following completion of the system design package, the detailed design effort must proceed with production of construction drawings and specifications. Further refinement of the system design is expected to occur through such activities as pilot trenching of proposed sewer routings, final selection of equipment, sub-surface investigations, etc.

The environmental sensitivities which were established in the development phases of the project should be maintained by the continued participation of the original project team throughout the detailed design and construction phases.

Construction

The selection of a construction contractor and construction scheduling/sequencing must recognize the need to maintain the operation of both the plant production facilities and the waste treatment facilities throughout the construction period.

LITERATURE CITED

1. "EPA National Pollution Discharge Elimination System Permit Regulations," Code of Federal Regulations, Title 40—Protection of the Environment, Subchapter D, Part 122.
2. "EPA General Provisions for Effluent Guidelines and Standards," Code of Federal Regulations, Title 40—Protection of the Environment, Subchapter N, Part 401.

Effectively Manage Stormwater in a CPI Complex

Segregation and minimization of contamination are key elements of a good stormwater management program.

Deepak Garg* and R. Benson Pair, Jr.,
The M.W. Kellogg Co.

*Mr. Garg is now with ENSR Consulting and Engineering.

Stormwater management systems are being closely scrutinized by regulatory authorities. Recent emphasis has been toward the minimization, segregation, and treatment of contaminated stormwater.

Because contaminated stormwater constitutes a significant peak load on a wastewater treatment facility, a large portion of the capital expenditure for an industrial wastewater treatment plant is related to stormwater treatment. There is, therefore, a strong economic incentive to carefully segregate and minimize contaminated stormwater.

The development of an effective stormwater management program for a modern chemical process industries (CPI) complex involves the following four steps:

1. Regulatory analysis.

2. Estimation of instantaneous and time-averaged stormwater flow rates.

3. Determination of stormwater segregation philosophy.

4. Design of stormwater collection, transfer, storage, treatment, and reuse facilities.

This article outlines several practical ways of segregating and minimizing contamination of stormwater, and discusses the design practices for stormwater collection, transfer, storage, treatment, and reuse systems.

Regulatory overview

Before the 1970s, it was common practice for stormwater surface runoff from CPI plants to be discharged without treatment. With the enactment of the Clean Water Act (CWA) in 1972 and the establishment of the National Pollutant Discharge Elimination System (NPDES), all aqueous wastes, including process wastewater and stormwater runoff, became subject to regulation. Other recently promulgated environmental legislation has tightened the control of the multimedia environmental risk associated with stormwater discharges. These regulations address issues of stormwater control related to point and nonpoint sources of pollutant loads to receiving bodies and the permitting requirements associated with such discharges, as well as leaks of toxic substances to the groundwater and emissions of volatile organic compounds (VOCs) from stormwater impoundment basins.

Rules promulgated by the U.S. Environmental Protection Agency (EPA) in 1992 *(1)* establish phased and tiered control requirements for stormwater discharges. These regulations specify permitting and minimum monitoring and reporting requirements, as well as the requirement that a facility develop and implement a Stormwater Pollution Prevention Plan (SWPPP). The SWPPP is intended to describe the control measures and practices that the facility will use to reduce pollutants in discharged stormwater. Source control and segregation of all process related or contact wastewaters are central to the SWPPP.

The Hazardous and Solid Waste Amendments of 1984 (HSWA) to the Resource Conservation and Recovery Act (RCRA) can impact stormwater management in several ways:

• The Toxicity Characteristic Leaching Procedure (TCLP) establishes that when a waste determined to be hazardous by this procedure is sent to an impoundment basin, the impoundment must meet minimum technology requirements (MTRs), which usually means double lining and leak detection systems;

• The Petroleum Refining Primary Sludge Listing establishes that stormwater basins receiving dry-weather flows (such as process wastewater or oily cooling water) would be regulated under RCRA and could be subject to MTRs; and

• The Land Disposal Restrictions and Variances establish that wastes determined to be hazardous must be disposed of in accordance with Best Demonstrated Available Technology (BDAT) standards and that impoundment basins used to store such wastes must meet MTRs.

Several air-pollution-control regulations (2–4) impact stormwater management programs indirectly. These regulations require the control of VOC emissions from process wastewater exposed to the atmosphere. If a combined stormwater sewer system is discharged to open management units, then some of the control requirements of these regulations may be triggered.

Most developing countries are establishing industrial stormwater management and discharge regulations that require similar controls as in the U.S., except in the area of stormwater segregation. There is usually no regulatory impetus to either segregate dry-weather flows from stormwater or to segregate various types of stormwater. However, some progressive grassroots projects are implementing some of the environmental control methods and segregation practices presented in this article that are beyond the minimum regulatory requirements.

Finally, recent environmental and economic forces have emphasized waste minimization and water reuse, driven by pollution prevention and water conservation initiatives. These practices are equally important in the U.S. and in the international arena. In some cases they are even more important outside the U.S. in countries where

water scarcity is an important consideration. Industrial stormwater, especially uncontaminated stormwater, when available, remains an important yet untapped source of water that has not been fully exploited by the industry. A related issue is the minimization of contaminated stormwater by adequate control of sources of contamination.

Estimating stormwater flows

The second step in the development of a stormwater management program is the estimation of instantaneous peak and time-averaged stormwater flow rates at various points in the collection, transfer, storage, and treatment system. There are various methods and models available for estimating stormwater flow characteristics, ranging from the simple

Recent environmental and economic forces have emphasized waste minimization and water reuse.

"Rational Method," to the complex and sophisticated "Storm Water Management Model" (SWMM) developed by the EPA.

The Rational Method calculates stormwater flow rate from a given area by:

$$Q = CIA$$

where: Q = stormwater flow rate, ft³/h; C = constant related to the type of surface (*e.g.,* grass, paved, dirt); I = design stormwater intensity, ft/h; and A = area contributing to stormwater flow, ft² .

The EPA SWMM, on the other hand, is a comprehensive mathematical model for simulating urban runoff water quality and quantity in storm and combined sewer systems. All aspects of the urban hydrologic and quality cycles are simulated, including surface and subsurface runoff, transport through the drainage network, and storage and treatment.

The choice of model and its level of sophistication are related to the level of

input data available for a particular project. This is especially true with regard to the data availability for cumulative rainfall characteristics.

For grassroots projects, typically, limited data are available when the conceptual plan for collection, transfer, storage, and treatment is developed. It is therefore very important to use the available data with good engineering judgment. A good simple approach for such applications relates cumulative rainfall characteristics with recurrence intervals and time of concentration (that is, the time for the entire area under consideration to be contributing to stormwater runoff) (5). This method can be used to size stormwater catch basins, sewer lines, lift stations, lift station pumps, stormwater storage volumes, and treatment capacity for grassroots projects, and in some cases even for revamp or upgrade projects.

For most upgrade projects at existing facilities, where the physical determinants of the stormwater management system are well-defined, a sophisticated analysis of stormwater flow rates and compositions is needed. The variation of flow characteristics with time at various system components such as sewer lines, lift stations, diversion boxes, and storage basins is also required. An example of such a project is the implementation of a segregated stormwater system at a facility that has a combined sewer system. In this case, a comprehensive model such as the EPA SWMM should be employed. A review of various stormwater estimation and management models and their applications is provided in (6).

STORMWATER SEGREGATION

The third step in designing a stormwater management program is defining a stormwater segregation philosophy. Segregation has become a key aspect of industrial stormwater management, driven by economics and environmental regulations.

Most stormwater pollution control regulations require the treatment of contaminated stormwater. Contaminated stormwater typically represents 20–60% of the design flow capacity of a CPI plant's wastewater treatment plant on an annual average basis. On an instantaneous basis, it can represent an even higher percentage of the inlet flow to the wastewater treatment facility. Minimization and segregation of the contaminated portion of a facility's stormwater runoff is a major goal of stormwater segregation.

In a combined sewer system, the mixing of dry-weather process wastewater with stormwater subjects the entire system to regulatory requirements. Thus, the other major goal of stormwater segregation is to segregate dry-weather process wastewater flows from stormwater collection, transfer, and storage systems.

Until approximately twenty years ago, most CPI plants had a combined sewer system for process wastewater and stormwater handling because that was the most cost-effective means of conveying these flows. During wet weather, all combined process water and stormwater from process units in excess of the treatment plant flow capacity was taken to impoundment basins for storage. This wastewater was later returned to treatment whenever it could be accommodated.

This approach presents two problems given the present regulatory framework: the potential migration of pollutants from the impoundment basins and the resulting contamination of groundwater, and the need to control the VOC emissions from the open impoundment basins. Regulations require the use of leakproof double-lined basins or, preferably, aboveground concrete or steel tanks to address the groundwater contamination problem, and the use of covers and emission control systems to mitigate VOCs. Both of these requirements are expensive considering the size of tanks and basins involved.

These potential costs can be reduced by carefully segregating the contaminated portion of the process unit stormwater runoff and sending it directly to treatment, and segregating all dry-weather

flows from stormwater. Stormwater that has the potential to be contaminated is sent to interim storage and returned for treatment if necessary.

Contamination sources

A large CPI complex may occupy several hundred to a few thousand acres. The process units usually occupy less than 20% of this area. The building blocks of the complex are the process units, utility systems, offsites and interconnecting facilities, and administrative facilities. The process units are typically clustered together, while the remaining support facilities are spread around to optimally serve the process units. It is important to note that tank farms, which include the tanks and tank dikes, can occupy 30% to 60% of the facility plot area.

Stormwater runoff can be classified into three categories:

• contaminated stormwater;
• potentially contaminated stormwater; and
• uncontaminated stormwater.

Contaminated stormwater is the runoff from known contaminated areas that requires treatment under most regulatory programs. Potentially contaminated stormwater may require treatment and thus needs to be tested before discharge to the outfall. This stormwater also has the potential for reuse if it is uncontaminated. The runoff from the uncontaminated areas of the site, such as roads, parking lots, and building roofs, is designated as uncontaminated stormwater.

The sources of contaminated stormwater are usually located within the process units. Pump pads for pumps in hydrocarbon service represent the most significant source of contaminated storm-

water, as there are hundreds of pumps in a typical plant that can contribute a significant flow of contaminated stormwater. Such sources as chemical injection systems, process unit battery-limit valving areas, as well as heat exchangers and vessels in some services, can also contribute to contaminated stormwater.

The process units are physically segregated from the nonprocess areas either by curbing or by grading and sloping of the surface. The remaining portions of the process units area that are not designated as contaminated should be regarded as potentially contaminated. There is always the potential of spills, leaking equipment, and leaking valves throughout the process unit. Although such spills and leaks are regularly washed down, there is always the possibility of generating contaminated stormwater from such areas when it rains.

In the offsites areas, pump pads, tank valving, and tank mixer areas are considered contaminated. Other areas such as tank dikes, solid waste areas, sulfur and coke stockpile areas, marine handling facilities, and chemical handling areas are considered potentially contaminated. The remaining areas of the site not designated as contaminated or potentially contaminated can be considered uncontaminated.

Recent experience with modern petroleum refining and petrochemical facilities has indicated that typically 1–5% of the facility area can be designated as contaminated, 25–60% of the area as potentially contaminated, and 35–74% as uncontaminated.

It should be noted that the potentially contaminated stormwater from tank dikes in the refinery tank farm areas is usually stored within the dikes during the rainfall event. This can represent 30–60% of the stormwater that falls on the facility. After the rain subsides, the stormwater is then tested for relevant parameters, such as chemical oxygen demand (COD), biological oxygen demand (BOD_5), and pH, and discharged directly to the outfall if uncontaminated, or to treatment if contaminated. The stormwa-

> *Minimization and segregation of the contaminated portion of a facility's stormwater runoff is a major goal of stormwater segregation.*

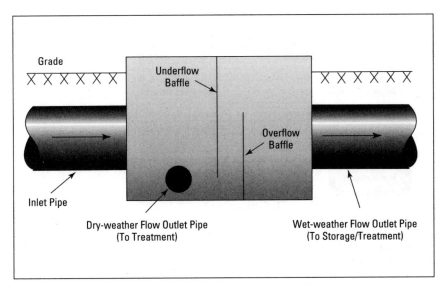

■ *Figure 1. A diversion sump routes washdown water and spills to treatment during dry weather.*

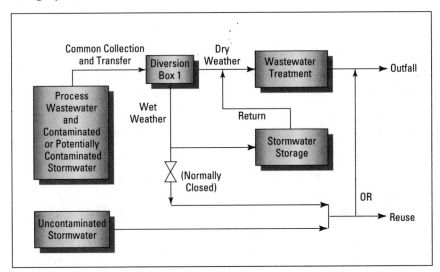

■ *Figure 2. A combined sewer system is not recommended for CPI plants.*

ter storage and surge capacity provided by tank dikes is an important consideration in any stormwater management plan.

Segregating dry-weather flows from stormwater

The first step in stormwater segregation is to segregate the dry-weather flows from stormwater. This type of segregation should be incorporated in all stormwater management programs.

Dry-weather flows from a typical CPI complex include process wastewater streams, utility blowdowns, and surface and equipment washdowns. These dry-weather streams include both continuous and intermittent streams.

It is preferable to segregate and hardpipe continuous streams directly to treatment. Intermittent streams and continuous streams that are not hard-piped can be effectively segregated from stormwater using an elevated hub collection system. Dry-weather flows are directly routed to drain hubs that are elevated approximately 6 in. above grade level. Once collected in the drain hubs, the dry-weather streams are conveyed via an underground or aboveground

sewer system to treatment. It should be noted that sending certain dry-weather wastewater streams to the plant's sewer can trigger various air-pollution regulations *(2–4)*.

Process unit surface washdowns and spills are dry-weather flows that have to be segregated from stormwater. Because these flows are generated at the unit surface, it is not possible to segregate them in the same manner as the dry-weather flows that are collected in raised drains. Washdowns and spills are collected in the surface runoff collection system. Thus, there is a need for a dry-weather diversion in the stormwater sewer system that routes these flows back to the process wastewater system.

This diversion is most effectively accomplished by means of a diversion sump (Figure 1), which is essentially an inline sewer box with underflow and overflow baffles. During dry weather, all flow entering the diversion box is conveyed to treatment. During wet weather, as the flow rate entering the box increases, the level in the box starts to rise. The height of the overflow baffle is set to a level where the water starts to overflow when the flow rate into the box reaches a predetermined level. This diverted flow is routed to storage and treatment. The underflow baffle aids in keeping free oil from overflowing to stormwater storage.

Segregating various types of stormwater

There are, in general, three conceptual methods of dealing with the contaminated and potentially contaminated stormwater. These methods are discussed below in order of increasing environmental benefit.

Combined sewer system. The first method, although not recommended, is currently still in practice in developing countries. There are several configurations that can be used with this approach, such as the one depicted in Figure 2.

All process wastewater, oily wastewater, and stormwater streams generated within the complex are routed through one sewer system to treatment and storage. There is a mechanism to segregate dry-weather flow from wet-weather flow to avoid overloading the treatment system. This wet-weather flow diversion is accomplished either by a hydraulic gravity-flow diversion box (Diversion Box 1 in Figure 2), or by pumps, depending on whether a stormwater holding basin or a tank is used for storage of wet-weather excess flow. During wet weather, however, the dry-weather flows can end up in the stormwater system, which is the major drawback of this system. The other major drawback of this method is that there is limited potential for reuse of the stormwater.

Combined contaminated and potentially contaminated stormwater. The second method, illustrated in Figure 3, is based on combining the contaminated and potentially contaminated stormwater systems at the source. This approach does not require segregation within process units by physical curbing.

Essentially all the stormwater that falls on the process units and other contaminated and potentially contaminated areas is collected and taken to a stormwater holding tank and brought back to treatment at a convenient time. The process wastewater streams are hard-piped or have a separate sewer to convey them to the treatment facility. The combined stormwater should all be considered contaminated.

This combined stormwater almost always has to be treated before discharge, unlike in the next approach, where the direct discharge of a substantial quantity of potentially contaminated stormwater to the outfall is often feasible. The option of reuse of some of this stormwater is lost by using this methodology.

Completely segregated system. The third approach, shown in Figure 4, handles contaminated and potentially contaminated stormwater separately. This is the most environmentally sound approach, and should be considered for implementation in a grassroots facility or in new process units within an existing facility.

Contaminated stormwater is segregated by means of physical curbing or grading and routed directly to treatment with

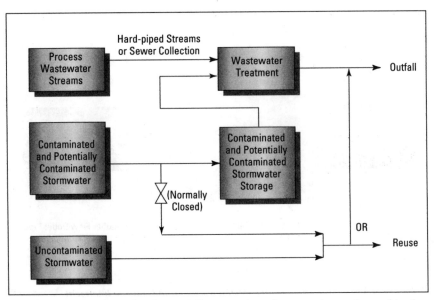

■ *Figure 3. Contaminated and potentially contaminated stormwater can be combined at the source.*

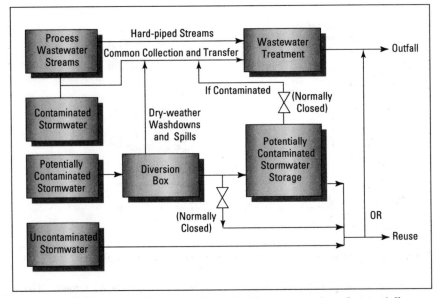

■ *Figure 4. The most effective segregation method keeps contaminated, potentially contaminated, and uncontaminated stormwater separate.*

the process wastewater. The potentially contaminated stormwater, which is also segregated by means of physical curbing, is routed to a holding tank (or basin), where it is stored, tested, and returned to treatment at a convenient time, if necessary. Spills that contaminate the process unit surfaces are collected in the same system as the potentially contaminated stormwater. A diversion mechanism is included to route the surface washdowns

and spills back to the process wastewater system. The decision to include a tank or a basin for holding the potentially contaminated stormwater is based on a consideration of economics, plot plan constraints, hydraulic constraints, and the regulatory climate.

This approach ensures minimum environmental compliance risk in the future, and has the maximum potential for water conservation by reuse of the potentially contaminated and uncontaminated stormwater. If implemented, it could obviate the need for major stormwater segregation capital projects in the future.

STORMWATER COLLECTION, TRANSFER, STORAGE, TREATMENT, AND REUSE

Once the stormwater segregation philosophy is determined, the final step in the development of a stormwater management program is to determine the method of collection, transfer, storage, treatment, and reuse.

There are two approaches to stormwater collection and storage — the surge-and-treat method, and the first-flush method. The choice of which method to adopt is important and has to be made early in the design of the stormwater management program.

In the surge-and-treat method, the stormwater conveyance system and the storage volume are sized to accommodate an entire design storm event. An example of a design storm event is a 24-h storm event with a 25-yr recurrence interval — that is, the largest rainfall expected during a 24-h period over an average 25-yr period. This stored stormwater is then sent to treatment at a rate that can be accommodated by the treatment plant. The surge-and-treat method results in large volumes for stormwater storage facilities, although the environmental risk associated with the potential discharge of contaminated stormwater to the outfall is minimized.

The first-flush principle assumes that only the initial stormwater runoff that flushes off contaminated or potentially contaminated surfaces is contaminated enough to require treatment. Typically, 5 to 20 minutes of first-flush at the peak rainfall rate is adequate to clean the surface. After this first-flush stormwater is collected for treatment, the subsequent surface runoff is discharged directly to the outfall.

When utilizing the first-flush principle in the sizing of stormwater holding volumes, it is important to note that the first-flush from all contaminated or potentially contaminated areas of the plant must be collected. Thus, the time of

Literature Cited

1. "EPA Administered Permit Programs: The National Pollutant Discharge Elimination System," Code of Federal Regulations, Title 40 — Protection of the Environment, Part 122, Subparts A, B, and C.
2. "Standards of Performance for VOC Emissions from Petroleum Refinery Wastewater Systems," Code of Federal Regulations, Title 40 — Protection of the Environment, Part 60, Subpart QQQ.
3. "National Emission Standards for Benzene Waste Operations," Code of Federal Regulations, Title 40 — Protection of the Environment, Part 61, Subpart FF.
4. "National Emission Standards for Hazardous Air Pollutants for Source Categories," Code of Federal Regulations, Title 40 — Protection of the Environment, Part 63, Subparts A, F, and G.
5. Elton, R. L., "Designing Stormwater Handling Systems," *Chem. Eng.*, pp. 64–68 (June 2, 1980).
6. Chieu, J., and S. Foster, "Improve Stormwater Management For Refineries, Part 1," *Hydrocarbon Processing*, pp. 73–79 (Dec. 1993).
7. Prosser, M. J., "The Hydraulic Design Of Pump Sumps And Intakes," British Hydromechanics Research Association, Bedford, U.K., and Construction Industry Research and Information Association, London (July 1977).
8. "Standards for Centrifugal, Rotary, and Reciprocating Pumps," 13th ed., Hydraulic Institute, Cleveland, OH (1975).

travel of the contaminated and potentially contaminated stormwater from the furthest contributing point in the facility to the storage system becomes significant. The holding volume of the storage system is sized based on the retention of 5–20 min worth of stormwater from this furthest point, plus all the stormwater from the nearer areas that enters the storage system during the time it takes the stormwater from the furthest point to reach storage. A retention time of 1–5 h of the plantwide stormwater based on this analysis is common.

Although the first-flush method does carry a higher risk of discharging contaminated stormwater to the outfall, if

designed correctly, it offers an effective and economic alternative to the surge-and-treat method. The first-flush method is applicable to all three methods of stormwater segregation discussed previously.

Basic configurations

When the dry-weather flows are combined with the contaminated and potentially contaminated stormwater streams (as in Figure 2), the dry-weather and wet-weather flows must be physically segregated to avoid hydraulic overload of the treatment system. This is done by means of a pumped or a gravity-flow diversion sump. One or several diversion sumps may be used depending on the size of the project, the plot plan, and elevations. The dry-weather flow lines from these sumps are headered together and dispatched directly to the treatment plant. The wet-weather flows are also headered together, but are taken to storage, then returned to treatment when convenient. Because of the potential for dry-weather flows to end up in the stormwater system, this storage should not be in an unlined basin.

A second diversion box may be used to route stormwater to the outfall once the storage volume is filled. This second diversion box is useful if the first-flush method is used to size the stormwater storage volume.

When contaminated and potentially contaminated flows are combined (Figure 3), a separate collection, transfer, and storage system must be designed for the combined stormwater. If the first-flush method is employed, the first-flush flow is transferred, by lift stations and pumps or by gravity flow, to a storage tank or basin; from there it is returned to treatment when convenient. The stormwater subsequent to the first-flush bypasses the storage system and goes directly to the outfall.

When all three types of stormwater are separate (Figure 4), a system for the collection, transfer, and storage of potentially contaminated runoff is required. Dry-weather flows and contaminated stormwater are transferred together directly to treatment in what is traditionally known as the oily wastewater or process wastewater system. The first-flush of potentially contaminated stormwater is transferred to a storage tank or basin, where it is tested and discharged to the outfall if uncontaminated or returned to treatment if contaminated. The surface runoff from the potentially contaminated areas subsequent to the first-flush is sent directly to the outfall.

Depending on whether a tank or a basin is used, the potentially contaminated stormwater is transferred from the collection catch basins by lift stations and pumps to a tank, or by gravity flow, if feasible, to a basin. Depending on the size of the storage facility and the treatment plant capacity, the potentially contaminated stormwater can be transferred back to treatment during dry weather. This way, the treatment plant is not overloaded and the storage tank or basin is emptied in a few days.

Collection and transfer systems

Regardless of which segregation scheme is employed, stormwater is typically collected in catch basins and transferred by underground gravity sewers either directly to a storage basin, or via a lift station to a storage tank. The design of the catch basins and sewer lines should be based on the peak flow rate estimated by one of the techniques discussed above.

If a lift station is used to transfer the stormwater, it typically employs large pumps (>100 hp motor). Instead of sizing based on retention time, it is recommended that the lift station be sized based on the approach presented in (7). This approach relates the working volume of the lift station to the maximum flow rate into the lift station, the design pump-out flow rate, and the design time between two consecutive starts of the lift station pump(s). The working volume is minimized if the design pump-out rate is two times the maximum flow rate into the lift station.

Care should taken to avoid pump motor overheating due to frequent consecutive starts. This criterion often determines the size of the lift station. For mo-

tors of the size under discussion, an allowance of 20–45 min between consecutive starts is a good practice.

Another important design consideration for lift stations is the provision of independent or redundant power sources for pumps sharing duty. Also, in the case of multiple pumps, it is good practice to sequence the starts such that the pumps experience even wear and individual pump motors have more time between consecutive starts. The physical characteristics of the lift station should be designed in accordance with Hydraulic Institute Standards (8).

Storage systems

The stormwater storage tank or basin is sized based on the surge-and-treat or the first-flush principles. Economics favor the first-flush sizing basis, as the required storage volume is minimized.

An aboveground concrete or steel tank is preferred over a below-grade impoundment basin, recognizing current and future environmental regulations. Tanks also occupy less area than basins and are more flexible in terms of plot location.

Transfer of stormwater to treatment

The stormwater return system and pumps are sized to empty the storage volume in approximately 3 to 15 days. A balance must be achieved between emptying the holding volume in time for the next storm event and limiting the pumping rate to within the treatment plant's capacity. Two or more pumps are provided to return the stormwater at varying rates depending upon the design of the storage system and treatment plant capacity.

Treatment and reuse of stormwater

Contaminated stormwater can be adequately treated by conventional primary and secondary treatment plants before its discharge to the outfall. If water reuse is a consideration, then the uncontaminated stormwater, or the potentially contaminated water after it has been found to be uncontaminated, should be reused prior

to mixing with the rest of the facility's wastewater for discharge to the outfall.

Typically, treatment to remove suspended solids, such as clarification and filtration, is adequate to allow this water to be reused as primary or secondary firewater, utility water, or cooling water during rainy weather. Other potential reuse options for this treated stormwater are irrigation of in-house facilities such as lawns, gardens, and landfarms, although these uses require distribution systems and are thus more appropriate to grassroots projects. **CEP**

D. GARG was an environmental engineer in the process engineering department of The M.W. Kellogg Co., Houston, TX, when he wrote this article. He is currently a project engineer with ENSR Consulting and Engineering, Houston, TX (713/520-9900; Fax: 713/520-6802). He has over seven years of experience in environmental process and system design, as well as in environmental compliance activities in the areas of wastewater and stormwater management and treatment, air-pollution control and permitting, and environmental impact assessment related to the hydrocarbon and chemical process industries. He received his MS in environmental engineering from the Univ. of Houston in 1988, and a BTech in chemical engineering from the Indian Institute of Technology, Madras, India, in 1986.

R. B. PAIR, Jr., is chief environmental technology engineer at The M.W. Kellogg Co., Houston, TX (713/753-2671; Fax: 713/753-6323), where he is responsible for the technical supervision on environmental projects, providing design guidance and engineering experience for all offsite utility areas. He has more than 20 years of experience in industrial wastewater treatment, hazardous waste management, and detailed design, plant startup, and operations, and is a recognized expert in the treatment of refinery and petrochemical wastewater. Prior to joining Kellogg in 1988, he was a branch manager for an environmental consulting firm. He holds bachelor's and master's degrees in chemical engineering from Rice Univ., and is a registered professional engineer in Texas. He is a member of the Water Environment Federation and a Diplomate of the American Academy of Environmental engineers; he is also a member of AIChE and a Director of the Environmental Division. He has authored numerous technical papers and reports.

Design Steam Strippers for Water Treatment

The high liquid-to-vapor ratios used and the possibility of fouling by organic and inorganic materials are among the challenges designers face.

José L. Bravo,
Jaeger Products, Inc.

A very prevalent water pollution problem is contamination by volatile organic compounds (VOCs). Many VOCs are only partially miscible with water, but in general they all present a certain solubility.

Some of the VOCs commonly found in water include benzene, toluene, xylene, naphthlene, acetone, and a wide range of chlorinated hydrocarbons. Contamination stems from such sources as gasoline leaks, solvent spills, and process spills and discharges. The problem of reducing VOCs in water applies to groundwaters, surface waters, and wastewaters alike; however, the origin of the water has some important design implications (which will be discussed later).

This article first reviews the basics of steam stripping. It then discusses the key issues and challenges faced by an engineer in designing a stripping system for water treatment (Table 1).

Why use steam stripping?

Dilute mixtures of organic materials in water can be concentrated by steam stripping. The end products of this operation are a clean water stream almost devoid of organic materials, and a highly concentrated organic stream suitable for recycle to a process or for disposal. The use of heat in the form of steam as a separating agent offers significant advantages over other methods, such as inert gas or air stripping.

Steam stripping for water cleanup is essentially a distillation process, where the heavy product is water and the light product is a mixture of volatile organics. These organics are present in the feed water in relatively small concentrations.

Steam stripping takes place at higher temperatures than air stripping, usually very close to the boiling point of water. Since the volatility of the organics is a very strong function of temperature, the high temperatures inherent in steam stripping allow for the removal of heavier, more-soluble organics that are not strippable with air.

Another advantage of steam stripping is that minimal secondary pollutants are generated. No off-gas treatment is needed. And, the only waste stream generated is a small amount of very concentrated organics, which are easily dealt with by incineration, biological treatment, or recycling back to the process.

Steam stripping can often achieve very high removals (more than 99%) and low effluent concentrations (below 5 ppb). It is the most economical removal technique at feed concentrations above 0.1% weight organics, and it is cost-effective at feed concentrations as low as 2 ppm. Depending on process needs, it can be operated at vacuum or pressure with little penalty. And, it can be made very energy efficient with heat recovery. However, a major drawback is that fouling is a continuous concern.

In summary, steam stripping is a good solution for wastewater streams that contain fairly soluble, semivolatile organics and where no off-gas stream is desired. On the other hand, steam stripping does necessitate the presence of steam (or process heat) and tends to be more capital-intensive than air stripping. Ideal settings for steam stripping are petroleum refineries and petrochemical and chemical plants.

What is steam stripping?

A wastewater stream is heated and put in intimate contact with steam in a packed or trayed tower. The combined effects of the steam and heat or temperature cause organic material to transfer from the liquid to the vapor phase. This material is then carried out with the vapor. As contacting proceeds down the tower, the wastewater becomes leaner in the organic material while the vapor phase becomes more enriched as it travels up the tower.

Steam is injected at the bottom of the tower to provide heat and vapor flow and the wastewater is fed at the top of the tower. Clean water leaves the bottom of the tower and the steam leaves the top heavily laden with organic material. The latter stream is condensed and processed further to separate the steam and organics. The net effect achieved in the steam stripper and condenser is that a contaminated wastewater and steam are injected into the tower and a clean water stream is obtained. A low-volume but concentrated water/organic mixture is also obtained as a byproduct.

The stripping tower employs either trays or packing to facilitate contact between the contaminated wastewater and the steam. Metal (stainless steel) random packings are suitable for most applications. Plastic (glass-reinforced polypropylene, polypropylene oxide, polyvinylidene fluoride, or polytetrafluoroethylene) random packings can handle acids. Metal (stainless steel or aluminum) structured packings provide higher efficiency or capacity. Stainless steel sieve trays may be used for fouling service. Column internals include distributors, redistributors, supports, and mist eliminators.

The configuration of a steam stripping unit can vary depending on the characteristics of the organic material to be removed and on what is to be done with it (that is, disposal or recycle). As a minimum, a steam stripping system will look like the unit depicted in Figure 1. It is important to note that heat recovery from the bottoms product is necessary for economical opera-

Table 1. Stripper design issues.

Process Characteristic	Recommendation
Use of air or steam	– Use steam if the organic is soluble or if recovery is feasible. – Steam could be more economical if air post-treatment is required.
Operating pressure for steam strippers	– As low as possible to minimize steam use and reduce fouling. – Limited by disposal of noncondensibles.
Fouling and foaming	– Use packed towers for foaming service; check for foaming tendencies under actual process conditions. – Steam strippers are prone to calcium fouling. – Air strippers are prone to iron and biological fouling. – Address water chemistry early in design and always consider pretreatment.
Vapor/liquid equilibria	– Consider interactions between components. – Use reliable data and activity models. – Consider dissolved gases and pH.
Mass-transfer efficiency	– Do not use conventional distillation efficiencies. – Use the *HTU/NTU* approach in packed towers. – Consider the effects of surface-active agents.
Materials of construction	– Fouling prevention technique could have a large impact on material selection. – Consider strippable gases such as H_2S, CO_2, and O_2.

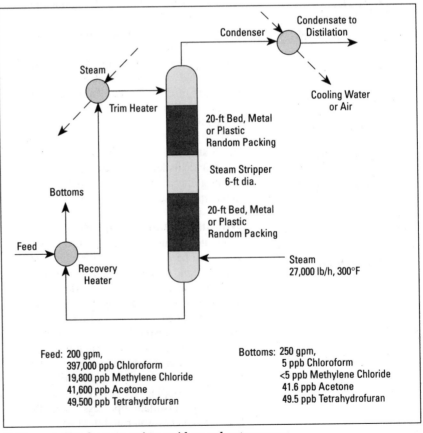

Feed: 200 gpm,
397,000 ppb Chloroform
19,800 ppb Methylene Chloride
41,600 ppb Acetone
49,500 ppb Tetrahydrofuran

Bottoms: 250 gpm,
5 ppb Chloroform
<5 ppb Methylene Chloride
41.6 ppb Acetone
49.5 ppb Tetrahydrofuran

■ *Figure 1. Simple steam stripper with no solvent recovery.*

tion. Operations at reduced pressure do not need recovery exchangers and operate at lower temperatures. The towers also tend to be a bit larger in vacuum operations.

Steam requirements for stripping vary with the operating pressure, the type of organic, and the degree of organic removal or recovery. Further steam requirements for heat balance purposes need to be accounted for as well. A very important consideration in the design of a steam stripper is the fact that the column must be capable of handling enough steam flow to operate without the recovery exchanger; this will be necessary during startup and when the exchanger is out of service for cleaning.

Some organic materials are not totally miscible in water and separate into a distinct organic phase when the concentration exceeds the solubility limit. Most aromatics and halogenated organics fall in this category. Steam stripping applications for these types of compounds can be very effective, since much of the concentration of the organic can be accomplished in a decanter, as indicated in Figure 2. In this case, the water layer is recycled to the stripping column for reprocessing.

The design of the decanter poses some interesting questions, since the water flow is generally significantly larger than the organic flow. Furthermore, in some cases (for example, benzene and toluene), the organic layer is the lighter of the two liquid phases. In applications involving halogenated organics, the organic liquid is heavier than water. Needless to say, good models to predict the phase behavior of the system in question are essential.

Different arrangements are needed when better organic recoveries are required from more dilute streams.

This can be accomplished using a separate recovery column (Figure 3a) or an integrated distillation/stripping column (Figure 3b) The selection between these two designs depends solely on the equipment sizing — the arrangement shown in Figure 3a is used when required steam flows are larger (that is, when the contaminants are less volatile).

Other variations on these flowsheets include the use of reboilers instead of direct steam injection, and operation at

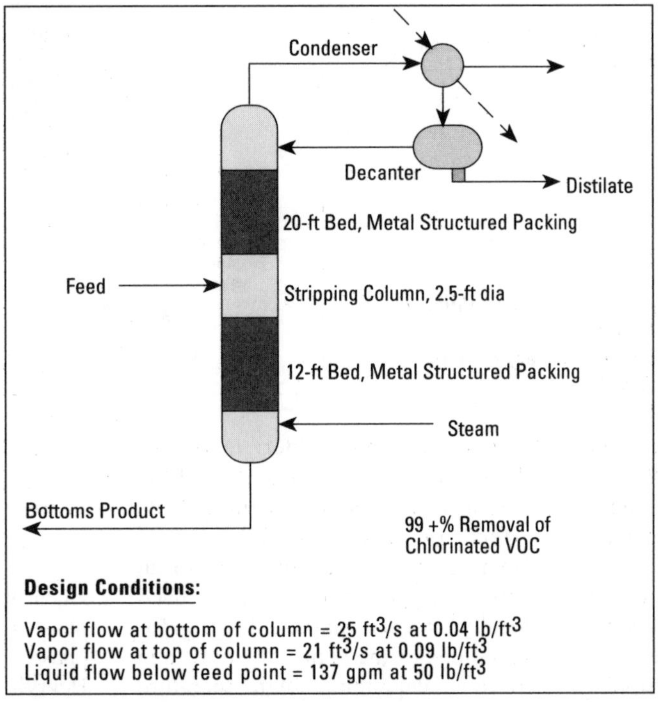

Figure 2. Steam stripper using structured packing; water (light phase) is recycled as reflux at top of stripper.

reduced pressure to lower operating temperature.

Important design considerations

Several aspects of the design of steam stripping systems are very crucial and not immediately obvious.

First, as in any distillation process, is the accuracy and reliability of equilibrium data. Steam stripping is a situation where the old reliable Henry's law (1, 2) needs to be used with care due to the broad concentration ranges,

high temperatures, extensive interactions between components, and the existence of two liquid phases.

The thermodynamic model of choice for steam stripping systems is one based on activity coefficients that can predict immiscibility. No model fits this function better than the nonrandom two liquid (NRTL) activity coefficient model (3). Pilot and laboratory tests to establish the adjustable parameters in the NRTL model for the mixture in question are advisable, but solubility and vapor pressure data can suffice as a good approximation.

Wastewaters can be very fouling, especially when the temperature is raised and inorganic salts precipitate. In typical steam stripping configurations, most of the fouling will occur in the recovery exchanger, so the system design must include provisions to allow for frequent cleaning.

In the absence of a recovery exchanger, the stripper will bear the brunt of the fouling. In such cases, the use of trays can avoid plugging even though packings would yield better performance. The use of sequestering agents is also a good solution for reliable and lengthy operation.

Materials of construction should be some grade of stainless steel or a high performance plastic due to the varied and changing nature of the water chemistry. Capital savings achieved by using lesser-quality materials of construction generally translate into severe problems and added expense later.

Startup of any steam stripper requires heating of the feed water to the operating temperature. This added heat has to be supplied in the form of steam at the bottom of the stripper. Design provisions must be made to accommodate this larger, but tempo-

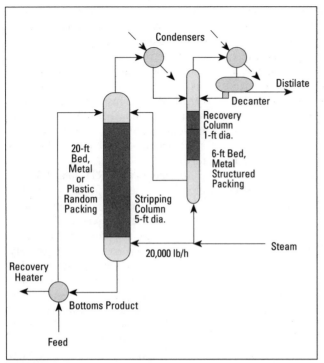

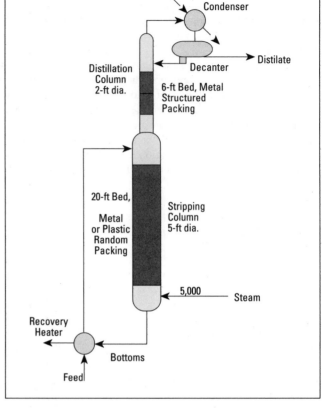

■ *Figure 3a (above). Steam stripper and recovery column for moderately volatile and immiscible systems.*
■ *Figure 3b (right). Steam stripper and recovery column combination for volatile and immiscible systems.*

rary, steam flow in the stripper. This capability is also desirable to allow for continued operation while cleaning of a fouled recovery exchanger takes place.

Design at low stripping steam rates is desirable since it reduces the downstream processing requirements. Optimum designs require stripping factors between 1.5 and 6. These stripping factors mandate more stages for separation and taller packed heights. Design under these conditions becomes very sensitive to the reliability of the equilibrium data and the mass-transfer models. And, in this regime, excellent packings and internals are necessary, and vendor experience in the design of steam stripping systems is extremely valuable.

Henry's law constant

The value of the Henry's law constant plays an important part in determining the required steam-to-water ratio in the stripper. It can also

be important in the determination of the number of transfer units and the height of a transfer unit through its effect on the stripping factor. Henry's law would not apply to an entire stripper, but it can be used locally at each stage, allowing for variations of the constant from stage to stage depending on temperature and concentration.

Reliable data on Henry's constants (H) are not easy to find, especially since these constants are dramatically affected by temperature and other solutes present in the water. In general, H increases with temperature and with the concentration of inorganic salts in the water. The effect of temperature should always be considered, whereas the effect of inorganic salts is usually neglected since this represents a conservative assumption. The effect of additional organic solutes, on the other hand, can be very important, as indicated in Table 2 for benzene. Reliable activity coefficient data and

models are indispensable in these cases.

Henry's law is usually expressed as:

$$yP = Hx \qquad (1)$$

where y is the mole fraction in the gas phase, x is the mole fraction in the liquid phase, P is the total pressure, and H is Henry's constant in pressure units (atm).

Table 2 lists some estimates of values of H for some common VOCs at ambient temperature. Note that the value of H would be much higher under steam stripping conditions. Also, these values are for ideal situations, where no interactions between solutes are present, and are for dilute conditions. These data appear to be conservative and can be used for design.

The value of H is strongly dependent on temperature because both the activity coefficient and the vapor pressure are a function of temperature. Extrapolation to different temperature requires good reliable models for vapor pressure and activity coefficients.

In more general terms, one can express equilibrium at low pressures (that is, for an ideal gas) by:

$$y P = g P^o x \qquad (2)$$

with

$$H = g P^o \qquad (3)$$

where g is the activity coefficient and P^o is the vapor pressure. This means that the temperature and concentration dependency of H can be described by g and P^o.

The NRTL model for activity coefficients was developed for partially miscible systems and appears to be the best for stripping applications. Its range of application covers air and steam stripping, and the interaction parameters for many of these VOCs are readily available (see (4)). When the actual binary parameters are not available, they can be inferred from mutual solubility data as provided in (5). As always, the Antoine equation for vapor pressure provides adequate temperature dependency.

Selecting operating pressure

The operating pressure of a steam stripper can have a large impact on efficiency and reliability. Lower pressures (vacuum stripping) mean enhanced volatility and lower operating temperatures. These two factors combined can mean in some cases less fouling by inorganic precipitation and lower heat-recovery requirements. Vacuum strippers can also be very energy efficient since the feed need not be heated to high temperatures. Furthermore, vacuum strippers offer the ability to use plastic internals at moderate temperatures, to deal with corrosive systems effectively.

On the other hand, vacuum strippers require the use of vacuum pumps or ejector systems. In pressurized stripping, the pressure of the stripper is determined by the pressure in the reflux accumulators, which will be "riding" on the pressure of the vent header. Pressure strippers operate hotter, but do not require a vacuum system.

Vacuum stripping is normally performed at around 2 psia to prevent inor-

Table 2. Typical values of Henry's constant for various VOCs at 20°C.	
Compound	**H, atm**
Carbon Tetrachloride	1,183
Chloroform	180
Methylene Chloride	125
Trichloroethylene	500
Perchloroethylene	800
1,1,1-Trichloroethane	200
1,1,2,2-Tetrachloroethane	20
Benzene	240
Benzene in the presence of 2%wt. isopropanol	15
Ethylbenzene	389
Dichlorobenzene	71
Methyl-Isobutyl Ketone	7.1
Methyl-Ethyl Ketone	1.7

ganic precipitation, whereas pressure stripping is often run at 5 to 10 psig.

The stripping factor

The stripping factor (S) is defined by:

$$S = m G_m / L_m$$
$$= (H/P) G_m / L_m \qquad (4)$$

where G_m and L_m are the molar flow rates of gas and liquid, respectively. The slope of the equilibrium line (m) is numerically equal to the Henry's law constant (H) expressed in atmospheres when the operating pressure of the stripper (P) is 1 atm.

The stripping factor is the most important design variable in a stripper. Higher stripping factors are conducive to more efficient stripping. The stripping factor for design purposes should always be greater than unity. In cases where reducing the amount of stripping medium is important, one can design for stripping factors between 3 and 6. The designer should always be aware of the reliability of the design value of H, since it has a major impact on the value of the stripping factor. This is why the use of S values below 3 is not recommended.

For steam stripping, in many cases the stripping factor can be set by the heat balance of the system. When cool water is introduced to the stripper, steam will be needed to bring it to the stripping temperature. The amount of steam required will be proportional to the degree of subcooling of the feed. In many cases, the steam required to heat the water far exceeds that required to strip the VOC. The designer should always keep in mind the steam requirements for heat balance purposes.

Trays or packing?

One of the early decisions that the engineer faces is whether to use a packed or a trayed column. This subject is explored at length in (6) and so won't be covered in detail here. Table 3 characterizes the applicability of trays and packings under a variety of process conditions.

This article will address some of the important guidelines that need to be observed when designing packed strippers. Many of the concepts are also applicable to trayed columns.

The use of high L/V ratios

The most economical stripper designs are those where the amount of stripping medium (air or steam) is minimized and the tower is designed to operate at the maximum possible liquid load. The diameter of the tower is usually dictated by the liquid flow. Stripping towers are not commonly designed using an "approach to flood."

Liquid loads in excess of 35 gpm/ft² are common and are coupled with gas F factors (gas velocity times the square root of gas density) as low as 0.5 ft/s•(lb/ft³)^0.5. This means L/V ratios as high as 50 on a mass basis.

The implications with respect to the selection of the proper predictive tools for column hydraulics and the design of liquid and gas distributors are significant.

Mass-transfer efficiency for packed towers

There are several ways to predict the mass-transfer performance of a given packing under the required operating conditions. In every case, the required height of packing will be the product of the number of transfer units (NTU) and the height of a transfer unit (HTU). The values of HTU and NTU to be used will depend on the stripping factor and the inherent efficiency of the packing. NTU is a variable that relates exclusively to the stripping factor and the degree of removal. HTU relates to the stripping factor, liquid load, and the packing efficiency.

The equation to determine the value of NTU_{ol} (number of transfer units referenced to the liquid phase) for a VOC stripper is:

$$NTU_{ol} = \left(\frac{S}{S-1}\right)$$
$$\times \ln\left[\left(1 - \frac{1}{S}\right)\frac{x_{in}}{x_{out}} + \frac{1}{S}\right] \quad (5)$$

where x_{in} and x_{out} are the inlet and outlet concentrations, respectively.

At values of $S = 12$ or above, Eq. 5 can be approximated by:

$$NTU_{ol} = \ln[x_{in}/x_{out}] \quad (6)$$

To determine HTU_{ol}, one can use a correlation or experimental data adapted to the conditions of the design and applicable to the packing being considered. Methods based on correlations can be fairly reliable if applied carefully to systems within the proven limits of the correlations. The best method currently available to the public is a modification of the Onda method that was developed by the Separations Research Program at the University of Texas at Austin (7).

The packed bed depth required to achieve a separation will then be:

$$H = HTU_{ol} \yen NTU_{ol} \quad (7)$$

Experimental data are difficult to obtain and difficult to validate, but they represent the best basis for design. It is always advisable to compare a design based on experimental data with a correlation method, as this will provide a good sense of the importance of the different variables on the design.

Typical design values of HTU_{ol} derived from performance data on random packings are shown in Table 4. The reader should understand that the selection of the proper value of HTU_{ol} for design should be done by the packing supplier, since a process guarantee is often associated with a design.

It should be noted here that the use of the height equivalent to a theoretical plate (HETP) in these types of applications is extremely tricky. Values of HETP for conventional random and structured packings can be several times those found in conventional distillation. This is because HETP is defined (as a distillation variable) in terms of the height of a transfer unit in the gas phase. HETP is a variable that is well-suited for systems where the resistance to mass transfer is in the gas phase. Stripping systems frequently exhibit the majority of the resistance to mass transfer in the liquid phase. Thus, when the values of HTU_{ol} are converted to HETP values,

one obtains HETP values in the 6- to 12-ft range. It should be noted that this effect is caused by the shift in resistance of the system and not by an inherent inefficiency in the packings.

Estimation of pressure drop is important in stripping applications, particularly in vacuum steam stripping. The bottoms pressure in the stripper has a pronounced effect on the bottoms temperature and on the volatility of the organic compound. Packed towers operate at significantly lower pressure drops than tray towers and are more desirable for low-pressure stripping applications.

Estimation of pressure drop and maximum capacity of the packed strippers is critical. The calculation methods presented in (8, 9) have proven to be the most reliable. Previously available methods described in the literature tend to fail in pressure drop and capacity predictions in steam stripping service

Table 3. Applicability of trayed and packed strippers.

Process Characteristic	Trays	Packings	Notes
Biological Fouling	A	A	Pretreatment required
Inorganic Fouling	A	A	Pretreatment required
Foaming	NA	VA	Packings recommended
Vacuum	MA	VA	Pressure drop crucial
High Removal	VA	VA	
Heat Recovery	A	VA	Pressure drop crucial
Corrosiveness	MA	VA	Plastic packings and internals
Turndown	A	A	Equal for trays or packing
Initial Cost	Lower	Higher	

Key: VA = Very Applicable; A = Applicable With Limitations; MA = Marginally Applicable But Not Recommended; NA = Not Applicable

Table 4. HTU_{ol} values for typical packings in VOC steam stripping.

Packing	S	Liquid Load, gpm/ft²	HTU_{ol}*, ft
2-in. Polypropylene Spherical Packing	6	25	2.9
	9	35	3.2
3½-in. Polypropylene Spherical Packing	6	25	3.4
	9	35	3.7
40-mm Stainless Steel Ring-Type Packing	6	25	3.2
	9	35	3.6

*HTU_{ol} values as indicated include a safety factor and can be used for design. They assume excellent liquid distribution and clean packing.

Literature Cited

1. Carroll, J. J., "What is Henry's Law?," *Chem. Eng. Progress*, **87**(9), pp. 48–52 (Sept. 1991).

2. Carroll, J. J., "Use Henry's Law for Multicomponent Mixtures," *Chem. Eng. Progress*, **88**(8), pp. 53–58 (Aug. 1992).

3. Renon, H., and J. M. Prausnitz, "Local Compositions in Thermodynamic Excess Functions for Liquid Mixtures," *AIChE J.*, **14**(1), pp. 135–144 (Jan. 1968).

4. Gmehling *et al.*, "Vapor-Liquid Equilibrium Data Collection," *Chemistry Data Series*, DECHEMA, Frankfurt (1977 to present).

5. Horvath, A. L., "Halogenated Hydrocarbons. Solubility-Miscibility with Water," Marcel Dekker, New York (1982).

6. Kister, H. Z., *et al.*, "How Do Trays and Packings Stack Up?," *Chem. Eng. Progress*, **90**(2), pp. 23–32 (Feb. 1994).

7. Bravo, J. L., and J. R. Fair, "Generalized Correlation for Mass Transfer in Packed Distillation Columns," *Ind. Eng. Chem. Proc. Des. Dev.*, 21, pp. 162–170 (1982).

8. Robbins, L. A., "Improve Pressure-Drop Prediction With a New Correlation," *Chem. Eng. Progress*, **87**(5), pp. 87–91 (May 1991).

9. Kister, H., and D. R. Gill, "Packing Capacity and Pressure Drop GPDC Interpolation Charts Atlas," Chapter 10, in "Distillation Design," McGraw-Hill, New York, pp. 585–652 (1992).

10. Kunesh, J., *et al.*, "Experimental Determination of Tray Efficiency Steam Stripping Toluene From Water in a 4-ft. Diameter Column," Presented at the Atlanta AIChE Spring National Meeting (Paper 90a), (Apr. 1994).

11. Chan, H., and J. R. Fair, "Prediction of Point Efficiencies on Sieve Trays. I. Binary Systems," *Ind. Eng. Chem. Proc. Des. Dev.*, 23, pp. 814–819 (1983).

12. Bonilla, J. A., "Don't Neglect Liquid Distributors," *Chem. Eng. Progress*, **89**(3), pp. 47–61 (Mar. 1993).

13. Bravo, J. L., "Effectively Fight Fouling of Packing," *Chem. Eng. Progress*, **89**(4), pp. 72–76 (Apr. 1993).

because the liquid-to-vapor ratio is generally significantly above the range of these older correlations.

Mass-transfer efficiency for trayed towers

When stripping operations are analyzed in terms of theoretical stage requirements, one usually finds that extremely high organic removals are possible in just a few theoretical stages. On the other hand, when one calculates tray efficiencies for these systems, tray efficiencies are often in the range of 25% to 40%. This has been corroborated by tests at Fractionation Research, Inc., on conventional sieve trays in toluene stripping service *(10)*.

As with the case of *HETP*s, tray efficiencies have been defined for application in normal distillation, where resistance is in the gas phase, unlike steam strippers, where the preponderance of the mass-transfer resistance is in the liquid phase. Low efficiencies in stripping systems are caused by the extremely large volatilities and stripping factors encountered and by the fact that efficiency is defined in terms of gas-controlled systems. Current tray efficiency correlations that take into account transfer in both phases can deal with this problem properly *(11)*.

Some pitfalls to avoid

Reliability of equilibrium data. The design of a stripper depends heavily on the value of the activity coefficient or the Henry's constant for the target VOC. The literature abounds with experimental values of Henry's constants, but unfortunately they do not always agree with other published values or with values apparent from field trials and installations.

The Henry's constant is a thermodynamic variable that depends only on temperature and composition. Many misguided efforts have tried to link the value of Henry's constant to mass-transfer performance by regressing values of H from actual stripping data. This is wrong and dangerous, since a fundamental thermodynamic value is held dependent on totally unrelated things

such as liquid distribution, packing shape and size, column levelness, gas distribution, instrument accuracy, and so on. Values of H derived in such manner should *never* be used for design, since they will prove unreliable in scale-up and will undoubtedly lead to wrong answers.

The correct procedure is to determine values of H from good experimental data on volatility and solubility and to determine column efficiency separately using the proper value of H.

Liquid and gas distribution. The performance of a stripper is in many case wrongly related only to the packing itself. In reality, the packing performs only as well as the initial liquid and gas distribution allows it to. Badly designed liquid distributors and inlet gas nozzles are the most common problems found in nonperforming strippers. Care should be taken to design and install proper distribution devices in the stripper.

Many stripper applications involve very large liquid loads, sometimes in excess of 30 gpm/ft², coupled with very low gas velocities ($F < 0.25$ ft/s•(lb/ft³)$^{0.5}$). In these cases, one finds a synergistic effect between liquid and gas distribution that can be disastrous in terms of mass-transfer performance. Uneven liquid distribution at very high loadings and liquid holdups produces large variations in void fraction across the bed. This causes severe gas maldistribution.

Care needs to be taken in the design of the liquid distributor for evenness of flow as well as geometric coverage. The use of pour point densities of 10 points/ft² in this application is normal. The real trick is in producing very low coefficients of variation for the liquid flow. The coefficient of variation is defined as the ratio of the standard deviation to the mean flow for the pour points in the distributor. A deterioration in performance at high loads can be expected if this coefficient exceeds 0.2 for a random sample of 20% of the pour points. The recommendations in *(12)* regarding liquid distribution are certainly pertinent to strippers and care should be exercised in the design and installation of liquid distributors.

Initial gas distribution is also critical, because strippers generally operate at low pressure drop levels. The best way to produce an even initial gas distribution is by the use of pressure drop. Any device that can direct the gas stream across the section with a pressure loss of about 0.5 in. of water will be adequate. Gas spargers and orifice trays designed to produce even flow at these pressure drops are recommended; these devices must be able to deal with the liquid downflow as well.

A common mistake in the design of stripping towers is that the gas inlet nozzle is positioned too close to the packed bed and the gas entrance velocity is too high. This produces an uneven gas velocity profile entering the bed. Also, the beams that support the packed bed can severely interfere with gas distribution when the gas inlet is too close to the bed. Because of these effects, a sound design will incorporate a minimum distance between the center of the inlet nozzle and the bed equal to half the column diameter.

Strippers operating at very high liquid-to-gas ratios can exhibit poor performance when liquid distribution is uneven in spite of having proper pour point density. This unique sensitivity to uneven flow out of the pour points can be attributed to the fact that strippers operate very close to the loading region by virtue of the very high liquid loads. Small changes in local liquid rates caused by uneven distribution can produce significant changes in local liquid holdup and in the effective void fraction of the irrigated packing. This results in extremely large variations in local gas flows to maintain pressure drop equilibrium with the ensuing deterioration of mass-transfer performance. In essence, uneven liquid distribution can cause severe gas maldistribution.

Misuse of safety factors in design. Many performance specifications for strippers include healthy safety factors in the inlet and outlet concentration requirements. Typically, the effluent concentration is set at the detection limit of the VOC in question and the inlet concentra-

tion is an absolute maximum that will very rarely (if ever) present itself.

Unfortunately, some mass-transfer device suppliers take advantage of this fact when presenting a design in a competitive situation. Their designs will be based on removals that are below the specified ones with the "hope" that the specified levels will never present themselves and the performance of the stripper will never be challenged.

Under-designs such as these give the false and dangerous impression that a packing is far better in performance than others. In reality, these suppliers are cutting corners at the expense of performance reliability of the stripper and are providing designs that do not meet the specified removal but meet the effluent characteristic only.

The user must be aware of this practice and protect against it by strongly requiring and verifying that all calculations and designs be based on meeting the specified outlet concentration given the specified inlet concentration.

Fouling and plugging of packings. Paradoxically, the high mass-transfer efficiency provided by the packing in a stripper promotes the deposition of insoluble metal oxides and salts and bacterial growth. Packings with high surface areas will be more efficient but will promote fouling as well.

There is no magic cure for fouling. The composition of the water, the irrigation and vapor rates, and the operating temperature have much more to do with how rapidly a tower will foul than the type of packing used.

If the contaminated water contains free iron or other minerals, the action of the stripping gas could cause some of these compounds to precipitate out and foul the packing media. Organic contaminants promote biological growth that accentuates the fouling problem. The reality is that all VOC strippers will eventually lose some of their efficiency and capacity due to fouling if the water is not pretreated before entering the tower.

The degree of fouling as well as the amount of time for the fouling to affect the performance of a stripper is a func-

tion of all of the above factors plus other unique characteristics of a particular site. It must also be noted that in many cases the fouling process is so slow that a contaminated site is essentially cleaned before fouling is a problem.

The best answer to the problem of fouling is a combination of good design and pretreatment. Pretreatment involves the continuous addition of a chemical to the water to keep the minerals from precipitating and to prevent algae buildup during the stripping process.

Also important to keeping fouling to a minimum are good maintenance practices, good monitoring of process conditions, and good overall process design. A fouling problem will not be resolved by trying a different packing unless important compromises are made in mass-transfer efficiency. Severely fouled packed beds are inefficient and cause high pressure drop. They can also be very dangerous, since support plates are generally not designed to handle the weight of packing heavily laden with inorganic salts. In some extremes, the weight of the packed bed can increase by a factor of ten or more as the packing fouls.

Ref. *(13)* describes methods and techniques to effectively deal with the problem of fouling in packed towers, with emphasis on air and steam strippers.

J. L. BRAVO is vice president for engineering at Jaeger Products, Inc., Houston, TX (713/449-9500; Fax: 713/449-9400). Formerly, he was program manager of the Separations Research Program at the Univ. of Texas at Austin, where he received an MS in chemical engineering. He has extensive experience in the design and troubleshooting of mass-transfer equipment and has published more than 30 articles in the areas of distillation, absorption, extraction, and stripping. In addition, he is the senior author of a book on separations technology, and an instructor for AIChE's Distillation in Practice course. He is a member of AIChE, and is a registered professional engineer in Texas.

Remove Organics by Activated Carbon Adsorption

Adsorption with granular activated carbon is a proven technology for process water purification. Here's how to evaluate and implement the technology.

Mark H. Stenzel,
Calgon Carbon Corp.

Activated carbon has a wide variety of applications in the chemical process industries (CPI), one of which is water and wastewater treatment. Activated carbon is especially effective at removing soluble organic compounds (SOCs) from both surface and groundwater sources. In fact, the 1986 Amendments to the Safe Drinking Water Act named adsorption with granular activated carbon as the yardstick by which other treatment technologies are to be evaluated for SOC control *(1)*.

The CPI have used activated carbon for diverse process water treatment applications, such as dechlorination of disinfected water supplies prior to process use, removal of trace organic contaminants for protection of resin beds, treatment of steam condensate for recycle, and removal of organic contaminants from surface or groundwater sources prior to plant use. Carbon adsorption also can be included in a wastewater treatment scheme for water reuse, where it can be employed as a point source treatment or as a final polish operation.

This article focuses on the use of granular activated carbon to remove specific compounds from water sources. It provides a methodology to evaluate adsorption, establish a process design, develop a system design, and estimate preliminary capital and operating costs.

■ *This two-stage carbon adsorption system treats up to 500 gal/min of groundwater, removing dissolved chlorinated solvents so the water can be used for potable and process applications.*

Principles of activated carbon adsorption

The removal of contaminants from the process stream occurs primarily by physical adsorption of the contaminants onto the surface of the carbon. This adsorption is due to naturally occurring attractive forces between the molecules on the carbon surface and those in solution. Physical adsorption is further enhanced by the lack of affinity of the contaminants for the solution.

Some organic contaminants are more adsorbable than others. Organic solvents, including chlorinated organic solvents such as trichloroethylene and aromatic solvents such as toluene, are adsorbable due to their low solubility in water. Higher molecular weight compounds, such as polynuclear aromatics, higher-molecular-weight amines, and surfactants, are also effectively adsorbed. Conversely, water-soluble compounds such as alcohols and aldehydes are poorly adsorbed. [Instead, oxidation technologies (2) can be used to destroy smaller water-soluble contaminants and ultrafiltration (3) can be used for effective removal of larger compounds.]

The unique nature of activated carbon greatly enhances the adsorption process. Activated carbon can be described as an amorphous form of graphite, with a random structure of graphite plates. The structure is highly porous, with a range of cracks and crevices reaching molecular dimensions. The larger openings (which make up about 25% of each particle's volume) function as transport pores through which the contaminants diffuse to the adsorption sites or pores (about 40% of the particle's volume). Adsorption takes place at the smallest gaps or openings, which are usually only a few molecules wide.

Activated carbon is used in both granular and powder forms. The powder form is usually added to a process stream and then filtered out downstream. The powder form can also be added to activated sludge processes to provide sites for bioactivity. The granular form is most commonly used for water treatment, where it is typically employed in a downflow fixed bed for flexibility.

Activated carbon is produced from materials with high carbon content, such as coal, peat, wood, or coconut shells. The coal-based raw material is the most common type of activated carbon. It is generally produced by grinding the coal, adding a binder, recompacting, and granulating, and then effecting thermal decomposition in the presence of steam to result in the carbon skeletal type particle.

The raw material may be an important factor in selecting an activated carbon for a given process. Lignite, subbituminous, and bituminous coals are used in activated carbon manufacture. Bituminous-coal-based activated carbon has the greatest hardness, abrasion resistance, and higher bed density, and lignite-based carbon has the least (4). These properties are important if the activated carbon is to be reactivated or if the fixed bed will require backwashing to remove filtered solids. The lignite-based material exhibits greater attrition. This results in lower yields during reactivation and loss of granules during backwashing, which may shorten the expected system life.

For industrial water treatment applications, the most common adsorption option is the downflow fixed-bed pressure system.

Bituminous-based carbon has the highest density and therefore has a larger number of small pores for adsorption per volume and a greater attraction for the contaminants. For removal of trace amounts of contaminants such as volatile organic solvents, the bituminous product provides the greatest capacity and attractive force. In addition, the more dense the particle of the starting carbon is, the more amenable it will be to reactivation and reuse. Thus, the bituminous product also benefits waste minimization either by higher capacity for contaminants or enhanced capability for reuse.

In addition to the physical adsorption of organic compounds, activated carbon removes free chlorine from water through a surface reaction. The free chlorine reacts with the carbon to form surface oxides:

$$C* + HOCl \rightarrow CO* + H^+ + Cl^-$$

where C* and CO* represent activated carbon and the surface oxide. This reaction occurs very quickly and effectively on all carbon surfaces.

Adsorption isotherm evaluation

In addition to knowing which compounds are adsorbable, it is often useful to conduct a series of tests to determine the applicability of adsorption for the specific water. The liquid-phase adsorption isotherm test is the basic preliminary evaluation tool for adsorption, and results of isotherm testing are widely available as reference information.

The isotherm is a batch test in which a fixed quantity of water sample is contacted with varying amounts of carbon. The carbon is pulverized to reduce the length of testing time required to reach equilibrium. For waters with trace amounts of contaminants, it is critical that equilibrium between the soluble and adsorbed condition be approached, as small errors would be amplified in the results. [Reference (5) provides further information on adsorption isotherm testing.]

Results of an adsorption isotherm are usually expressed in terms of the carbon's capacity for a given adsorbate at a specified equilibrium concentration. The Freundlich equation is used to express this relationship:

$$x/m = kC^{1/n}$$

where C is the concentration of unadsorbed compound left in solution or in equilibrium with the carbon and x/m is the amount of the compound adsorbed per unit weight of the carbon. The isotherm plot will approach a straight line on a logarithmic plot or a series of straight lines in the case of mixtures of differently adsorbing contaminants. On such a plot, k is a constant related to the adsorbability of the contaminant (capacity in mg/g at a concentration of 1 mg/L), and $1/n$ is the slope, or sensitivity to concentration, which reflects the ease or difficulty of removing that compound from solution. Some typical single component isotherm plots are shown in Figure 1, and Table 1 shows the Fruendlich constants for some typical organic compounds.

The isotherm or the Freundlich equation provides a good estimate of the capacity of carbon for the contaminant at equilibrium. Table 1 shows that water-soluble alcohols (butanol and phenol) and organic acids are poorly adsorbed and that the substituted aromatics are more easily adsorbable (as indicated by the intercept values) and that the latter remain adsorbable even at low concentrations (as shown by the effect of slope).

In summary, an adsorption isotherm will provide the following useful information:

• adsorbability of the compound(s) to determine whether further evaluation is warranted;

• equilibrium capacity to provide a basis for preliminary estimate of carbon usage; and

• changes in adsorptive capacity relative to contaminant concentration, which shows the effect of the adsorption process if contaminant concentrations are expected to change.

Dynamic adsorption testing

If isotherm data establish that carbon adsorption is a viable treatment process, additional information not readily available from the isotherm test should be obtained. Dynamic adsorption testing will provide the necessary design data, including the contact time required and the type of "break-

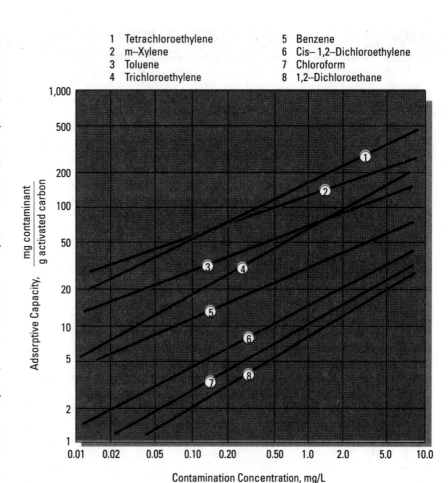

1	Tetrachloroethylene	5	Benzene
2	m–Xylene	6	Cis– 1,2–Dichloroethylene
3	Toluene	7	Chloroform
4	Trichloroethylene	8	1,2–Dichloroethane

■ *Figure 1. Single-component activated carbon adsorption isotherms.*

through curve." The breakthrough curve is the contaminant concentration exiting the bed over time as the carbon's capacity for the contaminant becomes fully used (that is, the concentration at which the contami-

nant "breaks through" the carbon and is no longer adsorbed).

The scaled column test is the most common and direct means to obtain the operating design information. The column test is set up with columns connected in series. Each column is filled with sufficient granular activated carbon to provide a specific contact time, typically 7.5–15 min per column. The hydraulic loading rate also can be close to the expected design, but it is not as critical as the contact time.

The column test should be run with all columns in operation for the best evaluation of contact time. The test can also be run as a staging study — two columns are on-line, and when the contaminant breaks through

Table 1. Typical Fruendlich isotherm constants.

	Intercept (k), mg/g	Slope ($1/n$)
Aniline	25	0.322
Benzene Sulfonic Acid	7	0.169
Benzoic Acid	7	0.237
Butanol	4.4	0.445
Butyric Acid	3.1	0.533
Chlorobenzene	40	0.406
Nitrobenzene	82	0.237
Phenol	24	0.271
Trinitrotoluene	270	0.111
Toluene	30	0.729

Table 2. Accelerated column test studies on actual water samples.

Case	Simulated Conditions Loading, (gal/min)/ft²*	Contact Time, min	Adsorbates Contaminant†	Concentration, mg/L	Breakthrough GAC Load, mg/g	Usage, lb/1,000 gal	Effluent, mg/L	Saturation Usage, lb/1,000 gal
1	3.8	18	Trichloroethylene	3.3	61	0.45	0.0001	
2	2.0	35	Benzene	0.024	0.5	0.4	0.002	
			Toluene	0.116	2.5	0.4	0.002	
			Xylene	0.630	13	0.4	0.001	
3	3.5	20	Methylene Chloride	0.007	0.014	5.8	0.001	5.2
			Chloroform	0.268	3.8	0.8	0.004	0.7
			Cis-1,2-Dichloroethylene	0.118	2.6	0.5	0.001	0.46
			Carbon Tetrachloride	39.8				
4	3.8	20	1,2-Dichloroethane	0.095	0.13	5.9	0.003	4.0
			Chloroform	0.223	0.37	5.0	0.010	4.0
			Carbon Tetrachloride	3.08	6.4	4.0	0.010	
5	1.3	100	Benzene	0.034	3.1	3.0	0.003	2.0
			Toluene	4.6	79.0	3.0	0.125	2.1
			Xylene	4.1				
6	1.3	44	Total Organic Carbon (TOC)	150			55	
			Benzene	10.2	7.7	11.0	0.002	
			Toluene	8.2	9.7	7.0	0.002	
7	2.6	22	TOC	15			15	
			Phenol	44	147	2.5	0.04	1.1
			Benzene	0.44	4.6	0.8	0.33	0.74

* 1,470 (gal/min)/ft² = m/s

† Contaminants are listed in order of breakthrough

the second column, the first is taken off-line and a third column is added after the remaining column as a new second stage. The scaled and full column studies each provide good data on carbon usage and breakthrough characteristics.

In evaluating water with trace amounts of contamination, however, the scaled column test will take a very long time, since it approximates actual operation. There may also be a problem with containing volatile contaminants in solution for the volumes and time required. To speed up the dynamic test, several researchers have developed minicolumn techniques, in which the carbon is reduced in size and the feed is pro-

vided under high pressure. Mathematical modeling based upon the particle diameter allows the results to be scaled up to a single breakthrough curve at a given contact time. This accelerated column test provides viable results in a matter of days instead of months (6, 7).

Table 2 shows the results of some minicolumn tests conducted on actual water samples (typical of those found at some industrial sites) containing trace concentration levels of contaminants (8):

• Cases 1 and 2 are low-usage situations where a single-stage adsorption system might suffice.

• Case 3 involves the presence of a weakly adsorbed contaminant,

methylene chloride, which requires a very high carbon usage rate. The system designer may want to evaluate other treatments to manage the methylene chloride and determine whether it will continue at these influent levels.

• Cases 4 and 5 show the benefit of using a two-stage system, allowing the first stage to approach saturation and obtain full carbon utilization.

• Cases 6 and 7 illustrate the effect of other organic compounds that may not need to be controlled by the treatment process. The adsorbability of these other compounds may influence the adsorbability of the specific compounds of interest.

In these examples, both the adsorbability of the compound to be removed and its concentration will determine when the compounds will exit the carbon bed if present with other compounds. Usually, specific testing will be required to evaluate waters with background organic compounds.

Adsorption system design

Adsorber design. After the contact time has been established and the evaluation of the breakthrough curve and usage rate has indicated whether a single-bed or staged system is preferred, the designer can select the type of adsorption system that best fits the application.

Adsorption systems generally are designed around three basic types of adsorbers. The gravity flow adsorber is often used for applications that have low contact times and high flow rates, such as municipal surface water treatment. The upflow pulse bed adsorber contains the breakthrough curve within a single adsorber, and small amounts of spent carbon are removed from the bottom with equal amounts of regenerated carbon added to the top to maintain the process. The upflow pulse bed is used for processes with long transfer zones and low flow rates, such as product purification for sugars or corn sweeteners.

For industrial water treatment applications, the most common adsorption option is the downflow fixed-bed pressure system, as shown in Figure 2. The water is introduced above the carbon bed, which is contained in a pressure vessel, usually without a distributor because the packed bed will serve to distribute the flow. The water flows down through the bed. Treated water is collected at the bottom of the bed by an underdrain consisting of slotted screen collectors and a header or collection system. Wedge-wire type slotted screens will retain the carbon

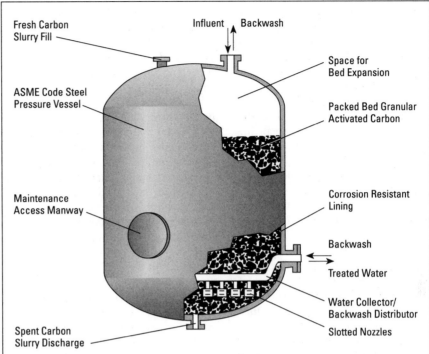

■ *Figure 2. Downflow fixed-bed adsorber.*

and allow water to pass out of the adsorber. The underdrain also serves as the backwash distribution, if necessary. Carbon is spent relatively evenly across the vertical cross-section due to the nature of flow through a packed bed.

The adsorbers are normally carbon steel pressure vessels that meet the American Society of Mechanical Engineers (ASME) pressure vessel codes. The interior of the adsorber has a corrosion resistant lining, because carbon in the presence of water will create a corrosion cell due to an oxidation-reduction differential. Also, due to the abrasive nature of carbon, the lining is typically a nominal 35 mils thick or greater. The lining may be an epoxy, fluoropolymer, or vinyl ester type coating, or rubber sheeting, depending on the liquid characteristics.

An advantage of the pressure adsorber system is that the water does not need to be repumped after treatment. For instance, many groundwater treatment systems will be placed between the well pump

and any downstream process or distribution system, and operate along with the well in an on-off mode.

The surface loading rates in a downflow adsorber can be as high as 10 gal/min per ft^2, depending on the vessel size and contact time. The pressure drop is dependent on the piping system and the type and depth of the granular activated column.

System components. A common downflow pressure adsorption system design will include two adsorbers in series operation. When the carbon in the first adsorber is spent, it is replaced with fresh carbon; what was the second stage is then moved to the first stage, with the fresher carbon in the final position. This design optimizes the use of carbon, as the second stage will continue to remove contaminants while the carbon in the first stage is more fully utilized.

The piping network for such an adsorption system allows either bed to be the first stage, and it allows isolation of either bed for carbon exchange or backwash operation. The granular activated carbon also can be moved in and out of the adsorber as a water slurry under water or air pressure. This slurry fill-and-discharge design requires a carbon discharge line at the bottom and a carbon fill line at the top.

In finalizing the system design, the engineer should estimate the optimum volume of carbon in the adsorber for commercially available fill-and-exchange services. The standard sized adsorber normally contains 20,000 lb of carbon (on a dry weight basis). This size adsorber has become an accepted design because spent carbon will contain water and adsorbate within the internal pore structure that cannot be drained, the weight of which is equal to that of the dry carbon. The total weight of spent carbon in one adsorber is thus approximately 40,000 lb, which approaches the limits for road transport. Smaller volumes may also have logical cutoffs, such as 10,000 or 2,000 lb, to accommodate other transport containers (*e.g.,* tote bins).

System costs. Complete adsorption systems consisting of two downflow fixed-bed adsorbers with all piping necessary for two-stage operation, backwash, and carbon exchange, are available from adsorption system suppliers. These systems generally will have lined ASME-code pressure adsorbers, corrosion resistant carbon transfer piping, and steel process piping.

Typical uninstalled capital costs for such systems are shown in Table 3. The two smaller systems can be provided as skid-mounted, preassembled units. The larger system would require additional field assembly.

Treatment trains. At times, the carbon adsorption system may be the sole treatment. These situations may include dechlorination of an incoming water source or treatment of relatively clean groundwater for trace organic contaminant removal. In such cases, the pressure adsorber concept is especially beneficial because no additional steps, such as collection, equalization, or repumping, are nec-

essary. The system designer only needs to assure that the water supply has sufficient pressure to accommodate the adsorption process, including the piping, the carbon bed, and downstream requirements.

Many times carbon adsorption is one unit operation in an overall treatment scheme. If the water being treated contains suspended solids, or if upstream unit operations include precipitation or settling steps for removal of metals or biological solids, a separate filtration step is needed. While granular activated carbon is an excellent filtration medium, the pressure drop may increase faster than in a design for in-depth filtration, such as a multimedia-type filter. Also, filter operations are usually designed for lower backwash requirements. If a carbon unit is operating without a prefilter, inclusion of backwash capability in the design is recommended.

Another unit operation that may be used as pretreatment to carbon adsorption is air stripping for removal of volatile solvents. In such cases, air stripping may remove the bulk of such contaminants prior to carbon adsorption. This leads to extended life of the carbon for removal of nonvolatile compounds or the remainder of the volatile compounds.

For complete removal of contaminants from the environment, carbon adsorption is often used to treat the off-gas from the air strippers. The change in treating vapor rather than liquid is often a cost-effective design, as carbon has higher capacities for organic compounds in the vapor phase than in the liquid phase *(9, 10)*.

Adsorption system operation

The operation of the downflow, fixed-bed adsorption system is relatively simple. The system can run unattended and can handle a wide variety of flow rates. As adsorption is taking place in the packed bed, the system also has great flexibility to accommodate changes in contaminant levels without risking the effluent objectives.

Normal monitoring includes periodic pressure drop readings and water sample analysis. In critical situations, the pressure drop can be alarmed and/or backwash operations automated.

The frequency and location of sampling depends upon the expected carbon bed life. Samples of both influent and effluent should be analyzed. In two-stage operation, interstage samples provide information on the characteristics of the breakthrough curve or indicate when the first stage has been fully utilized.

Other operations include backwash and carbon exchange. For backwash, uncontaminated water is introduced through the underdrain at a rate of about 12 gal/min per ft^2 to effect a 30% bed expansion. If a system will be backwashed, the bed should be backwashed upon installation. This initial backwash will classify the bed, so the bed will maintain the same position or classification after future backwashing.

In the carbon exchange procedure, the adsorber with spent carbon is isolated and pressurized, and the carbon exits the bottom of the adsorber as a carbon-liquid slurry. This slurry is

Table 3. Typical uninstalled capital costs for carbon adsorption systems with two adsorbers per system.

Carbon Capacity per Adsorber, lb	Typical Maximum Series Flow Rate, gal/min	Capital Cost (Including Carbon)
2,000	60	$50,000
10,000	250	$110,000
20,000	500	$165,000

directed to a receiving container or transport trailer, where the carbon is drained of all liquid except that portion held within the pores. This draining minimizes weight for shipment. The adsorber is refilled with fresh carbon in the same manner and placed back on-line.

Backwash or carbon exchange are normally infrequent batch operations. Accordingly, automation of these procedures is usually not necessary.

Adsorption systems have few moving parts, so maintenance costs are low. These costs can be estimated to be between 5% and 10% of the installed capital, depending on the number of carbon exchanges, the backwash frequency, and the water characteristics.

Carbon exchange

Carbon exchange, the removal of the spent carbon and resupply with fresh carbon, is often the most critical parameter of the adsorption system operation. Carbon exchange may be the predominant component of the operating cost, as well as the key to safe disposal of the organic contaminants removed from the process. A key advantage to the carbon adsorption process is that not only does it remove the contaminants from the water, but it also concentrates the contaminants and holds them within the carbon granule.

Spent carbon can be managed in a number of ways. It can be landfilled or incinerated as a solid waste. These two options often include liability and high cost concerns.

Coal-based granular activated carbon, however, is usually reactivated. The reactivation process drives off the more volatile organic compounds from the carbon surface of the equilibrium process and pyrolyzes and decomposes less volatile compounds held within the pores. Reactivation has proven to be a safe and cost-effective means to destroy adsorbed organic compounds and allow reuse of the carbon.

Thermal reactivation produces an activated carbon that is different from the original material because some of the smaller adsorption pores may be lost. The capacity for trace-level contaminants may be lessened in a reactivated product. If a process requires virgin-grade activated carbon, then the spent carbon should be replaced with virgin material and the used carbon reactivated for other uses. If the process can cost-effectively use a reactivated grade product, then a reactivated carbon should be used, so long as care is taken to ensure the proper quality of the reactivated product.

Thermal reactivation involves heating the spent carbon up to 1,800°F in a controlled atmosphere in a multiple-hearth furnace or rotary kiln. The process is equipped with off-gas treatment and spent and reactivated product handling. The handling and reactivation of granular activated carbon will cause product loss, which may be 10%–20% of the bituminous-coal-based carbon, and higher with softer lignite-based carbons.

In most cases an on-site thermal reactivation facility is not cost-effective until more than 10,000 lb/d of carbon are required. It is unlikely that treatment of process water will have usages this high, as the water will probably have low contaminant levels in order to make it a candidate for reuse.

A number of off-site reactivation suppliers have systems to take

M. H. STENZEL is manager, equipment product management, with Calgon Carbon Corp., Pittsburgh, PA (412/787-6809; Fax: 412/787-6676), where he is responsible for specification and marketing of the adsorption equipment product line. He has held a variety of technical positions during his 20 years with Calgon Carbon, including project engineering, process engineering, and marketing, and he has co-authored numerous articles on carbon adsorption and groundwater treatment. He received a BS in chemical engineering from Clarkson Univ. in 1969 and an MBA from Duquesne Univ. in 1989. He is a member of AIChE.

advantage of economy of scale and provide reactivation services and resupply of the proper grade of carbon. Typical costs are $0.80–$1.00/lb for liquid-phase virgin-grade carbon and $0.60–$0.80/lb of reactivated carbon. These costs include the exchange of the spent carbon but do not include freight to the off-site reactivation facility or delivery of the fresh carbon.

In situ regeneration processes, such as steam or hot gas treatment, have limited effectiveness in regenerating liquid-phase activated carbons. Such regeneration will restore some capacity to retain solvents, but it will not remove all of the adsorbed compounds from the adsorption pores. By leaving contaminants in the more effective adsorption sites, these processes reduce the ability of the carbon to remove contaminants to low levels. This effect is even more pronounced if nonvolatile compounds are also adsorbed in the pores. The only effective means of regenerating liquid-phase carbons, allowing essentially complete removal of contaminants in the pores, is through thermal reactivation.

System evaluation example

In-plant process water contaminated from a petrochemical operation is one CPI situation where carbon adsorption can be especially useful. This may be a wastewater or wash water that is to be reclaimed for plant use. This example illustrates how a preliminary evaluation might be conducted in such a situation.

The contamination is determined to be about 1 ppm benzene, toluene, and xylene (BTX) compounds, with a benzene concentration of 40 ppb. The water source delivers 400 gal/min to the plant for process water use. Case 2 in Table 2 can serve as a reference. It indicates that a BTX concentration of a little less than 1 ppm, with benzene at 24 ppb, had a usage rate of 0.4 lb carbon per 1,000 gal and a contact time of 35 min.

Because the example involves a benzene level 67% greater than that of the reference, the isotherm in Figure 2 is evaluated for concentration effects. The isotherms normally are not used for direct evaluation when a mixture of contaminants is present due to competitive effects that are not readily evaluated by the single-component isotherms. The isotherm can be useful, however, for evaluating changes in capacity with concentration, which in this case shows about a 25% greater capacity requirement as the concentration of benzene goes from 24 ppb to 40 ppb. The usage rate is influenced by two factors — the increase in the benzene concentration and the 25% increase in capacity for benzene (1.67/1.25 = 1.336). Thus, the usage rate can be expected to increase by 33.6%, to 0.53 lb/1,000 gal.

Finally, a two-stage system would be expected to further lower the usage rate, since it would allow saturation capacity of the carbon in the first stage. Two-stage systems improve usage by about a third, which in this case would lower the usage rate to approximately 0.35 lb carbon per 1,000 gal of water treated. The annual usage rate is estimated at 73,600 lb of carbon [calculated as 400 gal/min × 1,440 min/d × 365 d/yr × 0.35 lb/1,000 gal].

Based on these data, two 20,000-lb (carbon capacity) adsorbers operated in series would be a cost-effective treatment option. This design will provide 26 min total contact time, or 13 min per stage, which is adequate for this contaminant level. The two-stage system design will help to optimize and obtain the projected usage rate of 73,600 lb/yr and ensure effluent quality. Carbon exchange can be scheduled after breakthrough has been detected in the first stage. To allow for downtime and low usage periods, the annual carbon usage is estimated at 60,000 lb for estimating the operating costs.

The estimated installed capital cost of the adsorption system is $225,000. This includes $165,000 for the adsorption system and initial fill of carbon, $15,000 for installation and tie-ins, $20,000 for site work and foundation, and the remainder for site-specific work such as pump upgrade or backwash needs. For winterization, the system can be housed indoors or the exposed piping can be traced and insulated.

The annual operating cost elements are as follows:

• capital amortization (10 yr project life, 8% interest rate), $ 33,500;
• maintenance (8% of capital), $20,000;
• additional energy costs (7.0 bhp at $0.08/kWh), $5,000;
• carbon exchange (virgin-grade carbon at $0.90/lb), $54,000; and
• freight ($3,000 per exchange, 3 exchanges required), $9,000.

The total annual operating costs, therefore, are $121,500. This is approximately $0.46/1,000 gal treated. If the contaminant levels drop through process improvements or other design changes, the reduction in carbon usage rates can be estimated using the concentration/isotherm capacity relationship. **CEP**

Literature Cited

1. **Thompson, J. C.,** "The Safe Drinking Water Act," *Public Works,* pp. 123–125, 184–188 (Sept. 1986).
2. **Legan, R. W.,** "Ultraviolet Light Takes on CPI Role," *Chem. Eng.,* **89**(2), pp. 95–100 (Jan. 25, 1982).
3. **Bemberis, I. and K. Neely,** "Ultrafiltration as a Competitive Unit Process," *Chem. Eng. Progress,* **82**(11), pp. 29–35 (Nov. 1986).
4. **Deithorn, R. T., and A. F. Mazzoni,** "Activated Carbon; What It Is, How It Works," *Water Technology,* **9**(8), pp. 26–29 (Nov. 1986).
5. **Stenzel, M. H., and J. L. Fisher,** "Granular Activated Carbon for VOC Removal From Drinking Water Supplies," Proceedings of the American Water Works Association Conference, Kansas City, MO (June 1987).
6. **Rosene, M. R.,** *et al.,* "High-Pressure Technique for Rapid Screening of Activated Carbon for Use in Potable Water," Chapter 15 in "Activated Carbon Adsorption of Organics from the Aqueous Phase, Vol. 1," I. H. Suffet and M. J. McGuire, eds., Ann Arbor Science Publishers, Inc., Ann Arbor, MI (1980).
7. **Crittenden, J. C.,** *et al.,* "Predicting GAC Performance with Rapid Small-Scale Column Tests," *JAWWA,* **83**(1), pp. 77–87 (Jan. 1991).
8. **Stenzel, M. H., and W. J. Merz,** "Use of Carbon Adsorption Processes in Groundwater Treatment," *Env. Progress,* **8**(4), pp. 257–264 (Nov. 1989).
9. **Stenzel, M. H., and U. Sen Gupta,** "Treatment of Contaminated Groundwaters with Granular Activated Carbon and Air Stripping," *JAPCA,* **35**(12), pp. 1304–1309 (Dec. 1985).
10. **Adams, J. Q., and R. M. Clark,** "Evaluating the Costs of Packed-Tower Aeration and GAC for Controlling Selected Organics," *JAWWA,* **83**(1), pp. 49–57 (Jan. 1991).

Further Reading

Cheremisinoff, P. N., and F. Ellerbusch, eds., "Carbon Adsorption Handbook," Ann Arbor Science Publishers, Inc., Ann Arbor, MI (1978).

Susuki, M., "Adsorption Engineering," Chemical Engineering Monograph 25, Elsevier Science Publishing, New York, NY (1990).

Nyer, E. K., "Groundwater Treatment Technology," 2nd ed., Van Nostrand Reinhold, New York, NY (1992).

"Design and Use of Granular Activated Carbon — Practical Aspects," Proceedings of conference sponsored by American Water Works Association Research Foundation and U.S. Environmental Protection Agency, Cincinnati, OH, available from AWWA Research Foundation, Denver, CO (May 9–10, 1989).

Consider Fenton's Chemistry for Wastewater Treatment

Even though the chemistry was discovered a century ago, this technology is now being looked at for a variety of CPI wastewater-treatment applications.

Richard J. Bigda,
Technotreat Corp.

The chemical process industries (CPI) must treat wastewaters containing a wide variety of contaminants, ranging from toxic organics like phenol, benzene, other aromatics, formaldehyde, and amines, to inorganics such as sulfite, sulfide, mercaptans, and cyanide, to heavy metals such as hexavalent chrome. These wastewaters also have a wide range of concentrations and combinations of contaminants. The streams must be treated as inexpensively as possible and in a safe manner, preferably by processes that are easy to operate on-site and that require a minimum of labor and technical know-how. And, of course, the ultimate goal of this treatment is that the treated water meet all federal, state, and local discharge regulations.

One available wastewater treatment technology that few engineers seem to be familiar with is the Fenton reactor. In this advanced oxidation process, toxic wastewater is reacted with inexpensive ferrous sulfate catalyst and hydrogen peroxide in a simple, nonpressurized (typically batch) reactor to yield (if reacted to completion) carbon dioxide and water.

This article offers guidance on the use of this process by first explaining the mechanisms of Fenton's chemistry and then outlining how to apply it to industrial wastewater treatment.

The chemistry

One hundred years ago, H. J. H. Fenton discovered that by using a soluble iron catalyst and hydrogen peroxide, many organic molecules could be easily oxidized. No high pressures, no high temperatures, and no complicated equipment were necessary.

More recent research has demonstrated that the oxidation mechanism Fenton discovered was due to the reactive hydroxyl radical generated in an acidic solution by the catalytic decomposition of hydrogen peroxide. In the presence of ferrous iron, the peroxide is split into $OH^- + OH^\bullet$. Organic substrates are subject to free radical attack by the hydroxyl radical.

The completion of the oxidation is dependent on the ratio of hydrogen peroxide to organic, while the rate of oxidation is determined by the initial iron concentration and temperature. The amount of iron needed is low. However, the reaction is highly exothermic, so to prevent excessive heating, the peroxide must be added slowly.

The key to the process is the reactivity of the hydroxyl radical. It is twice as reactive as chlorine, with an oxidation potential between that of atomic oxygen and fluorine (Table 1).

Many chemicals can be oxidized by the Fenton reaction, as shown in Table 2.

When a complex mixture is being treated, the more highly reactive components, benzene or phenol for example, are oxidized more rapidly than less-reactive ones, such as chlorinated hydrocarbons. However, given sufficient time, the proper temperature, and the right H_2O_2 concentration, complete destruction can be achieved.

Meeting discharge regulations may not require total oxidation to carbon dioxide and water. Rather, partial oxidation to render the wastewater less toxic may be sufficient. For example, phenol concentrations can be reduced from 10,000 mg/L to less than 1 mg/L fairly quickly. (The reduction occurs in just a few minutes, but because of heat evolution, it is safer to add the peroxide slowly over a period of several hours.)

Limitations, of course, do exist. Some chemicals are not oxidized by hydrogen peroxide with an iron catalyst (Table 3). For example, organic acids such as acetic, maleic, and fumaric acids do not break down under the usual conditions (but formic, propionic, lactic, and other acids are oxidized easily).

In some cases, a seemingly nonreactive compound can be made to react by altering the pH or catalyst concentration. Many potential reactions have not yet been explored, and certainly treatability studies should be conducted to determine the optimum conditions. In addition, there may be situations where unwanted compounds are created, further illustrating the need for bench tests and good analyses before and after treatment.

For example, an operator was surprised to see acetone on the analysis report because no acetone was present in the initial wastewater, while isopropanol, which was in the wastewater, was not detected in the final analysis. What happened was that the alcohol was oxidized to acetone and the reaction stopped at that point. However, other ketones (dihydroxyacetone and methyl ethyl ketone) were destroyed by the OH·.

The reactor

Scale-up. If a bench test on the wastewater is successful, the procedure can usually be translated directly to a larger industrial-size unit with few problems.

One exception to this direct scale-up, though, is the reagent addition rates. In most bench-scale tests, all the reagents are usually added simultane-

Table 1. Oxidation power of common reactants relative to chlorine.

Fluorine	2.23
Hydroxyl Radical	2.06
Atomic Oxygen (Singlet)	1.78
Hydrogen Peroxide	1.31
Perhydroxyl Radical	1.25
Permanganate	1.24
Hypobromous Acid	1.17
Chlorine Dioxide	1.15
Hypochlorous Acid	1.10
Hypoiodous Acid	1.07
Chlorine	1.00
Bromine	0.80
Iodine	0.54

Table 2. Compounds that can be oxidized by the Fenton reaction.

Acids:	Formic	**Aromatics:**	Hydroquinone
	Gluconic	(con't)	para-Nitrophenol
	Lactic		Phenol
	Malic		Toluene
	Propionic		Trichlorophenol
	Tartaric		Xylene
			Trinitrotoluene
Alcohols:	Benzyl		
	tert-Butyl	**Amines:**	Aniline
	Ethanol		Cyclic Amines
	Ethylene Glycol		Diethylamine
	Glycerol		Dimethylformamide
	Isopropanol		Ethylenediaminetetraacetic
	Methanol		Acid (EDTA)
	Propenediol		Propanediamine
			n-Propylamine
Aldehydes:	Acetaldehyde		Explosives (RDX, or cyclonite)
	Benzaldehyde		
	Formaldehyde	**Dyes**	Anthraquinone
	Glyoxal		Diazo
	Isobutyraldehyde		Monoazo
	Trichloroacetaldehyde		
		Ethers:	Tetrahydrofuran
Aromatics:	Benzene		
	Chlorobenzene	**Ketones:**	Dihydroxyacetone
	Chlorophenol		Methyl Ethyl Ketone
	Creosote		
	Dichlorophenol		

Table 3. Chemicals that the Fenton reaction does not oxidize.

• Acetic Acid	• Methylene Chloride
• Acetone	• Oxalic Acid
• Carbon Tetrachloride	• n-Paraffins
• Chloroform	• Tetrachlorethane
• Maleic Acid	• Trichlorethane
• Malonic Acid	

ously. However, this is not feasible in a commerical system.

Scale-up of temperature control is also not straightforward. Lab trials may allow a rapid temperature increase, which speeds the reaction. But temperatures of large batches will rise more slowly, affecting reaction time and completion.

From bench tests, one can determine approximately how much catalyst and hydrogen peroxide are required and the optimum pH and oxidation-reduction potential (ORP) endpoint. For many chemicals, the literature indicates that the ideal pH for the Fenton reaction is between 3.0 and 4.0, and that the optimum catalyst-to-peroxide ratio (w/w) is 1:5. These pH conditions seem to be suitable for the full-scale reactor. However, usually much less iron is needed.

The initial concentration of oxidizable chemicals and the target final concentration will indicate the time and reagent quantities required to treat a batch. Experience acquired with different batch compositions will translate into more-efficient process control and give the operator confidence. However, always err on the side of caution — it is better to spend a bit more time treating a batch than to have an overheated reaction that can result in a spill over.

The process in the vessel. A batch Fenton reactor essentially consists of a nonpressurized stirred reactor with metering pumps for acid, base, a ferrous sulfate catalyst solution, and industrial-strength (35–50%) hydrogen peroxide. It is recommended that the reactor vessel interior be coated with an acid-resistant material because the Fenton reagent is very aggressive and corrosion can be a serious problem. Sensors to control the reaction include only pH, ORP, and temperature indicators. Ideally, and to reduce labor requirements, more-elaborate controls and safety features can be incorporated into the reactor system.

The vessel is first filled with the contaminated wastewater. The second step is to adjust the pH with dilute sulfuric acid. It is important to adjust the

Common CPI Applications for Fenton's Chemistry

Process Waters from the Manufacturing or Processing of:

- Chemicals
- Pharmaceuticals
- Insecticides
- Dyes and Inks
- Photochemicals
- Explosives (TNT, RDX)

Wastes from Petroleum Refineries and Fuel Terminals:

- Tank Bottoms
- Dilute BTX Streams
- Phenol Streams

Wastes from Specialty Chemical Applications:

- Plastics and Adhesives Manufacturing (Phenol, Formaldehyde)
- Wood Treating (Creosote, Copper)
- Paint Stripping (Phenol, Formic Acid)

Hazardous Wastewater Treatment

Groundwater and Soil Remediation

pH before the addition of the ferrous sulfate, because iron hydroxide forms at a pH of about 6, and the acid prevents the precipitation of the catalyst.

The third step is the addition of part of the catalyst dissolved in an acidic solution. As catalyst is added, the pH may shift, so this should be constantly monitored and adjusted as needed by adding either sodium hydroxide or acid. After some catalyst is mixed into the solution, hydrogen peroxide is slowly introduced, along with continuous catalyst supplements.

The reaction may be slow to start, particularly if the temperature is below 65°F. This is a critical period, since the initiation of the process may be sudden, especially if the peroxide concentration is high (for instance, due to continued injection of the reagent). This is why the peroxide *must* be added with patience while waiting for the temperature to rise above 80°F.

Automating the reactor. A commercial reactor can be automated to include several very useful features. By

using a computer or programmable logic controller (PLC), the treatment steps and parameters can be sequenced and controlled automatically. An instrumented unit is easily monitored and parameter readings can be recorded for review and to aid in regulatory monitoring and reporting. The only labor intensive operations involve reagent stocking and filtration.

The reactor can be programmed to shut off the peroxide and catalyst pumps if no significant change in the ORP has occurred in a specified time (for example, 20 minutes, depending upon the addition rate). An alarm can be built in to alert the operator that this step is taking too long.

The slow simultaneous addition of peroxide and catalyst allows the reaction to proceed slowly and the temperature and ORP reading to gradually increase. Peroxide addition can be further controlled by responding to a high ORP signal (about 350–400 mV). When the selected setting is attained, the pumps are shut off and time is allowed for the reaction to go to completion. If the ORP drops during this period, indicating peroxide consumption, reagent addition should be re-initiated. This continues until the ORP reaches equilibrium and does not call for more reagent within a certain period of time. At that point, the automated ORP controls will have ensured that the oxidation step is complete and (after confirmation by an analysis of the toxics) neutralization can take place. The reaction will be quenched by high pH, making it important to analyze the mixture to determine if the toxics are sufficiently low.

After the oxidation is complete, the pH is adjusted, usually to between 6 and 9, to precipitate iron hydroxide. After the oxidation step, other, more-conventional, treatment can be performed in the same batch reactor. For example, hexavalent chrome can be reduced, oils and solvents removed, and heavy metals precipitated. Adsorbent-like clays and activated carbon can also be added to remove trace amounts of other unwanted substances.

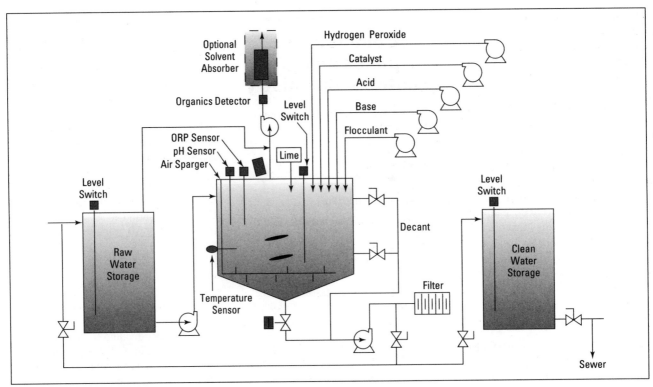

■ *Figure 1. Flow sheet for a commercial Fenton reactor.*

A commercial Fenton reactor, including the ancillary equipment necessary to carry out a complete treatment, is represented by the flow sheet in Figure 1. Included is a level sensor that performs several functions. To prevent overflow, it controls the feed pump and chemical pumps. If the level in the reactor drops, it shuts off the mixer. Finally, the level sensor also prevents accidental discharge, which could result if the operator leaves the bottom valve open when filling.

An exhaust vent prevents fumes from entering the work area, and if necessary, scrubbers could be attached to the vent. In most cases, at least the iron hydroxide must be removed by precipitation and filtration. Lime could be used for final neutralization and precipitation of metals. If the solution to be treated is very dilute, perhaps this step can be eliminated.

The peroxide and catalyst pumps are controlled by the ORP sensor/controller. ORP measurement gives a good indication of the completeness of the oxidation and prevents excessive per-

oxide use. For safety purposes, failure of the catalyst pump will shut off the peroxide pump. Both the acid and base addition are controlled by the pH sensor/controller.

In this particular system, volatile organic compounds (VOCs) can be removed by air sparging, the duration of which is controlled by an organics detector in the vent. The sparge is stopped when the exiting organic vapors reach a specified concentration level.

After neutralization, it may be desirable to add coagulation and flocculation agents. An adjustable pump can be used to inject a specified volume of solution to the batch. The final step is filtration for removal of heavy metals, dirt, and iron hydroxide from the water. The use of an air-operated diaphram pump is recommended to feed a filter press. Other types of filters could be used, depending on the volume and loading of the solid waste.

Preventing runaway conditions. The worst thing that can go wrong with a Fentor reactor is a runaway reaction, which can only occur if an ex-

cess of peroxide is present with a large amount of oxidizable material. Conditions of this sort can occur if the initial temperature is too low, if the pH is outside the optimum range, or if there is insufficient catalyst.

This situation can be prevented by the slow, careful addition of peroxide and catalyst, especially at low temperatures. (When the temperature rises above 130°F, more time should be allowed for heat to dissipate. Although this is not a critical temperature for the oxidation, it could present a danger to personnel.) Constantly monitor the pH and keep it within the optimum range for promotion of the reaction. Coupling the catalyst pump with the peroxide pump ensures proportionate levels of injection. If the catalyst flow is interrupted for any reason, the peroxide pump should be immediately stopped to prevent the buildup of excess peroxide.

Operating and capital costs

Capital costs for a Fenton reactor can vary from low for a basic no-frills unit, to moderate for a PC-controlled system with automatic process sequencing and control. A basic 1,000-gal unit with four simple on/off metering pumps activated by pH and ORP sensors, a manually controlled electric stirrer, and discharge pump would cost about $40,000. A computer-controlled system programmed to sequence chemical steps, control chemical addition, and record events could cost up to $100,000 for the same capacity.

Reactors with more sophisticated instrumentation obviously cost more than those with less instrumentation. Increasing capacity, though, has less impact on cost. For example, doubling the size of an instrumented reactor adds only 25% to the system price, whereas doubling a basic, noninstrumented unit may increase the price by 50–70%.

Wastewater compositions are very diverse and changeable, so each Fenton reactor needs some custom design work to provide the user with future flexibility and alternatives. Thus, the wastewater generator and equipment engineer should discuss optional features.

Operating costs depend primarily on reagent consumption. Labor and power costs are nearly constant for most batches — although a more-concentrated solution may require more time, the same amount of labor is expended using the automated system.

Figure 2 illustrates the operating costs of a Fenton reactor for reducing

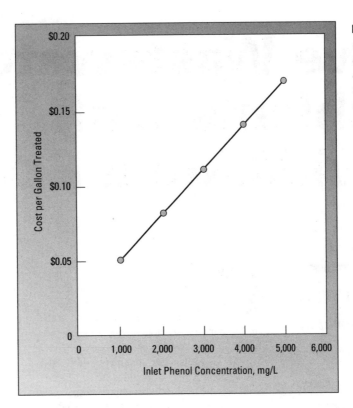

Figure 2. *Operating costs of treatment as a function of inlet phenol concentration.*

Further Reading

Potter, F. J., *et al.,* "Oxidation Kinetics of Aromatics in Washwater with Hydrogen Peroxide," presented at the AIChE Spring National Meeting, Houston, TX (Apr. 1989).

Sedlak, D. L., and A. W. Andren, "Oxidation of Chlorobenzene with Fenton's Reagent," *Environmental Science Technology,* 25(4), pp 777–782 (1991).

Walling, C., "Fenton's Reagent Revisited," *Accounts of Chemical Research,* 8(5), pp.125–131 (May 1977).

the concentration of a solution containing 1,000–5,000 mg/L of phenol to a residual of less than 1 mg/L. (This does not include capital cost, but does include the cost of chemicals, power, and labor.) For a 1,000 gal. reactor, power consumption is about 11 kW and labor is estimated at one hour per batch at $25/h.

A look ahead

Each year sees heightened interest in using hydrogen peroxide for environmental applications. Research indicates success in such new areas as remedial treatment of pesticide- and chemical-soaked soils, destruction of explosives such as TNT and RDX (cyclonite) in the wastewater from explosives manufacturing, and removing polyaromatic and polychlorinated hydrocarbons from sludges. Data are becoming available on the oxidation kinetics of many chemicals, which is helpful in the design of Fenton reactors. Much of the work is being done at Vanderbilt Univ., Washington State Univ., Virginia Polytechnic Institute, and by hydrogen peroxide producers,

including Solvay Interox Corp., FMC Corp., and Du Pont. The U.S. Environmental Protection Agency has been encouraging application research and development on this process through its Superfund Innovative Technology Evaluation (SITE) program.

New R&D work and increasing commercial applications for Dr. Fenton's discovery are reviving a chemistry to help solve the wastewater problems of the 21st century. Though he was a century ahead of his time, his technology is now proving to be a treatment of choice. **CEP**

R. J. BIGDA is president of both Technotreat Corp. and the consulting firm Richard J. Bigda and Associates, Tulsa, OK (918/445-0996; Fax: 918-445-0994). Technotreat was established to custom-build specialized industrial wastewater treatment systems for installation in the U.S. and overseas. He holds a BS from Wayne State Univ. and an MS from the Univ. of Michigan, both in chemical engineering, and is a registered Professional Engineer. He has authored many technical papers and reports, and has lectured at various engineering conferences. He is a member of AIChE and the American Chemical Society.

Solve Wastewater Problems with Liquid/Liquid Extraction

LLE is well-suited to the removal of trace organics from wastewater. This overview offers guidelines on process design and solvent selection.

Roger W. Cusack,
Glitsch Process Systems, Inc.

Chemical process industries (CPI) plants today are faced with more-stringent wastewater discharge regulations than ever before, with the probability of more restrictions on the horizon. Plant designers and operators involved in wastewater treatment often encounter particular difficulties when faced with organic compound contamination.

Liquid/liquid extraction (LLE) is a powerful separation technique that is finding wider application in the CPI to solve difficult environmental problems, particularly in the removal of trace organic compounds from wastewater streams. LLE is usually only applied when more conventional techniques such as steam stripping or distillation are not suitable. This is because LLE usually involves the introduction of a new component (the solvent) to the process, and this has obvious environmental implications.

Liquid/liquid extraction has been applied only sparingly in the past, because engineers are unfamiliar with its possibilities and how to employ this powerful technology. This article will review the advantages and limitations of LLE in general, and outline practical guidelines for wastewater applications. Particular emphasis will be given to the considerations that go into the selection of the solvent for the process.

When to use LLE

The most common method of treating wastewaters to reduce the level of organic contaminants is steam stripping, particularly when the contaminant's boiling point is lower than the boiling point of water. However, there are sometimes cases where stripping is not feasible. It is in these more difficult cases where LLE is often a viable alternative.

Table 1 lists some typical organic contaminants commonly found in wastewater. Those normally handled by steam stripping are found at the top of the table, and those better handled by LLE are at the bottom of the table.

Boiling point. The first key parameter to consider when deciding between steam stripping and liquid/liquid extraction is boiling point. If the boiling point of the contaminant is significantly below that of water, it is usually handled by stripping.

An example of a compound that can be controlled by steam stripping alone is benzene, which has been identified as a carcinogen and must be reduced to ppm or ppb levels in plant effluent waters. Since benzene boils at 80.1°C, and also forms a low-boiling-point azeotrope that boils at 69.4°C, it can be effectively reduced by steam stripping. The benzene-rich stream produced as an overhead from the stripper can be sent to incineration for final disposal, or recovered and recycled.

However, a boiling point below that of water is not the only consideration in choosing between stripping and extraction. Some organic compounds behave nonideally in aqueous solutions and form minimum boiling azeotropes with water. This allows them to be separated from water even though they boil higher than water.

For example, toluene boils at 110.8°C. But it also forms an azeotrope of 20.2% water and 79.8% toluene that boils at 85°C, allowing steam stripping to be used to remove toluene from wastewater.

Table 1. Some organic contaminants are best removed from wastewater by stripping, others by extraction.

STRIPPING

Organic Compound	Boiling Point, °C	Solubility, %	Azeotrope Boiling Point, °C	Azeotrope Water Concentration, %	Typical Reduction Levels
Methylene Chloride	40.0	2.0	38.1	1.5	< 50 ppb
Acetone	56.2	∞	—	—	< 50 ppb
Methanol	64.5	∞	—	—	< 50 ppb
Benzene	80.1	0.18	69.4	8.9	< 50 ppb
Toluene	110.8	0.05	85.0	20.2	< 50 ppb

EXTRACTION

Organic Compound	Boiling Point, °C	Solubility, %	Azeotrope Boiling Point, °C	Azeotrope Water Concentration, %	Typical Reduction Levels
Formaldehyde	−19.0	∞	—	—	< 800 ppm
Formic Acid	100.8	∞	107.1	22.5	<1,200 ppm
Acetic Acid	118.0	∞	—	—	< 800 ppm
Pyridine	115.5	57	92.6	43.0	< 10 ppm
Aniline	181.4	3.6	99.0	80.8	< 10 ppm
Phenol	181.4	8.2	99.5	90.8	< 10 ppm
Nitrobenzene	210.9	0.04 (approx.)	98.6	88.0	< 10 ppm
2,4-Dinitrotoluene	300.0	0.03	99–100 (est.)	90+ (est.)	< 10 ppm

LLE for high-boiling organics. Now consider the case of phenol. It, too, boils at a higher temperature than water (181.4°C) and forms an azeotrope that boils lower than water (90.8% water, 9.2% phenol, B.P. = 99.5°C). However, in this case, separation by stripping is not feasible because of the narrow difference between the boiling points of the azeotrope and water — just 0.5°C — and because the azeotrope takes so much water overhead with it.

For these reasons, phenol is almost always removed from wastewater by other techniques, usually liquid/liquid extraction, biological treatment, or adsorption onto carbon. Very often LLE is used upstream and in conjunction with these other processes to remove the bulk of the phenol. LLE typically reduce phenol levels to less than 10 ppm, and in pilot tests has reached levels of less than 20 ppb.

LLE for low-concentration acids. Also shown in the lower section of Table 1 is acetic acid, which has long

been recovered by LLE. It is present in the wastewater discharge of many chemical processes, particularly those which involve an oxidation reaction. The problem is that it is usually present at relatively low concentrations (0.5–3.0%).

Acetic acid boils at 118°C and does not form any azeotrope with water. Therefore, it cannot be stripped out. Since it is not toxic, very often it is treated merely by sending it to a biological treatment plant. However, with ever-tightening requirements for reduction of chemical oxygen demand (COD) in plant outflows, it has recently become economical for companies to remove and recover the acetic acid (plus formic acid, which is often a co-contaminant) before sending the stream to biological treatment.

Liquid/liquid extraction can be economically used to recover acids from even these low-concentration streams.

LLE for hydrogen-bonded compounds. A very curious component in the bottom section of the Table 1 is

formaldehyde. Like acetic acid, formaldehyde often appears in the wastewater from chemical operations as a byproduct of oxidation reactions. With its boiling point (–19°C) so much lower than that of water, it would appear to be easily stripped out of water.

However, formaldehyde actually is extremely difficult to remove from water once it has been dissolved. This is due to the hydrogen bonding phenomenon, which bonds the formaldehyde molecules very tightly to the water molecules by means of electrostatic charges.

LLE can be used to extract formaldehyde from wastewater down to levels of less than 1,000 ppm.

LLE for zero-volatility compounds. LLE has been used for many years for the recovery of primary metals, such as copper, nickel, and zinc. Thus, it is not surprising that it is finding applications today for the removal of metals or their organometallic derivatives from the discharges of process plants. These metals or metal

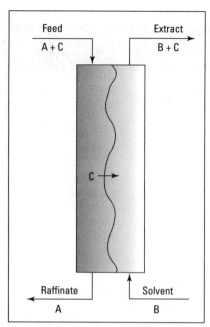

■ *Figure 1. Basic extraction process.*

compounds are usually the result of contamination from the catalysts used in the process. Since LLE works on the principle of chemical structure rather than volatility difference, it is ideally suited for solving these difficult removal problems.

Strengths of LLE

Some of the particular strengths of liquid/liquid extraction over competing technologies, such as carbon adsorption, wet oxidation, incineration, membranes, and biological treatment, are:

• *Ruggedness.* The liquid/liquid extraction column can handle the wide concentration ranges, chemical aggressiveness, and fouling characteristics typically encountered in CPI wastewater applications. In addition to the particular organic compounds that must be removed, these wastewater streams often contain insoluble oils, tars, and solids, which LLE columns can handle without expensive prefiltering or shutdowns for periodic cleaning.

• *Flexibility.* Liquid/liquid extraction can handle the wide range of flow rates encountered while still maintaining its efficiency. Typically, a well-designed LLE column can have a turndown ratio of 3:1 to 4:1 (that is, it can

be run effectively at 25–33% of its full design capacity) without a loss in extraction efficiency. This is important because wastewater stream flow rates can vary widely depending on plant operating conditions.

• *Recycle capability.* Unlike other processes that destroy the contaminant, LLE can recover the material for recycle. For example, in the manufacture of polycarbonate resins, water streams are produced that can contain up to 5% phenol. LLE can be used to recover this phenol and recycle it back to the reactors rather than destroying it. Likewise, LLE wastewater processes can be designed to recover acetic or formic acid pure enough for either reuse or resale on the open market.

• *Low energy requirements.* With the proper solvent, LLE is extremely energy efficient compared to other technologies (especially compared to wet oxidation processes using peroxide or ozone). This is because LLE is se-

lective and removes only certain components, whereas the oxidation processes destroy essentially all of the waste stream's COD, and this can increase the operating cost significantly.

LLE limitations

Of course, liquid/liquid extraction also has its limitations.

• *May introduce additional solvents.* LLE usually involves the introduction of a new solvent into the process plant. It is important to consider all possible solvents already in use at the plant for possible use in an LLE wastewater project. Be certain the company selling LLE technology has the capability to thoroughly test a wide range of solvents, including ones already present in the plant and processes before automatically specifying a new, and additional, solvent.

• *Selective, so may not reduce total COD.* Because LLE is selective and only removes certain components, it may not

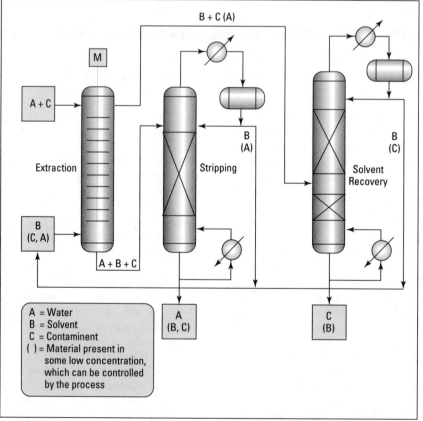

■ *Figure 2. Typical extraction process for removing contaminant C from water (A).*

reduce the total COD of the stream to the required level. Therefore, LLE is seldom used as a stand-alone treatment. Instead, it is often used in conjunction with other technologies to arrive at a total solution. The advantage of this approach is that the load on the final treatment plant is reduced significantly by the pretreatment of some streams.

Process fundamentals

Liquid/liquid extraction (Figure 1) is a mass-transfer process in which a feed material is brought into contact with a specially selected solvent, and this solvent removes (*i.e.,* extracts) some particular chemical compound(s) from the feed stream. The material being extracted is known as the *solute,* the solute-depleted stream leaving the LLE column is called the *raffinate,* and the solute-rich stream leaving the extractor is called the *extract.* The subscripts *F, S, E,* and *R* are used to refer to the feed, solvent, extract, and raffinate streams, respectively.

A typical extraction system is illustrated in Figure 2. For example, assume that water (A) is contaminated with phenol (C). By bringing this water stream into contact with a suitable solvent (B), it is possible to extract the phenol from the water phase into the solvent phase. The phenol is usually recovered from the solvent phase by distillation.

The important thing is how the streams are brought into contact. One of the liquids must be dispersed into the other in order to produce droplets and, therefore, interfacial area. This is necessary so that mass transfer can take place. The mass transfer can be either from the dispersed phase to the continuous phase, or vice versa.

A very useful chemical engineering tool for looking at this mass-transfer operation is the familiar McCabe-Thiele diagram (Figure 3). A complete discussion of this can be found in *(1).*

The Kremser equation. Actually, though, it is usually not necessary to make McCabe-Thiele plots to evaluate LLE. As long as the straight-line assumption holds (that is, mutual solubility of the feed and solute is negligible

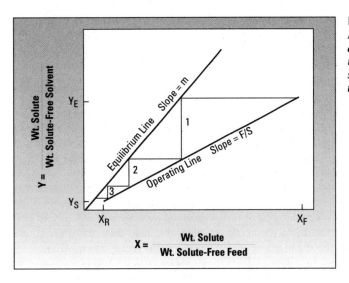

■ *Figure 3. A McCabe-Thiele diagram can be used to understand the extraction operation.*

[as is usually true in wastewater treatment applications]), the relationships among the variables can be represented by the Kremser equation:

$$N = \frac{\log\left[\left(\dfrac{X_F - \dfrac{Y_S}{m}}{X_N - \dfrac{Y_S}{m}}\right)\left(1 - \dfrac{1}{E}\right) + \dfrac{1}{E}\right]}{\log E}$$

where N is the number of theoretical stages, X_F is the concentration of solute in the feed, Y_S is the concentration of solute in the solvent, m is the distribution coefficient, which is equal to Y/X, X_N is the concentration of solute in the raffinate, and E is the extraction factor, which in turn is equal to $m(S/F)$ where S and F are the solvent and feed rates, respectively. All rates and concentrations must be expressed on a "solute-free" basis.

The equation has been solved graphically, as shown in Figure 4. To use this plot, first determine the degree of solute removal required as defined by the fraction unextracted, X_N/X_F. Once the distribution coefficient for the solvent is known, the solvent/feed ratio is arbitrarily varied to give various extraction factors. For each value of E, find on the abscissa the number of theoretical stages required. In this way, one can conduct preliminary screening of possible solvents, operating conditions, and economics before proceeding to pilot tests and equipment design.

As a general guideline, an LLE process should be run with a minimum extraction factor of 1.3. Below this value, the number of theoretical stages required become excessive and somewhat impractical. Most environmental extraction applications require anywhere from six to ten theoretical stages.

To illustrate the use of the Kremser equation, let us consider again the removal of phenol from wastewater. Several possible solvents for this operation are listed in Table 2.

Suppose that the stream contains 1,000 ppm phenol, which must be reduced to 1 ppm. Therefore, X_N/X_F is approximately 0.001. Referring to Figure 4, we see that at $E = 1.3$, more than 20 ideal stages are needed, at $E = 1.5$, 15 ideal stages are needed, and at $E = 2.0$, 9.2 ideal stages are needed.

To keep the stage requirements within a reasonable range, let us assume operation at an E of about 2.0. Thus, for each 100 lb/h of wastewater passing through the system, the following solvent rate (in lb/h per 100 lb/h of feed) would be required:

Toluene	101.5
Benzene	86.9
Isopropyl ether	6.9
n-Butyl acetate	2.8
Methyl isobutyl ketone	2.5

These results point strongly toward methyl isobutyl ketone (MIBK) as the preferred solvent. And, indeed, MIBK has been used extensively in phenol LLE applications.

Solvent selection

Because the solvent has such a pronounced impact on the overall process feasibility and economics, solvent selection is probably the most important aspect of any LLE process design.

The impact of solvent selection is evident when examining the overall flow sheet for a typical LLE process. There are three operations involved: (1) extraction, (2) water stripping to remove the solvent that dissolves in the water, and (3) distillation to recover the solvent for recycle. In a typical LLE plant, the extractor itself represents only about 10–15% of the capital investment and 5% of the operating cost. The balance of the investment and operating cost is associated with the water stripper and solvent recovery columns. Hence, from an economic standpoint, the importance of solvent selection is clear.

The "ideal" solvent. An ideal LLE solvent has several characteristics. Let us discuss these characteristics with respect to phenol removal, for which there are several commercially used solvents, as shown in Table 2.

• *High distribution coefficient.* A high distribution coefficient will minimize the solvent circulation rate.

Table 2 shows that the distribution coefficient for MIBK is much higher than that for toluene. As a result, phenol extraction with MIBK can usually be accomplished at a solvent/feed (*S/F*) ratio of 1:15, whereas toluene, due to its poorer distribution coefficient, requires an *S/F* of 2:1. Thus, the net effect on solvent circulation rate is a factor of 30. Since all of this solvent must be distilled for recycle back to the process, it is easy to see that MIBK is a much more energy efficient solvent.

• *Low solubility in water.* This is important because a solvent that is soluble in the water to any extent will have to be removed prior to discharge of the water. Unfortunately, it is a general (but not universal) characteristic of most solvents that the better or more-

efficient the solvent, the higher its mutual solubility with water.

This can be seen in the case of MIBK vs. toluene for phenol extraction discussed above. Although MIBK is much more efficient (by a factor of 30) when looking at the extract side, it is less efficient when looking at the raffinate side. MIBK is much more soluble in water than toluene (2.7% vs. 0.05%) and, therefore, must be stripped from the water, whereas the toluene raffinate could probably be discharged directly without further stripping. Despite this, the overall economics still greatly favor MIBK.

• *Low toxicity.* It is obvious that any solvent selected must have low toxicity, since some of it will necessarily remain in the discharge water, even if only at ppm or ppb levels. Also, the control of solvent vapors must also be considered in the process design.

For example, methylene chloride is a very powerful solvent for organics, but its low boiling point (40°C) indicates the potential for vapor emissions. Also, chlorinated organics are under increasing regulation from environmental authorities and are, therefore, seldom considered.

• *Low solubility of water in the solvent.* This is desirable because any water that dissolves in the solvent must probably be removed, usually by distillation. This aspect is particularly important in the recovery of compounds like acetic acid, which usually has to be recovered in glacial form (less than 0.5% water). The more water carried along with the extract stream, the more energy that must be expended to remove this water and reach the glacial point.

• *Chemical stability and safety.* These characteristics are important because they impact equipment design.

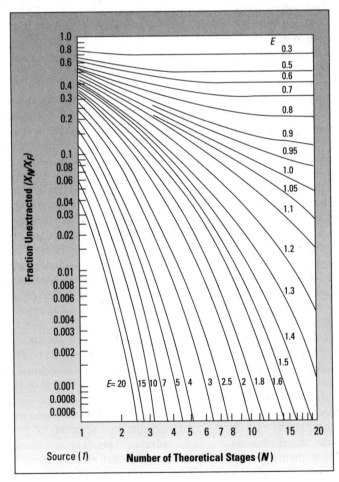

■ *Figure 4. Graphic solution of the Kremser Equation.*

Table 2. Possible solvents for phenol extraction.

Solvent	Boiling Point, °C	Distribution Coefficient	Solubility in Water, wt.%
Toluene	110.8	1.97	0.05
Benzene	80.3	2.30	0.178
Isopropyl Ether	68.5	29.0	0.90
n-Butyl Acetate	126.2	71.0	1.20
Methyl Isobutyl Ketone (MIBK)	116.6	80.0	2.70

For example, diisopropyl ether is a good solvent for phenol extraction. But it also has the tendency to form peroxides, which are explosive hazards, and for this reason would not normally be chosen for this process. Likewise, isopropyl acetate is also a possible solvent, but it has a tendency to undergo a hydrolysis reaction, producing isopropyl alcohol and acetic acid. In this case, not only is there a loss of solvent, but byproducts are produced that must be removed from the system.

• *Ease of recovery for recycle.* There are very few processes where the solvent is used on a once-through basis, and the ease of recovery for recycle should be considered. The most common method of recovery is distillation, and since this works on the principle of boiling point difference, it is important that there be a reasonable difference between the boiling points of the solvent and the solute being recovered.

• *Compatibility with the process.* The recovered material that is recycled to the process will often contain some traces of solvent. Thus, it is important to ensure that the solvent is compatible with the process, and to determine what, if any, levels of this new material are acceptable in the recycle stream.

In today's environment, there is always reluctance to introduce any new raw materials into the plant, since this usually involves obtaining additional permits. So, it is often advisable to try to find a possible solvent from among materials that are already being used in the process or somewhere else in the plant. It may not be the "best" solvent

from a purely LLE standpoint, but it may be the "best" from an environmental standpoint. It is usually better to accept some inefficiency in the extraction operation rather than risk the possibility of creating new process or environmental problems.

Process sequence

Once a possible solvent (or solvents) has been identified, the next step is to evaluate possible variations in process sequence and perform some preliminary process and economic calculations before proceeding to pilot-plant testing. Figures 5-8 illustrate several process sequencing options for a plant having three processes (Processes X, Y, and Z), each of which produces a water effluent with a particular contaminant (contaminants A, B, and C).

Typically these effluents would be combined into a single plant effluent stream for treatment, as shown in Figure 5. The first treatment is usually a stream stripper to remove the volatile components (in this case A and B). But if component C is something that does not strip out (*e.g.*, phenol), it must be handled by other means (such as LLE as shown here). The outlet from the LLE is then sent to final treatment (usually biological) before discharge from the plant.

The main disadvantage of this sequence is that the entire water stream passes through the LLE step. Since the solvent flow is proportional to the feed flow, this increases the solvent circulation rate and, therefore, the investment and operating costs.

A better arrangement is depicted in Figure 6, where the stream undergoing extraction is segregated from the other effluent streams. This reduces the size of the equipment and the energy requirements for the LLE step. This is often referred to as moving the LLE upstream in the process to treat the contamination at its source.

Both of these schemes (Figures 5 and 6) use a solvent external to the process. But as mentioned earlier, it is beneficial to try to use as a solvent a material that is already present somewhere in the process or plant to minimize problems with compatibility or permitting of new materials. This concept is illustrated in Figure 7, where a solvent used elsewhere in Process Z performs the extraction.

For example, assume again that contaminant C is phenol. As noted previously, the best solvent to extract phenol is probably MIBK. If MIBK is not currently used on-site but toluene is, toluene might be a better choice, even though it is significantly less efficient. In this case, it is the best solution from an overall plant-operation viewpoint.

Current plant design philosophy regarding wastewater treatment emphasizes water recycling. Because of the increasing costs associated with discharging wastewater, the practice of recycling water is becoming more and more common. Water is finally being recognized as one of the most valuable raw materials for the process, and therefore must be conserved to reduce costs.

This approach is shown in Figure 8. Here, effluent from each individual

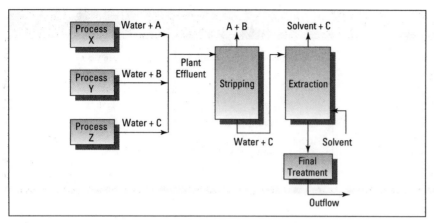

■ *Figure 5. In the past, process effluents have been combined into a single stream for treatment.*

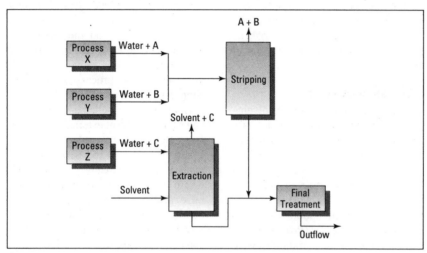

■ *Figure 6. A better approach is to segregate the stream undergoing extraction from the other stream.*

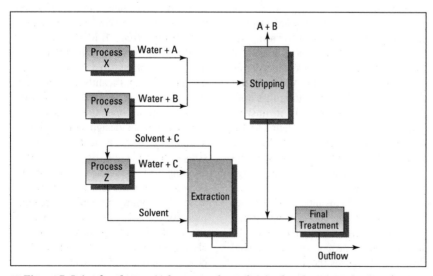

■ *Figure 7. It is often better to choose a solvent that is already present in the process or plant than to introduce an additional material.*

process is treated at its source and each process recycles its water, with a small purge to control quality. This not only saves on the cost of water discharge, but also reduces the size and cost of the final treatment step.

Pilot testing and scale-up

Unlike distillation processes, which can usually be designed from simulations alone, LLE is more complex and almost always requires some type of pilot-plant test to generate the necessary data for process design. This is especially true in environmental applications because the streams being processed can often vary widely in composition and contain trace quantities of other materials that affect the hydraulic capacity or efficiency of the process. Any pilot testing should *only* be done with actual plant materials, as synthetic blends will not reveal these problems. Also, the pilot tests should involve not only the LLE step, but also the stripping and solvent recovery steps, since all three must work together.

Extraction. There are many types of LLE devices available to accomplish the LLE step, including mixer-settlers, packed columns, sieve tray columns, agitated columns, and centrifugal units. A comparison of all of these is beyond the scope of this article, but the reader can find more information on this in *(2–3)*.

The application often dictates that the agitated liquid/liquid extractor be used for pilot testing. If the system requires a large number of theoretical stages, or requires fine droplets for mass transfer (and, thus, intense mixing), then a rotating impeller column is the proper choice. It has a high efficiency per unit of column height compared to other designs.

However, if the system tends to form stable emulsions upon mixing, a reciprocating plate extractor would be better. It mixes fluids by reciprocating plates rather than rotating impellers, producing somewhat larger droplet sizes but at the same time minimizing the possibility of emulsion formation. Other possible columns are the rotating

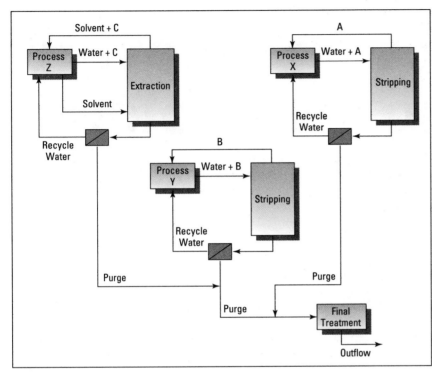

■ Figure 8. The best arrangement is to treat each process effluent at its source and maximize water recycling.

disc contactor and the pulsed extractor, but these are less often applied to environmental applications. If very large flow rates are involved and only a few theoretical stages are required, then a packed extractor may be the best choice. However, the packed extractor will have less flexibility and will be more susceptible to fouling.

After the type of extractor has been selected, the next step is to determine the setup of the experimental equipment and the test plan. For agitated

Literature Cited

1. **Perry, R. H., and D. W. Greene, eds.,** "Perry's Chemical Engineers' Handbook," 6th ed., McGraw-Hill, New York, pp. 15-14 to 15-18 (1984).
2. **Cusack, R. W., et al.,** "A Fresh Look at Liquid-Liquid Extraction, Parts 1–3" *Chem. Eng.,* **98**(2), pp. 66–76 (Feb. 1991), **98**(3), pp. 132–138 (Mar. 1991), and **98**(4), pp. 112–120 (Apr.1991).
3. **Godfrey, J. C., and M. J. Slater, eds.,** "Liquid-Liquid Extraction Equipment," Wiley, New York (1994).

columns, testing can be conducted in small (1–3-in.-dia.) columns. This has several advantages. First, it minimizes the amount of material required for the tests; tests can usually be completed with only one or two 55-gal drums of wastewater feed. Second, the small size allows equilibrium to be established quickly, which minimizes the time required to collect the data from the tests. A typical experimental program can be completed within one week. For packed columns, testing must usually be conducted in 4–6-in.-dia. columns, obviously requiring much more material and time to collect the data.

Solvent stripping. As mentioned earlier, most solvents have a significant solubility in the water phase, and therefore must be stripped from the water before it leaves the process, both for economic and environmental reasons.

The stripping tests are done using the raffinate from the LLE step. These tests are also conducted in small (2–3-in.-dia.) columns. Like the LLE step, the tests are usually run over a range

of feed rates and steam/feed ratios, and the performance measured at each condition. In addition to measuring the separation, look for any tendency toward foaming and any problems with fouling of the column internals. With these data, the stripper can be designed using conventional design techniques.

The stripping step is sometimes not required if the solvent used has very low solubility in water. In cases such as these, the water leaving the extractor can be discharged directly.

Solvent recovery. As mentioned earlier, one of the most critical aspects of any LLE system design is the solvent recovery step. The importance of this is often overlooked by novices to the LLE field because almost all of their attention is directed toward the LLE step and the selection of the solvent. But if the solvent cannot be effectively recovered for recycle back to the extractor, then the process will not be economically viable.

As with the solvent stripping tests, solvent recovery tests are usually conducted in 2–3-in.-dia. columns that are high enough to meet the product and recycle solvent specifications. Also like the stripping tests, look for foaming and fouling tendencies, in addition to separation characteristics.

From these pilot-plant data, the full-size solvent recovery column can be designed utilizing standard design techniques. **CEP**

R. W. CUSACK is vice president of Glitsch Process Systems, Inc., Parsippany, NJ (201/299-9350; Fax: 201/402-0335) He has been with the Glitsch organization for 12 years, and has been directly involved with extraction technology for the last decade. From 1990 to 1994, he held a managing director position for Otto York, NV, a European subsidiary of the Glitsch-held Otto York, Inc., where he was responsible for European marketing of all product lines, including solvent extraction. Previously, he was an engineering manager with Chem-Pro Corp. and worked in refinery and petrochemical design for Exxon Research and Engineering. He holds a BSChE and an MSChE from Manhattan College and is a member of AIChE.

Selecting Innovative Cleanup Technologies: EPA Resources

EPA offers many informational resources to aid in identifying and screening innovative technologies for waste site remediation.

Daniel M. Powell,
U.S. Environmental
Protection Agency,
Technology Innovation Office

The scope of the contaminated site cleanup problem in the United States indicates the need for more effective, less costly remediation technologies. However, the menu of routinely selected treatment options is minimal.

Existing technologies do not provide all of the answers to this dilemma. Although proven remedies are indeed effective in certain settings, they are limited by a number of factors. For example, incinerators can be very costly and difficult to site. In addition, many of the technologies used to date do not address some of the more complex problems faced at contaminated sites (such as mixed radioactive and hazardous wastes, or dense nonaqueous phase liquids [DNAPLs] in groundwater).

Although innovative solutions are being selected for a growing number of cleanup actions, several barriers hinder their routine use. The availability of adequate performance and cost data on such technologies is one of the greatest obstacles. These data are lacking mainly because only 14 of the 263 Superfund source control projects for which innovative remedies were selected have reached completion, the stage where performance and cost information become available.

This lack of information on the full-scale field use of innovative technologies is problematic for several reasons. First, it decreases the willingness of regulators and the public to accept new remedies for site cleanup. And, it discourages site managers from selecting innovative technologies for use at their sites, since data are not available to compare these technologies to more proven remedies.

The U.S. Environmental Protection Agency (EPA) encourages the use of innovative technologies, and by improving the availability of information to decision-makers at hazardous waste sites, the Agency hopes that they will be considered more often. Indeed, EPA offers a number of resources to site managers as they move though the decision-making process that leads to remedy selection.

This article describes the steps involved in researching the technology alternatives and the information available from EPA at each of these steps. It is not intended to be an authoritative guide to the remedy selection process. Rather, it can serve as a roadmap to help site managers navigate through the growing amount of information available on innovative treatment technologies. This roadmap should help managers categorize data available to identify and narrow the alternatives applicable to particular sites. Each publication description includes an EPA identification number, and the sidebar on p. 35 contains ordering instructions.

Steps in the remedy selection process

The following discussion of publications and databases is organized as a model reflecting the phases through which the hazardous waste site decision-maker progresses. In the initial remedy scoping phase, a site manager must identify and conduct a preliminary review of cleanup alternatives and locate the data available on each of them. This step involves both a screening of potential technologies and a search of the existing literature describing those technologies. In the feasibility study stage, the site manager must then review the information gathered and analyze its applicability to the particular site. Finally, the site manager must determine the site-specific applicability of the technologies identified in the first two phases, and, ultimately, develop a decision document (*e.g.,* a Superfund Record of Decision [ROD]) justifying the selection of the most appropriate alternative for that site.

■ *Soil and groundwater remediation occurs simultaneously in this integrated vapor extraction/steam vacuum stripping system. Photo courtesy AWD Technologies.*

In addition to these steps, some activities are interwoven throughout the selection process. The site manager should network among other professionals as well as identify and weigh the policy considerations involved in selecting remedies at hazardous waste sites. And, although treatability studies can occur in the latter stages of the technology selection process to help determine the applicability of a particular treatment process to a site-specific problem, EPA encourages site managers to conduct them as early in the process as possible.

The stages outlined above provide a convenient framework for describing information resources available from EPA to analyze innovative cleanup options. These resources are developed primarily by two offices within EPA — the Office of Solid Waste and Emergency Response (OSWER) and the Office of Research and Development (ORD). Within these organizations, the Technology Innovation Office, the Office of Emergency and Remedial Response, the Office of Solid Waste, the Office of Underground Storage Tanks, and EPA Laboratories provide numerous resources to assist site decision-makers with remedy selection. The rest of this article highlights the resources related to innovative technologies, focusing primarily on technologies available for source control and the treatment of contaminated soils.

Technology screening

The first step in the remedy selection process involves a general screening of the technologies available for dealing with the particular problem at the site. As a tool to scope innovative options, EPA, in conjunction with the U.S. Air Force, has developed the *Remediation Technologies Screening Matrix and Reference Guide* (EPA/542/B-93/005).

The first part of this document is a general screening matrix that compares technologies for soils, sediments and sludge, groundwater, and air emission/off-gas treatment according to a variety of parameters. These parameters include: the development status of the technology; contaminant groups treated; whether the technology is capital or operation-and-maintenance (O&M) intensive; whether it is used as part of a treatment train; residuals produced; whether it addresses toxicity, mobility, or volume; and long-term effectiveness and permanence. In addition, the matrix rates each technology (better, adequate, worse, inadequate information available, or not applicable) according to overall cost, time to complete cleanup, system reliability and maintenance requirements, awareness of the remediation consulting community, regulatory and permitting acceptability, and community acceptability.

The second portion of this document,

the reference guide, provides additional information to increase the usability of the matrix. It describes the basis of the ratings, gives more detailed information on the strengths and weaknesses of each technology, and includes citations to published information on each technology type.

In general, the *Screening Matrix and Reference Guide* is intended as a general reference to be used in the initial screening. It should not be used as the sole basis for remedy selection. A review of more detailed references along with an analysis of individual site conditions should follow this initial screening process.

To provide a fuller understanding of the menu of innovative technologies, EPA has developed several general technology survey reports giving extensive information on innovative technologies. One such resource is *Innovative Treatment Technologies: Overview and Guide of Information Sources* (EPA/540/9-91/002; PB92-179001). This is a compilation of information on treatment technologies available for use in the Superfund program.

The *Overview* includes sections on incineration, thermal desorption, soil washing, solvent extraction, dechlorination, bioremediation, vacuum extraction, *in situ* vitrification, and groundwater treatment. Each section contains: brief descriptions of the processes employed by the technology; summaries of its status, applications, strengths, and weaknesses; and, where available, facts on waste site characteristics that may affect performance. The document also lists reference materials and provides contacts within EPA, state agencies, and the contractor community who have experience with issues related to these technologies. Although the *Overview* is relevant at this initial screening stage, it

Ordering the publications

All of the documents discussed in this article are readily accessible to hazardous waste professionals, both public and private.

Many of the references are available, in limited quantities, free of charge. Documents with EPA/... numbers can be obtained from the National Center for Environmental Publications and Information (NCEPI); OSWER Directives are available from the Superfund Document Center.

When supplies are exhausted, a PB... number is assigned, and the publication must be ordered, for a fee, from the National Technical Information Service (NTIS).

To order, simply provide your request, including publication numbers, to one of the following organizations:

NTIS:
U.S. Dept. of Commerce
National Technical Information Service
Springfield, VA 22161
703/ 487-4650; Fax: 321-8547

NCEPI:
U.S. EPA
National Center for Environmental Publications and Information (NCEPI)
26 W. Martin Luther King Dr.
Cincinnati, OH 45268
Fax Orders: 513/891-6685
No telephone orders accepted

Superfund Document Center:
U.S. EPA
Superfund Document Center
401 M St., SW, OS-245
Washington, DC 20460
Attn: Superfund Directives
202/260-9760

should remain useful throughout the remedy selection process.

For site managers facing problems related to underground storage tanks (USTs) for petroleum products, *Technologies and Options for UST Corrective Actions: Overview of Current Use* (EPA/542/R-92/010; PB93-145589) summarizes treatment technologies used in state UST corrective-action programs. It also includes descriptions and operating parameters for a number of innovative technologies pertinent to UST-related contaminants. Finally, it gives examples of some of the state requirements that may affect the use of the technologies.

The Technology Innovation Office has also developed an electronic resource useful in screening innovative technologies and identifying technology vendors. The *Vendor Information System for Innovative Treatment Technologies (VISITT)* contains current information on the availability, performance, and cost of innovative treatments to remediate contaminated hazardous waste sites. The searchable database allows the user to find information on commercially available treatment processes based on a number of criteria, including contaminants, waste type, and waste source. It includes technologies at all levels of development — bench-, pilot-, and full-scale. This system enables vendors to notify the user communities of the availability of their technologies, and it enables site managers to determine what technologies may be available to treat the problems encountered at their sites.

VISITT Version 2.0 is now available. It profiles 231 technologies offered by 141 companies, 65% of which are available commercially at full-scale. Of the 231 technologies, 28 treat groundwater *in situ,* 164 treat soil, 77 treat sludge, 66 treat natural sediments, 32 treat solids, and 10 are off-gas treatments; 149 of the technologies treat volatile organics, 146 treat semi-volatile organics, 46 treat metals, and 28 treat other inorganics.

The *VISITT* system is distributed free of charge through EPA's National Center for Environmental Information (NCEPI) [see sidebar]. Information on the contents of the system along with details for ordering it can be found in the *VISITT Bulletin* (EPA/542/N-93/004).

EPA's Risk Reduction Engineering Laboratory has developed a series of *Engineering Bulletins* summarizing the latest information on specific treatment and remediation processes. These bulletins provide site managers with an understanding of the data and site characteristics necessary to evaluate (at the technology scoping level) the potential applicability of a technology to their particular problem.

As the program gathers new information on the technologies, addenda will be added to ensure that the documents remain up-to-date. The bulletins related to innovative treatment technologies published to date are listed in Table 1. In addition, EPA plans to publish two new bulletins, including one on *in situ* biodegradation.

Literature review

After the initial screening of available technologies, the next step in the selection process is a review of available literature on the potentially applicable options. Various resources are readily accessible, both searchable electronic media as well as the more generally used print media.

Electronic media. The foremost bibliographic database on hazardous waste site treatment technologies offered by EPA is the *Alternative Treatment Technology Information Center (ATTIC).* It contains information on biological, chemical, and physical treatment processes, solidification and stabilization, and thermal treatment technologies. The *ATTIC* system provides users with on-line access to several databases as well as an electronic bulletin board, a hot line (which allows searches for those without computer capabilities), and a repository of documents related to alternative and innovative treatment technologies.

The primary component of *ATTIC* is the *ATTIC* database, a searchable, bibliographic database providing abstracts on over 2,000 references. The on-line system also provides access to a treatability database, a message center, and a comprehensive calendar of technology-related events.

There is no charge for the use of *ATTIC* and it is available 24 hours a day, seven days a week. The on-line number for the *ATTIC* system is 703/908-2138, and the system operator/hotline number is 703/908-2137. The settings for on-line access to *ATTIC* are 8 data bits, no parity, and 1 stop bit.

A publication developed to facilitate

access to the large body of information related to innovative treatment technologies is *Accessing Federal Data Bases For Contaminated Site Clean-Up Technologies, Third Edition* (EPA/542/B-93/008 or PB94-144540). It contains a series of 23 profiles describing databases, expert systems, and electronic bulletin boards maintained by federal agencies. The listed systems contain information on innovative technologies or on completed demonstration projects. The publication explains the type of information each system contains, and notes the accessibility of the systems (restricted vs. open), hardware and software specifications, and agency contacts.

In addition to federally sponsored systems, various commercial databases and software can assist cleanup professionals in locating bibliographic information. To improve awareness of the available literature in the field of hazardous waste cleanup technologies, EPA has prepared a *Literature Survey of Innovative Technologies for Hazardous Waste Site Remediation: 1987-1991* (EPA/542/B-92/004). Compiled based on a search of com-

mercial databases (including CA Search, Compendex Plus, Energy Science and Technology Database, the National Technical Information Service (NTIS) Database, and Pollution Abstracts), this document is an extensive listing of technical literature related to innovative technologies, organized by technology type.

[*CEP*'s annual *Software Directory* is another source of information on available software. — Editor]

Print resources. EPA also provides printed materials useful in identifying technology references. *Selected Alternative and Innovative Treatment Technologies for Corrective Action and Site Remediation* (EPA/542/B-93/010) is a bibliography of EPA reports describing treatment technologies for hazardous waste sites. Published semi-annually, this brochure provides titles, document numbers, and ordering information for technology documents. The bibliography also includes details on EPA information systems relevant to site cleanup technologies.

A broader look at the reports and publications available from the federal government as a whole is provided by

Table 1. EPA *Engineering Bulletins* on specific remediation technologies.

Subject	Document Number
Solvent Extraction Treatment	EPA/540/2-90/013
Mobile/Transportable Incineration Treatment	EPA/540/2-90/014
Chemical Dehalogenation: Alkali Polyethylene Glycol (APEG) Treatment	EPA/540/2-90/015
Slurry Biodegradation	EPA/540/2-90/016
Soil Washing Treatment	EPA/540/2-90/017
In Situ Steam Extraction	EPA/540/2-91/005
In Situ Soil Vapor Extraction	EPA/540/2-91/006
Thermal Desorption Treatment	EPA/540/2-91/008
In Situ Soil Flushing	EPA/540/2-91/021
Chemical Oxidation Treatment	EPA/540/2-91/025
Supercritical Water Oxidation	EPA/540/S-92/006
Rotating Biological Contactors	EPA/540/S-92/007
Technology Preselection Data Requirements	EPA/540/S-92/009
Pyrolysis Treatment	EPA/540/S-92/010
Selection of Control Technologies for Remediation of Lead Battery Recycling Sites	EPA/540/S-92/011
Solidification/Stabilization of Organics and Inorganics	EPA/540/S-92/015

Federal Publications on Alternative and Innovative Treatment Technologies for Corrective Action and Site Remediation, Third Edition (EPA/542/B-93/007 or PB94-144557). This 42-page bibliography references reports describing federal research, evaluation, and demonstration of innovative treatment processes for hazardous waste sites. Over 400 reports are referenced and categorized by subject area, with listings for general survey reports as well as technology-specific topic groupings. The bibliography also provides document numbers and ordering information.

Another aid in the literature screening stage is the *Compendium of Superfund Program Publications* (EPA/540/8-91/014). Much more general and comprehensive than the bibliographies, this publication is the most complete source of information provided by EPA on Superfund documents, including fact sheets, directives, publications, and computer materials. The compendium also includes detailed information on how to order Superfund publications, most of which are available through the NTIS for a fee.

Due to the high level of interest in bioremediation, the Technology Innovation Office has developed the *Bioremediation Resource Guide* (EPA/542/B-93/004) to aid decision-makers in reviewing the applicability of bioremediation. This document provides access information on electronic resources and hot lines, cites relevant federal regulations, and abstracts pertinent print resources such as bibliographies, guidance documents, workshop proceedings, overview documents, study and test results, and test designs and protocols. Particularly handy is a detailed "bioremediation resource matrix," which compares the documents by technology type, affected media, and contaminants. The guide also provides detailed information on how users may obtain the listed publications.

Finally, EPA conducts annual symposia to provide information useful in the literature screening phase of the remedy selection process. The *Forum*

Table 2. Available abstracts from *Forums on Innovative Hazardous Waste Treatment Technologies: Domestic and International* conferences.

1st Forum, Atlanta, GA	EPA/540/2-89/055; PB90-268509
2nd Forum, Philadelphia, PA	EPA/540/2-90/009; PB91-145649
3rd Forum, Dallas, TX	EPA/540/2-91/016; PB92-233881
4th Forum, San Francisco, CA	EPA/540/R-92/081
5th Forum, Chicago, IL	Not Yet Assigned

on Innovative Hazardous Waste Treatment Technologies: Domestic and International is a national conference where hazardous waste professionals can exchange solutions to hazardous waste problems. The last forum, held in San Francisco, November 17–19, 1992, was attended by over 1,000 representatives from the U.S. and 25 foreign countries. Attendees heard 42 technical presentations describing domestic and international technologies for the treatment of waste, sludges, and contaminated soils at hazardous waste sites, focusing on physical/chemical, biological, thermal, and stabilization techniques. Abstracts and papers from this, as well as the first three forums, are available from EPA. Publication numbers for the abstracts are shown in Table 2. The next forum is being held this month (May 3–5 in Chicago), and abstracts from that conference will be available shortly.

Narrowing the options

The next step is the information review and analysis phase. By this point, the site manager has identified information on available alternatives, and can now begin to review the technical publications for more specific information on the options deemed relevant to the particular contamination problem at the site.

EPA has developed a number of resources that contain detailed information on the application of innovative treatment technologies. They can serve as a basis to determine whether alternatives identified in the previous stages warrant further consideration.

Two publications provide brief synopses of innovative technologies that

are in federal demonstration programs. First, the EPA Superfund Innovative Technology Evaluation (SITE) Program publishes an annual compendium of abstracts describing the innovative technologies or processes it has accepted into the SITE program. This publication, entitled *The Superfund Innovative Technology Evaluation Program: Technology Profiles, Sixth Edition* (EPA/540/R-93/526), contains profiles of 156 demonstration, emerging, and monitoring and measurement technologies currently being evaluated. Each profile provides an abstracted description of the technology, a discussion of its applicability to various wastes and media, an update on its development or demonstration status, demonstration results (if available), and contacts (for both the developer and at EPA).

EPA also publishes a compendium of federal demonstration and evaluation projects entitled *Synopses of Federal Demonstrations of Innovative Site Remediation Technologies, Third Edition* (EPA/542/B-93/009 or PB94-144565). Similar in format to the *SITE Technology Profiles*, this publication is a compilation of 112 abstracts describing demonstrations of innovative technologies conducted by federal agencies. Of the 112 projects, 32 are classified as bioremediation, 9 as chemical treatment, 17 as thermal treatment, 14 as vapor extraction, 16 as soil washing, and 24 as "other physical treatment." In addition, an appendix contains information on 14 demonstration projects involving incineration and solidification/stabilization. The abstracts profile both completed projects and planned demonstrations that are nearing implementation. Each profile includes the name and telephone number of a contact for the project.

The *SITE Technology Profiles* and the *Federal Synopses* provide brief abstracts on each technology. The full results of each specific project completed under the SITE Demonstration Program are incorporated in two documents, the *Technology Evaluation Report* and the *Applications Analysis Report.*

The *Technology Evaluation Report* is a comprehensive summary of the results of the demonstration. It gives a detailed review of the performance of the technology, as well as the advantages, risks, and costs for a given application. This report focuses on the needs of the site manager who is responsible for evaluating the applicability of the technology in relation to a specific site or waste situation.

The *Applications Analysis Report* (which will become the *Innovative Technology Evaluation Report* within the next year) is intended for use by decision-makers responsible for the selection and implementation of a remedy. It aids site managers in deciding whether the documented technology is viable for further consideration to clean up a specific site. In addition to providing an evaluation of the performance of the technology at the demonstration site, this report incorporates data from other projects to give an indication of the broader applicability of the tested technology. Since waste characteristic differences from site to site may affect a technology's success, it is necessary to examine data available from other field applications. In addition, EPA evaluates the applicability of each technology to sites and wastes other than those tested and studies the technology's likely costs in those situations, and the results from these analyses are included as well. Table 3 shows the *Applications Analysis Reports* now available.

An effort that should be particularly helpful in a detailed review of technologies is the *WASTECH* project. WASTECH is a multiorganizational

initiative (including AIChE's Environmental Div.) being coordinated by the American Academy of Environmental Engineers (AAEE) through a cooperative agreement with EPA's Technology Innovation Office. WASTECH is developing a series of eight engineering monographs on innovative cleanup technologies. The monographs are written by experts in each technology field and are exten-

sively peer-reviewed by government, academic, and professional organizations involved in site cleanup and technology development. Each monograph includes a comprehensive description of the technology or process, a description of suitable applications, an evaluation of the technology based on "a synthesis of available information and informed judgments," a description of the limitations of the process, and a

Table 3. Completed "Applications Analysis Reports."

Technology Type	Vendor	Report Number
Biological Technologies		
Biotreatment of Groundwater	Biotrol	EPA/540/A5-91/001; PB91-227983
Physical/Chemical Technologies		
Basic Sludge Extractive Treatment (BASIC)	Resources Conservation Co.	EPA/540/AR-92/079
Carver-Greenfield Separation Process	Dehydro-Tech	EPA/540/AR-92/002
Chemical Fixation/ Stabilization	Chemfix Technologies, Inc.	EPA/540/A5-89/011
Chemical Oxidation Treatment	Perox-Pure	EPA/540/AR-93/501
In Situ Stabilization	IWT/Geocon	EPA/540/A5-89/004
In Situ Steam/Hot Air Stripping	Toxic Treatments (USA)	EPA/540/A5-90/008
Integrated Vapor Extraction/Steam Vacuum Stripping	AWD Technologies, Inc.	EPA/540/A5-91/002
Membrane Filtration	SBP Technologies	EPA/540/AR-92/014
Membrane Microfiltration	DuPont/Oberlin	EPA/540/A5-90/007
Mobile Volume Reduction Unit	U.S. EPA	EPA/540/AR-93/508
Pneumatic Fracturing and Hot Gas Injection	Accutech	EPA/540/AR-93/509
Soil Recycling Treatment Train	Toronto Harbour Commissioners	EPA/540/AR-93/517
Soil Washing	Biotrol	EPA/540/A5-91/003; PB92-115245
Solidification	Hazcon	EPA/540/A5-89/001
Solidification	Soliditech, Inc.	EPA/540/A5-89/005
Solidification/Stabilization of Organics/Inorganics	Silicate Technology Corporation	EPA/540/AR-92/010; PB93-172948
Solvent Extraction	CF Systems, Corp.	EPA/540/A5-90/002
UV-Ozone Treatment for Liquids	Ultrox International	EPA/540/A5-89/012
Vacuum Extraction	Terra Vac	EPA/540/A5-89/003; PB90-119744
Wastewater Treatment System	PO*WW*ER	EPA/540/AR-93/506
Thermal Technologies		
Oxygen Enhanced Incineration	American Combustion	EPA/540/A5-89/008; PB90-258427
Cyclone Furnace Vitrification	Babcock and Wilcox	EPA/540/AR-92/017; PB93-122315
Flame Reactor	Horsehead Resource Development	EPA/540/A5-91/005; PB89-194179
Plasma Centrifugal Furnace	Retech	EPA/540/A5-91/007
Low Temperature Thermal Treatment (LT3)	Roy F. Weston, Inc.	EPA/540/AR-92/019
Infrared Incineration	Shirco	EPA/540/A5-89/010

prognosis of other processes or elements of processes that require further research before considering full-scale application.

The first two monographs, on thermal desorption and soil washing/soil flushing, were completed in late 1993 and are available from AAEE (130 Holiday Court, Suite 100, Annapolis, MD, 21401, 410/266-3311). The remaining six (on stabilization/solidification, vacuum/vapor extraction, thermal destruction, chemical treatment, solvent/chemical extraction, and bioremediation) are in progress. The completed set should be available by late summer or early fall.

Networking

As indicated earlier, the process of networking is not a one-time activity. Instead, networking should take place throughout the selection process. Networking involves communicating with other hazardous waste professionals to learn of their experiences and to tap their expertise. Cleanup decisions have already been made at many sites, and the experiences of those site managers can provide valuable lessons relevant to future remedy selection decisions. To facilitate networking, EPA offers several mechanisms.

The first mechanism consists of technology newsletters, which provide regular forums to publicize ongoing technology applications. EPA publishes three such newsletters on a quarterly basis.

The *Tech Trends* newsletter is self-described as "an applied journal for Superfund removals and remedial actions and RCRA [Resource Conservation and Recovery Act] corrective actions." Among the issues addressed by *Tech Trends* are new technologies, innovative uses of existing technologies, overcoming bureaucratic obstacles to the use of innovative technologies, and the applicability of innovative technologies used in the Superfund program to the RCRA corrective action and closure programs. The most recent issue is

February 1994 (EPA/542/N-94/001). *Bioremediation in the Field* is described as "an information update on applying bioremediation to site cleanup." It features articles on a number of topics, including results of projects at sites supported through EPA's Bioremediation Field Initiative, developments from laboratory and field tests, information on upcoming conferences and seminars, and a listing of relevant bioremediation publications available from EPA or NTIS. It also includes an up-to-date tracking table of sites where bioremediation projects have been completed, are in operation or design, or are being considered for use. The table contains site names, points of contact, contaminants and media treated at each project, project status, clean-up levels, and the specific treatment option chosen for each. This table can provide site managers with a starting point for further information on the applicability of bioremediation to additional sites. The table in the last issue of *Bioremediation in the Field* (March 1994, EPA/540/N-94/500) listed 148 bioremediation sites.

Another newsletter with a specific focus is *Ground Water Currents,* which focuses on issues affecting the development and use of innovative groundwater treatment technologies. Along with articles on field applications of such technologies, *Ground Water Currents* also provides information on groundwater research and regulatory issues affecting the development and application of technologies. The February 1994 issue (EPA/542/N-94/002) is the most recent.

An electronic bulletin board system, the *Cleanup Information (CLU-IN)* bulletin board, allows more timely, day-to-day communications among the clean-up community. *CLU-IN*'s features allow users to exchange messages (either with individual users or to large audiences), exchange computer files and databases, read bulletins on-line, or access several databases on-line. *CLU-IN* also includes a number of special interest group areas (or sub-bulletin boards) on groundwater cleanup, treatability studies, and training.

CLU-IN is available 24 hours a day, seven days a week, and there is no charge to use it. The on-line number for *CLU-IN* is 301/589-8366, and the telecommunications parameters must be set at 8 data bits, no parity, and 1 stop bit. *CLU-IN* also offers a help line staffed by the system operator at 301/589-8368, as well as several source documents, including an introductory flyer (*Exchanging Information on CLU-IN,* EPA/542/F-93/001), a users' guide (*Cleanup Information Bulletin Board System User's Guide,* EPA/542/B-93/002), and a one-hour, self-guided lesson for beginners (*A Guided Tour of CLU-IN,* EPA/542/B-93/003).

Several publications described earlier may also be useful for networking. Both the *Semi-Annual Status Report* and the SITE program's *Tech Profiles* provide contact names for the technology projects that they list. These contacts can provide valuable information on particular technologies. Also, *Accessing Federal Data Bases* lists other electronic systems available from other federal agencies that have communication capabilities.

Site-specific application

Technical assistance. After the options have been researched, a determination must be made as to the site-specific applicability of a particular technology. Several additional resources are available to provide support at this point in the selection process. *Technical Support Services for Superfund Site Remediation and RCRA Corrective Action* (EPA/540/8-91/091) is a directory of services within EPA to assist EPA project managers in solving specific technical problems. Also included are listings of salient publications and information on relevant automated information systems. Although many of the help lines are for EPA personnel only, a number of the resources listed are accessible to outside parties.

Another aid in identifying potential sources of assistance is *Innovative Hazardous Waste Treatment Technologies: A Developer's Guide to Support Services* (EPA/540/2-91/012). The primary purpose of this publication is to provide technology developers with a listing of resources available to them as they seek to advance their technologies from the proof-of-concept stage to commercialization. The booklet includes information on regulatory requirements applicable to the development of new technologies, sources of grant funding and technical assistance, and the identification of incubator, testing and evaluation, and university-affiliated research facilities that provide a host of development and evaluation services. Some of these services may also be useful to site managers interested in specific technologies.

Treatability studies guidance. The resources discussed thus far can aid a site manager in determining the applicability of a particular technology to site-specific contamination. However, of fundamental importance to remedy selection is the treatability testing process. Treatability studies provide a much clearer indication of whether a technology or treatment process is truly applicable to a specific contamination problem at a site.

To assist site managers in conducting treatability studies, EPA has developed a number of guidance documents that address both general and technology-specific issues. Tables 4 and 5 outline the available documents and fact sheets on treatability testing.

The goal: remedy selection

The resources described so far are intended to lead the hazardous waste professional up to the point of selecting a remedy at an individual site. EPA also provides support at the actual selection point.

Although the benefits of innovative technologies are often cited, site managers have little written guidance to support their decisions in selecting innovative technologies. In fact, some of the regulatory and administrative requirements may impede the use of innovative technologies. Furthermore, the fear of failure for unsuccessful applications of new processes has

Table 4. Treatability studies — general guidance.

Title	Document Number
Analysis of Treatability Data for Soil and Debris: Evaluation of Land Ban Impact on Use of Superfund Treatment Technologies	OSWER Directive 9380.3-04; PB90-258476
Conducting Treatability Studies Under RCRA — Quick Reference Fact Sheet	OSWER Directive 9380.3-09FS; PB92-963501
Guide for Conducting Treatability Studies Under CERCLA — Final	EPA/540/R-92/071A; PB93-126787
Inventory of Treatability Study Vendors, Draft Interim Final	EPA/540/2-90/003a
Regional Guide: Issuing Site-Specific Treatability Variances for Contaminated Soils and Debris for Land Disposal Restrictions (LDRs) — Quick Reference Fact Sheet	OSWER Directive 9380.3-08FS; PB92-963284
The Remedial Investigation Site Characterization and Treatability Studies	OSWER Directive 9355.3-01FS2 (Fact Sheet); PB90-274408
Treatability Studies Under CERCLA: An Overview, 12/89	OSWER Directive 9380.3-02FS (Fact Sheet); PB90-273970

Table 5. Treatability studies — technology-specific guidance.

Title	Document Number
Aerobic Biodegradation Remedy Selection	
— Interim Guidance	EPA/540/2-91/013A; PB92-109065
— Quick Reference Fact Sheet	EPA/540/2-91/013B; PB92-109073
Biodegradation Remedy Selection	EPA/540/R-93/519A
Chemical Dehalogenation	
— Interim Guidance	EPA/540/R-92/013A; PB92-169044
— Quick Reference Fact Sheet	EPA/540/R-92/013B; PB92-169275
Soil Vapor Extraction	
— Interim Guidance	EPA/540/2-91/019A; PB92-227271
— Quick Reference Fact Sheet	EPA/540/2-91/019B; PB92-231281
Soil Washing	
— Interim Guidance	EPA/540/2-91/020A; PB92-170570
— Quick Reference Fact Sheet	EPA/540/2-91/020B; PB92-231281
Solvent Extraction	EPA/540/R-92/016A; PB92-239581

also served to create an atmosphere that discouraged the selection of such technologies.

In the summer of 1991, OSWER recognized these shortcomings and issued *Furthering the Use of Innovative Treatment Technologies in OSWER Programs,* also known as Directive 9380.0-17 (available in fact-sheet format, OSWER Directive 9380.0-17FS) to emphasize the commitment of EPA senior management to the selection of innovative technologies. This Directive targeted both EPA site managers seeking support for their decisions to select innovative technologies and responsible parties needing a tool to negotiate the use of these technologies with reluctant EPA project managers. The Directive formally emphasizes OSWER's commitment to the use of innovative technologies, and indicates that efforts to further their application are viewed by senior management as a benefit that should be considered in remedy selection decision-making.

The Directive contains a number of initiatives in areas such as: funding for innovative projects; contracting issues; budgetary prioritization; treatability testing; use of innovative technologies in emergency response actions; testing at federal facilities; enforcement issues; use of innovative technologies in other OSWER programs (UST and RCRA corrective actions); and greater cooperation with private parties through the mechanisms available under the Federal Technology Transfer Act.

Reluctance on the part of management and regulators represents only one hurdle that site managers must clear in selecting innovative technologies at their sites. The public's lack of understanding and fear of being treated as "guinea pigs" represent another barrier.

To assist site managers in addressing the needs of surrounding communities to understand innovative technologies, EPA has prepared a series of *Citizen's Guides to* *Understanding Innovative Treatment Technologies.* They are written in predominantly nontechnical language and are available in both English and Spanish. Eight of the *Citizen's Guides* contain information on specific treatment technologies, one provides an overview of innovative treatment technologies, and another highlights success stories about locations where innovative treatment technologies have been applied. Table 6 is a listing of the available *Citizen's Guides.* CEP

Table 6. *Citizen's Guides* to understanding innovative treatment technologies.

A Citizen's Guide to ...	Document Number
Innovative Treatment Technologies for Contaminated Soils, Sludges, Sediments, and Debris	EPA/542/F-29/001 (English) EPA/542/F-92/014 (Spanish)
How Innovative Treatment Technologies Are Being Successfully Applied at Superfund Sites	EPA/542/F-92/002 (English) EPA/542/F-92/015 (Spanish)
Soil Washing	EPA/542/F-92/003 (English) EPA/542/F-92/016 (Spanish)
Solvent Extraction	EPA/542/F-92/004 (English) EPA/542/F-92/017 (Spanish)
Glycolate Dehalogenation	EPA/542/F-92/005 (English) EPA/542/F-92/018 (Spanish)
Thermal Desorption	EPA/542/F-92/006 (English) EPA/542/F-92/019 (Spanish)
In Situ Soil Flushing	EPA/542/F-92/007 (English) EPA/542/F-92/020 (Spanish)
Bioventing	EPA/542/F-92/008 (English) EPA/542/F-92/021 (Spanish)
Using Indigenous and Exogenous Microorganisms in Bioremediation	EPA/542/F-92/009 (English) EPA/542/F-92/022 (Spanish)
Air Sparging	EPA/542/F-92/010 (English) EPA/542/F-92/023 (Spanish)

Further Reading

U.S. EPA, *Cleaning Up the Nation's Waste Sites: Markets and Technology Trends,* Publication No. EPA/542/R-92/012, PB93-140762 (1992).

U.S. EPA, *Innovative Treatment Technologies: Annual Status Report, Fifth Edition,* Publication No. EPA/542/R-93/003 (Sept. 1993)

D. M. POWELL is an environmental protection specialist with the U.S. Environmental Protection Agency's Technology Innovation Office, Washington, DC, where he is primarily responsible for cooperative technology demonstration and information exchange endeavors with other federal agencies, the states, and private industry. Before coming to the TIO in 1990, he was a 1988 Presidential Management Intern with EPA, completing assignments with the Headquarters and Regional Superfund programs, the personal staff of U.S. Representative Michael Bilirakis (Florida), the Office of Underground Storage Tanks, the waste programs budget office, and the EPA Information Resources Management Office. He received a BA in political science and urban studies from Roanoke College in 1985 and a Master's of Public Administration from the Woodrow Wilson School of Government at the Univ. of Virginia in 1988.

Clean Up Hydrocarbon Contamination Effectively

Choosing the right technology involves a thorough site characterization and feasibility study based on the technical, economic, regulatory, and political issues peculiar to the site.

Gilbert M. Long,
ENSR Consulting
and Engineering

Hydrocarbon contamination of soil and groundwater is a serious health and environmental issue. Typical contamination sites include manufacturing plants, petroleum refineries, fuel and chemical storage facilities, gasoline service stations, and vehicle depots.

Unfortunately, there has been a dearth of useful information on the "how-to" aspects of the hydrocarbon cleanup process. Thus, remediation technologies have been improperly applied at some sites due to a lack of knowledge about pertinent science and engineering.

Effective remediation of a site contaminated with hydrocarbons requires a sound understanding of regulatory issues, technology options, and the site's hydrogeology. Many treatment options exist, but none is a panacea. The optimal solution for a site is based on a thorough site-specific characterization, followed by a feasibility study that evaluates treatment alternatives. The outcome will be a cost-effective solution that combines treatment alternatives best suited to the site.

This article outlines the technologies available for soil and groundwater remediation and how to select an appropriate technology from among them. [More detailed articles on some of these technologies are planned for future issues. — Editor]

Regulatory considerations

A practical understanding of regulatory issues and agency policies is a prerequisite to effective site remediation. Often more than one agency or set of statutes may govern cleanup. In the case of multiple regulatory sources, the most stringent requirements will govern site cleanup, with two notable exceptions: if the remediation is voluntary, it may be exempt from some requirements, and a risk assessment may sometimes justify relaxation of the usual regulatory mandates.

The major federal regulations for hazardous waste site remediation are the Comprehensive Environmental Response, Compensation, and Liability Act (CERCLA), the Superfund Amendments and Reauthorization Act (SARA), and the Resource Conservation and Recovery Act (RCRA). [For information on these laws, see (1).] Site owners may also need to contend with state and local regulations, which are often more stringent than federal mandates.

Permitting considerations also affect the selection of a treatment solution. In fact, permitting is a cost issue that should be evaluated early on. Many states require air permits to regulate air discharges from treatment systems, and groundwater treatment of a hydrocarbon-contaminated site can also affect existing permits.

Trigger levels. To complicate decision-making, the level of hydrocarbon contamination triggering remediation varies from state to state and between the saturated zone and the unsaturated (or vadose) zone (see Figure 1). Soil concentrations as low as 50 ppm total petroleum hydrocarbons (TPH) can necessitate site cleanup. To determine the levels for specific compounds, RCRA standards are compared to the results of toxicity characteristic leaching procedure (TCLP) tests on the soil. In many states, specific volatile organic compounds (VOCs) and base/neutral compounds have specific thresholds above which remediation is required. Other regulations may also apply, but they vary from state to state and from industry to industry.

In the case of contaminated groundwater, action levels depend upon whether or not the water may be used for human consumption. If so, drinking water standards apply. If groundwater is not intended for human consumption, maximum concentration limits (MCLs) under the Safe Drinking Water Act or less stringent regulations are generally applicable. Based on a site-specific risk assessment, treatment standards for both soil and groundwater may also be set or modified from normal regulatory targets.

Selecting remediation methods

Before selecting a treatment methodology, one must first understand the site's hydrogeology. Through an evaluation of the movement of contaminants through soil and groundwater (vertically, horizontally, or both), appropriate remediation techniques and goals can be identified.

A critical consideration in selecting a remedial technology is the "fate and transport" of contaminants. Mechanisms governing the fate of hydrocarbons in the environment, the impacts of various treatment processes, and the chemistry of individual contaminants must be understood to determine the ultimate destination of contaminants — and, in turn, the most effective cleanup technologies for a particular site. For instance, while a treatment method might be viable for excavated soils, the excavation process itself may generate toxic air emissions. On the other hand, an *in situ* process might solve the problem of air emissions, but prove slower or less effective in stemming long-term contaminant migration through groundwater.

In choosing a remediation technology, it is important to realize that there is no "cookbook solution." Each site poses its own unique challenges, which must be dealt with by answering such site-specific questions as:

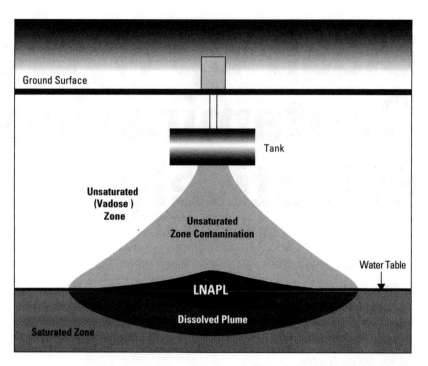

1. How much soil or groundwater requires cleanup?

2. Must contaminants be excavated and treated aboveground or can they be treated using *in situ* technologies?

3. How much time is available for cleanup?

4. Can air emissions and waste streams be minimized by combining treatment technologies?

5. How will cleanup (or no cleanup) affect site neighbors?

6. To what extent must a combination of technologies be pilot tested or otherwise demonstrated to agencies or committees that must approve their use?

7. Can technologies be combined to optimize treatment efficiency, meet cleanup standards, and minimize costs?

Today, many site owners are turning to on-site technologies. Benefits of on-site treatment include avoidance of transportation costs and liabilities, reduced human exposure to contaminants, and sometimes even exemption from permitting requirements. On-site treatment also minimizes long-term liabilities associated with hazardous waste storage and disposal as well as the risks involved with using an independent treatment facility where wastes may be commingled with materials from other sources.

■ *Figure 1. Soil and groundwater remediation zones.*

Dealing with LNAPL

Remediation of a site contaminated with hydrocarbons typically involves cleanup of soil and groundwater, as well as removal of light nonaqueous-phase liquids (LNAPL) or free product. LNAPL, which is often a major portion of the hydrocarbon mass at a site, serves as a primary source for groundwater contamination. Thus, the more LNAPL that is removed, the less water treatment is required.

To minimize costs, remediation should start with removal of the contaminant source and LNAPL. Once LNAPL has been eliminated to the greatest extent possible, soil and groundwater remediation can begin, along with cleanup of areas beyond the sources.

A number of techniques can accomplish LNAPL removal, including recovery wells, trenches, vapor extraction, oil-water separation, and chemical oxidation. The particular method chosen for LNAPL collection depends on site-specific conditions, such as soil permeability, depth to water table, and type of hydrocarbons present, as outlined in Table 1. Technology selection is essentially a tradeoff among these factors.

Generally, LNAPL is collected from wells and/or trenches, pumped into a collection or separation system, and then disposed of or recycled. To minimize contaminant migration, LNAPL recovery should be started quickly. The typical approach is to employ two pumps in a well. One pump lowers the water table near the LNAPL to create a cone of depression that serves as a sump to collect the water. The second pump collects the LNAPL and pumps it to the surface for subsequent incineration or recycling.

Soil permeability and the viscosity of the LNAPL determine the time required for LNAPL removal. LNAPL forms a free-floating layer on groundwater and exists in soil pores above the groundwater; due to

Table 1. Compatability of LNAPL recovery methods with various site conditions.			
Site Parameter	Recovery Well	Trench	Vapor Extraction
Soil Permeability	All Permeabilities	All Permeabilities	Medium to High
Depth to Water	Medium to Deep	Shallow	Shallow to Medium
Thickness of LNAPL	All Thicknesses	All Thicknesses	Thin (Sheen)
Weight of Hydrocarbon	Light to Medium	Heavy	Light
Permitting Time	Rapid	Rapid	Lengthy
Tidal Influence	Low to Medium	High	High
Impact on Groundwater Remediation	High	Medium	Low

Note: The terms used to describe site conditions are qualitative and are stated for comparative purposes only.

fluctuations in groundwater levels, LNAPL may also become trapped in soil pores below the measured groundwater table down to the lowest groundwater level. Standard collection systems can recover up to 60% of the LNAPL, and enhanced removal methods such as vapor extraction, steam injection, and air stripping may collect the remaining material from the soil pores.

Conventional LNAPL removal is among the least expensive cleanup methods available. Treatment of the water pumped to lower the water table is a major portion of the cost. Some companies have reduced overall site cleanup costs by reusing LNAPL as fuel at the site.

Remediating soil

Remediating contaminated soils lies at the heart of hydrocarbon cleanup. Indeed, it is useful to think of soil contamination as the source of contamination at these sites since groundwater will not generally stay clean until the soil is clean.

Because most hydrocarbon is trapped within soil pores or adsorbed to soil surfaces, the following factors will govern the selection of a cleanup process:

• the ability of the soil to transmit air and water through the unsaturated zone or water through the saturated zone (i.e., permeability);

• the ease with which the water table can be lowered to expose more soil to the unsaturated treatment zone; and

• the type of hydrocarbons present and their physical and chemical properties.

Soil treatment technologies. A range of treatment technologies is available to remediate soils. The proper mix of technologies depends upon soil permeability and hydrocarbon type, as shown in Table 2.

If most of the contaminated soil is below the water table, a pump-and-treat system or *in situ* bioremediation (or both) will often be the best cleanup alternatives. The pump-and-treat method involves the use of well pumps to carry groundwater to the surface and contain the contaminant plume. The water is then treated and discharged back into the ground or into a drainage system.

If soils are so impermeable that they do not permit sufficient water transport rates for pump-and-treat systems, thermal treatment can be used. If a pump-and-treat method is applied, however, some kind of *in situ* treatment makes sense whatever the permeability limitations.

Soil treatment costs typically run from about $75–$200/yd^3 for *in situ* or on-site treatment. Thermal methods are considerably more expensive if the material is considered hazardous. Of course, the type and

Table 2. Unsaturated soil treatment options for on-site remediation.

Contaminant Level	Soil Permeability to Water* (cm/s)		
	Low (<10⁻⁵)	Medium (10⁻⁵–10⁻²)	High (>10⁻²)
Fuel: Gasoline, Naphtha			
Low (100–500 mg/kg)	SVE, LT, TD	UZB, SVE, LT	UZB, SVE, LT
Medium (500–5,000 mg/kg)	SVE, LT, TD	UZB, SVE, LT	UZB, SVE, LT
High (>5,000 mg/kg)	SVE, LT, TD	UZB, SVE, LT	SVE†
Fuel: Diesel, No. 2 Fuel Oil, Kerosene, JP-1, Mineral Spirits			
Low (100–500 mg/kg)	SVE, TD	UZB, SVE, LT	UZB, LT, TD
Medium (500–5,000 mg/kg)	SVE, LT, TD	UZB, SVE, LT	UZB, LT, TD
High (>5,000 mg/kg)	SVE, LT, TD	UZB, LT, SVE, TD	LT, TD
Fuel: No. 4 or No. 6 Fuel Oil, Crude Oil, Tar, Residue Oils			
Low (100–500 mg/kg)	LT, I	LT, I	UZB, TD, I
Medium (500–5,000 mg/kg)	SVE, LT, TD, I, B	SVE, LT, TD, I, B	SVE, TD, I, B
High (>5,000 mg/kg)	SVE, LT, TD, I, B	SVE, LT, TD, I, B	SVE, LT, I, B

Legend:
B = Bioreactor
I = Incineration
LT = Land Treatment
SVE = Soil Vapor Extraction
TD = Thermal Desorption
UZB = Unsaturated Zone Bioremediation
* Saturated hydraulic conductivity.
† Other technologies may work if toxicity is not a problem.

extent of contamination as well as the volume of soil to be treated will determine the ultimate treatment costs at a site.

For less than 500 yd³ of soil, the most cost-effective remediation option may well be excavation and landfill disposal, if it is allowed for the site.

Remediating groundwater

Frequently, contaminated soil migrates into groundwater which, in turn, transports and spreads contaminants to a broader area. Groundwater remediation, therefore, must focus on removing dissolved hydrocarbons and hydrocarbons that may be trapped in the soil pores below the water table. Because groundwater typically moves very slowly with little mixing, soil contaminants must generally be redissolved before they are removed or destroyed.

Pump-and-treat technologies are standard procedures but they are very time-consuming. Years or even decades may pass before groundwater treatment standards are met. In some cases, currently enforced standards are simply not achievable using pump-and-treat methods alone. Effective source removal or active soil remediation can sometimes accelerate the process. For certain sites, *in situ* aquifer treatment technologies have also proven effective.

Before a groundwater recovery method or combination of methods can be selected, site hydrogeology must be thoroughly understood. Short-term pilot studies can be conducted to evaluate the impacts of groundwater withdrawals on flow patterns and contaminant migration. Groundwater flow and contaminant transport models can then be applied to simulate long-term impacts. Modeling is particularly useful for developing recovery strategies, optimizing groundwater and contaminant recovery, establishing project schedules, and estimating long-term operation and maintenance costs.

Groundwater treatment technologies. Various technologies are available for treating groundwater. Choosing one for a particular site involves understanding a variety of site-specific factors, such as the type, concentration, and depth of hydro-carbons in the water, the flow rate of water recovered from pumping, hydrogeologic features of aquifers, and fate and transport characteristics of both the aquifer and hydrocarbons. In addition, regulatory requirements such as the treatment standard for water discharge and the cleanup standard for the aquifer are also important.

Table 3 shows how such factors shape the final selection of remediation techniques. The discharge limits in the table are the limits prescribed in permits for water discharged from the treatment system to the environment or to a wastewater treatment system. Note that when the pumping system flow rate is less than 3 gal/min, use of carbon adsorption beds is generally the most cost-effective. Moreover, if the local sewage treatment plant can accept the stream without treatment, this approach may be the best alternative. This is, however, increasingly difficult to do.

The technologies

Hydrocarbon remediation technologies can be classified into four groups — physical, biological, thermal, and chemical treatment. Table 4 summarizes the classification scheme and the general applicability of the various treatment technologies.

Physical treatment technologies separate or concentrate treated materials but do not destroy or chemically modify them. By exploiting differences in physical characteristics such as density, vapor pressure, adsorption potential, or particle size, these methodologies extract particulate material from fluids, separate material into different phases, collect dissolved species, or remove vapors and liquids from solid and semi-solid wastes.

Bioremediation is a process that uses naturally occurring microorganisms in the soil to decompose contaminants. It was covered in an earlier

Table 3. Groundwater treatment options for on-site remediation.

Inlet Concentration	Benzene-Toluene-Xylene Discharge Limits			
	Low (<1 µg/L)	Medium (1–5 µg/L)	High (5–50 µmg/L)	Very High (>50 µg/L)
Water Flow Rate: 3–30 gal/min				
Low (<500 µg/L)	CBL	AS/CBV, CBL, UVO, B, SZB	AS/CBV, CBL UVO, SZB	AS/CBV, CBL
Medium (500–5,000 µg/L)	CBL, AS/CBL/CBV	AS/CBV, UVO, B, SZB	AS/CBV, CBL, UVO, SZB	AS/CBV, ISO
High (<5,000 µg/L)	AS/CBL/CBV, AS/CBL/CC	AS/CBV, AS/CC, B, SZB	AS/CBV, AS/CC, UVO, SZB	AS/CBV, AS/CC, ISO
Water Flow Rate: >30 gal/min				
Low (<500 µg/L)	CBL	AS/CBV, CBL, UVO,B, SZB	UVO, B, SZB	AS/CBV, B
Medium (500–5,000 µg/L)	AS/CBL/CBV	AS/CBV, UVO, B, SZB	AS/CBV, B, SZB, UVO	AS/CBV, ISO, B
High (<5,000 µg/L)	AS/CBL/CC	AS/CC, UVO, B, SZB	AS/CC, UVO, B, SZB	AS/CC, ISO, B

Legend:
AS = Air Stripping
B = Bioreactor
CBL = Carbon Bed, Liquid
CBV = Carbon Bed, Vapor
CC = Catalytic Combustion
ISO = *In situ* Oxidation
SZB = Saturated Zone Bioremediation
UVO = Ultraviolet Oxidation
/ Denotes combination of treatment technologies, primary followed by secondary treatment.

Table 4. Applicability of treatment technologies for site remediation.

	Surface Soils	Unsaturated Soils	Saturated Soils	Groundwater
Physical Treatment				
Air Stripping				✔
Carbon Adsorption Beds		✔		✔
Dissolved Air Flotation		✔		✔
Oil-Water Separation				✔
Soil Vapor Extraction	✔	✔		
Soil Washing	✔	✔		
Biological Treatment				
In situ Bioremediation		✔	✔	✔
Bioreactors		✔	✔	✔
Land Treatment	✔	✔		
Composting	✔	✔		
Thermal Treatment				
Incineration	✔	✔		
Thermal Desorption	✔	✔		
Chemical Treatment				
In situ Oxidation			✔	✔
Ultraviolet Oxidation			✔	✔

article *(2),* so this article focuses on two types of *in situ* bioremediation that were not covered previously.

In thermal treatment, heat is used to chemically or physically modify hazardous contaminants. In general, the process treats contaminated soils, sludges, and air, but not water, although aqueous streams containing RCRA-listed wastes may require incineration.

Chemical methods alter hazardous constituents in waste streams to reduce their toxicity or mobility or to create inert compounds from the source materials.

Soil washing

Soil washing, also known as soil flushing, removes water-soluble contaminants from soil. In one version, water is injected into the contaminated soil through a gallery system across the contamination zone to render the contaminants soluble. The water containing dissolved hydrocarbons is then recovered downgradient (that is, "downstream," except that there is not a discrete "stream" but rather a flow front) or in the center of the soil contamination. The contaminants are removed from the water, which is usually treated before reinjection.

A more complex application, depicted in Figure 2, entails loading excavated soils into a hopper. A screening device first removes rocks and other large objects. A solvent is added to the soils in a reaction vessel and contaminants are drawn into the solvent. Soils are then separated from the solvent via a combination of decanters, dryers, and other separation devices such as plate-and-frame filter presses. After contaminants are removed from the solvent for additional treatment or disposal, the solvent can be recycled.

Both organic and aqueous agents are used in soil washing, sometimes with additives to help mobilize contaminants. Surfactants may be very effective for organic contaminant removal, while acids are useful for metal removal.

Soil washing is most effective on sandy soils of moderate to high permeability and on soluble compounds at sites where cleanup standards are not very stringent. The technology is particularly feasible for water-soluble contaminants that will subsequently be removed from aqueous streams by means of temperature variation, extraction, or some other method. Soil washing is not usually done *in situ,* but rather is applied to excavated soils.

Organic contaminants, including volatiles, semivolatiles, and halogenated organics, are especially receptive to soil washing. The method has also proven effective in extracting organically bound metals, such as tetraethyl lead.

Soil washing does present some difficulties, however. Lower permeability soils and clay-like soils can pose significant handling problems and take longer to treat. Solids handling causes wear and tear on the equipment and problems in maintaining the soil in a slurry form. Careful system design is crucial since soil deposits in low-flow-rate areas of a treatment system can result in plugging, incomplete cleanup, and expensive maintenance. While the technology will quickly remove gross levels of contamination, it usually must be combined with another technique to meet current treatment standards. Finally, wash water must be treated before discharge.

Air stripping

Air stripping transfers volatile sparingly soluble compounds from water into the air. The method works on the premise that most volatile hydrocarbons are virtually insoluble in water and are attracted to air as a carrier.

A packed tower is usually the most efficient method of air stripping. Contaminated water is pumped from the ground and sprayed into the top of the tower. The water flows downward through the packing as air is fed into the bottom of the tower and blown upward. As the air and water pass through the packing layer, the hydrocarbons are transferred into the air.

To comply with emissions standards, the air stream leaving the tower may need further treatment to remove or destroy contaminants collected from the water. Vapor-phase carbon adsorption units or catalytic incinerators on the exhaust air outlet are typical treatment systems. Depending upon the treatment standard, treated water exiting the tower may be directly discharged to the environment or further treated with activated carbon.

Removal efficiencies can be greatly enhanced by preheating the air stream or otherwise raising the system temperature.

The most common problem with air stripping is scaling due to a buildup of insoluble metal salts, suspended solids, or biological growth on the packing media. Scaling slows the rate at which hydrocarbons are transferred to air. Tower packing usually requires cleaning or replacement once or twice yearly. If scaling occurs more frequently, periodic in-place tower flushing with chemical solutions may provide a temporary improvement. [For more information on dealing with packing fouilng, see *(3).*]

Oil-water separation

Oil-water separation is a pretreatment process that removes free-phase and emulsified oil, suspended solids, and compounds associated with the oil phase. Two processes,

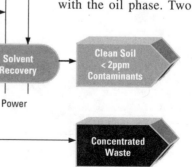

Figure 2. Flowsheet of a soil washing system.

dissolved air flotation (DAF) and gravity separation, are typically used. Both are fairly simple and inexpensive, feasible for removing oily contaminants in concentrated form, and highly effective (>90%) for polyaromatic hydrocarbon (PAH) reduction in suspended material.

In DAF, soil particles and oils float by attachment to fine air bubbles, which are released by depressurizing the water. The oily foam is skimmed off for recycling or disposal. DAF is best applied to waste streams in which the specific gravity of the material requiring separation is close to that of water. Because these particles settle very slowly or not at all, flotation and removal on the water surface is easier than natural separation.

In gravity separation, contaminated water is retained in a holding tank or coalescer. Materials with a specific gravity less than water float to the surface, while those with a greater specific gravity sink. Surface skimming and bottom collection systems in the tank then remove the separated oil. Gravity separation is typically used to remove free oil droplets larger than 150 microns.

Carbon adsorption

Carbon adsorption is a convenient process for collecting low levels of contaminants that can yield very low effluent concentrations in water and air streams. It is a highly efficient method for removing many hydrocarbons from carrier streams and is frequently used as a polishing step following some other treatment. Generally, a compound's adsorption is favored by increasing molecular size and aromaticity, and decreasing solubility, polarity, and carbon-chain branching. Some dissolved metals and inorganics, including arsenic, mercury, silver, chromium, and chlorine, have also shown good potential for carbon adsorption.

The carbon adsorption process bonds contaminants from an aqueous or vapor phase to the surface of the carbon, where they accumulate pending extraction or destruction. This phase transfer is the result of a contaminant's low affinity for the liquid or vapor phase and/or its strong attraction to the solid phase.

In carbon adsorption, carbon in granular or powder form is typically packed in a canister or tank. Generally, two or more of these fixed beds are partially filled with activated carbon and arranged in a series. Water to be treated flows downward through the carbon and exits at the bottom of the container. (A vapor-phase carbon system is similar except that contaminated gas flows upward and exits at the top of the container.)

Eventually, the carbon's surface becomes selectively deactivated as the active sites on the carbon become filled. When most of these sites are filled, hydrocarbon removal slows and "breakthrough" occurs as some of the hydrocarbons leave the bed with the water. To determine when breakthrough has set in, the exiting stream is sampled periodically.

G. M. LONG is a certified hazardous materials manager and general manager of ENSR Consulting and Engineering's Somerset, NJ, office (908/560-7323; Fax: 908/560-1688). He has 20 years of experience in the field of hazardous waste remediation and has special expertise assisting industrial companies with selecting and implementing a broad range of remediation technologies. He has designed, permitted, installed, and operated a variety of state-of-the-art technologies, including bioremediation and vapor extraction systems. He holds a BS in chemical engineering from Lafayette College and an MS in chemical engineering from the Univ. of Massachusetts.

Mr. Long is the primary author of ENSR's special report "A Guide to On-Site Remediation of Hydrocarbons." To obtain a free copy of that report, write on business letterhead to: ENSR Consulting and Engineering, Marketing Dept., 35 Nagog Park, Acton, MA 01720.

When breakthrough occurs, the lead (first) bed is removed and replaced with the second bed, and another bed is placed in the second position. The saturated carbon bed is regenerated on- or off-site or is disposed of entirely.

The cost and effectiveness of carbon adsorption depend on a variety of factors, such as the molecular weight and structure, solubility, and polarity of the compound being removed, and the pH and temperature of the aqueous phase. If a small amount of material is being collected, carbon adsorption is an inexpensive process.

[For a detailed discussion of carbon adsorption for water treatment, see (4). Carbon adsorption for recovering VOCs from air will be covered in the July issue. — Editor]

Soil vapor extraction

Soil vapor extraction, also known as soil venting and soil air stripping, allows remediation of volatile hydrocarbon contamination from the ground without the need for excavation. The process can also be used for above-ground soil piles. The approach is most useful for pure volatile chemicals and solvents (trichloroethylene, perchloroethylene, hexane, etc.) and volatile blended materials (gasoline, naphtha, solvent mixtures).

Soil vapor extraction is most effective after LNAPL has been entirely removed in high porosity soils that have a low water content and high permeability. If LNAPL is present, the low-surface-area LNAPL phase serves as a continuing source of contaminants to the air stream. The process is not very efficient in removing LNAPL.

In soil vapor extraction, a vacuum pump or blower moves air through the soil near the hydrocarbon contamination. As contaminated air is removed, cleaner air moves through the soil to replace it. This air movement also promotes microbial degradation of contaminants at many sites. The contaminant vapors are vented to the atmosphere or treated or destroyed in aboveground facilities. Sometimes the treated air is injected back into the ground to promote a circulation pattern.

The effectiveness of the method is determined by the Henry's Law constant for soils with moisture contents above approximately 15% and by the vapor pressure of the compounds in drier soils. Underground utilities or other channels that provide preferential flow paths for airflow, or tight soils that cannot be desaturated, may impede soil vapor extraction. Also, diesel fuel cannot be remediated through vapor removal alone because it is typically limited to no more than about 50% volatilization at practical vacuum levels. However, biological degradation will usually eliminate much of the remaining fuel if the soil moisture content is sufficient.

Saturated zone bioremediation

Saturated zone bioremediation treats soils and groundwater contaminated with biodegradable materials below the water table.

The key to success is careful control of hydrogeological, geochemical, and microbial activity. Pumping systems in wells, interceptor trenches, or other water recovery systems located at the leading edge of the contaminant plume control the groundwater flow near the contamination to capture contaminants and keep them from spreading. After treatment at the surface, water is generally reintroduced upgradient of the plume to establish a circulation loop.

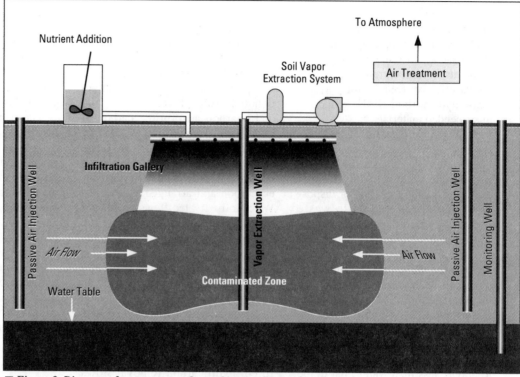

■ Figure 3. Diagram of an unsaturated zone bioremediation system.

The water may be reinjected above or below the water table depending upon the depth of contamination.

Before reinjection, the water is treated with nutrients, oxygen, and pH control additives to optimize subsurface bioremediation. Soil and water sampling ensures proper functioning of the bioremediation system, optimization of microbial activity, and protection of the groundwater aquifer.

Unsaturated zone bioremediation

Unsaturated zone bioremediation broadens the application of *in situ* bioremediation to soil between the surface and the water table. It is an especially useful technique for deeper regions where excavation is impractical and for soil contamination below buildings and other fixed structures. The method works best for fuels or hydrocarbon products that are not as volatile as gasoline and that cannot be vented from soil, but that are low enough in vis-

cosity to allow nutrients, air, and moisture to travel through the contaminated area. It can be applied to any site where permeable soils allow aqueous flushing of the unsaturated zone for nutrient and oxygen transport.

This type of bioremediation uses the microorganisms in the unsaturated zone to degrade hydrocarbon contaminants. After contamination has been located, an air movement system similar to a soil vapor extraction system is installed to provide fresh oxygen to the contaminated zone and remove volatile materials.

In typical applications, as shown in Figure 3, air withdrawal wells located in the middle of the contaminated zone draw fresh air from the perimeter. A liquid injection or filtration system covers the remediation zone with water containing nutrients for degrading hydrocarbon contaminants. Water may be added continuously or on a pulsed basis, but it should not saturate the soil or reach the water table. As an alternative, groundwater wells may be used to control water under the contaminated unsaturated zone and to recover and treat groundwater. The treated groundwater may be used for reinjection or infiltration. To prevent leaching of contaminants downward into the groundwater, the irrigation dosage must be carefully planned.

Catalytic combustion

Catalytic combustion burns contaminants in vapor streams before discharge to the atmosphere. In a process similar to that used to minimize exhaust gas hydrocarbons from automobiles, burned exhaust moves across a catalyst bed to complete a burning reaction. The technique treats pump discharges from soil vapor extraction systems, vapor streams from air stripping, collected air from land farming or composting within a building, or other process off-gases. Since it is essentially a permittable fume incinerator, catalytic combustion can destroy vapor-phase contaminants on-site.

However, it is important to realize that this technology is only for combustible contaminants. It is not effective for chlorinated materials at more than approximately 20% of the total hydrocarbon concentration. When chlorinated species are present, acid-resistant construction materials are necessary to minimize corrosion from the acid gases created by the chlorine. For low influent air concentrations, activated carbon may be more cost-effective in treating air stripper vapors. In addition, another limitation of catalytic combustion is that the catalyst is subject to fouling by sulfur and metals.

Because the initial costs for catalytic combustion are high, the process is more cost-effective for larger sites or for high vapor loadings that would require large amounts of activated carbon for treatment. In states that require operating permits for air discharges, a permit and process documentation may be necessary to meet regulatory requirements.

Incineration

Incineration, the controlled combustion of organic wastes at high temperatures (usually higher than 1,800°F), is frequently the most complete and permanent destruction technology for heavier hydrocarbons and chlorinated species. Incineration also eliminates most future environmental liability problems and greatly reduces waste volume. Through burning, it converts hydrocarbons in soils, sludges, liquids, or fumes to water and carbon dioxide.

Incineration can effectively treat just about any type of hydrocarbon. Though typically an expensive process (the estimated median cost is $1,000/ton), it may actually be the least costly alternative for treatment of heavy oils, tars, and some sludges.

Both on-site and off-site incinerators may be used. On-site incineration is usually best for large sites because meeting effluent limitations requires substantial pretreatment and exhaust treatment systems. Off-site incineration is frequently most useful for smaller sites to avoid the cost and time associated with permitting and equipment mobilization.

An incineration system includes a feed system, primary and secondary combustion chambers, and air pollution and ash control systems. Critical to safety is a raw material staging area for collecting and classifying materials and sorting out those materials that are not suitable for incineration. Rotary kilns and fluidized beds are the types of units typically used for soils.

In addition to high cost, the biggest disadvantage of incineration is the controversy surrounding potential health and environmental impacts. However, recent studies have shown that well-designed and -operated incinerators should not significantly affect human health and the environment. The local political climate generally determines the level of difficulty involved in securing approval for incineration.

Thermal desorption

A thermal process that uses lower temperatures than incineration (500° to 1,000°F), thermal desorption vaporizes most hydrocarbons from soil or sludge surfaces and recondenses them as liquids. They are then recycled if possible, or they are destroyed. Low-temperature thermal desorption is commercially available for removal of VOCs, semivolatiles, polychlorinated biphenyls (PCBs), and arsenic from soil. The technology is usually applicable where the materials to be treated are of intermediate molecular weight (fuel oils, mineral oils, lubricating oils) and the soil is relatively impermeable.

In thermal desorption, excavated soil is fed to a device that applies enough heat to volatilize and drive off the contamination. The contaminants are then burned in a secondary combustion chamber or separated from the gas stream by condensation or absorption. Components include a kiln, a feed preparation and treatment system, and condenser and vapor treatment systems.

Obtaining a permit for thermal desorption tends to be easier than for incineration. Thermal desorption seems to be more acceptable to the public because it does not burn anything. Also, costs are usually lower than for incineration but still higher than for *in situ* methods.

In situ chemical oxidation

In situ chemical oxidation decomposes or oxidizes hydrocarbons under controlled conditions. This technology is used when hydrocarbons are too concentrated or toxic to bacteria to allow direct bioremediation, but can be oxidized using air or peroxide. Soils must be sufficiently permeable for the air or peroxide solution to reach the hydrocarbons and for reaction products to move away from the area. Among the chemicals amenable to treatment are oxygenated solvents, aldehydes, organic acids, phenols, organosulfur compounds, and cyanides. The costs for *in situ* oxidation are about the same as for *in situ* bioremediation.

Note that the reaction rate for *in situ* oxidation must be carefully controlled. If the oxidation reaction reaches combustion rates, it can be dangerous. Furthermore, iron and manganese in the soil will compete with contaminants for oxygen. The technology is not feasible if LNAPL is present.

UV-enhanced oxidation

This technique chemically decomposes hydrocarbons in aqueous streams into carbon dioxide and water. Ultraviolet oxidation is usually most successful for low influent concentrations where very low effluent concentrations are required. Ultraviolet oxidation is not as cost-effective for influents in high concentrations nor can it easily handle contaminant spikes. The process offers on-site destruction of dissolved chlorinated hydrocarbons and does not create any secondary waste streams.

Ultraviolet oxidation works only on clear aqueous streams. The U.S. Environmental Protection Agency (EPA) lists ultraviolet oxidation as best available control technology (BACT) for treating PCBs, cyanides, and polyaromatic hydrocarbons in water.

In ultraviolet oxidation, hydrogen peroxide and ozone, typically employed as oxidants, are injected by themselves or together into the stream to be treated. The stream then moves through a bank of ultraviolet lamps to "activate" the oxidizers. This highly active solution rapidly breaks down the hydrocarbons into carbon dioxide and water (and chloride ions when chlorinated hydrocarbons are present).

Reaction rates differ for each chemical species under treatment, so the size of the system is usually geared to removing the slowest-reacting chemical at maximum concentration. A pilot test is generally required to properly size a unit.

Summing up

Hydrocarbon contamination of soil and groundwater exists at thousands of sites nationwide. Fortunately, a number of technologies are available for effective site remediation.

The key to choosing the right technology or combination of techniques is a thorough site characterization and feasibility study based on the technical, economic, regulatory, and political issues peculiar to the site. Depending on the technologies being considered and the site, the main considerations may include:

• concentrations of contaminants in the soil and/or groundwater;

• required clean-up level or final concentration;

• amount of soil and/or groundwater to be treated;

• soil characteristics, such as type, particle size, permeability, porosity, moisture content, pH, uniformity or stratification, sorption potential, chemistry, nutrient availability, and mineralogy;

• groundwater properties, such as flowrate, direction, hydraulic conductivity, pH, temperature, dissolved oxygen content, and oxidation/reduction potential;

• depth to groundwater and relative depth of contamination;

• physical and chemical properties of the contaminants, such as phase, specific gravity, solubility in water or other solvents, Henry's Law constant, boiling point, reactivity, and waste classification, as well as the presence of solids and oils;

• presence and properties of compounds that could interfere with treatment, such as suspended solids, foulants, or materials that are toxic or inhibitory to microorganisms;

• local permitting requirements and neighborhood constraints; and

• locations of underground utility lines, pipes, and trenches. **CEP**

Literature Cited

1. **Davenport, G. B.,** "The ABCs of Hazardous Waste Legislation," *Chem. Eng. Progress,* **88**(5), pp. 45–50 (May 1992).
2. **Bradford, M. L., and R. Krishnamoorthy,** "Consider Bioremediation for Waste Site Cleanup," *Chem. Eng. Progress,* **87**(2), pp. 80–85 (Feb. 1991).
3. **Bravo, J. L.,** "Effectively Fight Fouling of Packing," *Chem. Eng. Progress,* **89**(4), pp. 72–76 (April 1993).
4. **Stenzel, M. H.,** "Remove Organics by Activated Carbon Adsorption," *Chem. Eng. Progress,* **89**(4), pp. 36–43 (April 1993).

Further Reading

American Petroleum Institute, "A Compilation of Field-Collected Cost and Treatment Effectiveness Data for the Removal of Dissolved Gasoline Components from Groundwater," API, Washington, DC, Publication No. 4525 (Nov. 1990).

Hopper, D., "Cleaning Up Contaminated Waste Sites," *Chem. Eng.,* **96**(8), pp. 94–110 (Aug. 1989).

Sims, R. C., "Soil Remediation Techniques at Uncontrolled Hazardous Waste Sites," *J. Air Waste Management Assn.,* **40**(5), pp. 704–732 (May 1990).

U.S. Environmental Protection Agency, "Handbook on *In Situ* Treatment of Hazardous Waste-Contaminated Soils," EPA Risk Reduction Engineering Laboratory, Cincinnati, OH, EPA/540/2-90/002 (Jan. 1990).

U.S. Environmental Protection Agency, "Innovative Treatment Technologies: Semi-Annual Status Report," EPA, Technology Innovation Office, Washington, DC (first issued Jan. 1991).

Bioremediation — Why Doesn't It Work Sometimes?

Biological treatment of petroleum-contaminated soil has not always achieved the desired results. Here's how to evaluate a site's suitability for bioremediation.

**Robert Block,
Hans Stroo, and
Geoffrey H. Swett,**
RETEC

Biological treatment has rapidly become the technology of choice for remediation of soils contaminated by petroleum constituents. Since the mid-1980s, bioremediation has been used at more than 100 locations to cost-effectively remediate hundreds of thousands of cubic yards of contaminated soil. Two Superfund sites have been completely remediated using biological treatment, 25 other sites have treatment plans that include bioremediation, and bioremediation is being considered for more than 140 additional sites.

Numerous factors have contributed to the rapid and successful commercialization of bioremediation:

- low capital and operating costs;
- minimal specialized equipment requirements;
- the low profile of the technology; and
- availability of trained contractors to implement the technology on a fixed-price basis.

However, despite the excellent track record of bioremediation, during the past few years bioremediation was not successful at several sites. In retrospect, it is not surprising that these failures occurred. The project team broke the cardinal rule of remediation: *Always conduct some level of treatability work to guarantee that a unique condition does not exist that could render the chosen remedial alternative ineffective.*

Yet even after successful treatability testing, the lack of biological degradation in some cases was still puzzling. The constituents were supposedly biodegradable diesel fuel components. The same type of contaminated soils has been treated successfully at numerous other sites. The treatment process was the same, but bioremediation was not effective.

Testing identified other sites where bioremediation was unsuccessful for remediating petroleum constituents, and the factors that contributed to the failures were explored in greater depth. This article outlines a quick and inexpensive screening technique that allows one to determine whether bioremediation is practical and also provides an assessment of the time and cost factors. It involves four steps:

1. Site study.
2. Regulatory analysis.
3. Biological screening.
4. Treatability testing.

The methodology can be reduced to a set of decision trees to simplify the screening process. To obtain a copy of these, contact Geoffrey Swett at the telephone number on p. 45.

Site study

Bioremediation is appropriate for many sites. However, it is not a universal solution for the treatment of contaminated soils. The site screening process is critical in identifying candidate sites suitable for more detailed review.

Biodegradability of the contaminants. If the contaminants requiring treatment are not biodegradable, then the technology is not effective. While most petroleum constituents are biodegradable, rates of biodegradation can vary dramatically.

Variables include the source of the crude oil, refinery capabilities, and the blend of streams generated from crude distillation and downstream processing. This diversity makes exact chemical description of a specific petroleum product impossible. Therefore, one refiner's diesel may biodegrade in a significantly different manner than another's.

Normal, branched, and cyclic alkanes are the most abundant components of petroleum products. In general, normal alkanes are readily degradable. Branched and cyclic alkanes are also degradable; however, the rate of degradation decreases with increasing molecular size and complexity. For example, pyrene may take six times as long to treat as benzene.

The other major components of petroleum products are aromatic compounds, which are also biodegradable. However, like the branched and cyclic alkanes, the larger and more complex molecules take longer to biodegrade.

The literature contains a large quantity of data about the biodegradation of petroleum hydrocarbons. Much of the recent data lie in the hands of bioremediation process developers and contractors. These companies have shown that many compounds previously considered non-biodegradable can be cost effectively bioremediated. These developers and contractors should be contacted if literature data are insufficient to evaluate the feasibility of treating a particular contaminant.

Time requirements. Bioremediation typically takes two months to two years to complete, depending on the volume of soil requiring treatment and the availability of space. Many projects, particularly those related to real estate transactions, have severe time constraints for completion. For these sites, bioremediation may not be a viable option, and an alternative remediation technology should be considered.

Site conditions. Biological treatment of contaminated soil has particular site requirements. Soil must be handled on-site, and support facilities and equipment are needed. In order to deliver sufficient oxygen to the soil, space for tilling or installing vent pipes and blowers must be available for the duration of the project. If there are numerous buildings or obstruc-

tions on-site, *in situ* options may need to be explored. There should be sufficient access to affected areas.

Space constraints are not usually a major problem. In some instances, soils contaminated from underground leaking tanks have been bioremediated in as limited an area as a gas station.

Volume of materials to be treated. Bioremediation has certain fixed costs. In essence, the costs to conduct engineering and laboratory evaluations for 250 cubic yards or 3,000 cubic yards are similar. Due to these fixed costs, smaller projects are more costly on a per-yard basis. While the minimum volume can depend on the remediation options available, typically the minimum range is 250 to 500 cubic yards.

Costs of alternative remedial options. Bioremediation is only economically viable if it costs no more than other remedial options. Unfortunately, the costs of all remedial technologies are very site-specific, so there is no easy way to develop cost data for any given site. Site remediation planners should use screening cost estimates to narrow the number of alternative treatment technologies that warrant detailed evaluation.

An estimate of $25–$75/yd^3 for the bioremediation of petroleum hydrocarbons boiling at less than 650°F can be used for screening purposes. This screening cost estimate is also valid for soils contaminated with two- and three-ring polynuclear aromatic hydrocarbons. For other compounds it is necessary to check with a bioremediation contracting firm that has experience with the chemicals of concern to obtain a screening cost estimate.

The literature contains many sources that can be used to develop cost estimates for alternative remedial options. A particularly useful one is *(1)*.

Regulatory analysis

If a regulatory agency selects a treatment objective that cannot be achieved by bioremediation, then some other, frequently more costly, remediation technique must be employed. Usually, treatment objectives are developed on a case-by-case basis (although a few states are in the process of adopting state-wide standards). Thus,

R. BLOCK is a geotechnical engineer in Remediation Technologies, Inc.'s (RETEC's) Concord, MA, office. He has more than 16 years of experience in geotechnical and remediation engineering, and he has been responsible for the design and operation of numerous bioremediation projects across the U.S. at sites regulated under CERCLA and RCRA. This work included the design of the first Superfund bioremediation project to achieve treatment objectives in 1990. He received his MS in geotechnical engineering from the Massachusetts Institute of Technology.

H. STROO is manager of RETEC's Biotreatability Lab in Seattle and developed the Treatability Index Test procedures. He has more than 12 years of research and remediation experience in the use of microorganisms in the management of hazardous waste. He was responsible for the development and laboratory testing of a system for *in situ* biological treatment of creosote contaminated soil. In addition, he was responsible for the simulation of native microorganisms for the treatment of 6,500 cubic yards of contaminated soil at the first successful bioremediation Superfund site. He received his BS in soil science from Oregon State Univ., MS in agronomy from West Virginia Univ., and PhD in soil microbiology from Cornell Univ.

G. H. SWETT is a senior program manager at RETEC in Tucson, AZ (602/577-8323; Fax: 602/577-7455), specializing in the hazardous-waste management practices of heavy industry, including petroleum refining, petrochemicals, pharmaceuticals, and steel. He has more than 20 years of experience in the environmental management field. Prior to joining RETEC, he played a major role in the commercialization of UV/oxidation technologies. More recently, he has participated in the commercialization of RETEC's thermal desorption process. He has published numerous articles on UV/oxidation, bioremediation, and refinery oily sludge management. He received his BA in chemistry and mathematics from the Univ. of Denver and an MBA in finance from Golden Gate Univ.

numerous different approaches to setting treatment objectives have been used. The approach selected can be critical to the success of a bioremediation project.

Limiting regulations. Bioremediation frequently involves the introduction of an electron acceptor to increase oxygen availability. The amount of dissolved oxygen available in the subsurface can, in some instances, be most effectively increased with the use of hydrogen peroxide. There are also site-specific conditions that favor anaerobic biodegradation using anionic nitrate as an electron acceptor. In addition, it is routine to enhance biodegradation by adding inorganic nutrients (ammonia-nitrogen and orthophosphate) to injected groundwater.

These enhancements can be counter to state policies against chemical additions to groundwater. Some states have policies that prohibit the injection of chemicals into groundwater. These restrictions are in place to protect against aquifer degradation. However, in some instances, such restrictions can effectively forestall an otherwise applicable biological treatment scenario.

Statutory treatment criteria. Many regulatory agencies have selected total petroleum hydrocarbons (TPH) as the basis for establishing treatment requirements. Others have chosen a total of selected compounds, such as total benzene, toluene, ethyl benzene, and xylene (BTEX) or total polynuclear aromatics. In some cases, specific treatment objectives for individual compounds have been established. Each of these methods can establish barriers to successful bioremediation that do not necessarily provide additional protection to human health and the environment.

When no treatment criteria exist, the state or the U.S. Environmental Protection Agency (EPA) will either perform a risk assessment or require that the principal responsible parties (PRPs) perform a risk assessment for the site in question. Most PRPs will want to perform their own risk assessment, even when treatment criteria exist or when the state or the EPA also performs a risk assessment. Once risk assessments have been concluded, site assumptions can be discussed and treatment levels negotiated. In general, risk assessments are based on exposure pathways and concentrations of indicator parameters or leachable fractions of these constituents. [See *(2)* for further discussion of risk assessment.]

The leachable portion approach to setting treatment objectives is consistent with EPA's policy of establishing priorities based on relative risk.

TPH-based treatment objectives. One of the principal problems with using TPH as an indicator parameter is that there are no uniform national standards for measuring TPH. One lab's TPH may be different from another lab's TPH for the same sample.

All TPH analytical methods involve extraction of the environmental matrix with a solvent. While the solvent is very efficient at removing petroleum hydrocarbons, it also will remove nonpetroleum hydrocarbons, such as organic acids, esters, alcohols, and ketones. If these compounds are not removed from the extract before analysis, they will affect the results.

In the gravimetric TPH method, a solvent is used to extract the hydrocarbons from the soil. The solvent is then evaporated and the TPH concentration is determined based on the petroleum residue remaining. The problem with this technique is that many volatile petroleum components are evaporated with the solvent.

Infrared and gas chromatographic techniques are also available, but each of these has particular constraints. EPA Method 418.1 is an infrared technique involving measurement of the extract's maximum absorbance at 2,930 cm^{-1}, the absorbance of a linear carbon-hydrogen bond. This method is subject to interference from natural organics having linear carbon-hydrogen bonds. Additionally, this method relies on comparison of the absorbance of the environmental sample to a representative standard, which may be difficult or impossible to obtain.

More recently, a gas chromatographic technique has been employed whereby a flame ionization detector is used to identify the quantity of individual chemical constituents. Quantification with this method requires integration of the chromatogram and comparison with a known concentration of a standard presumed to be representative of the petroleum hydrocarbon in the sample.

While the analytical problems associated with using TPH for establishing treatment criteria are substantial, they pale in comparison to the risk assessment problems associated with TPH regulatory standards. Since TPH is a broad indicator of contamination, it should not be used to measure the risk the material presents to human health and the environment. Compounds with relatively high toxicity, such as benzene, are treated the same as compounds with low toxicity, such as hexane. Scarce cleanup dollars are often misallocated when TPH is used to establish treatment objectives.

The use of chemical-specific treatment objectives resolves the two major problems associated with TPH — analytical inconsistency and misallocation of cleanup dollars.

There are specific, accepted analytical techniques for most contaminants of concern. While these tests are more costly than TPH analysis, they all have a common methodology, and the results are generally comparable from lab to lab.

By relating cleanup objectives to individual contaminants, remediation dollars are focused on the chemicals with the largest potential risk.

Leachable-fraction-based treatment objectives. Measuring total levels of contaminants does not indicate the contaminants' availability to humans and the environment. Therefore, EPA established the toxicity characteristic leaching procedure (TCLP) *(3)*. There was a need to develop a method to measure the amount of contaminant available to the environment — just because a compound is present in the soil does not mean the material will leach out of the soil to adversely impact human health and the environment. Some soils leach contaminants very easily, while other soils bind contaminants up. Obviously, some contaminated soils pose a greater risk to the environment than other contaminated soils.

For example, we have compared TCLP values for individual polyaromatic hydrocarbon (PAH) compounds and total PAH in three soils before and after bioremediation. Despite wide variation in the three soils, losses of leachable PAH were greater than total losses, and this was true for all individual PAHs as well as the total PAH concentrations. In fact, for most PAHs, the final TCLP concentrations were below detection levels.

Compound-specific treatment objectives based on the leachable portion of the contaminants correspond more closely to the true risk to human health and the environment. This is particularly important in bioremediation projects, since much of the nonleachable fraction is also not available for biological degradation. Bioremediation generally yields far greater reductions in the leachable component. Furthermore, the leachable portion approach is consistent with the EPA's policy of using relative risk to establish priorities. EPA is increasingly concerned that the limited quantity of remediation funds are utilized in the most cost-effective manner.

Achievability of treatment criteria through bioremediation. The literature contains substantial data on the treatment levels that can be achieved by bioremediation. In addition, most of the recent data have been developed by bioremediation process developers and contractors. The literature data and process vendor data should be consulted to determine if the treatment objectives identified by the regulatory agency can be achieved through bioremediation.

Biological screening

Once a site has been identified as a suitable candidate for bioremediation, there are numerous biotechnology factors that will affect the success of the project.

Bulking or blending can often make inappropriate soil matrices amenable to bioremediation.

Soil/contaminant matrix. The soil/contaminant matrix at petroleum contaminated sites is a complex mixture of individual solid particles, liquids, and vapors. The primary solid components are particles comprised of various minerals in roughly spherical or platelet shapes. Organic matter and other solids are also present. Often, occluded droplets of petroleum product are present at these sites, randomly interspersed in the soil matrix. The primary liquid is water, and this water often contains a fraction of dissolved contaminants. The vapor phase is comprised primarily of air. Volatile petroleum contaminants will saturate the air, if they are present.

Soil texture significantly impacts bioremediation. Contaminants in soils with a high percentage of clays are generally biodegraded more slowly than the same contaminants in soils with lower concentrations of clays. There appear to be several possible reasons for this:

- Contaminants are adsorbed to the surface of the soil particles or organic matter;
- Contaminants are diffused into micropores of the particles; or
- Contaminants are trapped by surface tension as occluded droplets among the particles.

Sand and silt particles are generally spherical and have significantly less surface area than clay particles, which consist of flat platelets. Therefore, the potential for contaminant adsorption or diffusion is less with sandy soils than with clay.

Additionally, the sand and silt particles are less influenced by unbalanced electrolytic forces at the surface of the particles. Clay platelets inherently have unbalanced electrical charges at the surface with distinctly negative faces and positive edges. Surface force interactions coupled with the large specific surface areas of these particles serve to adsorb more contaminants to the clay particles.

Successful remediation of soils having higher clay contents depends on deflocculating the clay particles to reduce the surface tension effects and neutralizing the negative electrical field at the clay particle's surface.

Unfortunately, the existing database is not adequate to predict what percentage of clay will make bioremediation infeasible. At the current time, this factor is best evaluated by treatability tests.

Sorption of contaminants to organic matter also affects biodegradation. Many industrial sites have soils with a high concentration of organic matter, which can adsorb contaminants. It is difficult to remove the contaminants from the organic matter and bioremediation will, therefore, be impractical.

The sorption of contaminants onto organic material is likely to be a common explanation for the occasional failure of bioremediation to remove otherwise biodegradable contaminants. It is a difficult mechanism to prove or to predict without site-specific testing.

Bulking or blending can frequently be economic solutions to inappropriate soil matrices. If liquid-solids treatment is chosen as the biological remedy, then agents can be blended into the slurry mixture to effect desorption. Liquid-solids treatment involves contacting microorganisms with organic wastes in an aqueous suspension. Alternatively, if land treatment is the chosen alternative, bulking agents, such as manure or saw dust, can be tilled into the treatment zone to reduce the inhibitory effects of desorption.

Aging of the soil/contaminant matrix may also have a significant impact on the success of biological degradation. Microbial degradation of contaminants relies upon solubilization of the contaminant into the water phase where microorganisms can metabolize the contaminants.

Contaminants adsorb to soil particles as a result of electrical forces at the surface of the particle. Given increasing times of exposure of the contaminant to the soil mass, the contaminant migrates closer to the surface of individual soil particles, increasing the attractive force holding the contaminant to the particle. The impact of aging is to bond the contaminant more tightly to the soil matrix and reduce the ability of the contaminant to dissolve into the water phase.

If the contaminant does not dissolve into the water phase, it will not be biologically degraded. The microorganisms simply cannot reach the contaminant. The aging phenomenon not only reduces the mobility and bioavailability of the contaminant, but also reduces the risk of human or environmental exposure. This fact can be confirmed by evaluating the leachable fraction of contaminants. If the contaminants will not leach, they will not be available to microorganisms, humans, or the environment.

Data are not available to establish hard and fast rules concerning the effect of aging. As noted, testing for leachable fractions is one means to evaluate the possible effect of aging. Treatability testing will confirm the suitability of the soil for bioremediation.

Salinity. Salinity is one of the most common constraints on biologi-

■ *Soil pan testing replicates field conditions on a smaller scale.*

cal activity in contaminated soils, sediments, and sludges. Typically, the electrical conductivity (EC) of a saturated paste prepared from a solid sample is used as a guideline to salinity levels. Saline soils, for example, are classed as soils with a saturated paste EC greater than 4.0 mmhos/cm. Bioremediation can be successful in saline soils, but upper limits for effective biodegradation are generally in the range of 6 to 10 mmhos/cm.

Salts can be removed by leaching the soil in a prepared bed. Water with a low dissolved solids content is passed through the contaminated soil to remove the salts. The resulting leachate is collected and organic contaminants are removed. The treated leachate is then discharged to a publicly owned treatment works (POTW) or to a water body in accor-

dance with a National Pollutant Discharge Elimination System (NPDES) permit.

pH. A pH of between 6 and 8 is ideal for biological activity. However, there have been successful bioremediation projects outside of this ideal range. pH can be adjusted by adding lime, caustic soda, or acids. The benefits of pH adjustment should be weighed against the additional cost.

Toxicity. Toxicity can occur due to the organic or inorganic contaminants present in the soil matrix. Oily wastes can be toxic or inhibitory, either through direct toxicity or by restricting oxygen and water availability. Oil concentrations in excess of about 10% oil and grease have repeatedly been shown to inhibit biological activity. Metals and organic contamination may also be at toxic levels in soil. Pentachlorophenol and halogenated solvents are common organic toxins at many potential bioremediation sites. If methods to reduce toxicity are not practical, bioremediation often will not be successful.

One of the most cost-effective means to reduce toxicity is on-site blending. For example, soils with oil and grease concentrations in excess of 10% are blended with soils having concentrations less than than 10%. The resulting mixture can then be economically treated.

In some cases, it is possible to immobilize toxic constituents. Typically, this can be achieved by stabilizing toxic metals in the soil. Care needs to be taken in selecting the stabilization technique — one wants to stabilize toxic metals, not the organics, in the soil matrix.

Microbial populations. It is necessary to verify that a sufficient number of microorganisms capable of degrading the principal contaminants at the site are present. The total population and the numbers of specific degraders are determined in a bioremediation laboratory using standardized procedures.

Microbial counts in excess of 10^5 cells per gram are required for effective bioremediation. Sites typically contain 10^6–10^9 cells per gram.

If microbial counts are not sufficient, it is possible to increase the number of microorganisms. This can be done on-site in a fermenter or other suitable equipment. By using the naturally occurring organisms as a seed culture, one is assured of growing a population that is suitable for the on-site contamination. Since the microbes have been living in the contaminated soil, they are well-suited to the site environment.

In some cases, microbial populations can be increased by purchasing cultures grown in fermenters at a centralized facility. Microbes grown at a different location will need to become acclimated to site conditions (this can typically take 30–60 days), which reduces their cost-effectiveness in many situations.

Treatability testing

Finally, the screening procedure requires a certain amount of laboratory screening of the samples.

Initial testing. The first steps in this process we refer to as a Treatability Index Test, which includes an initial chemical, physical, and biological characterization. Typically, this analysis can be conducted for approximately $5,000 to $8,000 plus outside analytical lab costs. The specific tasks included in the Treatability Index Test are briefly described below.

The objectives of the initial chemical and physical characterization tests are to characterize the hydrocarbon contamination and assess potential limiting factors.

The organic constituent analyses for the compounds of interest at the site are conducted. The analyses should include testing for the leachable fraction of contaminant and total concentrations. In addition, the particle-size distribution is determined by sieve analysis.

Other analyses to be performed include organic matter content, moisture retention curves, total nitrogen, inorganic nitrogen, phosphorus, electrical conductivity (salinity), and pH. If necessary, the amount of base or acid required to adjust the pH to the optimal biodegradation range is measured. These analyses are needed to estimate nutrient requirements, establish target moisture contents during treatment, and assess possible physical or chemical problems that may inhibit biological activities.

The initial biological characterization is designed to determine if appropriate organisms are present, evaluate potential toxic or inhibitory conditions, and asseess the need for nutrient addition.

The initial numbers of microorganisms are measured. Also, the response of both total microbial numbers and the numbers of specific degrader organisms to a five-day incubation in stirred flasks under aerobic conditions with nutrient additions is determined.

If necessary for this test, pH is adjusted. Proven contaminant-degrading cultures can be added to samples in case these organisms are not present. These studies use respirometry to measure oxygen consumption rates at different loading rates.

Loading rates are varied and the results are used to assess the potential for toxicity and to set the loading rates for later bench-scale testing. The results also indicate whether the samples are toxic at higher loading rates. They also provide an initial estimate of the rates of treatment. The effectiveness of nutrient addition is determined as well.

For many sites the Treatability Index Test is sufficient to identify whether bioremediation is a feasible option. Additional tests are typ-

■ *A leachate collection system is being installed.*

ically needed to confirm the applicability and to obtain design data for those sites for which bioremediation has not been ruled out. The additional design and confirmatory data are best obtained from a liquid-solids treatment test or a solid-phase treatment test, or both. Not all situations require both tests, but experience indicates that at least one of the tests should be conducted at most sites.

Liquid-solids treatment test. The objectives of this testing are to measure the end-point achievable

■ *After biological treatment, such as in this portable tank system, API separator and dissolved-air flotation sludges can be disposed of on land.*

in a rapid test and assess biological activity in an optimized treatment system.

Duplicate stirred reactors are operated for a short period of time. The reactors are established under typical test conditions for optimizing activity in a slurry reactor with indigenous organisms. These conditions include near-neutral pH and nutrients added to target C:N:P ratios, with possible later additions as appropriate. Adequate mixing and aeration are provided to ensure aerobic conditions throughout the reactor. The amounts of neutralizing agents required to achieve and then maintain a neutral pH are recorded, since this information may impact project costs and material handling requirements.

Solid-phase treatment test. Since the cost of liquid-slurry treatment is higher than prepared-bed treatment, soil pan testing should be conducted to confirm the viability of a prepared bed treatment approach.

Soil pan testing consists of incubating contaminated materials in laboratory soil pans. The pans require less than one cubic foot of soil. The soil in the pans is tilled one to three times per week, depending on the compounds and nature of the project. The water content is maintained at near optimal levels by weekly watering. The laboratory

process replicates field conditions on a smaller scale.

Wrap-up

After examining all the reasons why bioremediation does not work, one may wonder if there are any sites at which it will work. The answer is yes. Our firm has conducted more than 200 treatability studies and 30 full-scale projects. More than 95% of our treatability studies demonstrated that bioremediation was viable. Indeed, bioremediation has saved numerous firms millions of dollars in cleanup costs.

However, the technology, while appearing straightforward and simple, must be applied carefully. A screening methodology to properly apply the technology, such as the one presented here, needs to be employed. The only full-scale failures occurred when the cardinal rule of requiring some form of treatability work before going to the field was violated.

We anticipate providing additional details in the future on how to apply the screening methodology, as our database continues to expand. **CEP**

Acknowledgment

The authors would like to acknowledge Dale Simmons, RETEC's technical editor, for her help in preparing this article.

Literature Cited

1. **U.S. Environmental Protection Agency,** "VISITT, Vendor Information System for Innovative Treatment Technologies, User Manual," EPA Office of Solid Waste and Emergency Response and Technology Innovation Office, Washington, DC, Publication No. EPA/542/R-92/001 (June 1992).
2. **Kolluru, R. V.,** "Understand the Basics of Risk Assessment," *Chem. Eng. Progress,* **87**(3), pp. 61–67 (March 1991).
3. Code of Federal Regulations, 40 CFR Part 261.24 and 40 CFR Part 261, Appendix II.

Further Reading

Bradford, M. L., and R. Krishnamoorthy, "Consider Bioremediation for Waste Site Cleanup," *Chem. Eng. Progress,* **87**(2), pp. 80–85 (Feb. 1991).
Brubaker, G., "Screening Criteria for *In Situ* Bioreclamation of Contaminated Aquifers," *The Hazardous Waste Consultant,* **7**(4), pp. 1–4 (July-Aug. 1989).
Devine, K., *et al.,* "Bioremediation Case Study Collection: 1991 Augmentation of the Alternative Treatment Technology Information Center (ATTIC)," U.S. Environmental Protection Agency, Bioremediation Action Committee, Data Identification/Collection Subcommittee, Washington, DC (Feb. 1992).
Linz, D. G., *et al.,* "The Influence of Soil Composition on Bioremediation of PAH-Contaminated Soils," *Remediation,* **1**(4), pp. 391–405 (Autumn 1991).
Long, G. M., "Clean Up Hydrocarbon Contamination Effectively," *Chem. Eng. Progress,* **89**(5), pp. 58–67 (May 1993).
Swett, G. H., "Bioremediation: Myths vs. Realities," *Environmental Protection,* **3**(4), pp. 22–26 (May 1992).
Torpy, M. F., *et al.,* "Biological Treatment of Hazardous Waste," *Pollution Eng.,* **21**(5), pp. 80–86 (May 1989).

A Screening Protocol for Bioremediation of Contaminated Soil

Jean A. Rogers

James M. Montgomery, Consulting Engineers, Inc., Mannheim, Germany

Dante J. Tedaldi

Bechtel Environmental, Inc., San Francisco, CA 94105

Michael C. Kavanaugh

James M. Montgomery, Consulting Engineers, Inc., Walnut Creek, CA 94598

A bioremediation treatability protocol for soil is presented which can provide feasibility study and remedial action engineers maximum information with respect to the viability and efficiency of bioremediation. The protocol, divided into two main phases, progressively evaluates the viability of biodegradation and the requirements for optimization of the process once implemented. Chemical and microbiological baseline conditions and the potential for contaminant degradation are assessed during Phase I screening. During Phase II the endpoint achievable and kinetics of the biodegradation reactions can be established with pan and slurry reactor tests for ex-situ systems or column tests for in-situ systems. Models are reviewed which can predict the rate of removal of organic constituents, and these data in conjunction with the Phase I and II may be used to estimate the potential time required to achieve cleanup standards, assess the relative importance of biological versus chemical removal mechanisms, and compare expected performance of alternative bioremediation methods.

INTRODUCTION

The screening protocol for evaluating and implementing bioremediation involves several distinct phases. Information about the contaminant and the contaminated media must be gathered during the site characterization and feasibility study stages. A treatability study should then be performed to develop information on the effectiveness of bioremediation for specific contaminants and media, and to optimize process parameters. Finally, contaminant removal rates and scale-up parameters should be considered before a project's design phase. A general bioremediation project timeline is shown in Figure 1. This paper presents a practical, phased approach for evaluating bioremediation as a cleanup alternative at hazardous waste sites, and presents the methodology for obtaining kinetic and equilibrium parameters from treatability study data.

SITE CHARACTERIZATION/FEASIBILITY STUDY ISSUES

A wide range of information should be collected about a potential bioremediation site during the site characterization and feasibility study phases. The data gathered at this point are critical to the evaluation of bioremediation as a viable remedial technology. The factors which should be examined include the chemical characteristics of the contaminants and the chemical, physical, and microbiological characteristics of the site [1, 2]. A summary of these factors is presented in Table 1 (after Dupont [3]).

The data on contaminant and site characteristics gathered during this phase are used to decide on two fundamental issues prior to the commencement of the treatability study and design phases of the project. First, the information gathered provides

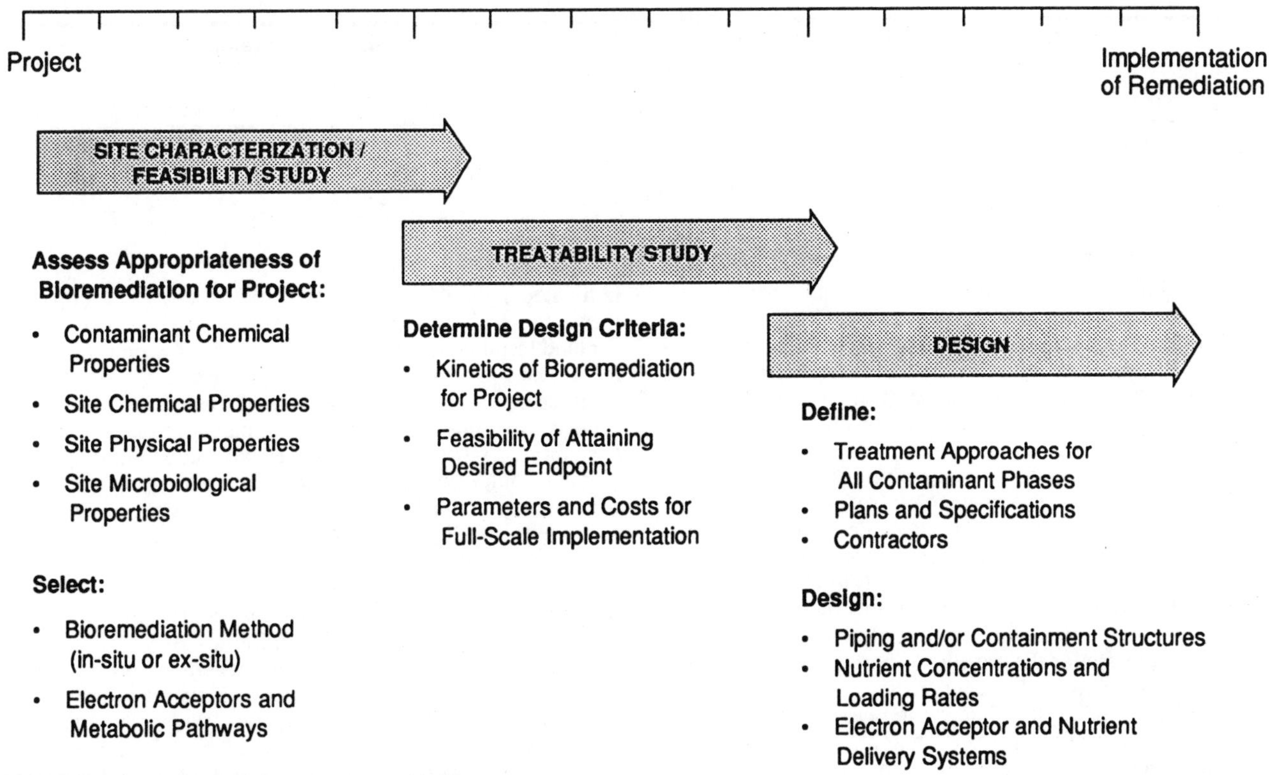

FIGURE 1. Bioremediation project timeline.

a basis for the selection of the most appropriate electron acceptor and redox environment (e.g., oxic/aerobic, anoxic/denitrifying, anaerobic/methanogenic [4, 5, 6]. The complete oxidation of a single organic chemical present at low concentrations in groundwater will typically require very large quantities of oxygen; often well beyond the ability of the system to naturally replenish the depleted O_2 supply. Thus, within these environments, addition of alternate electron acceptors or the addition of pure oxygen or hydrogen peroxide as a source of oxygen may be required to maintain the viability of the degradation process.

In cases where oxygen concentrations (either in soil gas or

Table 1 Summary of Important Site Characteristics

Soil/Groundwater Characteristics

  Texture, pH, nutrient availability, competing carbon sources/oxygen depleters
  Porosity, permeability, bulk density
  Organic matter/organic carbon content
  Cation exchange capacity, clay content
  Dissolved oxygen, redox potential, metals (Mg, Cu, Ni, total/dissolved Fe & Mn)
  Alkalinity, moisture content of soils
  Microbial population—total, contaminant-degraders

Site Characteristics

  Recharge rate/runoff potential/water balance
  Depth of water table
  Depth of contamination, areal extent of contamination
  Site/soil temperature
  Site surficial geology

Waste Characteristics

  Whole waste
    Existence of carrier fluid
    Carrier fluid chemical composition
    Carrier fluid density, viscosity
  Hazardous constituents
    Soil concentration, dissolved concentration
    Physical/chemical properties
      Vapor pressure, boiling point, melting point, solubility,
      Molecular weight, diffusivity
    Distribution in soil environment
      Soil/water, soil/air, air/water, carrier fluid/soil-air-water
    Degradation rate constants
      Biotic, abiotic

Table 2 Electron Acceptors and Redox Potential in Bioremediation Systems

Process	Electron Acceptor	Environment	Typical Redox Potential, mV	Order of Preference
Aerobic	O_2	Aerobic metabolism	+810	1
	NO_3^-	Denitrification	+750	2
Anaerobic	SO_4^{2-}	Sulfate reduction	-220	3
	CO_2	Methanogenesis	-240	4

in dissolved in water) are insufficient to maintain aerobic respiration, nitrate (NO_3^-), sulfate (SO_4^{2-}), iron (Fe^{3+}), and manganese (Mn^{2+}) can act as electron acceptors if the organisms have the appropriate enzyme systems. However, these reactions can occur only if the organic matter (contaminant or cometabolite) is present in a soluble and consumable form, the bacteria present have a suitable supply of nutrients to maintain the biochemical process, and temperature variations are not excessive. Electron acceptors in microbial processes, typical values for redox potential of various modes of microbial metabolism, and the order of preference for use by microorganisms (after Vogel [7]) are presented in Table 2.

The second principal task is to use the information to identify the general method (in-situ or ex-situ) of bioremediation that is most appropriate for the project. A thorough understanding of contaminant chemical characteristics (such as molecular structure, degree of substitution, vapor pressure, and partition coefficient) is critical to the selection and design of an appropriate treatment system. The complexity and structure of the organic material often determines the likelihood of the molecule being degraded. In general, most simple petroleum hydrocarbons, phenols, and lower ringed polynuclear aromatic hydrocarbons (PAHs) are degraded rapidly under aerobic conditions [8]. While a few chlorinated compounds can be used as primary substrates for growth many are transformed as secondary substrates. Less chlorinated compounds are more easily transformed by oxidation processes, while more highly chlorinated compounds are more easily transformed by reduction processes.

Furthermore, the estimated total mass in each phase (e.g., dissolved or solid), media chemical characteristics (including redox potential and organic carbon content), and site microbiological and hydrogeological characteristics should all be very well defined during the site investigation to ensure that treatment system design address the complexities of the site as well as those of the contaminants.

taminated soils typically rely on modifications and improvements to aerobic processes which were originally developed for land farming of petroleum wastes and for wastewater treatment. These processes consist of enriching an environment with a source of oxygen in the presence of the appropriate microorganisms to mineralize the contaminants to carbon dioxide and water. Ex-situ treatment systems include land treatment, biopiles/composting, and liquid/solids (slurry) reactors. In-situ systems include above-ground inoculation of extracted groundwater followed by injection after addition of nutrients and an electron acceptor; direct injection or infiltration of nutrients, electron acceptors and bacteria, and bioventing which seeks to stimulate biodegradation process through enhanced air flow (oxygen transfer) through the soil.

In-situ bioremediation offers the benefit of not requiring movement of contaminated soils to establish the appropriate conditions for contaminant degradation. Typically, nutrients, water and oxygen have to be supplied by injection wells to stimulate the indigenous organisms to metabolize the waste. In-situ bioventing may offer a less intrusive yet effective approach under certain circumstances. Implementation of ex-situ bioremediation avoids the difficulties imposed by hydrogeological constraints [9, 10]. In-situ systems are generally applicable where the hydrogeology of the site permits the transport of water, nutrients, and/or oxygen through the subsurface and permits the hydraulic containment of the contaminant [11]. In general, sites with conductivities greater than 10^{-4} cm/s and fairly homogeneous stratigraphy are good candidates for in-situ bioremediation [12, 13]. If the contamination is shallow and can be excavated easily, or site physical and/or chemical characteristics prohibit in-situ bioremediation, an ex-situ system may be preferred.

Bouwer [14] has summarized the favorable and unfavorable characteristics affecting the feasibility of in-situ bioremediation (Table 3).

BIOREMEDIATION METHOD

The method of bioremediation to be employed is chosen based on the information gathered during the characterization phase of a project. Conventional treatment processes for con-

COMPONENTS REQUIRED FOR BIODEGRADATION

The basic components required for the degradation process include the following:

Table 3 Favorable and Unfavorable Factors Affecting Bioremediation

Favorable Chemical and Biological Factors	Unfavorable Chemical and Biological Factors
Small number of organic contaminants	Numerous contaminants, or a complex mixture of inorganic and organic compounds
Non-toxic concentrations	Toxic concentration
Diverse microbial populations	Sparse microbial activity
Suitable electron acceptor condition	Absence of appropriate electron acceptors
pH 6 to 8	pH extremes

Favorable Hydrogeologic Factors	Unfavorable Hydrogeologic Factors
Granular porous media	Fractured rock
High permeability ($>10^{-4}$ cm/sec)	Low permeability
Uniform mineralogy	Complex mineralogy with high organic carbon content
Homogeneous media	Heterogeneous media
Saturated conditions	Unsaturated strata, or intermittently saturated conditions

- Microorganisms
- Terminal electron acceptor
- Carbon source
- Nutrients
- Water.

The biotreatability study evaluates the presence of these components, and to what extent these components need to be supplemented. Tests which can be performed encompass both chemical analyses necessary for basic soil and groundwater characterization, and engineered studies to assess optimization of parameters, and to determine kinetic and/or equilibrium measurements.

Information from mass balance studies, including laboratory screening, bench- and pilot-scale studies, is combined with information concerning site and waste characteristics in order to determine applications and limitations of each technology. Information obtained from treatability studies should be focused on identifying ultimate limitations to the use of a remediation technology at a specific site, which usually are related to 1) time required for cleanup 2) level of cleanup attainable, and 3) cost of cleanup.

Environmental factors which will affect remediation efforts include:

- Oxygen and nutrient availability
- Soil moisture content
- The pH of the soil, groundwater, and hazardous waste
- Soil structure and organic content
- Temperature
- Solubility of the pollutants
- Concentration of toxic compounds
- Concentration of contaminant-degrading microbes.

All of these factors should be assessed during treatability studies, and subsequently controlled during the bioremediation process.

Microorganisms

The availability and viability of microorganisms indigenous to the contaminated media are determined during treatability studies. It is generally desirable to enhance the microbial activity of indigenous organisms, rather than using exogenous organisms, because the indigenous organisms are already acclimated to the waste material. Additionally, it has been shown that exogenous organisms do not effectively compete with indigenous microorganisms. Only if the environment is sterile, or the present microbial population does not degrade the contaminant, should exogenous microorganisms be considered. Whatever types of degraders are selected, nutrient addition rates favorable to organisms which preferentially degrade the contaminants of concern must also be determined during treatability studies.

Substrate Requirements

Microorganisms require a primary substrate, which is the carbon and energy source, also termed the electron donor, in the redox reaction governing oxidation of hydrocarbons, for example. Ideally, the contaminant of concern serves as the primary substrate. If the contaminant of concern cannot be degraded as primary substrate, then an analog compound which acts as the primary substrate must be added, and the contaminant of concern can be cometabolized, or degraded as a secondary substrate. Additional substrate may also be required if contaminant levels are too low to support the microbial mass necessary for degradation.

Nutrient Requirements

Nutrient requirements for microorganisms have been established through extensive research for activated sludge processes in the wastewater treatment field. The nutrients include both macronutrients and micronutrients, and are based on the composition of cell matter. Macronutrients include nitrogen, phosphorus, sulfur, iron, potassium, calcium, magnesium, and manganese. Nitrogen and phosphorus are the major requirements in soil bioremediation systems because the soil itself generally provides the other nutrients which are needed in much smaller amounts. Oxidation-reduction reactions can be written for the synthesis and growth of the microbes in order to rigorously determine the mass of nutrients required for a given mass of contamination to be degraded based on the method of McCarty [15]. A rule of thumb ratio for C:N:P is 120:10:1 on a weight basis [16].

Oxygen Requirements

Oxygen requirements for aerobic systems can also be determined by mass balances, and for *in-situ* systems, hydrogen peroxide may be evaluated during treatability studies for use as an oxygen source. Hydrogen peroxide is cytotoxic to those species of microorganisms that do not possess the catalytic enzyme catalase that breaks down hydrogen peroxide to oxygen and water [17]. If hydrogen peroxide is being considered as an alternative oxygen source, the population of biodegrading organisms present should be tested for this enzyme in the treatability study stage.

FOCUS AND OBJECTIVES OF TREATABILITY STUDIES

There are two fundamental objectives of treatability studies, which define the scope of the study to be conducted. The primary objectives of the treatability protocol are:

- to rapidly and inexpensively evaluate the susceptibility of site soils to biological treatment, and
- to determine the rate and extent of treatment which can be achieved.

Secondary objectives include:

- to better understand site-specific chemical partition coefficients of soils for use in risk assessment and remediation efforts, and
- to provide insight regarding non-biological soil treatment options, e.g., soil washing.

A preliminary (Phase I) study is performed during the remedial investigation/feasibility study (RI/FS) stage, in order to obtain comparative information for technology selection. The intent of the Phase I treatability study is to determine if bioremediation is indeed an appropriate remedial activity, given the hydrogeological and contaminant characteristics. A Phase II treatability study is appropriate after bioremediation has been selected as the technology of choice, in order to provide design criteria for a full-scale remediation project. The Phase I and Phase II treatability studies can then be followed by field studies, if necessary, prior to full scale implementation.

Primary Elements of the Soil Treatability Protocol

While some elements of environmental fate mechanisms and required components for biological degradation can be quantified under strict laboratory conditions for some simple cases,

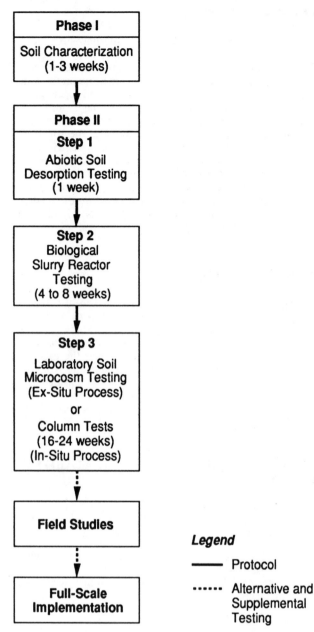

Phase I

Soil Characterization
(1-3 weeks)

↓

Phase II

Step 1
Abiotic Soil
Desorption Testing
(1 week)

↓

Step 2
Biological
Slurry Reactor
Testing
(4 to 8 weeks)

↓

Step 3
Laboratory Soil
Microcosm Testing
(Ex-Situ Process)
or
Column Tests
(16-24 weeks)
(In-Situ Process)

⋮

Field Studies

⋮

**Full-Scale
Implementation**

Legend

—— Protocol

······ Alternative and
Supplemental
Testing

FIGURE 2. Elements of the soil treatability test protocol evaluation.

tions. Slurry reactor configurations, used in Phase II studies, maximize contaminant mass transfer and ensure that the rate of desorption from soil to water is as rapid as possible. Slurry reactor and soil desorption test results are used to determine the relationship between the extent of contaminant soil desorption and the extent of biodegradation. Soil desorption testing results also provide an indication of the leaching potential of soil contaminants.

Results from this accelerated test protocol can then be applied toward field studies and full-scale implementation. Each of the steps in the treatability study protocol is discussed below.

Phase I—Soil Characterization

The objective of Phase I is to define the chemical, microbial, and physical characteristics of the soil. Chemical analyses are used to identify the chemicals of interest and to quantify their concentrations for a particular site soil. Statistical analysis can determine if soil collected for treatability evaluation is representative of the site (as defined by the site investigation). The chemical concentrations are also needed to help determine the amount of soil, nutrients, and other additives required for the soil desorption and slurry reactor tests. A complete gas chromatographic (GC) scan for volatile and semi-volatile organics should be conducted, as well as testing for metals if the soil has not previously been tested. In addition, the following tests should be conducted:

- Gas chromatographic analysis of volatile organic compounds and base/neutral extractable contaminants of concern
- Total organic carbon
- Ammoniacal nitrogen phosphorous
- Nitrate
- Potassium
- Sulfate
- pH
- Moisture content
- Redox potential
- Metals concentrations (Fe, Mn, Mg, Cu, Ni)
- Radioactivity (gross alpha and gross beta radiation), if suspected
- Alkalinity.

The characterization of soil and water chemistry indicates the contaminants of concern, possible metabolic modes for microbial activity, existing nutrients, and potential inhibitors.

Microbial Enumeration/Composition

Microbial characterization should include enumeration of total microbes and contaminant-specific degraders. Total microbes may be determined by the "Agar Plate Method for Total Microbial Count" as described by Clark [20]. The total microbial count should be compared to an estimate of the population present which will degrade the contaminant of concern. The preliminary screen for contaminant-specific degraders is performed in a medium that provides only the necessary inorganic nutrients required for microbial growth and no intrinsic carbon source. The contaminant supplement serves as the sole source of carbon. Cultures of selected potential biodegraders are prepared in supplemented and unsupplemented (control) media. Following incubation, the cultures are evaluated for the presence and relative abundance of microbial growth. Isolates demonstrating little or no growth are judged to be poor or non-degraders of the contaminant and should be eliminated from further study. The remaining isolates can be evaluated for the effects of environmental parameters or for performance in bench scale reactor systems [21].

Protocol for isolating and identifying contaminant-specific degraders should be developed in conjunction with a micro-

no one model or set of models can adequately simulate or predict biodegradation in soils because of the complexity of the system. The accelerated soil treatability protocol discussed below was developed and evaluated for the United States Environmental Protection Agency [16, 18] and was proposed earlier by Nakels and Smith [19]. The protocol is based on the premise that chemical contaminants must first desorb and diffuse from the soil and enter the aqueous phase before they can be assimilated by the bacteria and degraded.

The treatability study protocol is designed to evaluate 1) equilibrium sorption relationships, 2) sorption kinetics, and 3) biological oxidation. Protocol elements are shown in Figure 2. Phase I consists of complete characterization of the soil for chemical, microbial, and physical parameters. During Phase II the rate and extent of contaminant desorption which can be achieved is determined. The ability of the microbes to biodegrade the contaminants under aerobic or anaerobic conditions can also be assessed. These tests are not intended to simulate particular soil treatment processes, but rather are intended to examine the biodegradability of contaminants associated with site soils under optimal environmental condi-

biologist. There are many competent laboratories which can develop specific agars for enumeration of bacteria which degrade the contaminants of concern. Numerous other methods for determining viable counts of microorganisms have been developed recently. These include electron microscopy, viable counts, epifluorescence microscopy, and measurements of biochemical components. Readers are referred to selected references for additional information on newly developed microbial enumeration techniques [5, 22]. Results of total microorganisms and contaminant-specific degraders provide an indication of microbial activity for the soil for existing (unamended) site conditions. If viable populations exist, the potential for bioremediation has been established.

Another technique which establishes the viability of the existing population is respirometry. This method measures the CO_2 production of the microbes, and indicates the activity which is present. This technique is better suited to monitoring and process control than site characterization, because a real-time reading can be obtained. However, the respirometry technique does not distinguish between total organisms and contaminant degraders.

Toxicity Testing

In conjunction with plate counts, bioassays may be used to determine if toxic substances which may inhibit biodegradation exist in the soil matrix. The Microtox® assay is an aqueous general toxicity assay that measures the reduction in light output by a suspension of marine luminescent bacteria in response to an environmental sample. Bioluminescence of the test organism depends on a complex chain of biochemical reactions. Chemical inhibition of any of the biochemical reactions causes a reduction in bacterial luminescence. Therefore, the Microtox® test considers the physiological effect of a toxicant, and not just mortality.

Matthews and Bulich [23] describe a method of using the Microtox® assay to predict the land treatability of hazardous organic wastes. When plate counts yield low results for indigenous microorganisms, the Microtox® assay indicates whether the condition is controllable or not. If the low counts are due to the presence of toxins, bioremediation may be very complicated or infeasible. If toxins are absent, the microbial activity may be enhanced through nutrient addition.

Physical Characterization

Particle size distribution using a dry sieve method developed by Lambe [24] with points of clarification provided by ASTM [25] is used to interpret the results of desorption and slurry reactor testing, and to determine if *in-situ* bioremediation is feasible with respect to hydraulic considerations. Field hydraulic conductivity tests should also be performed by a hydrogeologist as part of a Phase I screening if *in-situ* biodegradation is being considered.

Phase II Testing

Step 1 Abiotic Soil Desorption Testing

Soil desorption tests are performed under abiotic conditions with the results used to compute site-specific soil/water partition coefficients (K_p) for chemicals of interest (i.e., site-specific desorption isotherms). Desorption tests can be conducted many different ways. Readers are referred to selected references to design an applicable protocol for the contaminant of concern [16, 26, 27, 28]. Experimentally determined partition coefficients provide a quick indication of the extent to which organic chemicals will leach off of the site soils. The extent to which chemicals desorb off of soils into solution can be directly correlated with their susceptibility to biodegradation. If a specific organic chemical is detected near its aqueous solubility, then it is expected that biodegradation will occur to some extent. If, on the other hand, a specific chemical is not measured in solution, then it is unlikely that bioremediation of this chemical will occur to any significant extent [29]. Thus, the ultimate effectiveness of bioremediation is strongly affected by the solubility of a contaminant to be degraded. Many common organic contaminants found in soil and groundwater are highly hydrophobic and nonpolar; thus they have relatively low solubilities in water. This fact can hamper the degradation of these compounds because degradation is most likely to occur in the aqueous phase where the bacteria are present and where assimilation of substrate and nutrients can easily occur. The role of sorption and desorption in the aqueous phase biodegradation process has been modeled by Annokee [30] and is presented in Figure 3. In this model, biological oxidation can

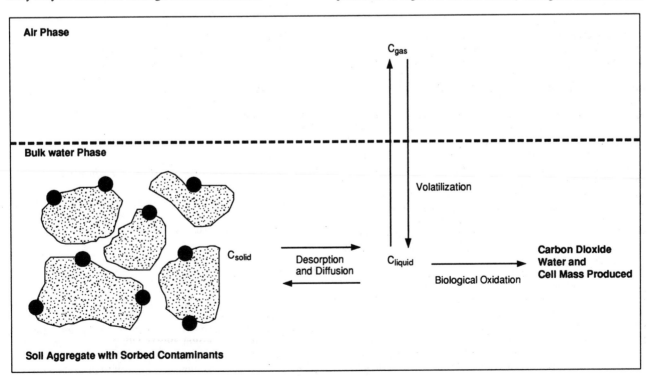

FIGURE 3. Role of desorption/diffusion in biodegradation process.

only occur if the compound within or on a soil particle desorbs and diffuses into the aqueous phase. Once in solution, volatilization can also occur. In many cases, desorption and diffusion of the contaminant into the aqueous phase may be the rate limiting step controlling both volatilization and biological oxidation [31]. These mechanisms control removal of contaminants in both *in-situ* and *ex-situ* biodegradation systems.

A great deal of research has been devoted to the facilitation of bioremediation through the enhancement of the aqueous solubility of organic chemicals. For example, Loehr [32, 33] observed that the loss rates of PAH compounds found at manufactured gas plant (MGP) sites were near zero and little affected by changing nutrient addition rates, water content, temperature, or pH. The test soils were non-toxic, and despite low loss rates of the compounds, significant populations of bacteria were present. Also, aqueous extracts of MGP soils indicated that PAH compounds in such soils were not soluble. This failure of bioremediation of PAHs was surprising as these compounds had been successfully biodegraded in numerous other laboratory and field studies [34]. It was postulated that the probable cause of this phenomenon was the biological unavailability of these compounds. The low solubilities of these compounds and their strong adsorption to the soil could lead to inaccessibility of these chemicals to microorganisms. As a result of these findings and corroborating research by others, many researchers have been investigating the use of surfactants and cosolvents to enhance the solubility of organic chemicals in the environment and thus, make these chemicals more available for bioremediation [35, 36, 37, 38].

Current research in this area has focused on several major areas:

- Surfactants or cosolvent additions are not required at all
- Surfactants are indigenous in the soil
- Surfactants are created by the microbes
- Excess mixing in above-ground reactors enhances chemical release
- Biotreatment is not required at all
- Surfactants or cosolvents remove contaminants from soil
- Design optimization
- Combined use of surfactants/cosolvents and bioremediation.

Large increases (orders of magnitude) in the solubility of individual compounds have been observed by these researchers through surfactant or cosolvent addition. However, because of the very low initial solubilities of many higher ringed PAHs (often less than 200 μg/L for the pure chemical) the resultant aqueous concentrations are still very low and biotreatment may have difficulty in proceeding. In such instances, even with enhancement of solubility and increases in the mass degraded, the mass of contaminant remaining on the soil may still be quite high and well-above regulatory limits. It is important to note that in all cases, these researchers reported that very high concentrations (tens of percent) of surfactants or cosolvents were required to achieve very small increases in solubility. Thus, even if surfactants and cosolvents are found to be effective for the enhancement of bioremediation, reagent costs as well as the ultimate environmental fate of the added reagents become important considerations which may detract from the applicability of these techniques.

Because aqueous solubilities within a chemical class (e.g., chlorinated aliphatics) can vary by several orders of magnitude, it is not possible to provide a broadly-based characterization of relative solubilities. Thus it is important to review the solubilities of contaminants of interest in bioremediation studies. From the solubility and desorption data, it may also be possible, based upon the hypothesis that mass transfer limitations dominate chemical fate in a soil matrix, to estimate the treatment endpoint for some chemicals.

Step 2 Biological Slurry Reactor Testing

Biological slurry reactors are operated to examine the ability of the microbes to biodegrade the desorbed organics under either aerobic or anaerobic conditions. Bioreactors maximize degradation rates by reducing or eliminating mass transfer limitations, thereby providing the equilibrium concentrations which will be achievable in the shortest period of time. Figure 4 provides a schematic representation of a common laboratory slurry reactor. Initial and final aqueous phase chemical concentrations are measured along with the concentration time profile of the chemical in the soil phase. The proper microbial environment (i.e., pH, nutrients, sufficient electron acceptor, and no toxic compounds) must be maintained through periodic monitoring and supplemental additives. The biodegradation tests can be completed with 4 to 8 weeks. In the event that necessary indigenous bacteria are not sufficient to biodegrade the aqueous phase organics, the protocol can involve bioaugmentation using cultured organisms.

Bioreactors can be operated in batch, fed batch, or continuous culture modes. The size of the reactor is dependent on the quantity of waste to be treated as well as the hydraulic retention time required to meet target levels. If the slurry reactor tests show statistically significant reduction in soil contaminants, then the technical viability of soil biodegradation will be established. Under slurry reactor conditions, soil desorption and solubilization of organic compounds is maximized, thus, biodegradation should also be maximized. The premise of the protocol is that if biodegradation cannot be achieved under slurry reactor conditions, then it is highly unlikely that biodegradation can be achieved in any other soil treatment process. Biological slurry reactor data quickly provides an estimate of the potential end point for biological treatment. With this understanding, it is possible to decide if further investigation is warranted to select the best process configuration to take advantage of the waste-specific desorption and biodegradation factors. Optimization of operating parameters can be quickly accomplished, such as temperature variation and surfactant addition.

After viability is established, a bench-scale or pilot-scale test is completed to determine the system's operating parameters, which are listed below [39, 40].

- Biological solids retention time
- Hydraulic retention time
- Total suspended solids
- Mixed liquor volatile suspended solids
- BOD, COD, TOC, nitrogen, and phosphorus removal efficiencies
- Sludge yield coefficients
- Dissolved oxygen uptake rates
- Surfactant addition requirements
- Nitrification potential investigations.

Step 3 Pan Microcosm Testing or Column Tests

Pan microcosm tests and/or column tests are conducted to determined the kinetics of the reaction, depending upon whether an *ex-* or *in-situ* approach is desired.

Pan Microcosm Testing

This procedure most closely emulates the land treatment process and represents the conventional test procedure for developing design parameters for the biological land treatment of contaminated soils. For the case of hydrocarbon-contaminated soil, the test is conducted by mixing contaminated soils with an appropriate amount of clean soil to produce a mixture that has oil and grease levels no greater than 1 to 2 percent by

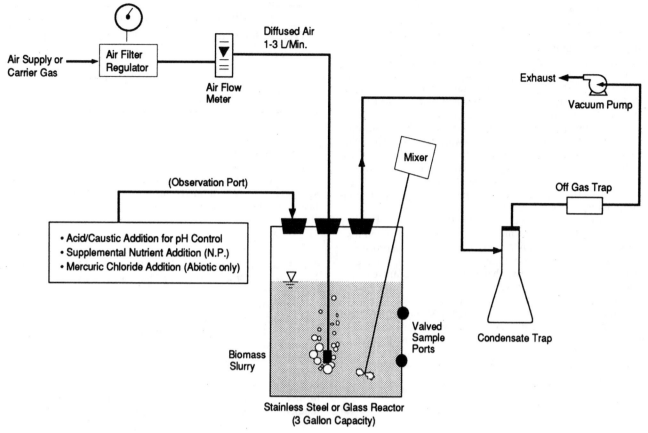

FIGURE 4. Slurry reactor process schematic.

weight. This mixture is placed in a pan reactor with typical dimensions of 20 cm by 30 cm with 10 cm depth [18]. Nutrients (i.e., phosphorus and nitrogen) and water are then added to the soil to enhance the activity of the indigenous bacteria, in amounts determined by the initial soil characterization. Periodically, the pan soils are mixed, and nutrient, pH, and moisture levels are measured, and if necessary, adjusted. Samples of the soils are taken at the start of the experiment, and every week for 2 to 3 months. The concentration-time profiles that are generated are used to estimate contaminant removal kinetics and the basic design parameters for the land treatment process. The pans must be operated for a 2- to 3-month period because this bioremediation process is not as aggressive as a slurry reactor in terms of maximizing soil surface exposure to biological activity.

Column Tests

This procedure simulates *in-situ* biodegradation of organic contaminants in the subsurface environment. A vertical column of contaminated soil is used, allowing a solution of water, microorganisms, and nutrients to flow through the system. Test and control columns are run. Hydraulic conductivity considerations can be assessed in this manner, as well as the potential for clogging due to biomass generation. Biofouling, common in *in-situ* systems, is caused by poor soil permeability and inhibits the flow of applied materials through the treatable soil matrix. The resultant biomaterial "plug" precludes homogeneous aqueous dispersion, saturation, and maintenance of biological activity, thereby negating effective biotreatment. During the course of the column studies, chemical, microbiological, and contaminant parameters of control and test systems are evaluated. Samples are subsequently evaluated for chemical and biological changes as well as terminal end procedure. If anaerobic or methanogenic conditions are being studied, metabolic pathways and potentially toxic by-products should be investigated [41, 42, 43].

Data Analysis

Following the collection of site characterization and contaminant treatability data, various data interpretation and modeling approaches are available to estimate the expected effectiveness of bioremediation for cleanup of soils contaminated with organic chemicals. Modeling approaches range from simple analytical equations to complex computer codes. The two most effective uses of models in remediation projects are for either screening purposes or for field validation activities. In this section, relatively simple analytical equations are presented which allow the rate and extent of contaminant removal to be estimated. These model results can be used to estimate the potential time required to achieve cleanup standards, assess the relative importance of biological versus physical/chemical reaction mechanisms, and compare expected performance of alternative bioremediation methods.

The data obtained from the slurry reactor study can be interpreted using the classical model of microbial growth, given by the Monod equation:

$$\frac{dS}{dt} = -\frac{kXS}{K_s + S} \tag{1}$$

Where:

S is substrate concentration, mg/L
X is biomass concentration, mg/L
K_s is the half-velocity coefficient, mg/L
t is the time in hours
k is the maximum specific substrate utilization rate, 1/hr.

The measurement of substrate and biomass with time will enable the quantification of K_s and k. The application of the Monod expression to soil bioremediation, however, has been limited in practice due to difficulty in quantifying X, which represents the active, contaminant-degrading microorganisms on a mass/volume basis and is a required parameter in the above second order equation. Many treatability studies rely on cultural techniques to estimate the abundance of microorganisms in soils, although these methods may detect less than ten percent of the amount actually present [22]. The application of this equation has traditionally been in slurry reactors, where activated sludge is dried and weighed to determine X. The presence of a solid phase in pan or column reactors complicates biomass determination. Non-traditional methods of determining X for soil/water systems are being investigated. Readers are referred to selected references for further information [5, 22]. Additional research is needed in the area of soil biomass quantification to provide a reliable method to distinguish between contaminant degrading the non-contaminant degrading cells.

In addition to problems associated with quantifying X, many of the assumptions upon which Monod kinetics are based may not be valid for *in-situ* applications or land treatment processes. These include the assumption that utilization rate is limited by a single enzymatic process, that there is a constant enzyme concentration, and that there is no change in the species distribution of the microbial community.

In view of the problems of applying the Monod equation in *in-situ* or land farming processes, approximations to the rigorous equation given above can be used. At high primary substrate concentration, the rate is a maximum and first order with respect to organism concentration, but zero order with respect to substrate concentration:

$$-dS/dt = kX \text{ with } (S \gg K_S) \qquad (2)$$

At low primary substrate concentration, the rate of utilization is first order with respect to both organism concentration and substrate concentration:

$$-dS/dt = (k/K_S)XS \text{ with } (S \ll K_S) \qquad (3)$$

The ratio of k/K_S is called the second order rate constant for biological degradation. This value is useful when considering the degradation of contaminants at low mg/L concentrations.

From these equations it is clear that X, or the biomass concentration should be increased in order to increase the rate of degradation. At a constant biomass level, one can determine the half-life of a compound by integrating.

$$C_0 - C_t = kX(t - t_0) \qquad (4)$$

where

C_t = concentration at time t,
C_0 = concentration at time $t_0 = 0$
for the half-life, $t_{1/2}$, set $C_t = 1/2C_0$

$$1/2C_0 = kXt_{1/2} \qquad (5)$$

$$t_{1/2} = C_0/2kX \qquad (6)$$

These concepts are illustrated by the following example, for this case assume:

$k = 0.1$ d^{-1} (a typical value)
$C_0 = 100$ mg/L, or 100 ppm
$t_{1/2} = 1,000$ hr, or about 40 days
then, X (biomass concentration) = 12.5 mg/L

The relatively slow rate of biodegradation is partially due to the very low biomass concentration and this factor would tend to limit in-situ degradation. This slow rate is further compounded by the fact that many degradable compounds may have half-lives of months to years under natural conditions.

Table 4 Volumetric Productivity Comparison

Type of Bioreactor	Volumetric Productivity (mg/L-hr)
Fermentor	5,000 to 25,000
Trickling filter	5 to 250
Activated sludge	10 to 100
Bioremediation	0.02 to 30

The extent to which biomass concentrations can be increased can be estimated by comparison to other biotreatment processes. In order to compare biodegradation with other bioprocesses it is useful to consider another way to express microbial kinetics in terms of volumetric productivity [44]. In the example just given, the rate of conversion could be stated as:

$$\text{Volumetric productivity} = \text{Rate/volume} = kX \qquad (7)$$

or, volumetric productivity $= (0.1 \text{ d}^{-1})(12.5 \text{ mg/L}) = 1.25$ mg/L-d. Other examples of biological reactions stated in the same terms are listed in Table 4.

One strategy to enhance biotreatment is to add a substrate that causes the biomass to increase *in-situ*. However, this is limited by the stoichioimetric constraints discussed above particularly for aerobic reactions. This same effect can also be obtained by adding externally grown biomass to the contaminated matrix, i.e., enrichment. In this case, biomass would be grown in a fermentor where it is much easier to optimize nutrient addition rates and environmental conditions such as temperature and pH. In a fermentor, biomass concentrations of greater than 25,000 mg/L can be achieved. The addition of this biomass to a contaminated matrix may increase the rate of degradation by orders of magnitude. However the rate will also decay as the bacteria die off if the substrate (and/or nutrient) concentration in the soil is insufficient for continued cell maintenance and growth. In addition, although the introduction of exogenous biomass (as recycle in the activated sludge process for conventional wastewater treatment) may be practical in open or batch systems, it is unlikely that biomass addition to groundwater systems will be practical due to problems associated with fouling of the porous media.

The minimum substrate concentration is also a useful parameter which can be obtained from slurry reactor data, [14] and is given as follows:

$$S_{min.} = \frac{K_s b}{Yk - b} \qquad (8)$$

Where:

Y = yield coefficient, mg bacteria/mg substrate
b = decay coefficient or death rate, day^{-1}.

S_{min} represents the minimum substrate concentration that can support a viable biomass community. If the substrate concentration is below S_{min} the organisms will not multiply or increase in concentration. Thus, if a contaminant enters the soil at a low level, the bacteria responsible for degradation will not increase substantially and the degradation will take place very slowly.

For example under aerobic conditions where $k = 0.1$ hr^{-1}, $K_s = 10$ mg/L, $Y = 0.5$, and $b = 0.005$ hr^{-1}, the S_{min} would equal about 100 μg/L. That is, degradation cannot achieve a concentration below 100 μg/L within a reasonable period of time. S_{min} is likely to vary by an order of magnitude or more above or below 100 μg/L, depending on the organism, the substrate, and the environmental conditions under which the organism grows. In any event, there appears to be a minimum value under steady-state conditions to which substrate concentration can be reduced when this substrate is the sole source of energy. However, it may possible to overcome this limitation by adding a non-toxic primary substrate which can act as cometabolite. This has the effect of increasing the rate of metabolism and overcoming the minimum concentrationi limitation. In this way it may be possible to achieve residual concentration levels of μg/L or even ng/L levels.

This determination is useful to estimate minimum concentrations which can be achieved without adding an additional primary substrate. The significance of determining S_{min} from treatability study data is that the treatability studies do not have to be carried out to their endpoint. Once sufficient data is collected to determine kinetic parameters, S_{min} can be predicted.

Although a contaminant being degraded as a primary substrate can not be reduced to concentrations below S_{min}, a contaminant acting as a secondary substrate may be removed to lower concentrations through cometabolism. In this case, utilization of the primary substrate determines the amount of biomass present in the system and the secondary substrate is degraded without contributing significant energy to the system. Kinetic parameters determined for bioremediation projects can be based on either primary or secondary substrate kinetics.

Recent efforts to describe substrate utilization in soil and subsurface environments are focusing on a biofilm concept as a basis for modeling [45]. Soil microorganisms are typically present as a biofilm on the soil particles. The change in substrate concentration, then, is a function of the mass transport of substrate, nutrients, and electron acceptor into the biofilm.

A comparison of the rates observed in the slurry reactor and in either the column or pan reactor will permit an evaluation of the rate limiting process. If the observed kinetics are similar in the two reactor configurations, the rate at which the microorganisms can metabolize the substrate limits the rate at which the substrate is used. Similarly, if the kinetics observed in the pan or column are much slower than those in the slurry reactor, then either the rate of desorption, mass transfer, or nutrient/electron acceptor availability limit the rate of substrate utilization. The latter case is typical of *in-situ* applications and suggests that the rate of disappearance of contaminant is typically related to the engineering of an appropriate system, rather than on the ability of the microorganisms to degrade the contaminant quickly. The column or pan reactor, being more representative of the system in which the remediation effort will occur, will give a closer approximation of the kinetics which can be expected in the field.

SUMMARY

The above treatability protocol is designed to provide maximum information with respect to the viability and efficiency of bioremediation, with a minimum of laboratory effort. The Phase I screening provides information with regard to chemical and microbiological parameters required to determine if bioremediation is a feasible technology given the site conditions. Once bioremediation has been selected as the appropriate technology, a Phase II screening should be performed. The Phase II screening provides information pertaining to the endpoint achievable (slurry reactors) or kinetics of the reaction (established with pan tests for *ex-situ* systems or column tests for in-situ systems). Data analysis can predict the rate of removal of constituents using one of the models discussed. Data from the preliminary screening and the kinetic predictions from the model are used to estimate the potential time required to achieve cleanup standards, assess the relative importance of biological versus chemical removal mechanisms, and compare expected performance of alternative bioremediation methods.

ACKNOWLEDGMENTS

The authors would like to acknowledge the indirect and direct contributions of researchers in the Environmental and Water Resources Program at the University of Texas at Austin especially Dr. Raymond C. Loehr, and the approach towards volumetric productivity advanced by Dr. Philip Stewart of the Center for Interfacial Microbial Processes at Montana State University.

LITERATURE CITED

1. McCarty, P. L., "Bioengineering Issues Related to In-Situ Remediation of Contaminated Soil and Groundwater," Presented at Reducing Risks from Environmental Chemicals through Biotechnology, University of Washington, Seattle, WA (1987).
2. Kobayashi, H., and B. Rittman, "Microbial Removal of Hazardous Organic Compounds," *Environmental Science and Technology*, **16** p. 170A (1982).
3. Dupont, R. R., "Applications of Treatability Studies in Management of Fuels/Petroleum Waste Impacted Soils," For Presentation at 84th Annual Meeting & Exhibition, Vancouver, B.C., (1991).
4. McCarty, P. L., B. E. Rittmann, and E. J. Bouwer, "Microbial Processes Affecting Chemical Transformations in Groundwater," *Groundwater Pollution Microbiology*, G. Bitton and C. P. Gerba (Eds.) J. Wiley and Sons, New York, N.Y. (1984).
5. Lee, M. D., et al., "Biorestoration of Aquifers Contaminated with Organic Compounds," CRC Critical Reviews in Environmental Control, **18** (1) pp. 29–89 (1981).
6. Babea, L., and D. D. Vaishnav, "Prediction of Biodegradability of Selected Organic Compounds," *Journal of Industrial Microbiology*, **2** pp. 107–115 (1987).
7. Vogel, T. M., C. S. Criddle, and P. L. McCarty, "Transformations of Halogenated Aliphatic Compounds," *Environmental Science & Technology*, **21** (8) pp. 722–736 (1987).
8. Bradford, M. L., and R. Krishnamoorthy, "Consider Bioremediation for Waste Site Cleanup," *Chemical Engineering Progress*, **87** (2) pp. 80–85 (1991).
9. Brown, R. A., R. D. Norris, and G. R. Brubaker, "Aquifer Restoration with Enhanced Bioreclamation," *Pollution Engineering*, pp. 25–28 (1985).
10. Brubaker, G. R., and E. L. Crockett, "In Situ Aquifer Remediation Using Enhanced Bioreclamation," In Proceedings of Hazmat 86. Atlantic City, Tower Conference Management, Glen Ellyn, IL (1986).
11. Raymond, R. L., M. D. Lee, and C. H. Ward, "Assessment of Potential for In-situ Bioremediation," In Proceedings of the DECHEMA Conference, Frankfurt am Main, pp. 30–31 (1989).
12. Thomas, J. M., and C. H. Ward, "In Situ Biorestoration of Organic Contaminants in the Subsurface," *Environmental Science and Technology*, **23** (7) pp. 760–765.
13. United States Environmental Protection Agency (EPA), International Evaluation of In-Situ Biorestoration of Contaminated Soil and Groundwater, EPA 540/2-90/012, Office of Emergency and Remedial Response, Washington, D.C. (1990).
14. Bouwer, E. J., "Bioremediation of Organic Contaminants in the Subsurface," in *New Concepts in Environmental Microbiology*, Ralph Mitchell ed., in Press.
15. McCarty, P. L., "Stoichiometry of Biological Reactions," Presented at the International Conference, Toward a Unified Concept of Biological Waste Treatment Design, Atlanta, GA (1972).
16. Sims, R. C., "Soil Remediation Techniques at Uncontrolled Hazardous Waste Sites," *J. Air and Waste Management Association*, **40** (5) pp. 704–732 (1990).
17. American Petroleum Institute, "Field Study of Enhanced

Subsurface Biodegradation of Hydrocarbons using Hydrogen Peroxide as an Oxygen Source," Publication No. 4448, Washington, D.C. (1987).

18. United States Environmental Protection Agency (EPA), "Guide for Conducting Treatability Studies under CERCLA," EPA/540/2-89/058, Office of Solid Waste and Emergency Response, and Office of Research and Development, Washington, D.C. (1989).

19. Nakles, D. V., and J. R. Smith, "Treatability Protocol for Screening Biodegradation of Heavy Hydrocarbons in Soil," In: *Proceedings of Hazardous Materials and Management Conference and Exhibition*, Atlantic City, NJ (1989).

20. Clark, F., "Agar Plate Method for Total Microbial Count," Methods of Soil and Analysis, American Society of Agronomy, Madison, WI, Vol. 2, pp. 1460–1465 (1965).

21. Kaufman, K., "Applied Bioremedial Technology," Presented at Hazmacon '89, Santa Clara, CA (1989).

22. Ghiorse, W. C., and D. L. Balkwill, "Microbial Characterization of Subsurface Environments," in *Ground Water Quality*, Ward, C. H., W. Giger, and P. L. McCarty eds., Wiley, New York, NY (1985).

23. Matthews, J., and A. Bulich, "A Toxicity Reduction Test System to Assist in Predicting Land Treatability of Hazardous Waste," in Hazardous and Industrial Solid Waste Testing: Fourth Symposium, STP-886, J. K. Petros, Ed., ASTM, Philadelphia, PA (1984).

24. Lambe, T. W., *Soil Testing for Engineers*, John Wiley and Sons, New York, NY (1951).

25. American Society of Testing and Materials (ASTM), "Standard Method for Particle-Size Analysis," ASTM D 422-63, (Reapproved 1972).

26. DiToro and Horzumpa, "Reversible and Resistant Components of PCB Adsorption-Desorption: Isotherms," *Environmental Science and Technology*, 16 (9) pp. 594–603 (1982).

27. Wu, S., and P. M. Gschwend, "Sorption Kinetics of Hydrophobic Compounds to Natural Sediments and Soils," *Environmental Science and Technology*, 20 (7) pp. 717–725 (1986).

28. Hamaker, J. W., and J. M. Thompson, *Adsorption of Organic Chemicals in the Soil Environment*, Vol. I, C. A. I. Goring and J. W. Hamaker, Eds., Marcel Dekker, Inc., New York, N.Y. (1972).

29. Torphy, M. F., H. F. Stroo, and G. Brubaker, "Biological Treatment of Hazardous Wastes," *Pollution Engineering* May 80–86 (1989).

30. Annokee, G. J., "Research on Decontamination of Polluted Soils and Dredging Sludges in Bioreactor Systems," Assessment of International Technologies for Superfund Application EPA/540/2-88/003, Washington, D.C. (1988).

31. Hamaker, J. W., and J. M. Thompson, "Adsorption,"

Organic Chemicals in the Soil Environment, Vol. I, C. A. I. Goring and J. W. Hamaker, Eds., Marcel Dekker, Inc., New York, N.Y. (1972).

32. Loehr, R. C., "Treatability Potential for EPA Listed Hazardous Wastes in Soils," EPA-600/2-89-011, Robert S. Kerr Environmental Research Laboratory, Ada, OK. (1989).

33. Loehr, R. C., "Bioremediation of PAH Compounds in Contaminated Soil," Second Annual West Coast Conference on Hydrocarbon Contaminated Soils and Groundwater, Newport Beach, CA (1991).

34. Park, K. S., R. C. Sims, and R. R. DuPont, "Transformation of PAHs in Soil Systems," ASCE *Journal of Environmental Engineering*, 116 (3) pp. 632–642 (1990).

35. Vedvyas, S. K., "Enhancement of the Solubilization of Polynuclear Aromatic Hydrocarbons Using Cosolvents and Surfactants," M.S. Thesis, University of Texas at Austin, Austin, TX (1990).

36. Groves, F., Jr., "Effect of Cosolvents on the Solubility of Hydrocarbons in Water," *Environmental Science & Technology*, 22 (3) pp. 282–286 (1988).

37. Ellis, W. D., J. R. Payne, and G. D. McNabb, "Treatment of Contaminated Soils with Aqueous Surfactants," EPA/600/S2-85/129. USEPA Hazardous Waste Engineering Research Laboratory, Cinncinnati, OH (1985).

38. Nkedi-Kizza, P., P. S. C. Rao, and A. G. Hornsby, "Influence of Organic Cosolvents on Sorption of Hydrophobic Organic Chemicals by Soils," *Environmental Science & Technology*, 19 (10) pp. 975–979 (1985).

39. Eckenfelder, W. W., *Industrial Water Pollution Control*, 2nd Edition, McGraw-Hill, Inc., New York, NY (1989).

40. Benefield, L. D., and C. W. Randall, "Biological Process Design for Wastewater Treatment," Prentice-Hall Series, Englewood Cliffs, NJ (1980).

41. Bragg, J. R., J. C. Roffall, S. McMillen, "Column Flow Studies of Bioremediation in Prince William Sound," Exxon Production Research Company, Houston, TX (1990).

42. Bouwer, E. J., and P. L. McCarty, "Transformations of Halogenated Organic Compounds Under Denitrification Conditions," *Applied and Environmental Microbiology*, 45 1295 (1983).

43. Gibson, S. A., and J. M. Suflita, "Extrapolation of Biodegradation Results to Groundwater Aquifers: Reductive Dehalogenation of Aromatic Compounds," *Applied and Environmental Microbiology*, 52 p. 161 (1986).

44. Stewart, P. S., D. J. Tedaldi, A. R. Lewis, and E. Goldman, "Biodegradation Rates of Crude Oil in Seawater," *Water Environment Research* (In Press).

45. Rittman, B. E., D. Jackson, S. L. Storck, "Potential for Treatment of Hazardous Organic Chemicals with Biological Processes," in Biotreatment Systems, CRC Press, Boca Raton, FL (1989).

How's Your "Pump-and-Treat" System Doing?

Use these techniques to evaluate the performance of groundwater recovery and treatment systems.

Gerald L. Kirkpatrick, P.G.
Environmental Resources
Management, Inc.

Groundwater recovery and treatment, or "pump and treat," systems are the most widely used technology for cleaning up contaminated aquifers. Although they are widely criticized for their inability to satisfy remedial goals, many of these programs are currently operating. Furthermore, this remediation method will continue to be selected for hazardous waste sites nationwide, at least in the near future, even though emerging and innovative technologies hold great potential.

The performance of these systems must be evaluated on an ongoing basis to prove that adequate protection of human health and the environment is being achieved through efficient operation of the system. In addition, system owners and operators need to provide supporting data that operating capital is being spent effectively through proper operation and maintenance of equipment. If necessary, assessment results can be useful in modifying the recovery system in order to meet the original performance goals.

This article discusses several methods used to analyze the long-term efficiency of a pump-and-treat system, such as:

• degree of compliance with an agreement with the regulatory agency(ies) (*e.g.,* Consent Order or Record of Decision);

• hydraulic demonstration of system performance;

• contaminant mass recovery calculation;

• system design/performance comparison;

• statistical evaluation of groundwater quality; and

• integration of several assessment methods.

Groundwater quality data will indicate the degree of protection of human health and the environment being achieved, which is a key objective, yet the equipment may be operating ineffectively and inefficiently, consuming capital. The degree to which a regulatory agreement is being followed may provide legal evidence of system success, but long-term operation may show that the system as originally designed is either inadequate or overdesigned. In such cases, modifying the system to improve its performance is appropriate.

Which recovery system assessment method(s) to apply depends on the type, amount, and quality of data available. No single assessment method provides a comprehensive evaluation of the overall success of a groundwater recovery and treatment system. Therefore, the use of an integrated approach combining several of the methods is recommended to obtain the most valuable information for evaluating the success of a pump-and-treat system. The methods outlined here should enable engineers and corporate management to hold meaningful discussions with their environmental consultant and regulatory agency. In addition, an independent, third-party, system evaluation can be valuable.

Plan a strategy

Before collecting any data to evaluate system performance, one must develop measurable remedial goals and identify a data collection strategy. The data are then used to assess how well the remedial goals are satisfied.

The primary goal of a groundwater remediation program is to restore the affected aquifer to a condition protective of human health and the environment. For example, for plume containment, the primary goal of a groundwater recovery system is to limit the expansion and migration of the contaminant plume and to provide protection of nearby receptors.

As mentioned, achieving remediation goals using pump-and-treat systems is assessed in several ways:

• *Compliance with a regulatory agreement.* Frequently, administrative orders require meeting certain performance criteria. These criteria are usually expressed in terms of compliance with water quality goals such as maximum contamination levels (MCLs);

• *Demonstration of hydraulic contaminant control.* Recovery system design is usually based on a straightforward extension of well hydraulic concepts. Assessment based on hydraulics depends on the acquisition of reliable field data and a reasonable interpretation of the data acquired. Included in this method is demonstrating that favorable plume geometry changes are taking place over time (that is, plume size decreases, stabilizes, or is otherwise altered to reduce risks).

• *Contaminant mass recovery.* The ratio between the estimated contaminant mass within the affected groundwater regime and the actual mass recovered can be calculated. Determining such a ratio relies on reliable initial estimates of the contaminant mass and accurate mass recovery records.

• *System design/performance comparison.* Using the initial system design as a "benchmark" for comparison with actual operation statistics may also be useful. This is appropriate for either groundwater volume data — comparing design recovery volumes to actual recovery volumes — or contaminant mass recovery data — comparing design contaminant mass removal rates to actual mass removal rates.

• *Statistical evaluation of groundwater quality.* Statistical analysis of long-term groundwater quality trends (which consider natural compound fluctuations) also provides an indica-

tion of recovery system success. Using time-dependent statistical techniques, it is possible to better quantify groundwater improvement or degradation.

• *Integration of assessment techniques.* Integrating several of these methods provides a more comprehensive evaluation of the total performance picture.

Now let's look at each of these techniques in more detail.

Compliance with a regulatory agreement

Records of Decision (RODs), Consent Orders, Best Demonstrated Available Technology (BDAT) standards, and certain regional permits may require that the recovery system operate until it attains specific MCLs or health-based risk standards. It is commonly stated by regulating agencies that these standards must be maintained in the groundwater regime on a permanent basis. Evaluating system performance by comparing exist-

The goal is to restore the aquifer to a condition that protects human health and the environment.

ing groundwater quality to ultimate groundwater quality goals is a common assessment method.

The data required to assess recovery system performance using this criterion are readily available for nearly all projects. Accurate analytical data from carefully designed networks of groundwater monitoring wells are required. The monitoring well networks may (and probably should) evolve as more information and understanding are attained, despite increased practical and administrative difficulties.

One way to assess the success or failure of the system is to prepare a quarterly summary that compares groundwater quality data, such as Table 1. This table should identify the regulated compound(s) used to determine groundwater quality, the initial concentration in the well, the current concentration, the agreed-upon quality standard, and the degree to which the goal has been attained. Validated analytical data and carefully placed monitoring wells are required to use this approach successfully.

Compliance with regulatory agreement is a useful assessment tool when system performance requirements are clear and well-defined. Only if quantitative performance standards are specified can this method yield meaningful information. Ambiguous phrases such as "until aquifer quality has been restored to useful benefit" will result in confusion and significant interpre-

Table 1. A quarterly reporting form can be used to compare current groundwater quality to compliance goals.

Monitoring Well	Benzene Concentration, ppb			Percent of Objective
	Initial	Current	Goal	
MW-1	121	100	5	18%
MW-2	100	65	5	37%
MW-3	107	88	5	19%
MW-4	115	102	5	12%
MW-5	90	87	5	4%
MW-6	8.5	9	5	—
MW-7	2.7	2.5	5	NA
MW-8	<0.5	<0.5	5	NA
MW-9	<0.5	<0.5	5	NA
MW-10	40	37	5	NA

tation challenges in determining whether the system is working as planned.

Demonstration of hydraulic control

An important and effective method for identifying successful recovery system performance, particularly where groundwater quality is of concern, is the assessment of hydraulic manipulation of groundwater by a network of injection and/or recovery wells. Hydraulic manipulation includes groundwater flow direction reversal or contaminant containment or both. This manipulation results in a decrease in contaminant plume spreading such that the aquifer and people who use it are better protected.

This method of assessment is interpretive and, therefore, somewhat subjective. However, it is possible for various parties to arrive at a common interpretation, because system design and operation is usually a straightforward extension of well hydraulic concepts. These concepts are in turn based on assumptions and limitations used in developing the conceptual model for the system.

This approach requires locating monitoring wells in positions that enable the hydrogeologist to map the potentiometric surface (that is, the theoretical groundwater surface under the ground) to determine which direction the groundwater flows. This mapping of the potentiometric surface defines plume hydraulic control, or the process of controlling which way the groundwater flows.

One way to exert hydraulic control is to pump water from wells at various depths and at various rates. Pumping water from an aquifer into wells results in a cone of depression at each well [Figure 1 *(1)*]. If a groundwater recovery system is designed properly, groundwater flow

will be so well controlled that contamination cannot migrate beyond the recovery well network. This is sometimes demonstrated by showing that the wells have overlapping cones of depression.

It is worth noting that installing monitoring wells to assess changes in groundwater quality may not reveal what hydraulic effects a recovery system operation will have on groundwater flow. Frequently, such hydraulic effects are difficult to identify, because uncertainties such as unpredictable nearby pumping centers, measurement and instrumentation error, mechanical equipment

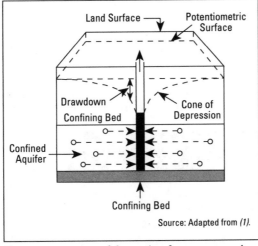

■ *Figure 1. A cone of depression forms as water is withdrawn from an aquifer into a well.*

failure, and complex geology complicate interpretations. In addition, it is important to monitor changes in area groundwater usage (new users, increased usage, or discontinued use) and account for these changes when interpreting the data.

Assessments of this type are usually completed once during the early stages of system testing and design (pump testing), and sometimes a second time during the initial full-scale operation. Unfortunately, aquifer uses change. In addition, reductions in well production capacities during the remediation program are not uncommon. For these reasons, an operator should consider a synoptic water-level

monitoring test, in which the water level in all wells is measured simultaneously, once every five years to assess what effects changes in the aquifer system and/or recovery equipment have on the efficiency of the recovery system.

Using these methods will identify plume geometry changes such that system success or failure will become apparent after a relatively short time — typically several quarterly sampling periods.

In general, demonstration of hydraulic control is a reliable assessment method for those cases where an adequate monitoring well network exists that enables one to identify the hydraulic effects of the recovery system. The method is a preferred choice when plume containment is a primary goal. However, this method may not be useful when hydraulic data are of suspect quality, such as when there are few measurement points or when there are inadequate historical records to compare pre-pumping to pumping conditions.

Contaminant recovery

Groundwater recovery and contaminant treatment systems include pretreatment influent sampling. Results of this sampling are useful for estimating contaminant mass removal from an aquifer. For a dissolved-phase plume, the total mass recovered is expressed in pounds or kilograms. The total contaminant mass is estimated by converting average influent concentrations and recovered water volumes to gallons (or liters) of influent contaminant recovered. Using the weight per unit volume, an operator can estimate the total system mass removal. Generally, such estimates should be made on a monthly basis.

For nonaqueous-phase liquids, such as petroleum hydrocarbons, a reasonable estimate can be made about the expected recoverable product volume. A simple calculation of the percent of recoverable product removed may, in some cases, be enough to assess system success or failure.

This assessment technique is particularly useful at sites requiring frequent influent and effluent sampling. Most remediation projects are subject to Resource Conservation and Recovery Act (RCRA) or National Pollutant Discharge Elimination System (NPDES) sampling and reporting protocols. When monitoring a large suite of compounds at the treatment unit, it is possible to report impressive contaminant mass removals. The technique is less valuable when monitoring only a few compounds present in low concentrations.

Performance vs. design

Basing the final design of a groundwater recovery and treatment system on technical information available prior to system operation is common practice. However, even the most carefully collected hydraulic, geologic, and engineering data will be insuf-

Experience has led to the development of an integrated approach to evaluating system performance.

ficient to predict with certainty the result of long-term groundwater withdrawal programs.

Predictive limitations aside, using the initial groundwater recovery mechanical system design as a benchmark against which actual operating statistics are compared is helpful. For this comparison to be of value, it is important to develop and implement a data collection program prior to system startup.

Table 2 illustrates a sample monthly reporting form that is useful when

recording the performance of mechanical equipment. Use of this form, which can be adapted to any of several spreadsheet software programs, will enable the operator to detect both long-term and short-term problems with the recovery system equipment. Monthly reports are preferable to quarterly reports so that operators can correct problems, should they arise, more quickly.

In addition to quantitative assessment of recovery volumes, the form includes failure codes to identify the source of significant operating and maintenance difficulties. This particular form has six categories of system difficulties:

• electrical problems — problems associated with electrical circuitry, as well as power surges and failures;

• well-specific mechanical problems — pump failures, pumping lift, changing groundwater conditions, and motor burnout;

• delivery system mechanical problems — delivery line (the line from the wellhead to the treatment system) ruptures, or other damage or blockage;

• chemical and biological fouling — iron, manganese, and sulfate bacterial problems resulting in loss of well yield and plugging;

• well maintenance — downtime as a result of routine maintenance; and

• other — vandalism, lightning strikes, vehicular damage, and other types of difficulties not accounted for in the categories noted above.

By analyzing the monthly summary reports, the operator can identify which wells have recurring problems and what types of problems they are, and can then determine how to correct them. The data can be compiled into narrative annual reports that include graphic presentations of monthly and cumulative production trends for each well in the recovery system, as shown in Figure 2.

By collecting these data, trends may be identifiable that indicate a gradual production rate decline, possibly indicating a slow deterioration of well performance. An operator can evaluate the

---

Table 2. This monthly reporting form can be used to document recovery system operation.

Project: Wood's Service Station
Location: Anytown, PA
Date of Report: 15 December 1991
Time Period: 1 November – 30 November 1991

Reported By: D. C. Manning
Checked By: C. F. Michaelson

Recovery Well	Operating Time, min	Groundwater Recovered, gal	Average Rate, gal/min	Original Design, gal/min	Percent of Original Design	Failure Codes
RW-1	40,300	443,520	11	25	44%	3 — see note
RW-2	40,185	1,245,735	31	30	103%	
RW-3	32,800	524,800	16	25	64%	1 — see note
RW-4	40,300	725,400	18	20	90%	
RW-5	37,500	2,700,000	72	70	103%	
RW-6	22,500	900,000	40	40	100%	

Operating Notes:

RW-1 — Plugged delivery pipe increasing backpressure. Line to be replaced next month.

RW-3 — Experienced recurring electronic failure in down-hole wiring. After examination of system it was determined that pump wires were shorting out. Wires were replaced and normal operation has resumed.

Failure Codes:
1. Electrical system failure
2. Mechanical problems
3. Delivery system mechanical problems
4. Biological fouling
5. Well maintenance
6. Other

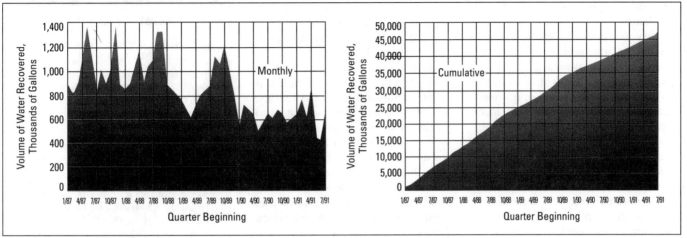

■ *Figure 2. Monthly and cumulative production trends can be charted to identify problems.*

cause of the decline and take corrective measures as appropriate.

This assessment technique is frequently ignored by engineers. The degradation of equipment performance is only identifiable if careful, complete operating records are maintained from the first day of operation. Otherwise, a reduction in well performance or increase in operating costs might go undetected until long after such conditions are reparable.

Statistical evaluation of groundwater quality

Several statistical approaches have been developed to assess temporal (*i.e.,* time-dependent) groundwater quality changes in an aquifer. Two of the most common are those presented in *(2, 3)*. Though not as well-known, comprehensive statistical software packages have also been developed (such as described in *(4, 5)*).

Assuming proper data collection and analysis, statistical programs can identify, with a reasonable degree of quantitative precision, whether a suspected trend in groundwater quality is actually present. Key groundwater monitoring wells and several indicator parameters are selected for analysis. The results may indicate improving, degrading, or stabilized groundwater

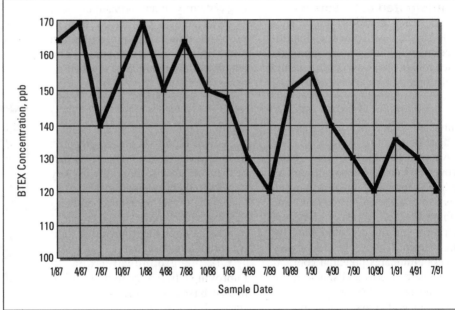

■ *Figure 3. Statistical analysis can show whether groundwater quality is improving, has stabilized, or is degrading.*

quality conditions. The quantified data trends can then be used to assess the effectiveness of the recovery system.

An example of a groundwater quality trend for benzene, toluene, ethylbenzene, and xylene (BTEX) compounds at a gasoline service station is shown in Figure 3. This time series plot is a simple *x–y* graph of the parameter measured, in this case BTEX concentrations in parts per billion, against time, here on a quarterly basis. In this case, groundwater quali-

ty with respect to BTEX appears to be improving. Though not shown here, there is a statistically identifiable decrease in compound concentrations over the life of the monitoring data.

Using statistical techniques for analyzing temporal groundwater quality data will detect trends in water quality parameters for a specific monitoring point, rather than comparing data to arbitrarily selected background wells. This approach may be especially useful in aquifers for which background concentrations cannot be easily defined — for example, those with a very low hydraulic gradient or that are tidally influenced.

A statistical assessment of groundwater quality data is most reliable when a few years' worth of quarterly data exist. It is a preferred assessment technique, because it is one of the best ways to quantitatively demonstrate environmental improvement and protection of human health and the environment. In some cases, regulatory personnel may require some education to feel comfortable with the resulting interpretation, but rarely is this assessment method unacceptable.

An integrated assessment

Experience at various project sites has led to the development of a multi-assessment, or integrated, approach, which makes use of available data and results in a comprehensive analysis of the system's operating success or failure. In addition, specific recommendations about potential improvements to the system can be formulated.

An independent review of the recovery system's operating history by qualified personnel is also encouraged. This can provide an objective outside opinion on the efficiency of system operation and the potential need for improvement. The independent review process eliminates potential conflict-of-interest concerns and should result in an unbiased evaluation.

The integrated assessment process begins with a preliminary data assessment that determines the amount, type, and quality of data available for evaluation. The ideal database for such an evaluation includes:
• the original groundwater recovery system design specifications;
• initial recovery well hydraulic data;
• quarterly groundwater quality data;
• quarterly groundwater level elevation data;
• monthly groundwater production and treatment volumes;
• well-documented system problems on a well-by-well basis; and

• monthly treatment system influent concentration data.

Rarely are all of these data available. If a system has been operating for years with only limited recordkeeping, it may be important to have the equipment manufacturer's specifications available (which would indicate how a particular piece of equipment should work). If a system has been operating for a short period of time and little groundwater quality data are available, it would be important to include contaminant mass recovery and hydraulic control information in the assessment process. In any case, it is important to develop an assessment plan that outlines the rationale for selecting or omitting certain system information.

After review of the assessment proposal, data collection, reduction and analysis are performed. The length of time required to conduct an assessment will vary with system complexity. Generally, it can be completed within 60 days of project authorization.

A final report should be prepared to summarize the assessment. As part of the conclusions, the following key issues should be addressed:
• Is the owner/operator complying with the agency order or agreement?
• Is the equipment operating as it was originally designed?
• Is the groundwater quality improving? Is the system attaining health-based groundwater quality standards or goals?
• Do any operating procedures or equipment require modification?

Answering these four questions is a basic goal of a groundwater recovery system assessment project. These questions can be most reliably answered by analyzing several categories of data in an integrated fashion and not relying exclusively on a single technique.

To properly assess the success of a recovery system, one can use a simple monthly reporting form that summarizes the relevant data for each of the assessment techniques described. Such a monthly reporting

Literature Cited

1. **Heath, R. C.,** "Basic Ground-Water Hydrology," U. S. Geological Survey Water Supply Paper 2200, U. S. Geological Survey, Denver, CO (1983).
2. **U. S. Environmental Protection Agency,** "RCRA Groundwater Monitoring Technical Enforcement Guidance Document," Publication No. OSWER 9950.1, U. S. EPA, Washington, DC (Sept. 1986).
3. **U. S. Environmental Protection Agency,** "Statistical Analysis of Groundwater Monitoring Data at RCRA Facilities — Interim Final Guidance," Publication No. PB89-151047, U. S. EPA, Washington, DC (Feb. 1989).
4. **Phillips, R. D.,** *et al.,* "WQSTAT II: A Water Quality Statistics Program User's Manual," Colorado State Univ., Fort Collins, CO (1989).
5. **Loftis, J.C.,** *et al.,* "WQSTAT II: A Water Quality Statistics Package," *Ground Water,* **27**(6), pp. 866–873 (1989).

G. L. KIRKPATRICK is a partner with Environmental Resources Management, Inc., an ERM Group member company, Exton, PA (215/524-3770; Fax: 215/524-7798), where he directs ERM's Hydrocarbon Remediation Services group. A certified petroleum geologist with more than 13 years of industrial and consulting experience in the applied geosciences, he holds a BS in geology from Muskingum College and an MSc in geology from Florida State Univ. He is the author of several technical publications and computer software programs, and he is a member of the Hazardous Materials Control Research Institute, the Society of Petroleum Engineers, and the Environmental Section of the American Association of Petroleum Geologists.

form would combine appropriate elements of Tables 1 and 2 and Figures 2 and 3. Combining these monthly reports into annual assessments of recovery system performance by an independent third-party reviewer is an important step in the evaluation process. **CEP**

Enhance Performance of Soil Vapor Extraction

Teaming soil vapor extraction with air sparging or steam injection can make SVE more effective or applicable where it otherwise would not be.

**David C. Noonan,
William K. Glynn, and
Michael E. Miller,**
Camp Dresser & McKee, Inc.

Hydrocarbon recovery as a means of soil and groundwater remediation has received considerable attention in the last few years as the shortcomings of groundwater pump-and-treat technologies have become more evident. A previous article *(1)* covered a wide range of *in situ* site cleanup technologies and provided guidance on how to choose among them. This article examines one of those technologies, soil vapor extraction (SVE), in more detail and explains how to improve the performance of SVE by combining it with air sparging or steam injection.

SVE is an *in situ* remediation technique whereby soil gas within the unsaturated contaminated soil matrix is extracted using an applied vacuum at one or more extraction wells (or trenches). Pressure gradients within the unsaturated (or vadose) zone induce convective air flow throughout the porous soil matrix, the extent of the vacuum influence determined by the soil properties. As the contaminated soil gas is removed, clean air from the surface is drawn into the contaminated zone and organic compounds are volatilized, depending upon their vapor pressure.

SVE has proven to be a very effective technology for the remediation of soil contaminated by petroleum hydrocarbons. This technology takes advantage of the highly volatile nature of many of the petroleum derivatives and of the relative ease of moving air through the unsaturated zone. However, SVE could find even more widespread application if:

• less volatile hydrocarbons could be partitioned from the soil to the vapor phase in the vadose zone, allowing their more complete recovery; or

• dissolved contaminants and residuals beneath the water table could be volatilized and transported into the vadose zone to be recovered by an induced vacuum, allowing SVE to be used for the treatment of contamination in the saturated zone and the unsaturated zone.

This article reviews two promising SVE enhancements — air sparging and steam injection. Air sparging injects air below the groundwater surface to promote the volatilization of volatile organic compounds (VOCs) from the groundwater into the vadose zone so that the VOCs can be removed via the SVE system. Steam injection injects steam into the vadose zone to increase the subsurface temperatures, thereby volatilizing organic compounds with high boiling points.

To facilitate the mass transfer of petroleum-based contaminants from the groundwater or beneath the water table (as a result of natural seasonal water table fluctuations, for example) to the vadose zone, several researchers have experimented with injecting air into the saturated zone using sparging techniques *(2, 3)*. Air sparging offers the promise of addressing both of the above conditions by adding air to the saturated zone where residual contamination has selectively partitioned onto the soils, has been trapped within the interstitial pore spaces, or exists in a dissolved phase. The injected air enhances volatilization by increasing the water-to-air surface area, and in some cases may induce upward migration of globules of product with migrating air bubbles.

To facilitate the removal of less

volatile and tightly sorbed constituents, other researchers have investigated steam injection *(4–6)*. The added heat provided by the steam enhances the volatilization of petroleum residuals that are sorbed to the soil. The steam front mobilizes the heavy residuals and volatilizes the light fractions. The vapors are removed by a vacuum and the condensate and oils are pumped out from the extraction wells. Enhanced volatilization and residuals migration can effect faster, more complete mass transfer, thereby speeding remediation and perhaps reducing costs.

However, these remediation techniques do have their drawbacks. They both may require significant expenditures for energy to establish positive displacement in the subsurface. And, there are questions as to the volume of residual that may be impacted by a single injection well. These factors can result in a lower overall effectiveness and increased costs.

Air sparging

Air sparging, also referred to as "*in situ* air stripping" or "*in situ* volatilization," is a treatment technology for removing VOCs from the saturated zone. Contaminant-free air is injected into contaminated groundwater to remove contaminants from the saturated zone and effectively capture them with an SVE system.

The use of an air sparging system results in a net positive pressure in the subsurface, which must be compensated for by the SVE system to prevent migration to previously uncontaminated areas. Without SVE, uncontrolled contaminated soil vapor flow may enter basements of nearby buildings, potentially creating an explosion or health hazard.

SVE without air sparging can remove contaminants from the saturated zone. However, the transport rates due to diffusion/dispersion of the dissolved contaminants in the

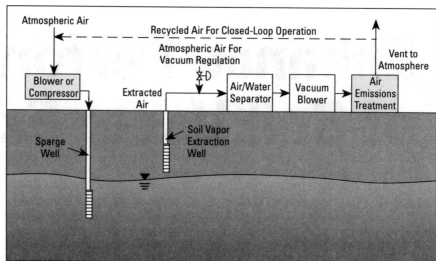

Figure 1. Typical air sparging system.

aqueous phase to the air-water interface limit the removal effectiveness. This rate of contaminant transport can be significantly increased by the addition of air sparging to an SVE system.

The effectiveness of the air-sparging/SVE system can be attributed to two major mechanisms — contaminant mass transport and biodegradation. Depending on the configuration of the system, the operating parameters, and the types of contaminants found at the site, one of these mechanisms usually predominates or can be enhanced to optimize contaminant removal.

The mass-transfer mechanism consists of movement of contaminants in the subsurface and eventual extraction via an SVE system. Contaminants adsorbed to soils in the saturated zone dissolve into groundwater. The sparged air displaces water in the soil pore spaces and causes the soil contaminants to desorb, volatilize, and enter the saturated zone vapor phase (SZVP). The mechanical action of the air passing through the saturated zone increases turbulence and mixing in the groundwater. Dissolved groundwater contaminants also volatilize into the SZVP and migrate up through the aquifer to the unsaturated zone. The SVE system creates a negative pressure gra-

dient in the unsaturated zone, which pulls the contaminant vapors toward the SVE wells.

Aerobic biodegradation of contaminants by indigenous microorganisms requires the presence of sufficient carbon source, nutrients, and oxygen. Air sparging increases the oxygen content of the groundwater, thus enhancing aerobic biodegradation of contaminants in the subsurface. The organic contaminants, especially petroleum constituents, provide the microorganisms with a carbon source. If the rate of biodegradation is to be significantly enhanced, nutrients such as nitrogen and phosphorus usually must be added to the contaminant zone. However, nutrient addition can cause excessive biological growth, which may cause significant fouling of the injection wells and thereby reduce the effectiveness of an air sparging system.

The design of an air sparging system involves selecting the well configuration, blower and compressor sizes, well design, and vapor treatment systems. In addition to the placement of the process equipment, proper gages and instrumentation are crucial for monitoring the effectiveness of the process and making adjustments as needed.

The aboveground components of a typical sparging system are

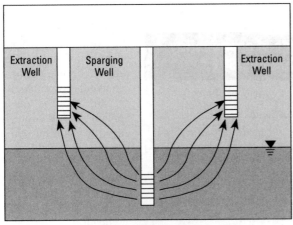

(a) Spaced Configuration

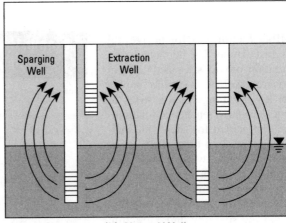

(b) Nested Wells

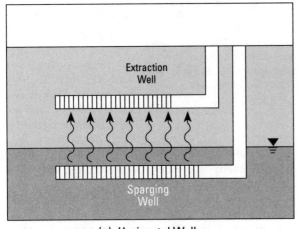

(c) Horizontal Wells

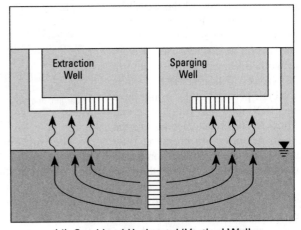

(d) Combined Horizontal/Vertical Wells

■ *Figure 2. Configurations for air sparging wells.*

shown in Figure 1. Its major components are:

- air blower or oil-free compressor;
- vacuum blower;
- air/water separator;
- air emissions treatment (such as activated carbon or catalytic oxidation);
- piping and valves; and
- instrumentation (pressure gauges, flow meters, and so on).

Perhaps the single most important design element of an air sparging system is the layout and construction of the well network. Various configurations, shown in Figure 2, have been used, each of which has its own unique advantages and disadvantages.

The placement of the air sparging and vapor extraction wells must take into account a wide range of groundwater, soil, and chemical properties of

the site. The physical conditions of the soil/groundwater matrix, such as depth to groundwater, soil particle-size distribution, soil stratification, and soil porosity, as well as the thermodynamic and transport properties, such as chemical/soil partitioning coefficients and pneumatic and hydraulic conductivity of the soil, will assist in determining the best well configuration for a particular site.

The spaced configuration (Figure 2a) is generally applied in a square grid pattern with the extraction well in the center and four injection wells at the corners. This pattern works well for sites with highly uniform sandy soils where an effective air flow pattern can be created between the injection and extraction points. The vertical wells are laid out throughout the site cov-

ering the zone of contamination. The spacing of the wells is based on the radius of influence of both the extraction and sparging.

Nested wells (Figure 2b) are extraction and sparging wells placed in the same borehole. The advantage of this configuration is that drilling costs may be reduced. The disadvantages are (1) care must be taken to properly grout the borehole to prevent short-circuiting of air and (2) the pressure gradient is primarily in the vertical direction. This configuration works well for sites with highly stratified silty soils where the vertical permeability is significantly less than the horizontal permeability.

Trenches or horizontal wells (Figure 2c) are formed by installing perforated pipe and gravel pack in a trench or by using new advanced drilling techniques for horizontal well installation. The horizontal configuration provides a more uniform pressure gradient at specific depths over a wider range than a series of vertical wells. Trenches are particularly well suited for sites with shallow aquifers less than 10 ft below grade. Horizontal wells are well suited for contaminant plumes resulting from leaking pipelines.

The radius of influence (ROI) around the sparging and extraction wells is the zone in which the vapor flow is induced toward the well. The ROI is determined by pressure gradients and/or changes in the chemical composition at distances away from the well. Soil permeability, among other factors, will affect ROI — soils with high permeabilities would have larger ROIs than soils with low permeabilities (all other factors remaining the same). Air injection pressure, flow rate, and the depth of injection below the water table will also affect the ROI.

The ROI is used to determine the well spacing and number of wells needed for the site. However, consideration should also be given to the travel time of the contaminant from the outer perimeter of the ROI to the extraction well. Restrictive travel times may impede the cleanup and therefore the well influences should be overlapped.

In general, high permeability sandy soils will result in a higher ROI and consequently higher flow rates than low permeability silty soils. The ROI for a sparging well can range from 10 to 100 ft from the injection point, while the ROI for an extraction well can range from 25 to 300 ft from the extraction point. The air injection flow rate is always less than the extraction flow rates in order to capture the injection air in the extraction system. Typical systems operate with injection air flows on the order of 10%–20% of the extraction flow rates.

Table 1. Range of applicability of air-sparging/SVE.*

Soil Types	Sand and Gravel; Fractured Bedrock
Contaminants	Benzene, Toluene, Ethyl Benzene, Xylenes (BTEX); Trichloroethylene (TCE); Tetrachloroethylene (PCE)
Depth to Groundwater	5–30 ft Below Grade
Vapor Extraction Wells	1–5 Wells
Air Sparging Wells	5–20 Wells
Cleanup Time†	2–24 Months

* Based on published information about 21 sites.
† Cleanup time refers to a significant reduction in groundwater concentrations from the initial air sparging operation.

The implementation of an air sparging system must also take into consideration the dynamic changes that may occur in the subsurface. The introduction of air below the water table will cause an increase in the groundwater elevation, which is also known as mounding. This effect, if not properly controlled, may cause further migration of contaminants away from the treatment area. Sparging can also cause dissolved minerals to precipitate, thereby impeding the flow of air through

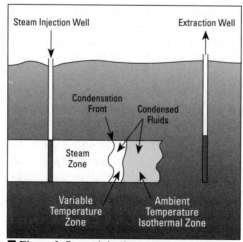

■ *Figure 3. Steam injection zones.*

the subsurface. Careful operation of the air injection rate or the use of nitrogen can avoid this problem.

This process has been applied at sites contaminated with gasoline and with chlorinated solvents. Table 1 summarizes the range of soil types, contaminant types, and construction and operating information for several applications.

Steam injection

Steam injection, also referred to as "soil heating" or "steam venting," is an *in situ* treatment technology for remediation of organic contaminants in the subsurface. Steam is injected into a contaminated subsurface to thermally recover volatile and semivolatile liquids in conjunction with water and vapor extraction.

Steam injection is coupled with an SVE system and a water extraction system in order to capture the contaminants that are liberated from the porous soil. The use of steam injection results in the migration of vapors in the steam zone and the flow of contaminant liquids ahead of the steam condensation front.

The effectiveness of the steam injection/recovery system can be attributed to two major mechanisms: vaporization of volatile and semivolatile contaminants, and displacement of liquids. Depending on the configuration of the system, the steam injection rate, additional operating parameters, and the types of contaminants found in the site, these mechanisms can be optimized for maximum contaminant removal.

As the steam is initially injected into the subsurface, the ambient soils remove the latent heat of vaporization from the steam and it condenses. Following additional steam injection, the steam condensate front moves outward from the injection point and

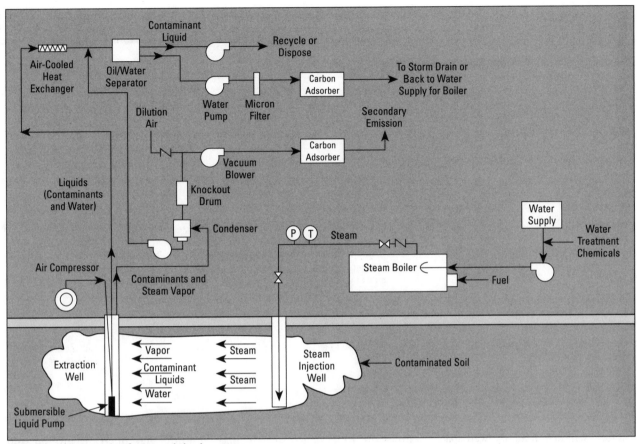

■ *Figure 4. Schematic of a steam injection system.*

an isothermal steam zone is evident. The zone beyond the steam condensate front is referred to as a variable temperature zone. The steam injection zones are shown schematically in Figure 3.

The volume affected by the steam zone depends on the steam injection rate. The system is generally characterized by a high vapor volumetric flow rate, typically in the range of 1,000–2,000 lb/hr, and temperatures can be raised to greater than 115°C.

Low-boiling-point liquids in the range of 90°C–150°C will generally be mobilized ahead of the steam condensate front in the variable temperature zone and accumulate in both the vapor and liquid phases. Organic contaminants with low vapor pressures (that is, C15 hydrocarbons and greater) may remain in the pore spaces within the isothermal steam zone. However, contin-

ued steam flow will subsequently evaporate these contaminants or enhance their migration toward the collection wells.

The design of a steam injection system involves selecting the well configuration, steam generator, blower and compressor sizes, well design, and water and vapor treatment systems. In addition, monitoring equipment (such as temperature probes), liquid content, and pressure gradients are crucial to determining the effectiveness of the process and evaluating adjustments to the system. A typical steam injection system is illustrated in Figure 4.

The removal of residual petroleum at a contaminated site can be accomplished over the entire contaminated area or sequentially in small areas. Although energy intensive to operate, a steam injection system is applied for only a fraction as long as conven-

tional remediation techniques — on the order of weeks as opposed to months or years for traditional remediation methods. In considering this process for cleanup of a site, a reasonable cleanup time must be estimated based on the specific site conditions, such as extent of contamination and soil permeability, in order to develop a comparable cost estimate with other remediation techniques.

The major factors affecting radius of influence of a steam injection system are soil permeability, steam injection pressure, and steam flow rate. In general, high permeability sandy soils will result in a higher ROI than low permeability silty soils. The ROI for a steam injection well can range from 25 ft to 100 ft from the injection point, and the ROI for an extraction well can range from 25 ft to 300 ft from the extraction point.

The ROI for the steam injection

Table 2. Summary of steam injection field studies.

Soil Type	Contaminants	Initial Soil Concentration	Final Soil Concentration	Cleanup Time	Comments
Sand above and below water table	Acetone; 2-Butanone; Chlorinated solvents; Benzene, toluene, ethyl benzene, and xylenes (BTEX)	2,000 mg/kg	12 mg/kg	140 hr	Pilot-scale. Closely spaced wells over small fraction of site.
Sand in the saturated zone	Diesel fuel	8,000 mg/kg	30% reduction*	5 mo*	Full-scale.
Interbedded sands, silts, and clay above and below water table	Solvents; Oil; Grease	Some pure product; Trichloroethylene at 40 mg/L in the groundwater			Full-scale. In design phase.
Sand and silt	Jet fuel	Pure product on the water table			Pilot-scale. In design phase.
Sandy gravel in the saturated zone.	Gasoline	Pockets of free product			Full-scale. In design phase.

* Remediation still in progress.

system will determine the well spacing and number of wells needed for the site. Based upon a square injection well grid, the maximum well spacing is the square root of two times the ROI. If a faster cleanup time is required, the injection wells should be spaced closer together than the maximum distance in order to heat the subsurface in less time. A model has been developed that, among other things, can determine the growth rate of the steam zone in a sandy soil. Based upon a steam injection rate of 1,000 lb/hr in a 10-ft layer of the vadose zone and a maximum ROI of 40 ft, the treatment zone is fully developed in about 16 days.

Steam injection has been used to remove gasoline and diesel oil in both the unsaturated and saturated zones. In general, the amount of steam required in the saturated zone is on the order of four to five times greater than that required in the unsaturated zone. This additional heat is required to physically displace, heat, and vaporize the groundwater. However steam injection cleanup times and costs are still only a fraction of those required for groundwater pump-and-treat systems.

The operation of a steam injection system begins with the simultaneous injection of steam and extraction of

D. C. NOONAN is an associate with Camp Dresser & McKee, Inc. (CDM), Cambridge, MA (617/252-8000; Fax: 617/621-2565). He has over 15 years of experience in soil and groundwater investigations and hazardous waste remediation; he has conducted research projects, site investigations, and feasibility studies for federal and state government and private industry. He received a BS in civil engineering from Northeastern Univ. and an MBA in business management from New Hampshire College. He has authored over 30 professional papers and co-authored two books. He is a registered professional engineer.

W. K. GLYNN is a senior chemical engineer in CDM's industrial and hazardous waste group, where he is responsible for the design and construction of waste treatment systems. He has developed full-scale designs of innovative treatment processes, including air sparging and soil vapor extraction systems, and evaluated the effectiveness of novel pilot-scale systems. He is also responsible for the treatability assessment of industrial and hazardous wastes to determine the most cost-effective treatment alternative. He holds a BS in chemical engineering from Tufts Univ. (1980) and an MS in chemical engineering from the Univ. of Kentucky (1984), and he is a registered chemical engineer in Massachusetts.

M. E. MILLER is an environmental chemist in CDM's hazardous waste division, specializing in the remediation of contaminated soil and groundwater and the fate and transport of organic and inorganic contaminants in soil, water, and air. Previously, he held a postdoctoral fellowship at Cornell Univ. Laboratory of Soil Microbiology, and was an Oak Ridge Universities postdoctoral scientist at the U.S. Dept. of Energy's Pittsburgh Energy Technology Center. He received a BA with high honors in chemistry from Swarthmore College in 1981, and his MS (1983) and PhD (1986) in physical and inorganic chemistry from Cornell Univ. He has published numerous refereed articles on topics from biodegradation to quantum mechanics.

liquids and vapors. During the first stage of operation, the subsurface is heated to the steam temperature as the steam front moves toward the extraction wells. After the injected steam breaks through to the extraction wells, steam injection continues until the contaminant concentrations approach the cleanup goals. At that point, steam injection is stopped while the vapor extraction system continues to operate. The continued SVE operation will result in the further vaporization of the residual contaminants in the pore spaces and the drying out of the treated soils.

Several sites have conducted limited trials with steam injection with very favorable results. Table 2 summarizes several field studies performed to date. The process is particularly applicable to high-molecular-weight petroleum products such as diesel fuel and jet fuel. In one case, the solvent concentrations in the soils were reduced by greater than 99% in less than six days.

Final thoughts

For removing VOCs from the saturated zone, the combination of air sparging and SVE has been shown to be more effective and have lower capital and operating costs than a traditional groundwater pump-and-treat system. For example, at one gasoline-contaminated site with groundwater benzene, toluene, ethyl benzene, and xylenes (BTEX) concentrations in the range of 50 mg/L, the capital cost was less than 50% and the annual operating costs were one-third of the costs for a traditional air stripping pump-and-treat system. Furthermore, air sparging systems have been shown to be effective at cleaning up a site in less than two years, while pump-and-treat systems often must operate for up to ten years or beyond.

For removing semivolatile organic contaminants from the unsaturated and saturated zones, steam injection combined with soil vapor extraction

Literature Cited

1. **Long, G. M.,** "Clean Up Hydrocarbon Contamination Effectively," *Chem. Eng. Progress,* **89**(5), pp. 58–67 (May 1993).
2. **Ardito, C. P., and J. F. Billings,** "Alternative Remediation Strategies: The Subsurface Volatilization and Ventilation System," Proceedings of the API/NWWA Conference on Petroleum Hydrocarbons and Organic Chemicals in Groundwater: Prevention, Detection, and Restoration, sponsored by the American Petroleum Institute and the National Water Well Association, Houston TX (1990).
3. **U.S. Environmental Protection Agency,** "Assessment of International Technologies for Superfund Applications," Publication No. 540/2-88/003 (1988).
4. **Lord, A. E., Jr.,** *et al.,* "Vacuum-Assisted Steam Stripping to Remove Pollutants from Contaminated Soils," U.S. Environmental Protection Agency, Publication No. 600/9-90/037 (1990).
5. **Hunt, J. R.,** *et al.,* "Nonaqueous Phase Liquid Transport and Cleanup, Part 1: Analysis of Mechanisms," *Water Resources Research,* **24**(8), pp. 1247–1258 (1988).
6. **Hunt, J. R.,** *et al.,* "Nonaqueous Phase Liquid Transport and Cleanup, Part 2: Experimental Studies," *Water Resources Research,* **24**(8), pp. 1259–1269 (1988).

Further Reading

Aines, R., *et al,* "Dynamic Underground Stripping Demonstration Project, Interim Progress Report, 1991," Lawrence Livermore National Laboratory (March 1992).

Anastos, G., *et al.,* "*In Situ* Air Stripping of Soil: Field Pilot Studies Through Implementation," *Toxic Hazardous Wastes,* **19**, p. 163 (1987).

Brown, R. A., *et al.,* "Use of Aeration in Environmental Clean-ups," Presented at Haztech International Pittsburgh Waste Conference, Pittsburgh, PA (May 14–16, 1991).

Kasbohm, P. L., *et al.,* "*In Situ* Soil Ventilation: A Case Study," Proceedings of the Ninth Annual Hazardous Materials Management Conference/International, Atlantic City, NJ (June 12–14, 1991).

Kasevich, R. S., *et al.,* "Remediation of Hydrocarbon Contaminated Soils Using Radio Frequency Heating," Presented at the 7th Annual East Coast Conference on Hydrocarbon Contaminated Soils, Amherst, MA (Sept. 1992).

Koltuniak, D., "*In Situ* Air Stripping Cleans Contaminated Soil," *Chem. Eng.,* **93**(16), pp. 30–31 (Aug. 18, 1985).

Marley, M. C., "Air Sparging in Conjunction with Vapor Extraction for Source Removal at VOC Spill Sites," Presented at the Fifth National Outdoor Action Conference, Las Vagas, NV (May 13-16, 1991).

Udell, K. S., and L. D. Stewart, "Combined Steam Injection and Vacuum Extraction for Aquifer Cleanup," Presented at Conference on Subsurface Contamination by Immiscible Fluids, Calgary, Alberta (Apr. 18–20, 1990).

Udell, K. S., and L. D. Stewart, "Field Study of *In Situ* Steam Injection and Vacuum Extraction for Recovery of Volatile Organic Solvents," Sanitary Engineering and Environmental Health Research Laboratory, Univ. of Calif., Berkeley, CA (June 1989).

has been shown to be more effective and requires less cleanup time than excavation and aboveground treatment. At a site contaminated with chlorinated aromatics with soil concentrations up to 5,000 mg/kg and groundwater concentrations in the range of 60 mg/L, an SVE/steam-injection system, although more expensive to install and operate than traditional cleanup methods, performed the cleanup in a fraction of the time required by the other techniques. In addition, the overall costs were 30% to 50% less than the cost for the more conventional approach.

Although these technologies are relatively new, they will continue to be implemented for the cleanup of petroleum contaminated sites and will have an enormous impact on future remediations. **CEP**

Soil Washing and Radioactive Contamination

Dirk Gombert

Westinghouse Idaho Nuclear Company, Inc.,
P.O. Box 4000, Idaho Falls, Idaho 83415-4000

INTRODUCTION

Soil washing, a treatment technique which may combine both physical and chemical processes to produce significant volume reduction of contaminated soils, is widely regarded as a panacea for the huge inventory of contaminated soils in the DOE Complex. While the technology has been demonstrated for organics and to some extent for metals, review of available publications on practical applications to radioactive sites indicates that most volume reduction is a product of unique circumstances such as screening or floating out non-soil materials containing most of the contaninants, or leaching contaminants (uranium or transuranic elements) that exist as anionic complexes [1] which are not held by the soil cation-exchange-capacity. In either case, the potential for success of the technology is extremely site and contaminant specific.

Environmental remediation techniques are based on the same principles used in mining, chemical manufacture, and water treatment. Nothing magic is implied by the environmental application; this new industry is just based on creative combinations of existing technology, with a little development to reach new goals. The constraints of waste and cost minimization have never been more critical.

This paper discusses the state of the technology and the unique challenges of treating radioactively contaminated soils, then focuses on how soil washing could be evaluated to provide definitive answers on when and where it should be employed. A logical, methodical approach must be designed to establish minimum acceptable criteria, determine what the controlling phenomena are, and then objectively evaluate whether a technology can potentially be applied to the problem. The Environmental Protection Agency's (EPA) guidance on soil washing treatability studies suggests a 50 percent reduction of contamination in particles over 2 mm as a reasonable cutoff for choosing soil washing for further development [2]. Once the decision has been made to attempt development, a systems approach is imperative to ensure a practical solutions. The EPA guidance also suggests a concept not so well recognized; "Residual risk, as applied to soil washing, assesses the risks associated with treatment residuals...sidestream and other treatment train processes should be included" [2]. Soil leachants cannot increase the toxicity of the soil product, interfere with downstream water treatment, or present an unacceptable hazard to remediation workers. This "big-picture" approach is frequently lost in the drive to develop the "magic bullet."

TECHNOLOGY

Mining Industry

A wealth of experience and knowledge is available in the mining industry on the recovery of metals. Commodity values have pushed retrieval efficiencies above 80 percent for ores containing parts-per-million (ppm) quantities of gold [3]. Crushing, screening, and grinding equipment comes in a wide variety of shapes and sizes. Particle size reduction is done to release individual mineral grains, remove surface contaminants, enhance chemical contact, and prepare material for particular applications. Though not typically viewed as a chemical engineering specialty, even Perry's Handbook devotes an entire chapter to these technologies [4].

Physical separation requires a distinct size or density difference between the material to be processed and the reject material. Where the species of interest exists in distinct particles, screening, settling, or flotation may be applied to provide a first cut, such that more cost-effective processing can be targeted at the more concentrated media. In some cases the particles may require separation from the host matrix or a depositional phase by grinding or attrition scrubbing. Settling and flotation may be facilitated by chemical additions that are species specific and cause or eccentuate density differences to promote settling or floating.

Acid leaching is typical of a chemical extraction to further concentrate the material for retrieval prior to purifying for its intended purpose. Acid is relatively inexpensive, and metals can be readily removed from other soluble salts, so the gross dissolution of other acid soluble compounds is acceptable. Once the metals are removed, the remaining salt bearing solution can be dried to a cake in an open evaporation pond.

Soil Washing

Application of these technologies to environmental restoration purposes entails a set of more confining constraints, while trying to satisfy more demanding goals. Not only do the contaminants include organics and inorganics, with cleanup criteria which may be orders-of-magnitude lower, but all effluents are subject to scrutiny for residual contamination. Surveys of soil washing as currently applied, particularly in Europe, show a key similarity in relying on size classification (physical soil washing) for the primary contaminant separation. Particles below a 63 to 74 micron cutoff range are separated and disposed of as hazardous waste [5, 6, 7]. Though chemical extraction may be employed in conjunction with the separation process, it is commonly accepted that fine particulate present in solution may resorb extracted contaminants, and it is not cost effective to continue to wash the fines. Soils containing more than 20 percent fines below the cutoff level, are generally not cost effectively treated with physical soil washing [5, 6, 7]. Thus soil washing typically provides volume reduction by a factor of 5–10, but does not actually reduce the toxicity of the contaminants.

While organic contamination is typically bound by the naturally occurring organic content of the soil, [8] or as coatings on larger particles, inorganic contaminants may be bound through any combination of several mechanisms including ion exchange, chemical and physical adsorption, precipitation, isomorphic substitution, and agglomeration. Adding to this complication is the extremely site and species specific nature of the problem. A chemical soil washing (extractive) flowsheet that works well to remove gasoline from a moist, humic soil would have a slim chance for success removing cadmium from an arid calcereous soil. Even a system proven successful for removing the same element from the same soil could have difficulty due to speciation, as is the case for trivalent versus hexavalent chromium. Applying extractive soil washing technology to inorganic removal requires a quantum leap in sophistication from the relatively simple application of heated water and surfactants which can be quite effective for organic contaminants. Even now satisfactory results beyond size classification cannot be obtained for many metals in many environments.

Radioactive Contamination

Adapting this technology to radionuclide contaminated soils presents a much greater challenge. At the Idaho National Engineering Laboratory (INEL), for example, some contaminated soils and sludges are one-third to three-quarters material below 50 micron [9, 10]. In addition, the contaminants may not exist as individual particulate; being derived from aqueous solutions, many radionuclides contaminate soils on an atomic level. The contaminants are not limited to transition and heavy metals, but also include alkalies and alkaline earth fission products that are chemically analogous to natural soil constituents which make up percent levels of the soil matrix. It is obviously questionable whether soil washing is applicable to these contaminants. More questionable is the ability to reach the radionuclide cleanup goals even if the conditions are ideal.

"How clean is clean?" takes on a special significance when radionuclides are discussed because instrumentation exists to measure radioactivity to extraordinarily low levels, and the risks projected by the cancer initiation models in use by the EPA have no lower bound [11]. To date, below regulatory concern (BRC) levels have not been quantified. The waste disposition criteria for radionuclides in commercial landfills permitted under the Resource Conservation and Recovery Act (RCRA) have defaulted to background values, which vary by location, but are very low everywhere, particularly for fission products. Continued placement of radioactive soils in Department of Energy (DOE) controlled landfills is complicated by the frequent association of radionuclides with regulated heavy metals and organics (mixed waste). The technology exists to treat most soils to the extent that they will pass the toxic characteristic leach procedure (TCLP) criteria under RCRA, but treatment may be complicated if not precluded entirely by radioactive contamination. Chemically extracting metals to meet the TCLP requirements such that soils may be placed in landfills may leave a significant non-leachable metal content in the soil, but radionuclide limits are not simply a function

of leachability. The very existence of radionuclides presents a potential cancer hazard as a function of proximity due to the non-contact effects of ionizing radiation. Obtaining contact handling standards of 5 millirem per hour for soils contaminated with cesium-137 requires contaminant reduction to parts-per-billion (ppb) levels. Treatment to eliminate all significant risks to allow free release, that is lower than the typically acceptable 10^{-6} levels under the Comprehensive Environmental Response, Compensation, and Liability Act (CERCLA) for remediatiion efforts, [12] can require decontamination down to part-per-trillion (ppt) levels in the soil [13]. This may require innovative extractive processing following physical separation.

Were it not for the potentially huge environmental gains to be made by successfully developing soil washing for remediating radionuclide bearing soils, the concept would have already been dropped. To develop an adequate knowledge base to definitively judge the usefulness of the technology requires some basic research in how the contaminants are bound, so flowsheets can be targeted at the controlling mechanisms. Without this knowledge, trial-and-error methods may never give the technology a comprehensive evaluation. It is very possible to free a species from one mechanism, while making it susceptible to another, with no net release from the soil. To obtain release, gross dissolution of an unnecessarily great fraction of the soil using acid may result in magnifying the waste disposal problem. Without the basic knowledge on what binds the contaminant, it may only be fortuitous to develop an acceptable strategy.

EVALUATION

Fixation Mechanisms

A logical evaluation program should begin to establish the insight necessary to either support site-specific soil washing flowsheet design or abandon the technology. As in a chemical process, where the rate-controlling step is identified so the design can fully exploit the kinetics, the key to evaluating a decontamination process is understanding how the contaminants are fixed. The strong retention of radionuclides in some soils was once considered an asset because it served as a final barrier to contamination of ground water. This attribute is now a liability to returning contaminated soils to a near pristine condition for uncontrolled release. Understanding the fixation mechanism(s) is necessary to developing an effective strategy to induce release.

Many inorganic contaminants are held in soils by physical adsorption or cation exchange due to the electrical charge on soil particles, particularly clays. Cations are attracted to the broken edges of silicate layers to stabilize the negative charge imbalance that occurs where the silica tetrahedra are incomplete. The electroneutrality may also be upset by isomorphic substitution of an ion of lesser charge. For example, a trivalent aluminum or iron ion may substitute for a tetravalent silicon ion leaving the matrix with a net negative charge [14]. This effect can be strongly influenced by pH; as pH is increased, the ability of water to stabilize the matrix is reduced, and the cation exchange capacity of the soil is increased [14]. Strontium has been shown to be less strongly bound by clays at lower pH values, probably due to reduced dissociation at hydroxyl sites, and the net positive charge induced by aluminum and iron oxides [15]. Similar to specialty ion exchange resins, some materials favor specific ions due to differences in relative affinity, charge spacing, or steric hinderances. Note that some contaminants exist in solution as anionic complexes or compounds and are held by analogous, but opposite forces. Some materials can be fixed in either manner depending on the soil redox chemistry and pH.

Contaminants can also be bound through precipitation from a super-saturated solution, coprecipitated with another material, or physically bound in an agglomerate due to the characteristics of another compound. As a solution evaporates, it eventually becomes super-saturated, and the salts crystallize on whatever solid surface to which they are exposed. Exposed to water again, these salts are typically redissolved, however, if they precipitate in pores where flow cannot be established, or as a new relatively insoluble compound, dissolution may be drastically limited. The local chemical equilibrium may also be controlled by pH such as in the carbon dioxide/carbonate/bicarbonate system which can cause fixation over a very sharp pH gradient around a carbonate mineral such as limestone [16].

Hydrated metal oxides are well known for their tendency to concentrate heavy metals. Oxides of aluminum, manganese, and iron readily hydrate to form a semi-continuous layer which may occlude other cations of particles [17, 14]. Their effect may be disproportionate to their concentration in the soil because they tend to form thin coatings on particles, yielding a tremendous surface area for fixing contaminants. Chemically or physically removing these layers would generally act to the benefit of removing a contaminant, but the layer may cover or block pore spaces or ion exchange sites, and removal may open these other sites for active fixation. Though these new sites would probably fix less contamination than that fixed by the hydrated oxides, the degree of decontamination realized may be lessened by this effect.

Organic materials may also bind contaminants due to biological activity, stabilization by free organic ligands such as polyelectrolytes, or charge neutralization with organic acids [16, 18]. Highly humic soils are typically acidic which tends to keep metals soluble, but also generally have a high cation exchange capacity (CEC) due to the activity of these other phenomena [14].

Over a long enough time contaminants may also substitute in a mineral lattice, becoming a part of the undissolved mineral fraction of the soil. Ironically, though this material is all but unavailable to the environment biologically, if radionuclides are involved, much investment may be necessary to solubilize them for removal.

Mechanism Identification

Certainly wet-screening the material to determine if a significant fraction may be discarded without treatment is a sound start. The contaminant distribution may be altered in favor of additional physical separation by controlled attrition of surface deposits. Flotation may be necesary to completely separate the fine clay and organic particles from the more coarse material. Complete segregation of size fractions is essential to evaluate contaminant distribution and the potential for physical volume reduction.

Assay of the soil fractions will not only identify which fraction(s) on which to focus, but analyses of other typical soil constituents may also provide initial insight into any correlations between the contaminants and compounds in which they could be fixed. These analyses will also provide the basis for understanding the soil chemistry, e.g. acidity, alkalinity, organic content, and fines fraction. For a small amount of soil, or a relatively straight forward decontamination, simple trial-and-error with the most widely used or most probably effective cleaning agents such as water (hot or cold), detergents, or possibly acid of caustic may be completely satisfactory to evaluate the efficacy or soil washing. For more complicated contaminants such as soil analogues, particularly with radionuclides where long term disposition and control drives waste minimization more than short-term costs, a more insightful strategy may be necessary. The more learned about the contaminant retention in the soils, the more likely an efficient contaminant-specific release flowsheet will be developed. Not all trial and error can be eliminated from the R&D program. However, developing a hypothesis on the principal retention

mechanism(s) for a species in a particular matrix will provide a basis for organizing an experimental strategy.

The next step may include a surface analysis if possible. Though the contamination could penetrate into the individual particles over time, the contamination was introduced from an external source, and therefore should reside primarily on the surface. Elemental mapping of the surface with a scanning electron microscope (SEM) may corroborate correlations indicated by the chemical assays, or it may indicate less obvious correlations such as the disproportionate metal fixing capability of the hydrated metal oxide films described earlier. While electron spectroscopy for chemical analyses (ESCA) is expensive, it may give direct indication of the compounds fixing the contaminants of concern.

Though significant work has been done to quantify the relative retention characteristics of contaminants in soils, most of this research has been in support of migration modelling [19]. Little work has been found in the environmental literature on studies dedicated to defining how radioactive contaminants are retained. What research there is, relates to decontamination of low-level wastes (Cs^{137} and Sr^{90}) by ion exchange on clays and silico-titanates [20, 21]. The function of clay is described above. The titanates are of interest because these materials can be engineered for ion-selectivity by controlling lattice spacings to function in a manner similar to molecular sieves.

Contact with many commercial vendors of soil washing technology has explained the dearth of published data; virtually all testing data with RCRA regulated species is protected as proprietary, and little or no work has been done with radionuclides because of the regulatory morass which makes profit questionable.

Much can be learned, however, from bioavailability studies on heavy metals in soils [16, 22, 23, 24]. Sequential leaching experiments are designed to destroy one fixation mechanism at a time, progressing from relatively mild to increasingly more aggressive treatments. Chemical analyses of both the solids and the leachate between steps suggest the mode of fixation for each contaminant. The finesse is in designing a treatment that will efficiently eliminate one mechanism, without attacking the remaining matrix. Not only is this ideality impossible to fully attain, it can be masked by the refixation of the contaminants in the remaining solids. In addition to providing step-wise yield data for the contaminants of concern, the chemical analyses can help to adjust the extraction procedure by providing verification of how completely a mechanistic compound is removed, and some insight into the unavoidable attack on the remaining mechanisms. Though imperfect, this technique can be used to deduce the most probable fixation mechanism so that it may be targeted by a flowsheet.

Table 1 summarizes some of the reagents listed in the references cited above to perform selective leach experiments. This is by no means a comprehensive list, and reference to the original publications and bibliographies contained therein is recommended for evaluating which treatment is most applicable to a given soil. The leachants are listed by mechanism/phase attacked, in the order of the typical sequential design. The first leach should be dedicated to physical adsorbtion and ion exchange, the second, carbonates, etc.

Examples of some of the leachant limitations and experimental suggestions gleaned from the literature are as follows:

Some of the ion exchange solutions may also attack calcium carbonate either due to the acetate complexing the calcium, or the ammonium ion neutralizing the carbonate [16]. Running the extraction at slightly over pH 8 rather than in a neutral solution may mitigate the complexing effect, and using the sodium salt may eliminate the neutralization problem [16]. In addition to pH, the time and temperature of the extraction should be held constant, experiment to experiment, to minimize data variability [16].

Reducible metal fractions may be altered during sample preparation, such as exposure to air for samples from reducing environments [16]. Chelating, and complexing effects of leach

Table 1 Mechanism Specific Leachants

Target Mechanism	Leachant
Adsorption/Ion exchange	0.05 M $CaCl_2$ $BaCl_2$, Triethanolamine, pH 8.1 $MgCl_2$, pH 7 NH_4OAc, pH 7, 8.2 NaOAc
Carbonates	CO_2 NaOAc/HOAc buffer, pH 5
Reducible Oxides	Acidic Hydroxylamine in Ammonium-Oxalate Buffer Hydroxylamine/Acetic acid 0.1 M NH_2OH/HCl, 0.01 M HNO_3, pH 2 0.2 M Ammonium Oxalate, 0.2 M oxalic acid, pH 3 Ammonium Oxalate/Oxalic Acid with UV Sodium Dithionite-Citrate Buffer Hydroxylamine/HCl, Acetic Acid
Organic/Sulfides	H_2O_2, NH_4OAc, pH 2.5 30% H_2O_2, 0.02 M HNO_3, pH 2, 85°C Organic Solvents 0.1 M NaOH, H_2SO_4 Potassium pyrophosphate
Residual	HNO_3 (conc.), 180°C Aqua Regia, HF, Boric acid HF/$HClO_4$

solution anions may also yield confusing results for metals [16]. Reprecipitation may occur due to pH swings at particle interfaces, or in the solution due to loss/gain of carbon dioxide [16]. Adjusting the pH up to 3, and decreasing the contact time to 15 minutes and the reagent concentration to 0.025 M may make the hydroxylamine hydrochloride leachant even more specific, removing the more readily reduced manganese oxides while leaving the iron oxides virtually intact [25].

Using sulfide bearing reagents may cause low solubility metal-sulfides to be lost from solution [16]. Care should also be taken in using dithionite because of inherent zinc contamination, and residual extractant can cause plugging problems in atomic absorption equipment [23].

Sodium and potassium chlorides have also been used to displace strontium and cesium [26]. Calcium chloride was shown to effective displace strontium which might be expected from their chemical similarity, but potassium seemed to be far more effective at displacing cesium than was sodium [27]. This may be due to the relative natural affinities predicted in the Hofmeister Series [28], or some degree of steric hinderance as in the case of a molecular sieve.

Chelants have also been used to extract heavy metals from soils [29], but are notably absent from Table 1. These materials enhance solubility, but provide no mechanism specific attack, and therefore are not particularly effective for a sequential leach.

The spacial variability of soils cannot be overstated. Just as each mechanistic study is unique to the location and contaminants, multiple samples from the same site may vary enough to mask trends in experimental data. Care must be taken to eliminate unique characteristics between samples. Adequate blending of the initial source material for experiments to ensure homogeneity is essential.

CONCLUSIONS

There is no question that there is a need for R&D to minimize the amount of waste the DOE must control for years to come. Though trial-and-error testing is quick and relatively inexpensive, it provides little or no insight for technology transfer to

other locations. A mechanistic approach yields enough data to help explain how a process works or why it should not be considered further. With this level of understanding, it will be easier to defend the evaluation as comprehensive to local communities. Finally, developing flowsheets to minimize waste, by minimizing the amount of soil dissolved, is cost effective.

Timely treatability studies are necessary to evaluate the potential for technologies such as soil washing for application to readionuclide contamination. The goal of the research should be to provide adequate insight into the physical/chemical processes involved to determine if soil washing should be considered for remedial action with radioactive soils. Soil assays, surface analyses, and sequential leaching strategies are tools available to provide the knowledge base needed to make informed decisions. Hopefully, the recommendation to the public to drop, selectively apply, or regard soil washing as a true "magic bullet" will be based on science.

LITERATURE CITED

1. **Grant, D. C., and E. J. Lahoda,** "Remediation of Uranium and Radium Contaminated Soil using the Westinghouse Soil Washing Process," *Proceedings of the 7th Annual DOE Model Conference on Waste Management & Environmental Restoration*, Oak Ridge, Tennessee (1991).
2. EPA, "Guide for Conducting Treatability Studies Under CERCLA: Soil Washing, Interium Guidance," United States Environmental Protection Agency, Office of Emergency and Remedial Response, EPA/540/2-91/020A (1991).
3. **Clark, M. L., et al.,** "Montana Tunnels-start-up Efficiencies Improved, Expansion Considered," *Mining Engineering*, 41(2), pp. 87–91 (1989).
4. **Perry, R. H., and C. H. Cholton,** Chemical Engineer's Handbook, McGraw-Hill, New York, NY, Chapter 8 (1973).
5. **Pheiffer, T. H., et al.,** "EPA's Assessment of European

Contaminated Soil Treatment Techniques," *Environmental Progress*, **9**(2), pp. 79–86 (1990).

6. EPA, "Innovative Technology: Soil Washing," United States Environmental Protection Agency, Office of Solid Waste and Emergency Response, PB90-274184, Fact Sheet Number 9200.5-250FS (1989).

7. EPA, "Assessment of International Technologies for Superfund Applications," United States Environmental Protection Agency, Office of Solid Waste and Emergency Response, EPA/540/2-88/003 (1988).

8. Dragun, J., "The Fate of Hazardous Materials in Soil Part 3," *Hazardous Materials Control*, **1**(5), pp. 24–43 (1988).

9. Del Debbio, J. A., "Sorption of Strontium, Selenium, Cadmium, and Mercury in Soil," *Radiochimica Acta*, **52/53**(Part 1), pp. 181–186 (1991).

10. Miller, J. D., *et al.*, "Particle Characterization of Contaminated Soil," prepared for the U.S. Department of Energy Idaho Operations Office, EGG-WTD-9736 (1991).

11. EPA, "Risk Assessment Guidance for Superfund Volume I Human Health Evaluation Manual (Part A)," United States Environmental Protection Agency, Office of Solid Waste and Emergency Response, EPA/540/1-89/002 (1989).

12. 55 FR 8848 (March 8, 1990).

13. Figueroa, Ines Del C., *et al.*, "Interim-action Risk Assessment for the Test Reactor Area (TRA) Warm-Waste Leach Pond Sediments (OU-2-10)," EGG-WM-9622 (1991).

14. Dragun, J., "The Fate of Hazardous Materials in Soil Part 2," *Hazardous Materials Control*, **1**(3), pp. 41–65 (1988).

15. L'annunziata, M. F., and W. H. Fuller, "The Chelation and Movement of Sr^{89}-Sr^{90} (Y^{90}) in a Calcareous Soil," *Soil Science*, **105**(5), pp. 311–319 (1968).

16. Calmano, W., and U. Forstner, "Chemical Extraction of Heavy Metals in Polluted River Sediments in Central Europe," *The Science of the Total Environment*, **28**, pp. 77–90 (1983).

17. Jeene, E. A., "Controls on Mn, Fe, Co, Ni, Cu, and Zn Concentrations in Soils and Water: the Significant Role of Hydrous Mn and Fe Oxides," Advances in Chemistry Series 73, American Chemical Society, Washington, D.C., pp. 337–387 (1968).

18. Mortensen, J. L., "Complexing of Metals by Soil Organic Matter," *Soil Science Society Proceedings 1963*, pp. 179–186 (1963).

19. Del Debbio, J. A., and T. R. Thomas, "Transport Properties of Radionuclides and Hazardous Chemical Species in Soils at the Idaho Chemical Processing Plant," prepared for the U.S. Nuclear Regulatory Commission under Department of Energy Idaho Operations Office, WINCO-1068 (1989).

20. Lacey, W. J., "Use of Clays as Ion Exchange Material to Remove Radioactive Contaminants from Water," American Chemical Society 137th National Meeting, Cleveland, OH (April 5–14, 1960).

21. Anthony, R. G., *et al.*, "Selective Adsorption and Ion Exchange of Metal Cations and Anions with Silico-titanates and Layered Titanates," Gulf Coast Hazardous Research Center's Fifth Annual Symposium on Emerging Technologies: Metals, Oxidation and Separation, Lamar University, Beaumont, TX (February 25–26, 1993).

22. Mclaren, R. G., and D. V. Crawford, "Studies on Soil Copper, I. The Fractionation of Copper in Soils," *Journal of Soil Science*, **24**(2), pp. 172–181 (1973).

23. Tessier, A., P. G. C. Campbell, and M. Bisson, "Sequential Extraction Procedure for the Speciation of Particulate Trace Metals," *Analytical Chemistry*, **51**(7), pp. 844–851 (1979).

24. Gibson, M. J., and J. G. Farmer, "Multi-step Sequential Chemical Extraction of Heavy Metals from Urban Soils," *Environmental Pollution/Series B*, **11**(2), pp. 117–136 (1986).

25. Chao, T. T., "Selective Dissolution of Manganese Oxides from Soils and Sediments with Acidified Hydroxylamine Hydrochloride," *Soil Science Society Proceedings 1972*, pp. 764–768 (1972).

26. Fuller, W. H., *et al.*, "Contribution of the Soil to the Migration of Certain Common and Trace Elements," *Soil Science*, **122**(4), pp. 223–235 (1976).

27. Miller, J. R., and R. F. Reitemeier, "The Leaching of Radiocesium Through Soils," Soil Science Society Proceedings 1963, pp. 141–144 (1963).

28. Kunin, R., Ion Exchange Resins, Robert E. Kreiger Publishing Company, Huntington, New York, p. 26 (1972).

29. EPA, "Cleaning Excavated Soil Using Extraction Agents A State-of-the-Art Review," prepared for the Environmental Protection Agency, Cincinnati, OH by Foster Wheeler Enviresponse, Inc. PB89-212757 (1989).

INDEX

Low-NOx burners, nitrogen oxides control, 45, 47, 57-60

M

MACT standards, 30, 128

Maintenance modifications, minimizing air toxics by, 29

Marine vapor control systems, 118-124

Mechanical seals
    fugitive emissions from, 144-149, 162-163
    selection, 164-167

Membrane separation, 10, 32-33

Mining industry, soil remediation, 308

Mist scrubbers, 23

Monitoring
    biomonitoring, 211-217
    compliance assurance monitoring, 67
    continuous emissions monitoring, 67-73
    parametric monitoring, 67
    path monitoring, 71-72
    point monitoring, 70, 71
    predictive monitoring, 67

N

Nitrogen oxides control
    burner modifications, 57-60
    selective catalytic reduction, 50-56
    technology selection, 43-49

Nitrogen usage, minimizing, 28-29

Non-catalytic reduction, selective, 45, 48, 50-56

O

Odor control
    mapping an effective strategy, 74-80, 81-87
    technologies, 80, 88-96

Odor minimization, 80, 88-89

Off-stoichiometric combustion, nitrogen oxides control, 45, 46-47

Oil coalescing, wastewater treatment by, 194, 196

Oil/water separation, hydrocarbon remediation, 271, 272-273

Operating permit, building flexibility into, 125-129

Oxidation reactions
    hydrocarbon remediation, 271, 275-276
    for VOC control, 18-20
    wastewater treatment by, 199, 201-202, 245-249

P

Packed bed scrubbing, 22

Packing, valve packing for fugitive emissions control, 150-155, 158-160

Pan microcosm testing, 290-291

Parametric monitoring, 67

Path monitoring, 71-72

Plasma technology
    hazardous air pollutants control, 39-40
    for VOC control, 10, 11

Point monitoring, 70, 71

Pollution control see Air pollution control; Emission control; Site remediation; Wastewater treatment

Polymeric adsorbents, for hazardous air pollutant control, 10

Polytetrafluoroethylene, valve packing, 150-155

Precipitation, wastewater treatment by, 199-200

Predictive monitoring, 67

Process changes, minimizing air toxics by, 24-29

Process emissions, 26

Process heaters, 90

Prompt NO, 44

Pumps
    EPA equipment leak standard, 133-134
    fugitive emission control, 162-163, 164-167
    mechanical seals, 144-149, 162-163, 164-167

Q

Quality improvement program, EPA equipment leak standard, 131, 134-135

Steam recovery, solvent recovery by adsorption on activated carbon, 97-101
Steam stripper, design, 229-236
Steam stripping, wastewater treatment by, 175-176, 197-198, 229-236
Steam venting, with soil vapor extraction, 304
Storage emissions, 26
Stormwater management, 218-221, 222-228
Stormwater segregation, 223-227
Stratified factors, fugitive emissions estimating, 141
Stripping
    liquid/liquid extraction, 257
    wastewater treatment by, 175-176, 194, 197-198
Supercritical extraction, activated carbon, 104
Supercritical oxidation, wastewater treatment by, 202-203

T

Tank breathing, emissions from, 3
Testing
    ambient testing, 61
    compliance testing, 62-63
    engineering testing, 63
    guarantee testing, 63
    pan microcosm testing, 290-291
    stack testing, 2, 62-66, 73
    stationary source testing, 61-66
    whole-effluent toxicity testing, 170
Thermal desorption, hydrocarbon remediation, 271, 275
Thermal incinerators, 89
Thermal oxidation, for VOCs control, 18-20
Thermal regeneration, activated carbon, 103-104
Thermal treatment
    hydrocarbon remediation, 271, 275-276
    of wastewater, 202-204
Total petroleum hydrocarbons (TPH), 279
Toxicity characteristic leaching procedures (TCLP), 280
Toxicity identification evaluation (TIE), 213-216
Toxicity reduction evaluation (TRE), 211-217

Trickling filters, wastewater treatment by, 203, 207

U

Ultra-low-NOx burners, for NOx control, 58
Ultraviolet oxidation
    for hazardous air pollutant control, 33-34
    hydrocarbon remediation, 271, 276
    for VOCs control, 10
Unit-specific correlation method, fugitive emissions estimating, 143
Unsaturated zone bioremediation, hydrocarbon cleanup, 274-275

V

Vacuum operations, emissions from, 4
Valves
    EPA equipment leak standard, 132-133
    fire safety, 157-158
    packing, for fugitive emissions control, 150-155, 158-160
Vapor destruction systems, in marine vessels, 120-121
Vapor-phase advanced oxidation, for odor control, 93-94
Vapor recovery systems, in marine vessels, 121-122
Vent condensers, 28
Volatile organic compound (VOC) control, 9, 10, 16, 25
    biofiltration, 10, 31-32, 94-95, 111-117
    current practices, 9-11, 38, 114
    equipment leaks, 130-138
    fugitive emissions control
        mechanical seals, 144-149, 162-163
        process equipment improvements, 156-163
    future trends, 12-15
    odor control, 88-96
    regulations, 11-12
    selecting a strategy, 16-23
    steam stripping, 229
    wastewater treatment, 229-236, 237-244
Volatile organic compounds (VOCs)
    groundwater remediation, 301-307

solvent recovery by carbon adsorption, 97-101

W

Wastewater
    characteristics, 191-192
    effluent and residual disposition, 168-169
    reuse, 180-181, 183-190
    zero discharge, 177-182
Wastewater treatment
    biomonitoring, 211-217
    carbon adsorption, 196-197, 237-244
    developing an effective strategy, 168-176, 209-210
    Fenton reactor, 245-249
    liquid/liquid extraction, 198-199, 250-257
    oxidation process, 199, 201-202, 245-249
    steam strippers, 197-198, 229-236
    stormwater management, 218-221, 222-228
    technologies in use, 191-210
    water reuse following, 183-190
Water reuse, 183-190
Water/steam injection, for nitrogen oxides control, 45, 47-48
Wet air oxidation, wastewater treatment by, 202-203
Whole-effluent toxicity testing, 170

Z

Zeolite adsorption
    for hazardous air pollutant control, 34
    solvent recovery carbon adsorption, 101
Zero discharge, 177-182